D0560368

Principles of Digital Computer Design

Principles of Digital Computer Design

Volume I

ABD-ELFATTAH M. ABD-ALLA
The George Washington University
Washington, D.C.

ARNOLD C. MELTZER
The George Washington University
Washington, D.C.

Prentice-Hall, Inc., Englewood Cliffs, New Jersey

Library of Congress Cataloging in Publication Data

ABD-ALLA, ABD-ELFATTAH M. (date)
Principles of digital computer design.

Includes bibliographies and index.
1. Electronic digital computers—Design and construction. I. Meltzer, Arnold C. (date), joint author. II. Title.
TK7888.3.A26 621.3819'58'2 75-12572
ISBN 0-13-701524-0

10 9 8 7 6 5 4 3 2 1

Printed in the United States of America.

PRENTICE-HALL INTERNATIONAL, INC., *London*
PRENTICE-HALL OF AUSTRALIA, PTY. LTD., *Sydney*
PRENTICE-HALL OF CANADA, LTD., *Toronto*
PRENTICE-HALL OF INDIA PRIVATE LIMITED, *New Delhi*
PRENTICE-HALL OF JAPAN, INC., *Tokyo*
PRENTICE-HALL OF SOUTHEAST ASIA (PTE.) LTD., *Singapore*

Contents

4 Codes 76

5 Switching Algebra and Logic Gates 96

Preface

The reason for this book is that the authors could not find a book to their liking for teaching a course entitled Digital Computer Design. Although many books exist in this field, many have a superficial depth on the subjects which constitute the area. Other texts digress into teaching software languages such as FORTRAN or application areas for computers. We have tried to stay in the area of digital computer *design*. We rely on other texts to convey the needed information for teaching software and applications. We have not tried to depict the entire field in a single volume, but present the fundamental principles of computer design in this text. The authors intend to write a second volume where the more advanced concepts of computer design will be developed. Therefore, this book is not a survey but develops the material in depth. It should be of use to the advanced undergraduate student, graduate student, or practicing computer specialist who wishes to learn how to design digital computers. The book has been written for use by programmers, systems analysts, computer scientists and electrical engineers. Many of the people in the above categories will never produce hardware designs, but they should be thoroughly familiar with the concept of the hardware in order to achieve their responsibilities in the overall computer system design.

The purpose of this book is to present the principles underlying the design of digital computers and to utilize these principles in the design of a small digital computer using microprogramming.

This book is not written as a first introduction to digital computers. In fact, it assumes that the reader has had some previous contact with digital computers, at least by the programming of these machines in a "higher level" language such as FORTRAN, ALGOL or PL/1. Although the text presents chapters on number systems, codes, and switching theory, they are very brief and intended only as an introduction to those subjects for the reader who lacks knowledge in those areas. The authors feel that the proper background for this text is:

1. A knowledge of programming using a high level language.
2. A knowledge of switching theory design.

This book is used in the second half of a one year senior level design course. The first half of the year is a course in Switching Circuit Design. The second half uses the material of Chapters 3, 6, and 7 through 12 of this text. The text is also used in a one semester graduate course which presents the entire text without the prerequisite course in switching theory. For those who wish to emphasize the system aspect of this material, the text is suitable to be used for two semesters. The only prerequisite would be a knowledge of programming. In that case, the material of Chapters 1 through 7 could be presented during the first semester and the remainder of the text in the second semester. The authors have presented the material in courses which have students whose backgrounds are extremely diverse. Both engineers and programmers have benefited from the text in these integrated courses.

Chapter 1 introduces the digital computer, some of the jargon used to describe the machine, the place of people in its design and a brief history of computing.

Chapter 2 presents the concept of number representation in various base systems. Conversion of numbers among the different bases is developed and the different representations for signed numbers is presented.

Chapter 3 develops various algorithms for performing arithmetic operations using signed binary numbers which are compatible with an electronic hardware implementation. The authors feel that separating the algorithm from the actual hardware design permits the reader to concentrate on only one new concept instead of trying to master both algorithm and design simultaneously. The hardware design to implement the algorithms is presented in Chapter 8.

Chapter 4 presents an alternative method of representing numbers in binary form by the use of codes. The instructor who prefers to teach codes along with number representation can present the material of this chapter before presenting Chapter 3. This chapter also introduces the concepts of error detection and correction and the use of codes to present alphabetic characters as well as numerics.

Chapter 5 gives a brief introduction to the design of combinational switching circuits using Boolean algebra. Only one minimization technique is presented, i.e. the Karnaugh map. The chapter also presents a brief review of some electronic implementations of logic gates in order to familiarize the nonengineering student with the technical terminology. The chapter is intended as a review for those who have a knowledge of switching theory.

Chapter 6 develops the concepts of various types of flip-flops and their use as storage registers. It introduces the concept of the register transfer equation which is used as the fundamental design tool throughout the remainder of the book. Sequential circuit theory and register transfer equations are used to develop the functional circuits which are used to construct the subsystems of the digital computer. The logic design of counters, shift registers, encoders, decoders, adders, pulse generators, and timing distributors are presented.

Chapter 7 introduces the architecture of the Von Neumann stored program digital computer. It describes various addressing methods and instruction formats. It presents the basic concepts of machine language programming and discusses the ideas of software-hardware tradeoffs. It introduces the instruction formats and architecture of several different computers which have been constructed during the past twenty-five years. It concludes with a brief description of the software systems needed by current large scale digital computers in order to function properly.

Chapter 8 is an in-depth development of the logic designs of the arithmetic unit for various number systems and data representations. Fixed-point binary arithmetic is implemented for both serial and parallel structures. Decimal arithmetic units are designed with three different implementations. A floating-point arithmetic unit is designed and the design of the units which produce logical operations is considered. The design for addition, subtraction, multiplication, and division is considered for each implementation. Techniques to increase the speed of arithmetic units are discussed.

Chapter 9 presents the theory and design of on-line storage. The magnetic core is introduced and developed into complete core memories using both 3D and 2D structures. The electronics and timing of the main memory subsystem are presented. The fundamental structures of LSI electronic memories are introduced. The characteristics of the magnetic drum and both the head per track and moving head magnetic disk are developed in depth.

Chapter 10 introduces the concepts of data structures, channels, and I/O devices. It starts with the development of the I/O structures for large digital computers introducing both the hardware and software needed to make this subsystem function. A detailed description of simple channels follows the bird's-eye view of the I/O system. It then presents the structure of the DMA channel. The chapter concludes with the design characteristics of various I/O devices such as paper-tape readers and punches, card readers, line printers, CRT terminals and magnetic tape units. Some readers may argue that magnetic tape units are storage and should be presented in

Chapter 9. We feel that magnetic tape is used as I/O and is not on-line storage. However, the instructor who wishes to present the section on magnetic tape along with the material of Chapter 9 can do so.

Chapter 11 develops the techniques used in the Control Unit of a digital computer. It is based on a knowledge of Chapter 7 and therefore that chapter should be understood before attempting Chapter 11. It presents the timing and sequencing of the various instructions and phases of the computer. It depicts various structures used in computers and shows how timing signals are derived and distributed. Finally, it presents the fundamentals of microprogramming for computer design.

Chapter 12 is the detailed design of a small general purpose digital computer. The machine is not a trivial academic exercise, but contains forty-seven instructions. The first part of the chapter shows the design from the viewpoint of the programmer's manual. The middle of the chapter details the register transfer equations for the computer and the architecture of the machine. The last part of the chapter shows the microprogramming of each instruction and the detailed logic design of the machine.

ABD-ELFATTAH M. ABD-ALLA
ARNOLD C. MELTZER

Washington, D. C.

1

Computer Uses and Application

This text is written to show a methodology for the design of modern digital computers. This methodology is the result of many man-years of scientific and engineering research and development by what is now a large and important industry. Because of this effort, digital computers have reached a level of performance and reliability which has enabled them to contribute to the very existence of our high-level technical society. Without these machines much of our way of life would be entirely different. Many of man's greatest achievements in medicine, space technology, production, process control, mathematics, electronics, and economics have been achieved with the use of the electronic digital computer. This industry is still very young and growing; new ideas and methods which will permit man to achieve a better way of life using this "information-amplifying" machine are still to be developed.

1.1 CHARACTERISTICS OF ELECTRONIC COMPUTERS

Before we delve into the detailed design of such a machine, we should consider some of its overall characteristics, structure, and uses. This will enable us to have an objective for our design. In this chapter we shall present such an overview.

We are talking about an electronic machine, that is, one whose central functions are performed by electronic processes using electronic circuitry. Why use an electronic machine? The answer is speed. The great achievements of the electronic computer are due to its relative speed in computing as compared to its users or surroundings.

In this book we shall be describing electronic *digital* computers. Are there other types of electronic computers? Yes—electronic *analog* computers. An analog computer works with continuous time electronic signals, while a digital computer has discrete time electronic signals. Because of the technical problem of accurately measuring a continuous time-varying electronic signal, the digital computer can achieve a much greater accuracy than the analog computer. Another difference in these two types of machines is their capability to store information for use in solving a problem. The digital computer has a large memory and can store vast amounts of numeric information for use while solving a problem, while the analog computer has a very limited amount of storage capability. Although these two types of computers are quite different, they also complement each other very well. The result has been the marriage of these machines into a third type of electronic computer called the electronic *hybrid* computer. This machine has the characteristics of both the digital and analog computer.

The primary characteristic of the digital computer has been its speed. This speed is achieved by building it with electronic components. These components are utilized as counting and comparing circuits and operate in a switching mode between two states. Thus, the machine operates with binary circuits and therefore in order to achieve high speed usually performs its arithmetic with a binary number system. This binary feature, which has been forced upon us by the limitations of our electronic components, has permitted the methodology of Boolean algebra to be used to design the logical circuits needed by digital computers. It is assumed that the reader of this text is familiar with the theory and practice of switching algebra and the logical design of digital circuits. A brief review will be given in Chapter 5 in order to establish a notation.

Another feature of the electronic digital computer is the concept of the stored program. Since the advantage of the machine is its speed, it must have both its instructions, i.e., what it is to perform, and its data stored internally in its memory prior to its actual operation on the data. Then it can perform the instructions on the data at electronic speeds. This stored program concept has led to a whole new profession called *programming*.

From the above description we could call the machine a "stored program binary electronic digital computer." During the remainder of this text we shall just call it "digital computer," or sometimes shorten it to "computer."

Although the digital computer is fast, it must also be reliable if it is to be of any practical use. That is, it must not "drop" a number, or interchange numbers, or "forget" a number which it is supposed to remember. The ability of the computer to perform reliably is a function of its design and the reliability of the electronic components of which it is built. It is the ability of these components to perform fast and reliably which has made the digital computer a practical machine.

In the chapters which follow we shall first study ways of representing information in a form that is compatible with digital computers. We shall study the methods of arithmetic which lend themselves to electronic implementation. In Chapter 5 we shall review the fundamentals of logic design, logic building blocks, and electronic implementation of logic. In Chapter 6 we shall use the knowledge gained from the first five chapters to design the subblocks needed in digital computers. In each of the last six chapters we shall design a particular subsystem of the digital computer, culminating in the complete design of a small-scale computer.

1.2 STRUCTURE OF A DIGITAL COMPUTER

Before we get to the "nitty-gritty" of a digital computer, let us look at the subsystems of the machine. This will be a preliminary peek so that we shall be acquainted somewhat with the machine while we study what goes into its design and construction. As stated in the previous section, the computer has a memory capability to store data and instructions. It performs these instructions many of which are arithmetic in nature. It must also be able to display the solutions to the problems which it computes and be able to accept data and instructions from humans prior to their execution and use. Thus, we have a simple block diagram of a digital computer as shown in Fig. 1.1.

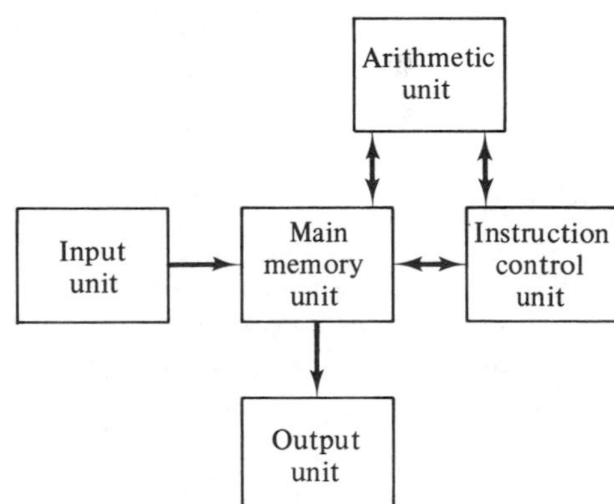

Figure 1.1 A simple block diagram of a digital computer.

1.2.1 Arithmetic Unit

The arithmetic unit of the digital computer performs the operations of addition, subtraction, multiplication, and division. It also performs certain logical operations such as logical-AND, logical-OR, and negation. The arithmetic operations are performed in the binary number system and may be either in integer or fractional binary form or scientific (exponential) binary form.

1.2.2 Instruction Control Unit

The instruction control unit of the computer obtains the instructions which are stored in the memory, interprets the instructions, and sequences the operation of the rest of the machine to perform the instruction. The control unit is the "symphony conductor" of the computer. The control unit contains registers which hold the instructions and some data which are presently being used. It contains a clock by which it maintains the time relationships needed for synchronization.

1.2.3 Main Memory Unit

The main memory unit contains the instructions and the data which are being processed by the control unit and arithmetic unit. The sequence of instructions which the control unit executes is called a *program.* A program is written by a programmer and placed in a form which is interpretable by a computer, such as punched cards. Once the entire program has been stored in the computer's memory and the data on which it is to operate are also inserted, the computer processes the information according to the instructions of the program at an electronic speed determined primarily by the rate at which it can fetch instructions and data from its main memory. The memory is also used to remember intermediate results obtained from computations during the execution of the program.

1.2.4 Input Unit

The input unit reads the instructions and data into the main memory of the computer where they are stored until they are processed. The unit reads only in a medium which is compatible with the computer. Some input devices are punched-card readers, punched-paper-tape readers, magnetic character readers, optical character readers, and typewriters. The first four devices require special preparation of the input information before the computer can read it. Thus, for example, the data must first be punched onto cards before the cards can be read by the computer.

1.2.5 Output Unit

The output unit presents the results of the computation to the user in a form compatible with the user and not as electric signals in binary code as it is represented within the machine. Some common output devices are card-punching machines, magnetic tape units, typewriters, oscilloscopes, and high-speed line printers. The first two are not compatible with humans but can be used as inputs to another computer. The last three can usually print or display alphabetic and numeric characters to be read by humans. The typewriter or teletype unit can be used for both input to the computer from the keyboard and output from the computer when the machine locks out the keyboard and prints.

1.3 USES OF THE DIGITAL COMPUTER

1.3.1 Scientific Applications

Science and engineering use mathematics as a tool to express the physical and biological laws of nature. By manipulating mathematical symbols and equations new inventions and better ways of doing things can be developed. Since the digital computer can manipulate numbers at electronic speeds, it is another valuable tool for science and engineering. The evaluation of some mathematical expressions may require many hours when performed manually, but when performed by a computer these laborious calculations may require only seconds. Many times it is necessary to calculate the same mathematical expression over and over with different sets of parameters. Even if a single evaluation takes only a few minutes, the repetition many times may take hours or days if performed manually. The computer is especially well suited for repetitive-type problems and can save time and money in performing this task. Many times an engineer is trying to determine an optimum solution to a problem or design. He generally formulates the problem in mathematical terms and proceeds to use known techniques to obtain an optimum solution. This may be difficult or impossible and also very time consuming. Instead of attempting to obtain an optimum point analytically, he could program a search technique on the computer and allow it to determine the optimum solution. Thus, we see that the solution to very time-consuming or extremely difficult problems may be more easily and less expensively solved by the use of the digital computer.

Although most digital computers can only add, subtract, multiply, and divide in performing mathematical operations, the use of *numerical analysis* to achieve higher mathematical functions such as evaluation of integrals, transcendental functions, and solutions to differential equations has enabled the computer to solve complex mathematical problems.

One of the first problems solved by electronic digital computers was the solution of the ballistic paths of artillery shells. Over the years this has developed into the ability to use digital computers to determine satellite and rocket trajectories and to guide a man to the moon and back.

1.3.2 Business Applications

The largest user of computers is not the scientific community but the business community. Many of the problems encountered in business are more easily solved and solved faster by a computer. Of course, they can be solved at a lower cost by computers or the businessmen wouldn't use them. Some of the business problems which are now solved with the use of computers are

1. Producing the *payroll* for large businesses.
2. Maintaining an *inventory* of parts produced by the business and also parts needed by the company.
3. Determining the *optimum ordering* point or plant production for new parts so as to maximize profits and minimize costs.
4. Maintaining up-to-date *personnel files* of the employees so that promotions and job placement can be done more quickly and efficiently.
5. Automation of *orders* and *sales* to determine both billing to customers and costs. This can progress to an automated (computerized) *accounting system.*
6. The generation of reports which summarized the activities of the company so that management can make decisions based on timely information.
7. The maintenance of a complete *data base* which contains all the data and information needed to run the company—a combination of items 1 through 6.

The use of data bases, which are becoming more prevalent, has led to the growth of a field called *information retrieval.* Although not strictly limited to business applications, information retrieval deals with obtaining the correct information from a data base to answer the needs and questions of one of its users.

1.3.3 Use of Digital Computers in Real-Time Control Systems

The ability of the digital computer to make fast and precise calculations has made it possible to incorporate them as the decision-making component of large *real-time* control systems. The computer has decision-making capability because it can compare two numbers and determine what should be performed based on this comparison. Since the computer is incorporated in a control loop, it must be fast enough to "keep up" with the remainder of the equipment in the system. A computer which can do this is considered to be operating in real time. Let us examine some examples of real-time computation.

In the area of manufacturing the digital computer has seen an increasing use as a real-time controller. For instance, in the petroleum industry the computer can control the refineries to optimize the outputs for specific quantities to fill sales orders. Thus, the business aspects of the computer combine with its scientific applications in a real-time control of a business.

Another manufacturing application in real time is the control of the routing of automobile parts to specific locations on the assembly line so that an automobile can be built without "holding up" the line. This routing problem of getting the right part to the right place at the right time is extremely complex, but the proper solution decreases the costs of cars by increasing the productivity of the assembly line.

Within the aerospace industry computers are used on board spacecraft and within rockets to guide the craft. These digital computers, usually of a special-purpose design, are in communication with large computers on the earth. Together they provide the decision-making capability in real time to guide the rocket.

A classic example of a large real-time digital control system is in air defense. In this system, radar data gathered from a large network are sent via communication channels to a centralized digital computer. The computer calculates the total situation of aircraft and missiles in the area. It displays a picture of this situation to the military for the human decision about interception. If the decision is made, the computer via communication links guides the interceptor missiles to their targets.

A language problem exists in most digitally controlled systems. Most of the input information from transducers is in the form of analog signals such as displacement, temperature, or pressure. Also most of the equipment which is to be controlled requires analog quantities as inputs, such as shaft rotation, speed, and valve openings. Since the digital computer operates with digital signals and not analog signals, a conversion must be performed on input to and output from the digital computer. The device which converts analog signals into digital signals is called an *analog-to-digital converter*; likewise, the device which converts digital signals to analog signals is a *digital-to-analog converter*. A description of both of these devices will be found in Chapter 10.

The subsystem components of the digitally controlled system are

1. The analog transducers, which obtain the measured information from the system to be controlled.
2. The analog-to-digital converters, which transform the measured signal into a form compatible with the digital computer.
3. The digital computer, which performs arithmetic and logical operations on the measured data to determine the decisions for response to the system.
4. The digital-to-analog converters, which transform the output signals of the computer back to a form that the activating devices will understand.
5. The activating devices, which control the system.

1.3.4 Types of Digital Computers and Their Operation

Digital computers may be subdivided into two broad categories: (1) *general-purpose digital computer* and (2) *special-purpose digital computer*. The general-purpose digital computer has a sequence of instructions formulated by a programmer and inputted to the machine to be stored in its main memory unit. The machine can execute a different sequence of instructions at another time by inputting these instructions into the main memory and then executing them. The set of instructions which the general-purpose digital computer can execute is usually quite large, and the machine is versatile in being able to perform many different types of *algorithms* to solve all types of problems. Since the sequence of instructions can be changed, the machine is flexible in its uses. The general-purpose computer can process the payroll for a company in one job and then have a new program inputted which determines the integral of a function in the next job.

The special-purpose digital computer has a fixed sequence of instructions permanently wired into its memory. This type of machine can be made lighter, smaller, and faster, and made to consume less power than a comparable general-purpose machine. However, if the instructions need to be changed in a special-purpose digital computer, it must be rewired. Special-purpose machines are used where weight and power consumption are critical, such as in spacecraft. This type of machine can solve only a narrow class of problems because of its fixed sequencing of instructions (fixed program). In this text we shall concentrate on the design of general-purpose digital computers.

General-purpose digital computers can operate in several modes. If only one program is in the memory and being executed, this is called *monoprogramming*, and this one program can call on all the resources available to the computer at any time it needs them. This is usually the way that a real-time control system would operate. Another way to monoprogram is to run each program through the computer in a sequential manner. This method is called *batch* processing. Another mode of batch processing is to have more than one program in the memory of the computer and to switch between the programs on a scheduled basis. When a program is completed another program takes its place in memory. This mode of operation is called batch *multiprogramming*.

In recent years, the large digital computer system has been designed to communicate with a user at a terminal. The user has a dialog with the computer, and the computer responds with answers to queries or computations. This system is called *interactive*, with the user being connected to the system via a terminal and communication line. Terminals connected in this manner are called *on-line*, as opposed to card punch machines, which are not connected to the system and are called *off-line*. Computer systems which provide on-line, interactive computing usually operate in a *time-sharing* mode.

In this mode the user is a research worker or an engineer or a businessman but not generally a trained programmer. The system provides him with a conversational language by which he communicates with the computer. He is one of many users all of whom think they have exclusive use of the computer when actually they each see only a small piece of the computer or all of the machine for only a short time. If the computer is sufficiently fast in responding to the requests of each user, then he does not realize that there are other users concurrently on-line. A time-shared computer must be fast enough to give adequate response to the on-line terminal users. These users have teletype keyboards or oscilloscope displays with which they converse directly with the machine in a special language. There are no input or output media conversions, such as punched cards, between user and machine. Although the system is large, fast, and expensive, if sufficient users can maintain a high rate of use of the facilities in the time-shared mode, then the cost per user is reduced, and much computing can be obtained per dollar.

The time-shared systems have been effective on many applications: computer-aided design or use of the interaction to perform complicated design of aircraft, lens systems, electronic circuits, etc. It has been one of the most cost-effective applications. A special-purpose system such as an airline reservation network has been another practical time-shared system, and most airlines now employ such systems. A large data base employing information retrieval methods using terminals uses the computer in the time-shared mode to respond to all the inquiries from its users.

1.4 DIGITAL COMPUTER DESIGN PROCEDURES

1.4.1 Computer Design

Just what procedures are needed to design a digital computer system? How does one begin to design such a system? *One* does not design a computer; many people are required to design and build such a complicated system. Among some of the types of people are *system architects.* It is the function of these people to decide on the broad specifications of the system. They develop these specifications by studying the markets to which they are trying to sell the machine to be built. Since the computers that we are going to study are general-purpose, they will fit into many markets. It is the job of the system architects to compromise the design so that the computers can be used by the largest number of people at the least cost of construction. Thus, the results of this group of people are a set of specifications which tell in some detail what the machine should do and how it should do it.

At the other end of the spectrum of computer designers are the engineers and physicists who design the electronic circuits and components. These people design the electronic packages which can be used to perform the processing of the electronic signals in the computer. We can call these people *circuit designers*.

Between these two groups, the system architects who try to develop broad system specifications and the circuit designer who develops the detailed packages of the computer, lies the *logical designer*. This group of people takes the specifications from the system architects and the packages of the circuit designers and produces the interconnection diagrams which determine how the components will be put together to realize the specifications. Of course, none of the three groups works in isolation from the others. They must overlap each other and interact in a feedback environment. One cannot design a fast computer using slow electronic components. Thus, the man in the middle, the logical designer, must have sufficient knowledge of electronic component design and breadth enough to understand specifications and cost/performance trade-offs to be able to design the best computer for the price.

A fourth type of designer who works very close with and in some ways is a part of the system architects is the *system software designer*. This group of people develops the programs which are needed in order to allow the various parts of the computer to operate. In fact, system software designers work on a special set of programs which are usually called the *operating system*. These programs are supplied with the computer by the manufacturer and permit the user to adequately utilize the computer without having to write all the required programs for himself.

1.4.2 System Architects

The system architects, in order to develop design specifications for a computer or family of computers, must consider the environment into which these computers will fit. Are the machines predominantly designed for the high-speed scientific market or the small business company or the real-time spacecraft control or the large-data-base information retrieval market? The answer will affect the types and speeds of the instruction sets that the computers and their associated input and output equipment will have.

For example, a computer which is designed to solve differential equations will need a high-speed exponential notation arithmetic unit (i.e., FORTRAN REAL variables) and only moderate secondary storage and input/output equipment. On the other hand, a computer system designed for a large business data base does not need high-speed scientific computation capability but needs a fast decimal arithmetic unit, a large amount of secondary storage to hold the data base, and a large number of terminals connected to it for input and output. These two systems will have quite different design specifications.

Of course, the system architect must be aware of the electronic components available or potentially available, i.e., in research, to build the system. Since the components will be integrated circuits, there will be a standard family of circuit packages available. He cannot design specifications which are beyond the capabilities of the available electronics.

To some extent, the system architect must be clairvoyant. He must anticipate how future markets will develop and must anticipate their computing needs, for it will do no good to design the specifications for a computer based on today's market, since it takes 3 to 5 years before the computer will be designed, built, and delivered.

1.4.3 Circuit Designers

The circuit designer designs and builds the electronic components from which the computer will be constructed. His components should be fast, cheap, small, and reliable; require a small amount of power; and be universally usable. Of course, he can only build circuits which approximate all these characteristics. Some of the above characteristics are in direct conflict, since it is very difficult to build fast circuits which do not use larger amounts of power than slower ones. Thus, the circuit designer must compromise his designs to achieve a balance of the above characteristics for a given price.

The circuit designer must cooperate with the logic designer. He must design a family of electronic packages which have the features which the logic designer can use to build a computer. He must build logical packages which are sufficiently fast such that when the logic designer uses them to implement a function or instruction in a machine the overall time of the operation will be sufficiently short for the machine to meet the market requirements.

To build such components, the circuit designer must be fully cognizant of the state of the art in digital circuitry. He will be designing integrated circuits and must be fully aware of the properties of electronic materials and and techniques for the fabrication of these materials so as to produce components which will meet the design criteria of the computer. Above all, his electronic components must meet a reliability specification. If a computer is to be useful and profitable in today's marketplace, it must have a long time between electronic component failures. To achieve a high reliability, the circuit designer must devise production methods and testing methods for his electronic packages.

It should be quite clear from the above description that no single individual has all the necessary capabilities to handle the job of the circuit designer for a large digital computer manufacturer. What we have described is really the job of many people who collectively design, produce, and test the electronic components needed to build a family of digital components. These people may be called circuit designers or component designers.

1.4.4 Logical Designers

The logical designer must take the electronic packages produced by the circuit designers and determine how they should be connected together to produce a digital computer which will meet the specifications of the system architects. Of course, there are usually several ways to make such connections. The trick is to do it for the least cost. This requires an insight into the feasible alternatives and a methodology to determine which design will meet the objectives of the minimum amount of logic, fabricating manpower, and production line costs. Again, his design must be reliable, just as the circuit designer's components must be reliable. He must make his design out of the fewest possible number of different electronic package types so that the costs of fabricating the electronics can be minimized.

It is assumed by the authors of this text that the reader has some previous knowledge of logical design and the various techniques for the minimization of logical circuits—that if the reader is an electrical engineer or a student in that field, he will have a course in the electronics of switching circuits. Thus, these aspects of the work of a logical designer will not be emphasized in this book. Instead in this book we shall develop the insight needed by the logical designer to evaluate alternative computer architectures and designs to obtain a working, reliable, and cost-effective digital computer.

1.4.5 Software Designers

A fundamental part of the design of a digital computer is the determination by the system architects of the instruction set of the machine. The choice of instructions will determine the type of applications for which the computer will be best suited. Following this basic decision by the system architects, the software designer must develop the necessary operating system software so that the machine will perform many of its operations without the purchaser having to write all the operating system programs for the machine. Many of the operations which a computer can perform are due to a combination of the software supplied in the *operating system* and the hardware delivered as the computer. Thus, although the machine has its fundamental instruction set built into the hardware, it requires the skillful programming of these instructions into programs which will permit the full potential of the computing system to develop. It is the job of the system software designer to develop the specifications of the operating system programs so that the hardware subsystems will be able to perform their jobs with an optimum efficiency and thereby have a high *throughput* for the computer. Throughput is the ability of the system to produce useful work.

System software designers must therefore be skilled programmers who can understand the fundamental concepts of machine language programming and the relationships of the hardware subsystems both to each other and to the needs of the user.

1.5 A BRIEF HISTORY OF THE DIGITAL COMPUTER

1.5.1 Mechanical Calculators

4000–3000 B.C.—The development of the abacus. The abacus is a mechanical device with beads which can move along a wire. The positions of the beads on each wire determine the value of a digit. A skilled operator can add, subtract, multiply, and divide on this simple device faster than most mechanical calculators can perform these operations.

1642—Blaise Pascal developed a rotating wheel mechanical calculator with automatic carry between digits for addition and subtraction of decimal numbers.

1671—Baron von Leibnitz extended the Pascal calculator so that it could multiply and divide. This required additional gearing and was similar to contemporary mechanical desk calculators.

With a mechanical desk calculator, the operator pushes keys to enter a number and a key to specify an operation. The number is then entered into a register or subtracted from or added to a register. Thus, the limiting factor in the speed of such a device is the rate at which the operator can enter data. To overcome this limiting speed factor Charles Babbage, a professor of mathematics at Cambridge University, during the beginning of the nineteenth century, proposed his "analytical engine." This machine was composed of two parts: a store to hold data until they were needed and a mill to operate on the data by instructions. Although Babbage obtained some funds from the British government to finance his engine, he could not complete it because of technical difficulties.

1890—Herman Hollerith, a census employee, conceived the idea of using holes punched in a card to store census data to be processed by machines. These machines were mechanical in nature. Hollerith's ideas were adapted by manufacturers and have been developed over the twentieth century into extensive card-processing equipment which was used prior to the introduction of electronic business computers.

1892—Oldher mass-produced the mechanical desk calculator for use by business. These calculators were crank driven and were limited to human speeds for entering data. In 1930, these machines became powered by electric motors but were still limited in speed.

1937—George Stibitz at the Bell Telephone Laboratories began building a relay computer called the Complex Calculator which was installed in 1940. The Bell Laboratories developed several other relay computers during the 1940–1950 period. The Bell IV computer was many times faster than the Mark I computer and the Bell V built in 1946 was about 200 times faster than Howard Aiken's Mark I.

1937—Howard Aiken of Harvard University conceived the Mark I Automatic Sequence Controlled Calculator, which was built in cooperation with International Business Machines Corporation (IBM). The machine became operational in 1944 and was a general-purpose mechanical digital computer. It was not a stored program machine but kept its programs on punched paper tape which could be looped through the input units. It also used plugboards and switches for sequencing instructions. It performed division in about 60 seconds, while today's computers perform division in less than a microsecond.

1944—Delivery of the Mark I.

1946—Construction of the Bell V.

1948—The Harvard Mark II Aiken Relay Computer was completed and installed at the Naval Proving Grounds at Dahlgren, Virginia.

1948—The IBM 604 Electronic Calculating Punch became available. This machine could read punched cards and perform arithmetic operations. The results were punched out on cards. The machine was programmed with a plugboard and was not a stored program machine. It had over 1400 vacuum tubes to perform arithmetic in electronic registers.

1949—The Northrop Corporation joined an IBM 604 to an IBM 400 series accounting machine to produce what became the IBM Card Programmed Calculator (CPC). This was not a stored program machine but could execute programs which were read from punched cards at the rate of 150 cards per minute. Each instruction on a card called a plugboard sequence of arithmetic operations and could also punch an output card or print a line on the accounting machine.

1.5.2 Electronic Computers

1.5.2.1 First-Generation Vacuum Tube Machines

1943—A proposal was made by a team of engineers and mathematicians headed by J. P. Eckert and J. W. Mauchly of the Moore School of Electrical Engineering of the University of Pennsylvania to the Ballistic Research Laboratories of the U. S. Army Ordnance Corps to build an all-electronic vacuum tube calculator for use in computing trajectories and firing tables for artillery shells. This computer, called the ENIAC, Electronic Numerical Integrator and Calculator, contained over 18,000 tubes and was completed in 1945. It was used as a computer until 1958. The machine was not a stored program computer but was programmed by means of patch boards and switches; the data were maintained in its vacuum tube flip-flop registers. This machine was about 30,000 times faster than the relay Mark I machine.

1945—Delivery of the ENIAC.

1945—Dr. John von Neumann proposed the general-purpose stored program binary electronic digital computer as a successor to the ENIAC. This machine, called the EDVAC, Electronic Discrete Variable Computer, also to be built for the U.S. Army, was started at the Moore School in 1946. The completion of the machine was delayed because Professors Eckert and Mauchly left the university to start their own company. It was not completed until late in 1950.

1949—The EDSAC, Electronic Delay Storage Automatic Calculator, at Cambridge University, performed its first computations. This machine, which was started in 1947 by Professor Maurice Wilkes, was the first stored program electronic digital computer to perform. It used mercury delay lines for its memory.

1950—The SEAC, Standards Eastern Automatic Computer, was the first stored program computer to be put into operation in the United States and was used for over 10 years by the National Bureau of Standards in Washington, D. C., which built this computer. It used mercury delay lines for its memory.

1950—The ERA 1101, built by Engineering Research Associates of St. Paul, was the first computer to utilize a magnetic drum for main memory to replace the mercury delay lines. It had 16,384 words of storage. Many different computers utilizing magnetic drums for main memory were constructed during the period 1950–1955.

1951—The first UNIVAC, Universal Automatic Computer, was delivered. This machine was contracted by the National Bureau of Standards for delivery to the Bureau of the Census for use with the data of the 1950 census. It was designed by the Eckert-Mauchly Computer Corporation. However, before the first UNIVAC was delivered, the company became the Eckert-Mauchly Division of Remington Rand Corporation. This computer used mercury delay lines as its memory and was the first commercially available stored program electronic digital computer.

1953—The IBM 701 computer, a large-scale scientific computer using a Williams electrostatic tube memory backed up with a magnetic drum, was delivered. This machine had parallel binary arithmetic and was much faster than the UNIVAC for scientific computations.

1953—The IBM 650 magnetic drum computer was announced. The drum stored 2000 10-digit words and rotated at 12,500 revolutions per minute. It had IBM cards for input and output but used stored programs, which were inputted from the cards. This machine was the sequel to the CPC mechanical type calculator. Because of the existing card market which IBM dominated, over 1000 IBM 650 computers were sold.

1953—A coincident-current magnetic core memory was installed on the Wirldwind computer at M.I.T. This computer was designed for a real-time application and used .5-microsecond circuitry.

1956—The IBM 704 scientific computer was delivered with a 12-microsecond magnetic core memory and exponential multiplication capability.

1.5.2.2 Second-Generation Transistor Machines

1958—The National Cash Register computer (NCR 304) was delivered. It was the first commercial transistorized digital computer.

1960—The IBM 1401 was delivered, and many thousands of this small transistorized computer were sold.

1960—The TRANSAC S-2000 manufactured by Philco Corporation was delivered. This was a large-scale binary scientific stored program transistorized digital computer with magnetic core storage.

1960—The CDC 1604 was delivered by Control Data Corporation. This is a transistorized machine with a 48-bit word length.

1960—The IBM 7090 was delivered. This is a transistorized scientific computer with a 2.18-microsecond magnetic core memory and over 32,000 36-bit words of storage.

1960–1961—The UNIVAC LARC and the IBM 7030 (STRETCH) computers were delivered. These rival machines were very large transistorized computers for their time. The LARC used surface barrier transistors, while the STRETCH used drift transistors. The LARC had a 4-microsecond memory, and the STRETCH had a 2-microsecond memory. Only two LARCs were delivered and only seven STRETCHs were delivered. These two machines provided the test vehicles for many features found in the third-generation machines.

1964—The CDC 6600 computer was delivered to the AEC Livermore Laboratory in California. This machine is more than three times as powerful as the IBM 7030 computer. It can execute, on the average, more than 3 million instructions per second. It achieves this effective speed by its parallel architecture, which has multiple arithmetic and logical units and 10 small computers used for input/output operations. This machine achieves its speed through a high degree of parallelism and overlap.

1.5.2.3 Third-Generation Integrated Circuit Machines

On April 7, 1964, IBM announced a series of six new computers called System 360. These machines used "emulators" to provide compatibility with the previous IBM machines. These machines used hybrid integrated circuits.

The line of machines now constitute models 20, 30, 40, 44, 50, 65, 67, 75, 85, 91, and 195. Thousands of these machines have been delivered and have become the de facto standard of the industry.

Shortly after IBM announced its System 360, RCA announced its Spectra 70 series of machines. These machines were almost completely compatible with the IBM 360 line. The RCA computers used monolithic integrated circuits. In 1971, RCA Corporation announced the abandonment of the commercial general-purpose digital computer market after having lost many millions of dollars in trying to compete with IBM.

Another third-generation line of computers was manufactured by the General Electric Company. The G.E. 600 series was scientific in nature but not very successful financially. In 1970, the G.E. computer division was sold to the Honeywell Computer Division to form Honeywell Information Systems.

Burroughs Corporation, which had produced second-generation machines, now offered a line of computers topped with its 5500, 6500, 7500, and 8500 machines. Univac, which lost its initial lead in the computer industry to IBM, produces a scientific computer UNIVAC 1108 and its 9000 series of computers. In 1968, National Cash Register announced its Century Systems of business computers. Many smaller companies have announced successful third-generation computers. In particular, Digital Equipment Corporation has delivered many thousands of minicomputers and introduced an entirely new concept to the computer industry.

In 1970, IBM announced a new series of computers called System 370. These machines use monolithic integrated circuits and are compatible with System 360 programs. They give more performance per dollar than the 360 series and are designed for both scientific and business applications. IBM has announced seven machines in the line—115, 125, 135, 145, 158, 168, and 195. Burroughs Corporation has announced a new line of machines, the 5700, 6700, and 7700 machines. Univac has announced its newer machine, the UNIVAC 1110, and CDC has announced its CYBER 70 series of machines. Honeywell Information Systems has announced a successor to the G.E. line of machines, the 6050, 6060, 6070, and 6080 systems.

1.5.3 Software Developments

Even prior to the operation of the first stored programmed digital computer, EDSAC, in 1949, programming had developed on mechanical calculators. At first these were often-used routines which were kept on the plugboard machines. When a particular process needed to be performed, that particular programmed plugboard was retrieved from the library of plugboards and inserted into the machine. These plugboards certainly were not *software* since they were wired boards, but they were the forerunners of the programming which has evolved in stored programmed machines.

By 1951, a prewritten set of generally used algorithms called *subroutines* was being used on the EDSAC computer. There was a library of such subroutines which the programmer could use and therefore not have to recode. These subroutines were in the language of the machine, since programming for the earliest machines was done in the binary coding of the machine.

In the early years of electronic digital computers, about 1952, it was realized that programming in machine language had some fundamental but routine aspects to it. The programmer had to keep track of the address of each piece of data he wished to use and also remember the various codes for the operations he wanted to perform. To obviate this chore, *assemblers* were developed. These programs permitted the programmer to use *mnemonics* in place of the operation codes and addresses. When his *source* program was translated (assembled) into the machine code, the assembler assigned absolute codes and addresses in place of the mnemonics.

During the early days, about 1953, of the UNIVAC I computer, Dr. Grace Hopper and others conceived the idea of allowing the computer to *translate* and *compile* programs which were written in languages which more nearly represented mathematics. She developed three *problem-oriented languages*, Math-matic, Flow-matic, and the A2 compiler, which were the forerunners of today's higher-level languages.

In 1957, Dr. James Backus of IBM developed a problem-oriented compiler called FORTRAN (from FORmula TRANslation). This language has gained wide use for scientific programming. Another language whose specifications were first published in 1958 was ALGOL (from ALGorthmically Oriented Language). Although this language is not used widely in the United States, it has been used extensively in Europe. ALGOL is also a scientific programming language. To help with business data processing, the Department of Defense organized a committee to develop a problem-oriented language for business programming. The language, which was first announced in 1960, is called COBOL (from COmmon Business-Oriented Language) and is widely used today. Many other problem-oriented languages have been developed over the years.

In addition to languages which have been developed to make the programming of the digital computer easier, *operating systems* have been developed to make the job of the computer operator easier and to increase the efficiency of the computer system. The need for such a master program was realized in order to maintain a smooth flow of jobs through the system, utilize the system resources to their fullest potential, call programs and subroutines from the storage library as needed, maintain a catalog of programs which were stored and their location in storage, perform accounting procedures so that users could be billed for their use of the computer, help to determine if a part of the computer was faulty, help programmers debug their programs, and perform various utility functions needed for proper programming.

In 1961, the users of the IBM 709 formed a committee to develop an operating system for this machine. This users' group, called SHARE, produced the Share Operating System, SOS. Since that time most large computers have had operating systems supplied by the manufacturers. The system supplied with the 360 line of computers is called OS/360 and has continued to grow and develop over the period from 1964 to the present, until it has grown so large that there is probably no single individual who understands all the coding and interrelationships of the entire operating system.

REFERENCES

BOWDEN, B. V., *Faster Than Thought.* London: Putnam & Co. Ltd., 1953.

BURROUGHS CORP., *Digital Computer Principles,* 2nd ed. New York: McGraw-Hill, 1969.

GOLDSTINE, H. H., *The Computer from Pascal to von Neumann.* Princeton, N.J.: Princeton University Press, 1972.

MAISEL, H., *Introduction to Electronic Digital Computers.* New York: McGraw-Hill, 1969.

ROSEN, S., "Electronic Computers: A Historical Survey," in *Computing Surveys,* Vol. 1, No. 1. New York: Association for Computing Machinery, March 1969.

———, ed., *Programming Systems and Languages.* New York: McGraw-Hill, 1967.

SAMMET, J. E., *Programming Languages: History and Fundamentals.* Englewood Cliffs, N.J.: Prentice-Hall, 1969.

STIBITZ, G. R., "The Relay Computers at Bell Labs," *Datamation* (April 1966), pp. 35–44; (May 1967), pp. 45–49.

TAYLOR, A. E., "The Flow-matic and Math-matic Automatic Programming Systems," *Review of Automatic Programming,* Vol. 1. Elmsford, N.Y.: Pergamon, 1960, pp. 196–206.

WARE, W. H., *Digital Computer Technology and Design,* Vol. 1. New York: Wiley, 1963.

WILKES, M. V., D. J. WHEELER, and S. GILL, *The Preparation of Programs for an Electronic Digital Computer.* Reading, Mass.: Addison-Wesley, 1951.

2

Number Systems

The history of mankind's development parallels man's ability to develop the mathematics and computational capability to describe his achievement. Of course his earliest and possibly his greatest achievement is the concept of numbers and counting. The earliest records of such a development were tally marks on the walls of caves. This counting led to the use of the decimal system, which is a natural outgrowth of our 10 fingers. The next great step forward came with the use of the Arabic symbols 0 through 9, instead of tally marks, and the method of value based on the positional notation. If you do not believe that this is true, try to add and multiply using Roman numerals.

With the use of positional notation, the Arabic numerals, and the concept of zero, arithmetic operations became much easier. This led to the continuous advancement in the field of mathematics. The mechanical calculator helped with the calculations but is limited to human speed. The electronic digital computer developed the speed capability needed to advance man's computational ability to keep in step with his mathematical conceptional ability. The use of this electronic machine has required man to use other number systems besides the decimal system. These concepts have become so important that we are teaching them to grammer school students in the "new math."

2.1 POSITIONAL REPRESENTATION OF POSITIVE NUMBERS

The most commonly used number systems are the positional systems. The term *positional systems* means that the positions of the various digits indicate the significance to be attached to each digit. This property allows compact notations of expressing quantities. But for the positional system to be useful, there must be general agreement about the significance of the various positions. We know this is true in the familiar decimal system.

In a positional number system, a number N is usually expressed by the form in Eq. (2.1), which is equivalent to the form in Eq. (2.2),

$$N = (d_{n-1}d_{n-2} \cdots d_1d_0 \cdot d_{-1}d_{-2} \cdots d_{-m})_r \tag{2.1}$$

$$N = d_{n-1} \times r^{n-1} + \cdots + d_0 \times r^0 + d_{-1} \times r^{-1} + \cdots + d_{-m} \times r^{-m} \tag{2.2}$$

where the coefficients d_0, d_1, etc., represent the digits; r the radix, or base; n the number of integral digits; and m the number of fractional digits. The coefficients $d_0, d_1, \ldots,$ etc., are chosen from the set $0, 1, \ldots, r - 1$. Conventionally, if the base is ten, the subscript r may be omitted from Eq. (2.1).

For example, if we write $(1011.01)_2$, this means the following sum:

$$1 \times 2^3 + 0 \times 2^2 + 1 \times 2^1 + 1 \times 2^0 + 0 \times 2^{-1} + 1 \times 2^{-2}$$

In general, the number of permissible digit symbols in a number system is called the radix or base. Expression (2.1) represents a number by a series of digits. The leftmost digit d_{n-1} is called the most significant digit (MSD) of the number, and the rightmost digit d_{-m} is called the least significant digit (LSD).

2.2 THE DECIMAL SYSTEM

The most familiar number system is the decimal system. Any decimal number is constructed from the ten digit symbols 0, 1, 2, . . . , 9. Consequently, the base of the decimal system is ten. When we count, we use these symbols consecutively: 0, 1, 2, . . . , 9. Then, after exhausting all available symbols we place the symbol 1 in a new position and repeat the cycle in the first position: 10, 11, 12, . . . , 19. Then we increase the digit in the second position by 1 and repeat the cycle in the first position. After exhausting all available symbols in the second position, we create a third position and so on. Counting in any other number system follows the same procedure; the only difference is the length of the cycle.

2.3 APPROPRIATE NUMBER SYSTEM FOR COMPUTER ARITHMETIC

The decimal system is the most familiar system for human use. However, selection of the proper system of representing numbers and alphabetic characters within the machine is determined by the consideration of economy, speed, and reliability. Electrical means of implementing coded information are exclusively used in computers, because they offer greater speed and economy. If the decimal system is used to code information, we need ten different electrical signal levels to distinguish between the ten different digits of the decimal system. Now consider a system with only two different digits, that is, the binary system. Using the binary system will require only two different signal levels to represent coded information. Moreover, looking into today's technology, we find that the simplest, cheapest, and most dependable devices for digit representation are binary elements. There is no indication of the development of comparable "decimal" devices, that is, a device that can represent on its output line 1 of 10 different signals which are easily distinguishable and stable. This does not mean that the computer has to deal only with the binary number system. Humans can use the decimal system to communicate with the machine, but individual digits or groups of digits will have a binary representation. Different ways of representing nonbinary digits in binary forms will be discussed in Chapter 4.

2.4 THE BINARY SYSTEM

The radix of the binary system is two, and consequently any number, when represented, in the binary system consists exclusively of the binary digits 0 and 1. The term *binary digit* is often referred to as *bit*. In this text both the terms bit and the binary digit refer to the same thing. For example, the binary number 10101.11 has 7 bits or 7 binary digits. Arithmetic operations in the binary system are performed according to the same general rules as decimal arithmetic. Usually computers add only two binary numbers at a time, and consequently the size of the carry does not exceed the largest admissible symbol in the system, a 1 in this case. Since there are only two different digit symbols in the binary system, the addition and the subtraction tables are very simple. Single-digit binary addition and subtraction are shown in Tables 2.1 and 2.2, respectively.

Table 2.1 *Single-Digit Binary Addition*

Addend digit	*Augend digit*	*Sum*	*Carry*
0	0	0	0
0	1	1	0
1	0	1	0
1	1	0	1

Table 2.2 *Single-Digit Binary Subtraction*

Minuend digit	*Subtrahend digit*	*Difference*	*Borrow*
0	0	0	0
0	1	1	1
1	0	1	0
1	1	0	0

2.5 OTHER NUMBER SYSTEMS

Although in theory we can have number systems with any radix, few number systems are of practical value. The number systems that are most frequently used in digital computers are the binary, octal, decimal, and the hexadecimal. Table 2.3 shows some number systems with their radix.

Table 2.3 *Some Number Systems*

Radix	
2	Binary
3	Ternary
4	Quaternary
5	Quinary
8	Octonary or octal
10	Decimal
12	Duodecimal
16	Hexadecimal

Number systems, other than the decimal and the binary, which are of some interest for computer use are the quaternary, octal, and hexadecimal systems. Note that a quaternary, an octal, or a hexadecimal digit corresponds to a combination of 2, 3, and 4 bits, respectively. The base of any of these systems is larger than the base of the binary system. This property sometimes is used as a convenient means of writing a long binary number. The relationship between these particular larger radix systems and the radix of the binary system makes the conversion very simple. For example, $(101011101011)_2$ can be written as $(5353)_8$, since $(101)_2 = (5)_8$ and $(011)_2 = (3)_8$. As will be shown in the subsequent sections on conversion, the binary-octal, binary-quaternary, and binary-hexadecimal conversions can be done by inspection. The symbols for the digits for some of the tabulated base systems which are shown in Table 2.3 are in Table 2.4.

Table 2.4 *The Digits of Some Number Systems*

Decimal radix (ten)	*Binary radix (two)*	*Octal radix (eight)*	*Hexadecimal radix (sixteen)*
0	0	0	0
1	1	1	1
2		2	2
3		3	3
4		4	4
5		5	5
6		6	6
7		7	7
8			8
9			9
			A
			B
			C
			D
			E
			F

2.6 RADIX CONVERSION

As long as there is more than one number system in use, conversion of numbers from one system to another is important. For example, while it is always convenient for people to deal with the decimal system, the digital computer might be a binary computer. In such a case, decimal-to-binary conversion is to be performed on input information, while binary-to-decimal conversion is necessary to obtain decimal output.

Since the most frequently used number systems in digital computers are the binary, octal, hexadecimal, and decimal systems, the five kinds of radix conversions of interest are binary-octal, binary-hexadecimal, binary-decimal, octal-decimal, and hexadecimal-decimal conversions.

2.6.1 Conversion to Decimal

A number written in any number system could be converted into the decimal using Eq. (2.2). For example, $(6734.62)_8$ can be evaluated in decimal notation as

$$\begin{aligned}(6734.62)_8 &= 6 \times 8^3 + 7 \times 8^2 + 3 \times 8^1 \\ &\quad + 4 \times 8^0 + 6 \times 8^{-1} + 2 \times 8^{-2} \\ &= 6 \times 512 + 7 \times 64 + 3 \times 8 + 4 \times 1 \\ &\quad + 6 \times .125 + 2 \times .015625 \\ &= 3072 + 448 + 24 + 4 + .75 + .03125 \\ &= (3548.78125)_{10}\end{aligned}$$

Repeated multiplication by radix is more convenient, especially for machine implementation. This method will be illustrated below, first for integral numbers and then for fractional numbers.

Let N be an integral number with five digits in a number system with radix r:

$$N = d_4 r^4 + d_3 r^3 + d_2 r^2 + d_1 r^1 + d_0 r^0$$

The above expression can be written as

$$N = \{[(d_4 r + d_3)r + d_2]r + d_1\}r + d_0$$

The above conversion process can be described as follows: Multiply the most significant digit by the radix r. Add the second most significant digit to the product to form the first partial product (M_1). Multiply the first partial product by the radix r and add the third most significant digit to form the second partial product (M_2). Continue the process until the least significant digit is added. After adding the least significant digit, the resulting number is the decimal equivalent of the number N, if the evaluation is done in the decimal system.

Example 2.1

Let $N = (101101)_2$, and denote the partial products by M_1, M_2, M_3, and M_4:

$$\begin{aligned}&\text{Most significant digit} = 1 \\ &M_1 = 1 \times 2 + 0 = 2 \\ &M_2 = 2 \times 2 + 1 = 5\end{aligned}$$

$$M_3 = 5 \times 2 + 1 = 11$$
$$M_4 = 11 \times 2 + 0 = 22$$
$$N = 22 \times 2 + 1 = 45$$

Example 2.2

Convert $N = (\text{A6B3})_{16}$ into decimal:

$$\text{Most significant digit} = 10$$
$$M_1 = 10 \times 16 + 6 = 166$$
$$M_2 = 166 \times 16 + 11 = 2667$$
$$N = 2667 \times 16 + 3 = 42675$$

For a fractional number, repeated division by radix is used. Consider a number system with radix r, and let N be a fractional number in that system with five fractional digits as below:

$$N = d_{-1}r^{-1} + d_{-2}r^{-2} + d_{-3}r^{-3} + d_{-4}r^{-4} + d_{-5}r^{-5}$$

To convert to decimal, by repeated division, the above expression can be written as

$$N = r^{-1}[d_{-1} + r^{-1}(d_{-2} + r^{-1}\{d_{-3} + r^{-1}(d_{-4} + r^{-1}d_{-5})\})]$$

The above conversion process is similar to the integral number conversion, except multiplication by the radix r is substituted by division by the radix and the process starts with the least significant digit and stops after adding the zero to the left of the radix point.

Example 2.3(a)

Let $N = (0.504)_8$; convert to decimal:

$$\text{Least significant digit} = 4$$
$$R_1 = (4 \div 8) + 0 = 0.5$$
$$R_2 = (0.5 \div 8) + 5 = 5.0625$$
$$N = (5.0625 \div 8) + 0 = 0.6328125$$

Example 2.3(b)

Convert $N = (0.10101)_2$ into decimal:

$$\text{Least significant digit} = 1$$

$$R_1 = (1 \div 2) + 0 = 0.5$$
$$R_2 = (0.5 \div 2) + 1 = 1.25$$
$$R_3 = (1.25 \div 2) + 0 = 0.625$$
$$R_4 = (0.625 \div 2) + 1 = 1.3125$$
$$N = (1.3125 \div 2) + 0 = 0.65625$$

Example 2.3(c)

Convert $N = (0.\text{F62B})_{16}$ into decimal:

$$\text{Least significant digit} = \text{B} = 11$$
$$R_1 = (11 \div 16) + 2 = 2.6875$$
$$R_2 = (2.6875 \div 16) + 6 = 6.16796875$$
$$R_3 = (R_2 \div 16) + 15 \simeq 15.3855$$
$$N = (\text{R}_3 \div 16) + 0 \simeq 0.96159$$

2.6.2 Decimal-to-Radix *r* Conversion

To convert an integral number from the decimal system into a system of radix r, the *division by radix* method is used. This is done by successive division of the given decimal number by the new radix r. In each division, the remainder represents a new digit of the converted number. The remainder from the first division gives the least significant digit of the new number. Then the quotient resulting from the first division is divided by the radix again to yield a new quotient and a new remainder. The process continues until the quotient is less than the radix r. When this happens, the next division gives a zero quotient and a remainder which becomes the most significant digit of the number written in radix r.

Example 2.4

Convert $N = (3964)_{10}$ to octal:

$$\begin{array}{r|r l}
8 & 3964 & \\
8 & 495 & d_0 = 4 \\
8 & 61 & d_1 = 7 \\
8 & 7 & d_2 = 5 \\
 & 0 & d_3 = 7
\end{array}$$

The remainders d_0 to d_3 give the equivalent octal number 7574.

Example 2.5

Convert $N = (45)_{10}$ to binary:

$$\begin{array}{rl}
2\underline{|45} & \\
2\underline{|22} & d_0 = 1 \\
2\underline{|11} & d_1 = 0 \\
2\underline{|\ 5} & d_2 = 1 \\
2\underline{|\ 2} & d_3 = 1 \\
2\underline{|\ 1} & d_4 = 0 \\
0 & d_5 = 1
\end{array}$$

The equivalent binary number is 101101, in agreement with the number in Example 2.1.

For conversion of fractional decimal numbers, the repeated multiplication by r method is used. This is done by first multiplying the fractional decimal number by the radix r. The resulting integral part is taken as the most significant digit of the new number. The fractional part is again multiplied by the radix. The result of this second multiplication generates an integral part and a fractional part. The integral part is the second digit of the new number, and the fractional part is multiplied again by r. This process continues until the fractional part of the multiplication is zero.† The above conversion process is illustrated by the two examples below.

Example 2.6

Convert $N = (0.6328125)_{10}$ to octal:

$$\begin{array}{rl}
0.6328125 \times 8 = 5.0625 & d_{-1} = 5 \\
0.0625 \times 8 = 0.5000 & d_{-2} = 0 \\
0.5000 \times 8 = 4.0000 & d_{-3} = 4
\end{array}$$

The digits d_{-1}, d_{-2}, and d_{-3} give the equivalent octal number as 0.504.

Example 2.7

Convert $N = (0.4)_{10}$ to binary:

$$\begin{array}{rl}
0.4 \times 2 = 0.8 & d_{-1} = 0 \\
0.8 \times 2 = 1.6 & d_{-2} = 1 \\
0.6 \times 2 = 1.2 & d_{-3} = 1 \\
0.2 \times 2 = 0.4 & d_{-4} = 0
\end{array}$$

†See Example 2.7 for a further discussion.

$$0.4 \times 2 = 0.8 \qquad d_{-5} = 0$$

$$0.8 \times 2 = 1.6 \qquad d_{-6} = 1$$

In this example, the fractional part of the multiplication process will never be zero. This is indicated by the starting of a new cycle at 0.8. The value of N is 011001 . . . . There is no finite exact representation of 0.4 in the binary system. In general a finite fraction in one number system cannot always be exactly represented by a finite fraction in another number system.

2.6.3 Binary-Octal Conversion

Since $8 = 2^3$, a binary number can be very easily represented in the octal system. For an integral binary number, simply divide it into groups of 3 bits each starting from the least significant bit first. Then for each group the equivalent octal number is obtained.

For a fractional binary number the above method applies if we start grouping from left to right, i.e., with the most significant 3 bits first.

Similarly, in octal-binary conversion each octal digit is replaced by the equivalent three binary digits.

Example 2.8

Convert $N = (11010111.01101)_2$ into octal:

$$\begin{aligned}(11010111.01101)_2 &= (011)(010)(111).(011)(010) \\ &= (327.32)_8\end{aligned}$$

Example 2.9

Convert $N = (327.32)_8$ into binary:

$$\begin{aligned}(327.32)_8 &= (011)(010)(111).(011)(010) \\ &= (11010111.01101)_2\end{aligned}$$

2.6.4 Binary-Hexadecimal Conversion

The relation between binary and hexadecimal systems is similar to that of the binary-octal relation. Since $16 = 2^4$, an integer binary number can be converted into hexadecimal by simply dividing the binary number into groups of 4 bits starting from the least significant bit first. For each group the equivalent hexadecimal is obtained. For fractional binary numbers the same method applies but with the grouping starting from the leftmost digit and proceeding to the right.

The hexadecimal-to-binary conversion is done by replacing each hexadecimal digit by the equivalent four binary digits.

Example 2.10

Convert $N = (1011111.01101)_2$ into hexadecimal:

$$\begin{aligned}(1011111.01101)_2 &= (0101)(1111).(0110)(1000)\\ &= (5\text{F}.68)_{16}\end{aligned}$$

Example 2.11

Convert $N = (\text{D}57.68)_{16}$ into binary:

$$\begin{aligned}(\text{D}57.68)_{16} &= (1101)(0101)(0111).(0110)(1000)\\ &= (110101010111.01101)_2\end{aligned}$$

2.6.5 General Radix Conversion

A number written in any number system, such as base N_1, can be converted to any other number system, such as base N_2. To do this, one must be able to represent the numbers of base N_1 in base N_2 notation and perform the mathematics in base N_2. Human beings do not readily handle mathematics in a general base such as N_2. Therefore, it is easier to perform a conversion from base N_1 to base 10 (decimal) and then convert from decimal to base N_2, i.e., use decimal notation as an intermediary. This is not needed in theory, but because of human error it is used in practice.

Example 2.12

Convert $(121)_3$ into base 7:

$$(121)_3 = 1 \times 3^2 + 2 \times 3^1 + 1 \times 3^0 = 9 + 6 + 1 = 16_{10}$$

$$(16)_{10} = (?)_7$$

$$\begin{array}{rl} 7\underline{|16} & \\ 7\underline{|\ 2} & d_0 = 2 \\ 0 & d_1 = 2 \end{array}$$

$$(16)_{10} = (22)_7$$

Therefore,

$$(121)_3 = (22)_7$$

2.7 NEGATIVE NUMBER REPRESENTATION

In ordinary arithmetic a negative number is represented by a minus sign followed by the magnitude of the number. The above representation could be implemented in a machine by simply storing a single binary digit associated with each number to indicate whether a given number is negative or positive. This is signed-magnitude representation. There are two other ways of representing numbers which are often used. One is called the true complement or radix-complement, and the other one is the radix-minus-one or diminished-radix-complement.

2.7.1 Signed-Magnitude Representation

An unsigned number N is represented by the digit sequence $(d_{n-1}, d_{n-2}, \ldots, d_1, d_0 \cdot d_{-1}, \ldots, d_{-m})$. Let us now assign the symbol d_n to the leftmost position of the sequence representing a number and arbitrarily assume that a 0 in this position indicates a positive number and that a 1 indicates a negative number. A possible interpretation of the new sequence representing the signed number N is

$$N = (1 - 2d_n) \sum_{i=-m}^{n-1} d_i r^i$$

In this case d_n indicates whether the number is positive or negative without affecting the magnitude of the number, which is still represented by $d_{n-1}, \ldots, d_{-m}$. With the above notation, 1526 represents -526, and 0526 represents $+526$. For binary numbers, 10110 represents -0110, and 00110 represents $+0110$.

2.7.2 Complement Representation

For a number system with radix r, let the radix-complement of a number N be denoted by $\bar{\bar{N}}$ and the diminished-radix by $\bar{N}$. Then $\bar{\bar{N}}$ and $\bar{N}$ are defined below:

$$\bar{\bar{N}} = r^n - N$$

$$\bar{N} = r^n - N - r^{-m}$$

where n is the number of integral digits and m the number of fractional digits in the number N. The radix-complement of $d_{n-1}, d_{n-2}, \ldots, d_1, d_0, d_{-1}, \ldots, d_{-m}$ can be formed by simply subtracting each digit d_i from $r - 1$

and then adding r^{-m} to the number. For integral numbers, $r^{-m} = r^{-0} = 1$; and for fractional numbers or mixed point numbers, adding r^{-m} corresponds to adding 1 at the least significant position of the number. For example, the radix-complement for 2457 is 7543, and the radix-complement of 653.72 is 346.28.

The diminished-radix-complement of a given number is formed by simply subtracting each digit from $r - 1$. For example, the diminished-radix-complement of 2457 is 7542, and the diminished-radix-complement of 653.72 is 346.27. Likewise, the diminished-radix-complement of the binary number 0101 is 1010, and the diminished-radix-complement of 0101.110 is 1010.001.

In the decimal number system, the radix-complement is called the 10's complement, and the diminished-radix-complement is called the 9's complement. Similarly, the radix-complement of a binary number is called the 2's complement, and the diminished-radix-complement of a binary number is called the 1's complement.

2.7.3 Signed Number Representation

To represent both positive and negative numbers, a sign bit is added to the left of the number digits. According to our previous notation, this digit would be d_n. A 0 in the sign bit position will represent a positive number, while a 1 in the sign position will represent a negative number. The other digits of the number are called the number digits. The signed numbers can be in one of three different forms:

1. Signed-magnitude representation
2. Radix-complement representation
3. Diminished-radix-complement representation

Both of the complement representations are used for negative numbers only. Positive numbers in any of the complement representations will not appear complemented. They will appear in their true form. Hence, a positive number will appear in the same form in all three representations.

A negative number in signed-magnitude representation is represented by a 1 in the sign position, while the number digits hold the magnitude of the number. In the radix-complement, a negative number will be represented by a 1 in the sign position, while the digit numbers represent the radix-complements of the magnitude. For illustration, Table 2.5 shows the three representations of some signed numbers in the binary system. The leftmost bit (the underlined bit) is interpreted as the sign of the number.

***Table 2.5** Representation of Signed Numbers*

	Number		
Representation	(+11)	(−11)	(−5)
Sign-magnitude	0,1011	1,1011	1,0101
Radix-complement	0,1011	1,0101	1,1011
Diminished-radix-complement	0,1011	1,0100	1,1010

The advantages of the complement representation of signed numbers lie in the fact that subtraction of a positive number can be performed by addition of a negative number. Since negative numbers can be represented easily in the complement form, the addition and subtraction algorithm is easier in the complement representation than the corresponding algorithm with signed-magnitude representation. This point will become clear in the next chapter.

REFERENCES

Chu, Y., *Digital Computer Design Fundamentals.* New York: McGraw-Hill, 1962.

Flores, I., *The Logic of Computer Arithmetic.* Englewood Cliffs, N.J.: Prentice-Hall, 1962.

Richards, R. K., *Arithmetic Operation in Digital Computers.* New York: Van Nostrand Reinhold, 1955.

Scott, N. R., *Electronic Computer Technology.* New York: McGraw-Hill, 1970.

Stein, M. L., and W. D. Munro, *Introduction to Machine Arithmetic.* Reading, Mass.: Addison-Wesley, 1972.

PROBLEMS

2.1. Convert the following integer binary numbers into decimal:
(a) 110111 (b) 111000 (c) 010101 (d) 101010

2.2. Convert the following fractional binary numbers into decimal:
(a) 0.1010 (b) 0.11 (c) 0.001 (d) 0.11001

2.3. Convert the following binary numbers into decimal:
(a) 101010.011 (b) 1111110.011001

2.4. Convert the following decimal numbers into binary:
(a) 25 (b) 69 (c) 102 (d) 135 (e) 284

2.5. Convert the following fractional decimal numbers into binary:
(a) .8750 (b) .593 (c) .03125 (d) .370

2.6. Convert the following decimal numbers into binary:
(a) 276.53125 (b) 99.4375 (c) 53.5625

2.7. Convert the following numbers as indicated:
(a) $(125.3)_8 = (?)_{10}$ (b) $(119)_{10} = (?)_8$

2.8. Convert the following numbers as indicated:
(a) $(1011010)_2 = (?)_8$ (b) $(1011101.0110)_2 = (?)_8$
(c) $(372.57)_8 = (?)_2$ (d) $(4757)_8 = (?)_2$
(e) $(54F2)_{16} = (?)_2$ (f) $(10110101110.011)_2 = (?)_{16}$

2.9. Convert the following octal and hexadecimal numbers:
(a) $(673.124)_8 = (?)_{16}$ (b) $(26153.7406)_8 = (?)_{16}$
(c) $(306.D)_{16} = (?)_8$ (d) $(2C6B.F2)_{16} = (?)_8$

2.10. Obtain the 1's and 2's complement of the following binary numbers: 101010, 110010, 00001, 010011.011, 00000, and 01110.110

2.11. Obtain the 9's and 10's complement of the following decimal numbers: 13578, 0990, 10000, 9090, 0000, 3721.99, and 5637.99

3

Arithmetic Operations

Arithmetic operations involve the process of data manipulation. The basic arithmetic operations to be described are shifting, addition, subtraction, multiplication, and division. Ordinary procedures to perform the above arithmetic operations prove generally unsuitable for digital computers, and hence new arithmetic algorithms suitable for machine implementation will be developed.

In Chapter 2, we discussed three representations of binary numbers, signed-magnitude, signed 2's complement, and signed 1's complement. In this chapter, algorithms for addition, subtraction, multiplication, and division will be developed utilizing these number representations. Algorithms to perform these arithmetic operations are to be developed without presentation of the hardware components required for their implementation. Details of the computer subsystem, that is, the arithmetic unit, will be described in Chapter 8.

In Chapter 2, we talked about numbers having their radix point at the extreme right, the extreme left, or anywhere between the digits of the number. The first case refers to integers; the second case corresponds to fractions. Both integers and fractions are called fixed-point numbers. The third case refers to numbers with both integral and fractional components and are referred to as floating-point numbers. Machine arithmetic is done either in fixed-point mode or floating-point mode. Fixed-point, fractional arithmetic is the most widely used in digital computers. In this mode, the radix point, although it does not physically exist inside the machine, is considered to be to the left of the most significant digit of the number digits or, in other words, just to the right of the sign bit of a signed number.

In Sections 3.1–3.4, only fixed-point arithmetic is considered. Floating-point arithmetic is treated in Section 3.5 of this chapter.

The position of the radix point does not have any effect on the result of addition and subtraction. Therefore, the methods of addition and subtraction which will be described in the subsequent sections apply to both integers and fractions. The position of the radix point is very important in the case of multiplication and division. Multiplication of fractions has two specific advantages. First, no overflow is expected (underflow might occur); second, the location of the radix point remains unchanged after multiplication. The difference between integer division and fractional division is treated when division is discussed.

3.1 SHIFTING OF SIGNED BINARY NUMBERS

In a positional number system, shifting the number to the left or to the right by one position with respect to the radix point is equivalent to multiplication or division by the radix, respectively. For example, the binary number 0,0101. (+5) when shifted to the left one place gives 0,1010., which is $5 \times 2^1 = 10$. Similarly, shifting the same number one position to the right results in 0,0010.1, which is 5/2, assuming that we are not restricted to a fixed number of digits. If we maintain the same number of digits, then the right shift yields 0,0010., and any odd number will be scaled down, thus losing some accuracy, which is similar to roundoff in division. For signed numbers, multiplication or division by 2 does not change the sign of the numbers. Therefore, a shifting algorithm for signed numbers should always keep the sign unchanged. Shifting to the left or to the right will always create a vacant position which has to be filled with either a 0 or a 1. The correct choice of adding a 0 or 1 depends on the sign of the number and also on the type of negative number representation.

A positive number has the same representation whether in signed-magnitude, 1's complement, or 2's complement. This representation consists of a 0 in the sign bit followed by the magnitude of the number. To multiply by 2, shift the number to the left, leaving the sign bit unchanged and enter a 0 in the rightmost position. The result is correct.

For a negative number, with the sign bit represented by a 1 for all three representations, the shifting algorithm depends on the particular representation. In all three representations, the number digits are shifted without changing the sign bit. The new digits to be entered to the left or to the right of the number digits (to fill the vacant positions) depend on the type of representation of negative numbers. The added digits to fill the vacant positions in the various number representations are given in Table 3.1, and the shifting process is illustrated in Examples 3.1(a) and 3.1(b).

Table 3.1 Added Digits for Shifting Signed Binary Numbers

Number representation	Added digits
Positive number	Add 0s
Negative number in signed-magnitude representation	Add 0s
Negative number in 2's complement representation	Shifting left, add 0s Shifting right, add 1s
Negative number in 1's complement representation	Add 1s

Example 3.1 (a)

This example shows the shift to the left or to the right of the positive signed binary number A = 0,1101. The added digits are shown in parentheses.

$A = 0{,}1101 = +13$	before shifting
$A \times 2^{1} = 0{,}1101(0) = +26$	by shifting one position to the left
$A \times 2^{2} = 0{,}1101(00) = +52$	by shifting two positions to the left
$A \times 2^{-1} = 0{,}(0)110.1 = +(\frac{13}{2}) = +6.5$	by shifting one position to the right
$A \times 2^{-2} = 0{,}(00)11.01 = +(\frac{13}{4}) = +3.25$	by shifting two positions to the right

Example 3.1 (b)

This example shows the shift to the left or to the right of the negative binary number A = −17 in the three different representations of negative numbers. The added digits are shown in parentheses.

1's Complement representation:

$$A = 1{,}01110 = -17$$
$$A \times 2^{1} = 1{,}01110(1) = -34$$
$$A \times 2^{2} = 1{,}01110(11) = -68$$
$$A \times 2^{-1} = 1{,}(1)0111.0 = -\tfrac{17}{2}$$
$$A \times 2^{-2} = 1{,}(11)011.10 = -\tfrac{17}{4}$$

2's Complement representation:

$$A = 1{,}01111 = -17$$
$$A \times 2^{1} = 1{,}01111(0) = -34$$
$$A \times 2^{2} = 1{,}01111(00) = -68$$

$$A \times 2^{-1} = 1,(1)0111.1 = -\tfrac{17}{2}$$
$$A \times 2^{-2} = 1,(11)011.11 = -\tfrac{17}{4}$$

Signed-magnitude representation:

$$A = 1,10001 = -17$$
$$A \times 2^{1} = 1,10001(0) = -34$$
$$A \times 2^{2} = 1,10001(00) = -68$$
$$A \times 2^{-1} = 1,(0)1000.1 = -\tfrac{17}{2}$$
$$A \times 2^{-2} = 1,(00)100.01 = -\tfrac{17}{4}$$

Notice that in the above two examples the added digits follow the rules in Table 3.1. The result is the exact product or quotient only if the number of digits in the numbers are allowed to increase as in the examples. If the number of digits are fixed, shifting to the left leads to losing the leftmost bits and the result is equal to the product only if the lost digits represent zero magnitude. In shifting to the right, dropping the rightmost digits amounts to less accuracy if the lost digits are of nonzero magnitude.

3.2 BINARY ADDITION AND SUBTRACTION

Addition and subtraction are the most frequent arithmetic operations performed by a machine, since most other arithmetic operations can be performed by a process of addition, subtraction, and shifting. The computer subsystem where arithmetic operations are performed is referred to as the arithmetic unit. Although having a built-in subtractor as well as an adder as part of the arithmetic unit seems to be necessary, many computers have either adder(s) or subtractor(s) built in. In general, any piece of hardware which can add signed numbers is capable of subtraction. For this reason, if we talk about algebraic addition, subtraction capabilities are implied, and the subtraction operation is not in any way different from the addition of signed numbers.

Addition and subtraction of two signed binary numbers depends on the representation of negative numbers. The addition and subtraction methods corresponding to each of the three negative numbers representations are considered separately in the following subsections.

3.2.1 Signed-Magnitude I

In this section, we shall assume that all the numbers are stored in the machine in the signed-magnitude representation. After the addition or the subtraction is performed on the two numbers, the resulting number has to be in the signed-magnitude form before it is stored back in the machine. Let the two signed binary numbers be A and B and their sum C, where each number is of fixed length, consisting of the magnitude of the number plus a sign bit. In other words, let A, B, and C have the form below,

$$A = (A_0, A_1, A_2, \ldots, A_n) \quad \text{and} \quad A^* = (A_1, \ldots, A_n)$$
$$B = (B_0, B_1, B_2, \ldots, B_n) \quad \text{and} \quad B^* = (B_1, \ldots, B_n)$$
$$C = (C_0, C_1, C_2, \ldots, C_n) \quad \text{and} \quad C^* = (C_1, \ldots, C_n)$$

where A_0, B_0, and C_0 are the sign bits of A, B, and C, respectively. The number digits, which in this case represent the magnitude of the number, consist of the rightmost n bits and are denoted by A*, B*, and C*. The restriction of fixed-length words implies that C has the same length as both A and B. That is, the sum of A and B does not exceed n bits. If the sum does exceed n bits, this will be called an overflow.

Let us begin by reviewing the paper and pencil steps of adding B to A. These steps are described below.

Step 1. Compare the two sign bits of the two numbers. If they are the same, do step 2. If they are different, do step 3.

Step 2. Add the magnitudes to form the sum C. The sign of the sum is the same as the sign of both A and B.

Step 3. Since the signs are different, we subtract the number with the smaller magnitude from the one with the larger magnitude. The sign of the result is the same as that of the number with the larger magnitude.

Subtraction can be performed by simply adding the negative of the number to be subtracted. In other words A − B can be substituted with A + (−B). Hence, in this section no special treatment for subtraction is necessary other than changing the sign of B prior to addition.

In the above steps for addition and subtraction, there was no concern about fixed-length words, and consequently there is no overflow problem. For machine arithmetic with fixed-length words, overflow must be checked. In this case the overflow could occur only if both numbers have the same sign. The check for overflow is considered in step 2 before attaching the sign to the sum. If there is an overflow after adding A* and B*, it can be detected by a 1, in the sign bit position of the sum C.

An algorithm for addition and subtraction is shown in Fig. 3.1. This algorithm follows the same steps for paper and pencil steps described above, except for the provision of an overflow check. When A and B are of different signs, a magnitude comparison and a possible change of operands are performed. For machine implementation, this could be time consuming, and

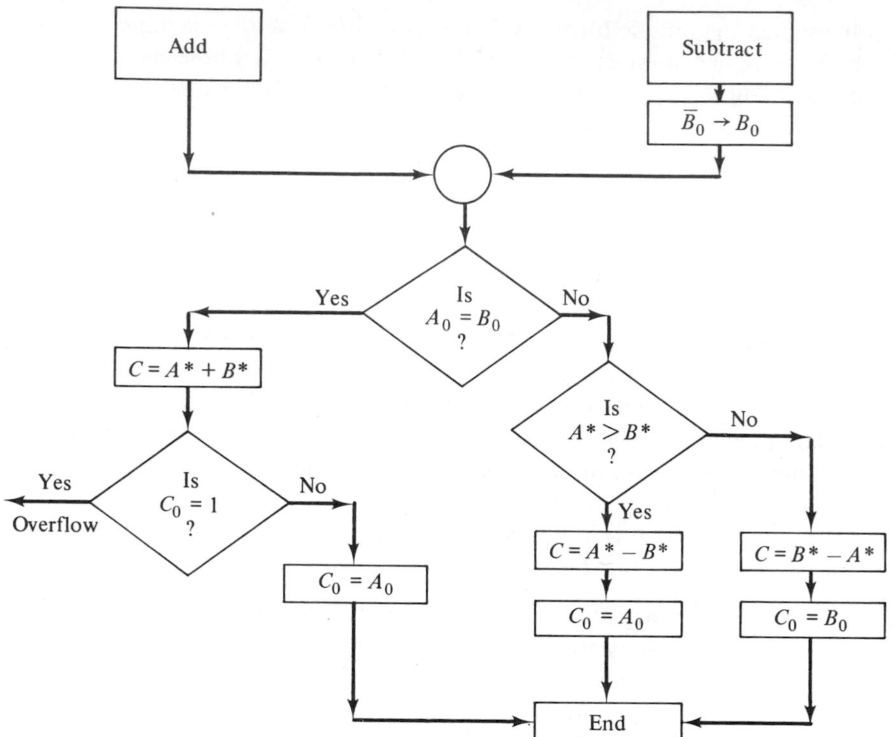

Figure 3.1 Addition and subtraction algorithm for numbers in signed-magnitude representation I.

a modification of the algorithm to eliminate this is possible. Such a modified algorithm is described in the subsequent paragraphs.

Let us assume that we shall always either add B* to A* or subtract B* from A*. In the case of both A and B having the same sign, no change in the previous algorithm is necessary. If A and B are of different signs, B* is subtracted from A* without prior comparison. There are two possible cases.

Case 1: A is positive, and B is negative (either B is originally negative and the process is addition or B is originally positive and the process is subtraction). This case is characterized by $A_0 = 0$, $B_0 = 1$.

a. A* is larger than B*, the difference A* − B* is positive, and consequently no borrow is generated. Hence, A* − B* corresponds to the correct value of C*, and the sign C_0 must be taken as A_0.

b. If A* is less than B*, the difference C = A* − B* is negative. As we recall, the importance of the complements lies in the fact that they can represent negative numbers. Hence, C* in this case is expected to be in 2's complement form. This is indicated by the appearance of a borrow in the sign bit. Therefore, if a borrow is generated, the result is negative and the correct magnitude C* is the 2's complement of A* − B*. The sign must be taken as $C_0 = \bar{A}_0$, where $\bar{A}_0$ is the complement of A_0.

Case 2: A is negative, and B is positive. That is, $A_0 = 1$, $B_0 = 0$.

a. If A* is larger than B*, the difference A* − B* is positive, and consequently no borrow is generated in the sign bit. Hence, C* = A* − B* and $C_0 = A_0$.

b. If A* is less than B*, the difference A* − B* is negative, and hence a borrow is generated in the sign bit. Hence, C* = the 2's complement of A* − B* and $C_0 = \bar{A}_0$.

In summary, if the signs of A and B are different, perform A* − B*. If no borrow is generated in the sign bit, C* = A* − B* is in correct form and the correct sign is $C_0 = A_0$. If a borrow is generated, C* is in 2's complement form and $C_0 = \bar{A}_0$. Figure 3.2 shows this modified algorithm of the addition and subtraction for numbers in signed magnitude representation. If a borrow is generated as a result of performing A* − B*, this borrow appears

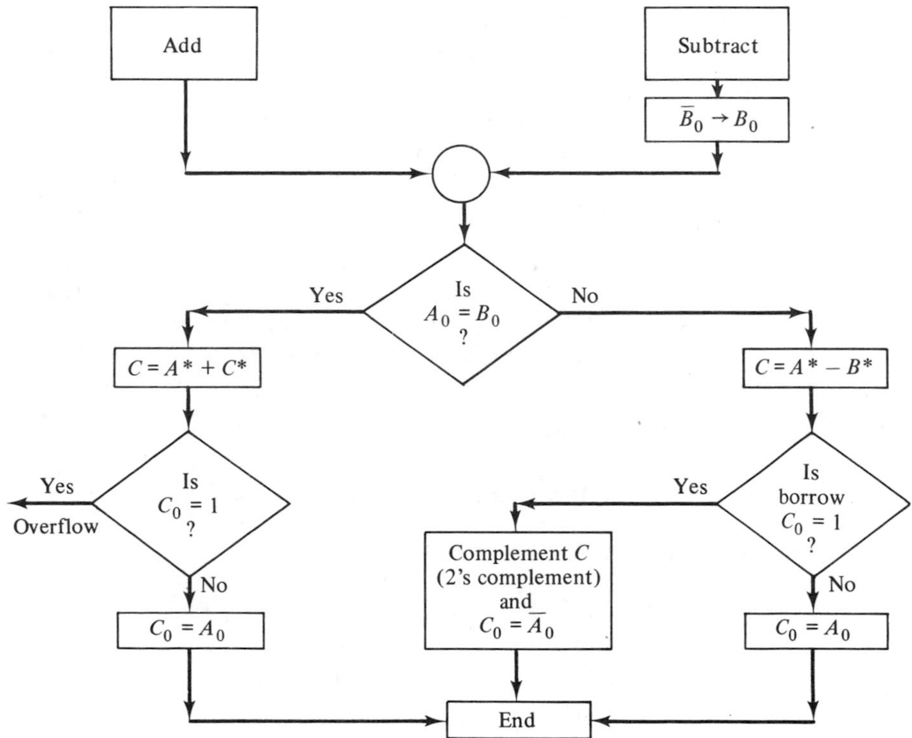

Figure 3.2 Addition and subtraction algorithm (modified) for numbers in signed-magnitude representation I.

in the position C_0. Hence, the check for borrow generation is done by checking C_0 immediately after the subtraction $A^* - B^*$, and then the correct sign for the results is inserted in the C_0 position.

Example 3.2

This example shows the addition of two 5-bit numbers (plus a sign bit). Only the cases where A and B have different signs are considered. The cases of A and B having the same sign is straightforward.

Case 1: A is positive, and B is negative.

a. For $A = +13 = 0{,}01101$ and $B = -11 = 1{,}01011$,

$$A^* = 01101, \quad B^* = 01011, \quad A_0 = 0, B_0 = 1$$

$$A^* - B^* = 01101$$
$$-01011$$
$$00010 \quad \text{no borrow (i.e., } C_0 = 0)$$
$$\therefore C^* = A^* - B^* = 00010$$
$$C_0 = A_0 = 0$$
$$\therefore C = 0{,}00010$$

b. For $A = +11 = 0{,}01011$ and $B = -13 = 1{,}01101$,

$$A^* = 01011, \quad B^* = 01101, \quad A_0 = 0, B_0 = 1$$

$$A^* - B^* = 01011$$
$$-01101$$
$$1\ 11110$$
$$\uparrow \text{ borrow (i.e., } C_0 = 1)$$
$$\therefore C^* = \text{2's complement of } (A^* - B^*) = 00010$$
$$C_0 = \bar{A}_0 = 1$$
$$\therefore C = 1{,}00010$$

Case 2: A is negative, and B is positive.

a. For $A = -13 = 1{,}01101$, $B = +11 = 0{,}01011$

$$A^* = 01101, \quad B^* = 01011, \quad A_0 = 1, B_0 = 0$$

$$A^* - B^* = 01101$$
$$-01011$$
$$00010 \quad \text{no borrow (i.e., } C_0 = 0)$$
$$\therefore C^* = A^* - B^* = 00010$$

$$C_0 = A_0 = 1$$
$$\therefore C = 1{,}00010$$

b. For A = −11 = 1,01011 and B = +13 = 0,01101,

$$A^* = 01011, \quad B^* = 01101, \quad A_0 = 1, \quad B_0 = 0$$

$$\begin{aligned} A^* - B^* = \; & 01011 \\ & \underline{-01101} \\ 1\; & 11110 \end{aligned}$$

$\uparrow$ borrow (i.e., $C_0 = 1$)

$$\therefore C^* = \text{2's complement of } (A^* - B^*) = 00010$$
$$C_0 = \bar{A}_0 = 0$$
$$\therefore C = 0{,}00010$$

To implement the above algorithm, we must have both an adder and a subtractor. The adder is used if both numbers have the same sign, while the subtractor is used when the two numbers have different sign bits.

The algorithms to be introduced in the subsequent sections require only an adder. Other algorithms which make use of only a subtractor exist but are not presented here.

3.2.2 1's Complement

In signed 1's complement representation, a negative number is represented by a 1 in the sign bit and the 1's complement of its magnitude in the number digits. Since A − B is equivalent to A + (−B), the subtraction operation can be replaced by complementation and addition. Hence, in developing an algorithm for addition and subtraction, we need only be concerned about algebraic addition of signed numbers in 1's complement representation.

Let A and B be two signed binary numbers in 1's complement representation and let C be their sum,

$$A = (A_0, A_1, \ldots, A_n)$$
$$B = (B_0, B_1, \ldots, B_n)$$
$$C = (C_{-1}, C_0, C_1, \ldots, C_n)$$

where A_0, B_0, and C_0 are the sign bits of A, B, and C, respectively. To add A and B, we simply add all the bits of A and B including the sign bits. If A is positive, it is stored as the value of its magnitude A^*, where $A^* < 2^n$. If A is negative, it is stored as $2^{n+1} - A^* - 1$.

There are three possible basic cases which might arise:

Case 1: When A is positive and B is positive, the addition of all the bits including the sign bits is straightforward. Since both numbers are positive, their sign bits are 0s, and the sum of these two bits is a 0, unless a carry from the most significant bit (MSB) position occurs. Hence, a 1 in the C_0 position indicates an overflow, and the C_{-1} position will always contain a 0.

Case 2: When a positive number A is added to a negative number B, their sum can be written as

$$C = A^* + 2^{n+1} - B^* - 1 = A^* - B^* + 2^{n+1} - 1$$

There are two situations:

a. If $A^* > B^*$, the sum is expected to be positive and equal to $A^* - B^*$. Recall that any positive number is characterized by $A^* < 2^n$, while 2^{n+1} represents a 1 in the C_{-1} position. A 1 in the C_{-1} position represents an end-around carry. If this end-around carry is added to the sum while it is dropped from the C_{-1} position, this results in

$$C = A^* - B^* + 2^{n+1} - 1 + (1 - 2^{n+1}) = A^* - B^* > 0$$

Since $A_0 = 0$, $B_0 = 1$, this single-bit addition of A_0 and B_0 results in a 1 in the C_0 position. But since we have an end-around carry generated in the C_{-1} position, this means that this carry is propagated from the C_1 through the C_0 into the C_{-1} position. Consequently, C_0 will contain a 0, which is the correct sign.

b. If $A^* < B^*$, the correct sum must be negative and equal to $B^* - A^*$. The sum can be written as

$$A^* - B^* + 2^{n+1} - 1 = 2^{n+1} - 1 - (B^* - A^*)$$

Since $B^* > A^*$, $(B^* - A^*)$ is positive, and consequently $2^{n+1} - 1 - (B^* - A^*)$ represents the 1's complement of $A^* - B^*$, which is the correct result.

Case 3: When both A and B are negative numbers, their sum can be written as

$$\begin{aligned} C &= (2^{n+1} - A^* - 1) + (2^{n+1} - B^* - 1) \\ &= (2^{n+1} - 1) + (2^{n+1} - (A^* + B^*) - 1) \end{aligned}$$

The 2^{n+1} shown in the left parenthesis represents an end-around carry. Adding the end-around carry, as in Case 2, results in the following value of C:

$$C = [2^{n+1} - (A^* + B^*) - 1]$$

The above expression represents $(A^* + B^*)$ in 1's complement, which is the correct result. Again in this case an overflow is possible, and if it occurs, it will be indicated by a 0 in the C_0 position.

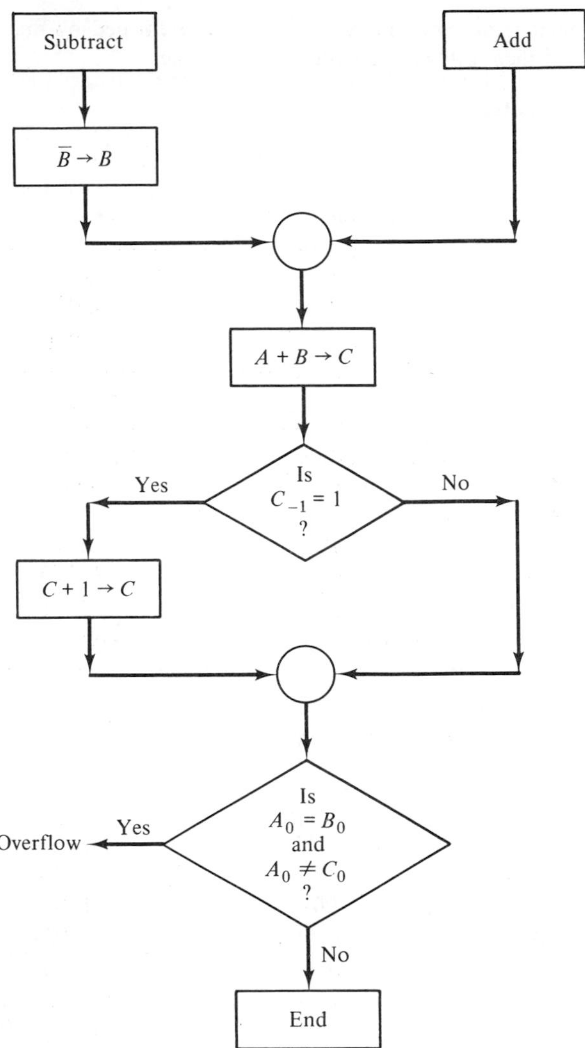

Figure 3.3 Algorithm for addition and subtraction in 1's complement.

A summary of the addition and subtraction algorithm in 1's complement is shown in Fig. 3.3. Notice that if the result of the addition or subtraction is negative, then it is in 1's complement form.

Example 3.3

In this example, we shall consider four bit plus sign numbers and apply the previous algorithms to illustrate the different cases of the addition of signed numbers in 1's complement representation. Subtraction is performed by first converting it into addition and then following the addition example.

Case 1: A is positive, and B is positive.

	Signed-magnitude notation		*1's Complement notation*
A =	+0011	=	0,0011
B =	+0100	=	0,0100
	+0111		0,0111

	Signed-magnitude notation		*1's Complement notation*
A =	+1011	=	0,1011
B =	+1100	=	0,1100
	1 0111 (carry)		1,0111 (overflow)

Case 2: a. A is positive, and B is negative, A* ≥ B*.

	Signed-magnitude notation		*1's Complement notation*
A =	+1001	=	0,1001
B =	−0100	=	1,1011
	+0101		1 0,0100
			→ 1 end-around carry
			0,0101

	Signed-magnitude notation		*1's Complement notation*
A =	+0011	=	0,0011
B =	−0011	=	1,1100
	+0000		1,1111 (negative 0)

b. A is positive, and B is negative, A* < B*.

	Signed-magnitude notation		*1's Complement notation*
A =	+0011	=	0,0011
B =	−1100	=	1,0011
	−1001		1,0110

Case 3: A is negative, and B is negative.

	Signed-magnitude notation		*1's Complement notation*
A =	−0100	=	1,1011
B =	−0111	=	1,1000
	−1011		1 1,0011
			→ 1
			1,0100

3.2.3 2's Complement

When negative numbers are represented in the 2's complement system, the addition and subtraction algorithm is similar to that in the 1's complement system. However, the 2's complement system has the advantage of not requiring an end-around carry during addition. Using similar mathematical steps as those used in the different cases of 1's complement, the reader can easily justify that in the 2's complement system the sum of the numbers will be correct if the carry in the C_{-1} position is disregarded whenever it is generated. Therefore, the algorithm for addition and subtraction in 2's complement is the same as that shown in Fig. 3.3 for 1's complement except that the end-around carry (C_{-1}) check is not needed and hence can be omitted. Again, if the two binary numbers in 2's complement have a negative sum, the result will be in 2's complement form. Example 3.4 illustrates the addition and subtraction in 2's complement representation.

Addition and subtraction in 2's complement representation requires negative numbers to be in 2's complement. The 2's complement of a number can be obtained by first obtaining the 1's complement, which is a bit-by-bit complement, and then adding 1 to the least significant bit of the register. In parallel machines this process is time consuming, and hence it is considered a disadvantage when the 2's complement is compared to the 1's complement. However, in serial machines, obtaining the 2's complement could be performed without any extra time. More on this point is presented in Chapter 8.

Example 3.4

In this example, the addition of signed numbers in 2's complement representation is illustrated for the corresponding different cases discussed in 1's complement.

Case 1: This case is identical to Case 1 in the 1's complement.

Case 2: a. A is positive, and B is negative, $A^* \geq B^*$.

	Signed-magnitude notation		*2's Complement notation*
A =	+0111	=	0,0111
B =	−0011	=	1,1101
	+0100		1 0,0100
			carry is discarded

	Signed-magnitude notation		*2's Complement notation*
A =	+0101	=	0,0101
B =	−0101	=	1,1011
	+0000		1 0,0000
			carry is discarded

b. A is positive, and B is negative, $A^* < B^*$.

	Signed-magnitude notation		*2's Complement notation*
A =	+0100	=	0,0100
B =	−1000	=	1,1000
	−0100		1,1100

Case 3: A is negative, and B is negative.

	Signed-magnitude notation		*2's Complement notation*
A =	−0011	=	1,1101
B =	−0100	=	1,1100
	−0111		1 1,1001
			discarded

3.2.4 Signed-Magnitude II

The signed-magnitude algorithm presented in Section 3.2.1 requires, for its implementation, both an adder and a subtractor. Moreover, the previous sections have indicated that algebraic addition in 1's or 2's complement notation is less complicated than in signed-magnitude. Another simpler algorithm for algebraic addition in signed-magnitude which uses only an adder is described below. A flow chart is given in Fig. 3.4.

Let us consider the algebraic addition of two numbers A and B in signed magnitude representation. There are four possible cases:

1. (+A) + (+B).
2. (+A) + (−B).
3. (−A) + (+B).
4. (−A) + (−B).

The most straightforward solution for the first and the last case is to add the magnitude of A and the magnitude of B. Since the numbers are in signed-magnitude, the addition of the magnitudes gives the correct magnitude of their sum. For either case the sign of the sum is the same as the original sign of any of the operands, and a carry into the sign bit indicates an overflow.

In Case 2 or Case 3, a possible solution is to complement (2's complement) the magnitude of the negative number and add it to the magnitude of the positive one. Case 2 is investigated below.

Case 2:

$$A^* + \overline{\overline{B}}^* = A^* + 2^n - B^* = 2^n + (A^* - B^*)$$

where $\overline{\overline{B}}^*$ represents the 2's complement of the magnitude of B. If A^* is greater than B^*, then $(A^* - B^*)$ is positive and 2^n represents an end carry (note that the number digits are of length n). However, the expected result is positive, and thus if an end carry occurs, the magnitude of the sum is correct and its sign is positive. When A^* is less than B^*, the above relation can be written as

$$A^* + \overline{\overline{B}}^* = 2^n - (B^* - A^*)$$

and since $(B^* - A^*)$ is positive, $2^n - (B^* - A^*)$ represents the 2's complement of $(B^* - A^*)$ Hence, if no end carry occurs, the result of the addition is the correct sum but in 2's complement form. For the sum to be in signed-magnitude, recomplement (2's complement) and attach a negative sign.

Case 3 can be investigated in a similar way. In fact, interchanging A^* and B^* in the above discussion leads to the same action. This algorithm does not require subtraction.

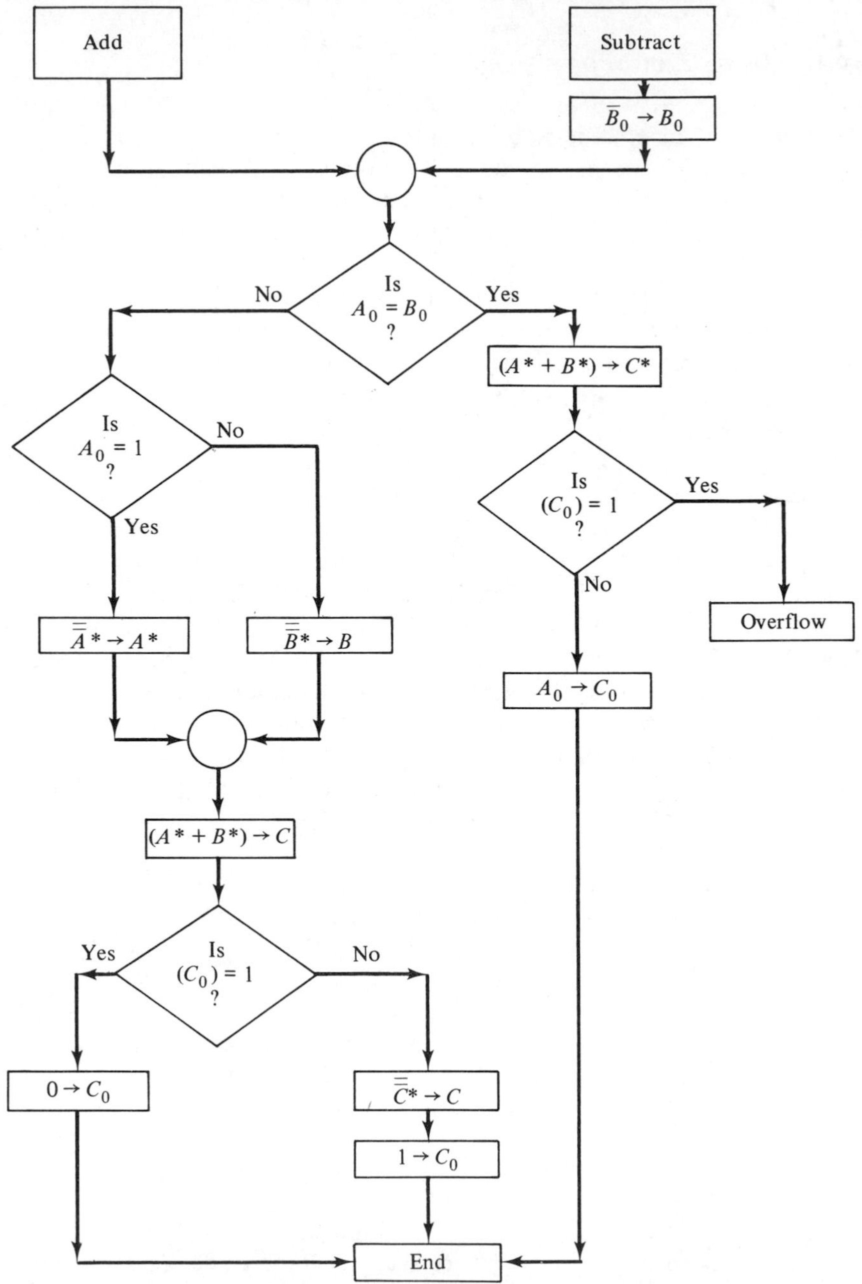

Figure 3.4 Flow chart for algebraic addition and subtraction in signed-magnitude II.

Example 3.5

In this example, addition and subtraction using the signed-magnitude II algorithm is illustrated for all the possible cases.

Case 1: A is positive, and B is positive.

a.

$$\begin{array}{lr} A = 0{,}0100 & A^* = 0100 \\ B = 0{,}1001 & B^* = \underline{1001} \\ & C = A^* + B^* = 1101 \end{array}$$

b.

$$\begin{array}{lrl} A = 0{,}1001 & A^* = \quad 1001 & A_0 = 0 \\ B = 0{,}1011 & B^* = \quad \underline{1011} & B_0 = 0 \\ & C = A^* + B^* = 1\ 0100 & \end{array}$$

$C_0 = 1$ indicates an overflow,

Case 2: A is positive, and B is negative.

a.

$$\begin{array}{lr} A = 0{,}0111 & A^* = \quad 0111 \\ B = 1{,}1001 & \bar{\bar{B}}^* = \quad \underline{0111} \\ & C = A^* + \bar{\bar{B}}^* = 0{,}1110 \\ C_0 = 0; & \therefore C = \overline{\overline{0{,}1110}} = 1{,}0010 \end{array}$$

b.

$$\begin{array}{lr} A = 0{,}0111 & A^* = \quad 0111 \\ B = 1{,}0011 & \bar{\bar{B}}^* = \quad \underline{1101} \\ & C = A^* + \bar{\bar{B}}^* = 1{,}0100 \\ C_0 = 1; & \therefore C = 0{,}0100 \end{array}$$

Case 3: A is negative, and B is positive. This case is identical to Case 2 with A and B interchanged.

Case 4: A is negative, and B is negative.

$$\begin{array}{lr} A = 1{,}0111 & A^* = \quad 0111 \\ B = 1{,}0011 & B^* = \quad \underline{0011} \\ & C = A^* + B^* = 0{,}1010 \end{array}$$

3.3 BINARY MULTIPLICATION

The multiplication of two signed binary numbers depends on the representation of the negative numbers. Contrary to addition and subtraction, discussed in Section 3.2, multiplication of signed binary numbers in the signed-magnitude representation is less complicated than multiplication of signed complement numbers. Hence, in designing an arithmetic unit, the choice of negative number representation is a compromise between these factors.

The simplest and most straightforward method of multiplication (similar to paper and pencil multiplication) of signed binary numbers is known as multiplication by repeated addition. This method, as described in the following subsection, is directly applicable to signed binary numbers in the signed-magnitude representation of numbers. When the machine uses signed complement representation, the negative numbers must be transformed to a signed-magnitude representation first. Other implementations of multiplication by repeated addition which apply directly to signed complement numbers are the following:

1. Burks-Goldstine-von Neumann method.
2. Robertson's first method.
3. Robertson's second method.
4. Booth's method.

High-speed multiplication methods will be described in subsequent sections. The details and variations of all these methods depend on the negative number representation used; the size of the product to be derived, i.e., single-length or double-length; whether integers or fractions are to be multiplied, and whether a parallel or serial computer is under consideration.

3.3.1 Multiplication by Repeated Addition

This method is used for numbers in signed-magnitude representation. Let us first review the simple paper and pencil multiplication of two signed binary numbers. The multiplication steps are shown in Fig. 3.5 for the two binary numbers A = +1011 and B = −1001.

```
      +1011    multiplicand
×     −1001    multiplier
-------------------------------
       1011    first partial product
      0000     second partial product
     0000      third partial product
    1011       fourth partial product
-------------------------------
   −1100011    product A × B
```

Figure 3.5 Paper and pencil multiplication.

In the paper and pencil multiplication, we simply form all the partial products and then add them in the proper position. The sign of the product is positive whenever the signs of the two numbers are equal, i.e., whenever both A and B are positive or both A and B are negative. Otherwise the sign of the product is negative. However, since it is easier for the machine to add two numbers at one time, the partial products must be accumulated individually toward the final results. This leads to the repeated addition method. The

```
    1011    multiplicand
    1001    multiplier
-----------------------------------------------
00000000    start with 0 result at beginning
   +1011    first partial product
--------
00001011    result after adding first partial product
  +0000     second partial product
--------
00001011    result after adding second partial product
  0000      third partial product
--------
00001011    result after adding third partial product
+1011       fourth partial product
--------
01100011    result after adding fourth partial product
            = product (A × B)
```

Figure 3.6 Example of multiplication by repeated addition.

steps for the multiplication by repeated addition are rewritten, for the same example, in Fig. 3.6. The sign bit of the product is obtained as the logical exclusive-OR of the two sign bits of A and B. Therefore, only the number digits of A and B are involved in the repeated addition process, while the sign bit of each number is used to form the sign of the product.

When the multiplicand and the multiplier have the same fixed length, their product is of double-length. If only a single-length product is required, roundoff is used to reduce the number of digits representing a number. The algorithm for the multiplication by repeated addition can be stated as

1. Start with accumulated product equal to zero.
2. Inspect the multiplier bits individually, starting with the least significant bit.
3. Add the multiplicand to the accumulated product if the multiplier bit is a 1; otherwise add 0s.
4. Shift the multiplicand by 1 bit to the left.
5. Go to step 2 and repeat the loop until all the partial products are added.

Detailed implementation for the multiplication by repeated addition for both parallel and serial computers together with some different register layouts is described in Chapter 8. The previous algorithm could be modified slightly to increase the speed of multiplication. Higher speed could be achieved by skipping over the zero digits of the multiplier. This simply means that if the multiplier digit is 0, then do not add the 0s. So for 0 in the multiplier, just shift the multiplicand and start to inspect the next bit of the multiplier. This modification is known as *skip over zero* or *shift across zero*, and the multiplication scheme which uses this method is known as multiplication by repeated addition and shift across zero.

3.3.2 High-Speed Multiplication Techniques

One idea for speedup techniques which has been proposed has turned out to be of little value. The basic idea is that the two operands could first be examined before the multiplication takes place to determine which has fewer 1s and interchange the multiplicand and multiplier, if necessary. The operand with the fewer 1s is used as the multiplier, thus decreasing the number of additions. Upon analysis it was found that as the number of bits per word becomes larger the improvement in multiplication time diminished toward zero. This is due to the fact that considering a large sample of numbers to be multiplied, the average number of 1s will approach $n/2$ for each number. For these multiplier-multiplicand pairs, the above technique will present no advantage. For small n, the actual time saved is significantly diminished by the extra time required for comparison and possible interchange.

The multiplication by repeated addition, as described in the previous subsection, required n additions and n shifts for n bit $\times$ n bit numbers. Shifting across zero reduces the number of additions by an average of 50%. Still the average number of additions required to form the final product is approximately $n/2$, which implies an average multiplication time proportional to n^2, assuming that the addition of two n-bit numbers is proportional to n. The implication is that multiplication can be a long process if long-length operands are used. With the multiplication by repeated addition in mind the two general approaches to decrease the overall multiplication time are

1. Use a more sophisticated multiplier decoding technique so that fewer addition cycles are required to form the final product.
2. Provide a faster adder, to reduce the time for each addition cycle.

Two of the techniques which have been developed as a result of the first approach are listed and briefly discussed below:

1. Multiplication by addition and subtraction.
2. Multiplication by uniform multiple shifts.

Multiplication by Addition and Subtraction Just as we were able to shift across a series of 0s, shifting across a series of neighboring 1s is also possible. If the multiplier contains a series of neighboring 1s, the individual additions can be replaced by a single addition and a single subtraction. Suppose that the multiplier is the binary number 011111; this requires five additions of the multiplicand according to the repeated addition scheme. However, an identical result could be obtained by performing

$$(\text{Multiplicand} \times 100000) - (\text{Multiplicand} \times 000001)$$

Now, this scheme requires only one addition and one subtraction. The combined idea of shifting across 0s and shifting across 1s is illustrated below.

Multiplier:	0 0 1 1 1 1 0 0 0 1 1 0	
Shifting across 0s:	+ + + + + +	add in these positions
Shift across 1s:	+ − + −	

The rules of multiplication by addition and subtraction can be expressed as follows:

Inspect the multiplier bits, starting with the least significant bit, and take the following actions:

1. If you encounter the first 1 in a series of 1s, then subtract the multiplicand from the accumulated product.
2. If you encounter a single 1, add the multiplicand to the accumulated product.
3. If you encounter the first 0 after a series of 1s, then add the multiplicand to the accumulated product.

In applying the above rules, the multiplicand has to be shifted an appropriate number of places, as in the normal multiplication procedure. This could mean that for each multiplier bit checked the multiplicand is shifted one place to the left. But much time can be saved if provisions are made for variable-length shifts. Hence, in this multiplication method the multiplicand could be shifted a number of places corresponding to the number of the multiplier digits inspected at any one time. Although this is a time-saving procedure, compared to a fixed shift by one place for every bit in the multiplier, it requires more hardware to implement. A compromise method, to gain speed and still maintain less hardware complexity, is provided by the multiplication using uniform multiple shifts.

The multiplication by addition and subtraction can be further speeded up. There are some instances where an isolated 0 in the multiplier occurs. In such an instance, the algorithm results in an addition followed by a subtraction which could be replaced by a single subtraction, as illustrated below:

Multiplier:	0 0 1 1 1 1 0 1 1 0
Shift across 0s:	+ + + + + +
Shift across 1s:	+ − + −
Single zero:	+ − −

Multiplication by Uniform Multiple Shifts This method uses shifts of uniform size to avoid the complexity of implementing a variable shift scheme. This is described below for the case of uniform shifts of two. Similar rules could be derived for a higher number of bits per shift.

Consider the case where the multiplier bits are to be inspected in groups of two at a time. A possible set of rules is given below:

Multiplier	*Action*
00	Nothing
01	Add 1 × multiplicand
10	Add 2 × multiplicand
11	Add 3 × multiplicand

This set of rules requires the addition of up to three multiples of the multiplicand. The first two rules are easy to implement. The third rule (add 2 × multiplicand) could be implemented easily assuming that we have a provision to shift the multiplicand one place to the left and then add it. The third rule needs two additions. Practically any addition of any multiple of the multiplicand which is a power of 2 (i.e., 0, 1, 2, 4, 8, etc.) is easy to implement. Therefore, it is preferred to have a set of rules which involves addition or subtraction of only multiplicand multiples which are a power of 2. Adding three multiples of the multiplicand is equivalent to subtracting the multipli-

Table 3.2 *Multiplication by Uniform Shifts of Two Positions*

Marker	*Multiplier group*	*Action*
0	00	Nothing
0	01	Add 1 × multiplicand
0	10	Add 2 × multiplicand
0	11	Subtract 1 × multiplicand and set marker to 1
1	00	Add 1 × multiplicand and clear marker to 0
1	01	Add 2 × multiplicand and clear marker to 0
1	10	Subtract 1 × multiplicand and set marker to 1
1	11	Set marker to 1

cand and then adding four multiples of the multiplicand. This is equivalent to subtracting the multiplicand, shifting it two places (multiply by 4), and remembering to add it when inspecting the next higher-order bit group of the multiplier. A possible set of rules which uses a marker to remember to add is given in Table 3.2. The inspection of the multiplier is done 2 bits at a time, starting with the least significant pair of bits. In each step, depending on the multiplier pair of bits, an action is taken, and the multiplicand is shifted two places to the left. The above set of rules (actions) is simply implemented by shifts and adds or subtracts. This technique is slower, when compared to the previous technique of multiplication by addition and subtraction, but eliminates the complexity of the variable shift implementation implied by the previous technique.

3.4 BINARY DIVISION

Just as multiplication can be performed by a series of additions and shifting, division can be performed by a series of subtractions and shifting. In this section, three methods of binary division are described:

1. Comparison method.
2. Restoring method.
3. Nonrestoring method.

While the comparison method employs a repeated subtraction technique, the restoring and nonrestoring methods utilize both addition and subtraction. The restoring method is a special case of the comparison method, where the comparison is performed by subtraction. Thus, the restoring division is more suitable for machine operation when compared to the comparison method.

The comparison and the restoring methods described here apply only to numbers in the signed-magnitude representation. The nonrestoring method applies to both the signed-magnitude and the 2's complement representations. There are equivalent methods which apply to the 1's complement representation, but they will not be presented.

3.4.1 Comparison Method

This method employs the technique of repeated subtraction and, as described, applies to signed-magnitude representation. First let us consider the paper and pencil division of binary numbers. This is illustrated in the following numerical example:

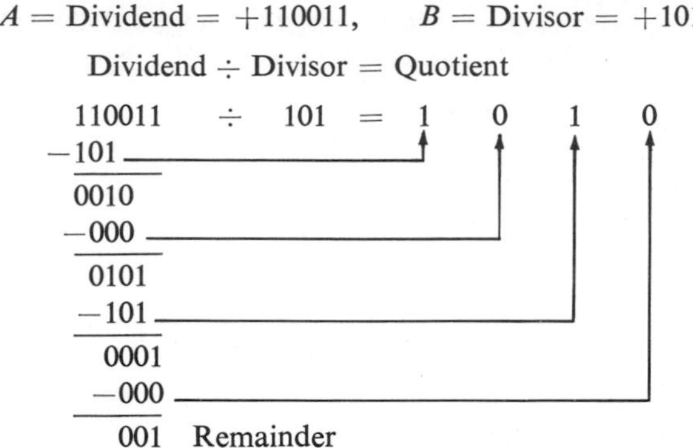

The bits of the quotient are obtained in serial fashion, 1 bit at a time. First compare the divisor with the most significant bits of the dividend. If the divisor is smaller than or equal to the high-order bits of the dividend, then the quotient bit is 1, and the divisor is subtracted from the dividend to form a partial remainder. Otherwise the quotient bit is a 0, and 0s are subtracted (actually, subtraction of 0s is not necessary). In either case the divisor is shifted one position to the right and compared with the new partial remainder (or modified dividend). This process continues until all the dividend bits are processed.

The paper and pencil division by comparison, as described above, is not suitable for machine implementation. Fixed-point machine arithmetic is performed either with fractional operands and generates a fraction result or with integer operands in which the result is an integer. Another restriction is that the machine must avoid both the division by 0 and the division of 0 by any number. In the first case, the result is infinity, which cannot be represented within the finite operand size of the machine. Dividing 0 by a finite number will always have a 0 result, and consequently it is a trivial operation and is not worth wasting the machine time. Hence, for fixed-point arithmetic, the range of the operands must be restricted, and machine division is better described either in terms of integers or fractions.

Machine division requires the derivation of a quotient with fixed word length. The operands have fixed word length. Let us recall that the product of two numbers of a fixed length is twice the length of the operands. If this is the case, then we talk about a double-length product, unless the product is scaled to a single length. Similarly, the dividend could be of double-length or single-length, and in either case a single-length quotient is to be derived.

1. *Division of Integers.* In the division of integers the dividend and divisor must satisfy the condition $0 < |\text{divisor}| \leq |\text{dividend}|$. This condition is necessary in order to have an integer quotient and also to make the division by 0 forbidden. Usually, before the division starts, the above condition is checked, and in case of violation, the machine gives a division-fault signal and the programmer has to scale his operands to satisfy the above condition. The dividend must not be too large compared to the divisor; otherwise an overflow occurs.

If there is no division-fault, the division is performed by a repeated comparison, subtraction, and shift. The last comparison of the operands yields the least significant bit of the quotient, which has a unity weight. Therefore, in this last comparison, the dividend and the divisor must be aligned; that is, the digits of equal weight must occupy the same position. To achieve this alignment and derive an n-bit quotient, division must start with the divisor shifted to the left $n - 1$ positions. Each time a quotient bit is derived, the divisor is shifted to the right one position and the process continues. This guarantees that when the last quotient bit is derived, the two operands are aligned. An equivalent method used by digital machines is to start with the divisor shifted to the left $n - 1$ positions, relative to the dividend, and then instead of shifting the divisor to the right, the dividend is shifted to the left. For a double-length dividend the comparison is done between the high-order bits of the dividend (or modified dividend) and the divisor. In the case of a single-length dividend, the dividend is extended to double-length by adding a full length of leading zeros. The algorithm for the division of integers by comparison is summarized in the following paragraph.

An overflow check can be achieved by comparing the high-order bits of the dividend with the divisor shifted n positions to the left. The dividend is shifted one position to the left, and the first bit of the quotient is obtained by comparing the high-order bits of the modified dividend with the divisor. If the modified dividend is larger than or equal to the divisor, the quotient bit is a 1, and the divisor is subtracted from the modified dividend; otherwise the quotient bit is a 0, and no subtraction is needed. The modified dividend (partial remainder) is shifted to the left one position, and the second comparison takes place. This process continues until the n bits of the quotient are obtained. The division of integers for a double-length dividend and a single-length dividend are illustrated in Examples 3.6(a) and 3.6(b), respectively.

Example 3.6(a)

$$A = \text{Dividend} = 1011$$

$$B = \text{Divisor} = 0010$$

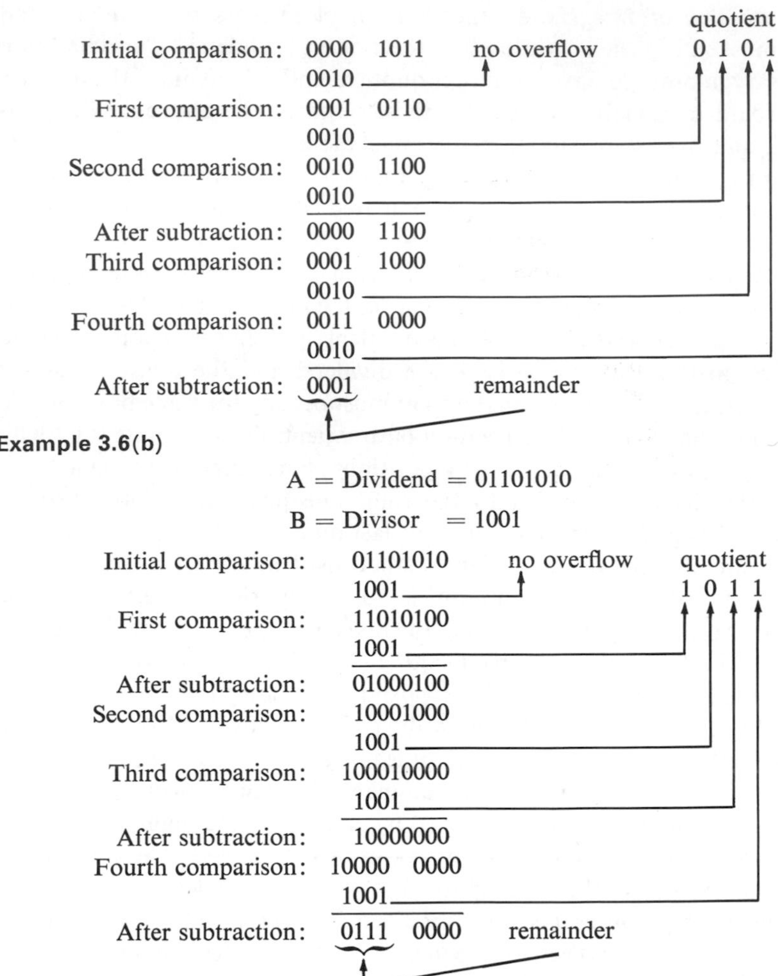

quotient
Initial comparison: 0000 1011 no overflow 0 1 0 1
0010
First comparison: 0001 0110
0010
Second comparison: 0010 1100
0010
After subtraction: 0000 1100
Third comparison: 0001 1000
0010
Fourth comparison: 0011 0000
0010
After subtraction: 0001 remainder
Example 3.6(b)
A = Dividend = 01101010
B = Divisor = 1001
Initial comparison: 01101010 no overflow quotient
1001 1 0 1 1
First comparison: 11010100
1001
After subtraction: 01000100
Second comparison: 10001000
1001
Third comparison: 100010000
1001
After subtraction: 10000000
Fourth comparison: 10000 0000
1001
After subtraction: 0111 0000 remainder

2. *Division of Fractions.* For fractions, the binary point is considered to the left of the most significant bit of the operands and a quotient is to be derived, with the binary point also to the left of the most significant bit. To obtain a quotient which is a fractional number and avoid a division-fault, the following relation must hold:

$$0 < |\text{Dividend}| < |\text{Divisor}|$$

If the above condition is violated, a division-fault signal is usually generated by the machine. Also, the dividend must not be too much smaller than the divisor; otherwise the quotient loses most of its significance.

Let us consider the division of two fractions. If we start with the operands aligned, the first comparison of their magnitudes leads to the quotient bit of unity weight. Hence, this initial comparison can be used to check the validity of the division; that is, if the quotient bit is a 1, then the operands are not scaled properly, and this 1 indicates an overflow. If the quotient bit is a 0, then the operands are in valid form, and the division can proceed. The first quotient bit (most significant bit of the fraction) is obtained by shifting the dividend to the left one position, and then the divisor is compared with the high-order bits of the dividend. The division process continues exactly the same way as in the division of integers. Hence, the division of fractions differs from that of integers only in the initial relative position of the operands. While for fractions the division starts with the operands aligned, the division of integers starts with the divisor digits shifted n positions to the left relative to the dividend digits of equal weight. The derivation of the quotient bits for each case is the same. A flow chart for the division by comparison which applies to both integers and fractions is shown in Fig. 3.7. In this flow chart it is assumed that the relative position of the divisor to the dividend is initially adjusted as described above.

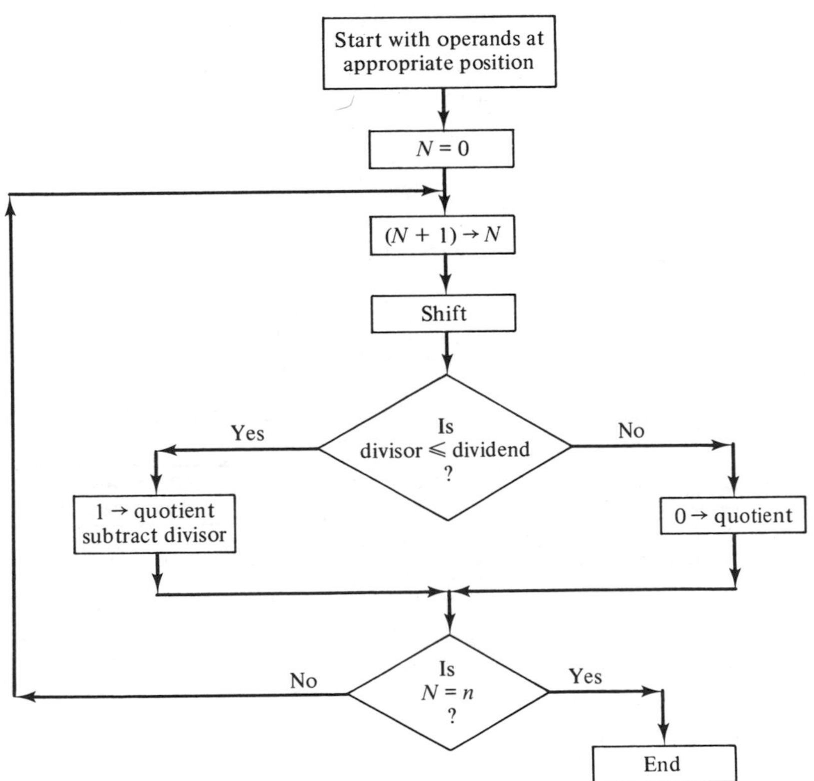

Figure 3.7 Division by comparison.

3.4.2 Restoring Method

The comparison method requires a comparator which is capable of comparing the dividend (or modified dividend) and the divisor. In most machines, instead of building a comparator, the already available adder or subtractor is used to do the comparison. If the comparison is done by subtraction followed by addition of the divisor, then the division is called *restoring* division. Hence, the restoring method of division is the same as the comparison method, except that each comparison which leads to a negative modified dividend is replaced by a subtraction followed by an addition of the divisor to restore the modified dividend to a positive number. The quotient bit

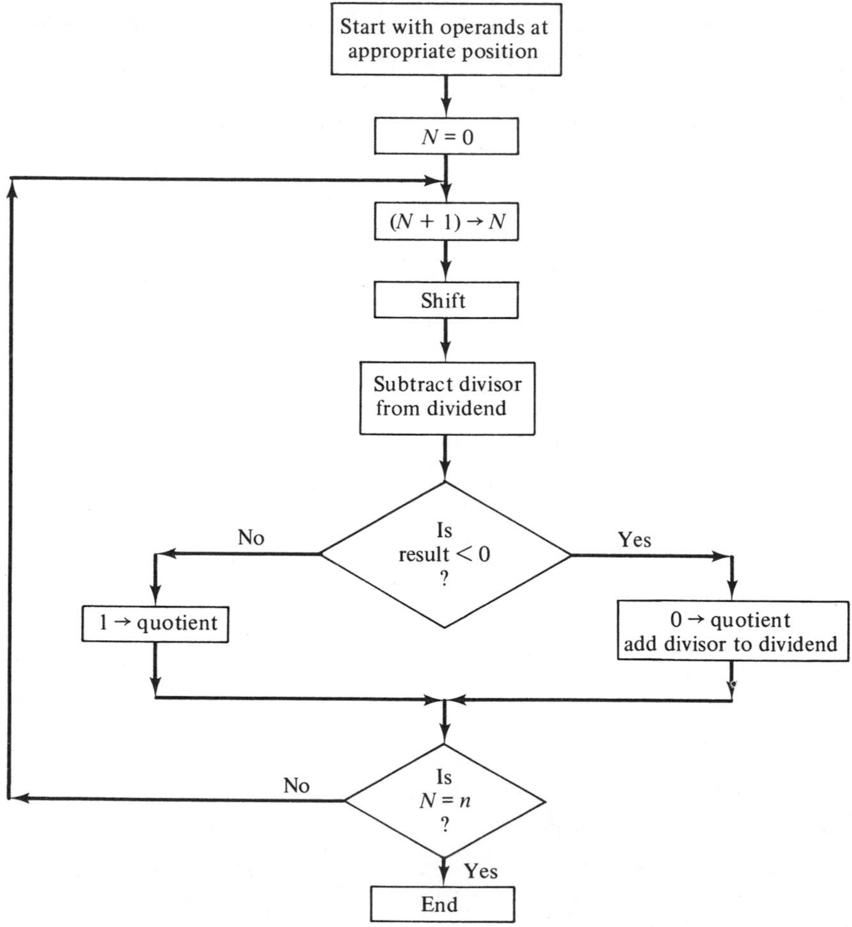

Figure 3.8 Restoring division.

for this comparison is a 0. The flow chart for the restoring division is shown in Fig. 3.8, and the algorithm is illustrated in Example 3.7.

Example 3.7

A = Dividend = 1001

B = Divisor = 0010

All numbers are shown in signed-magnitude notation.

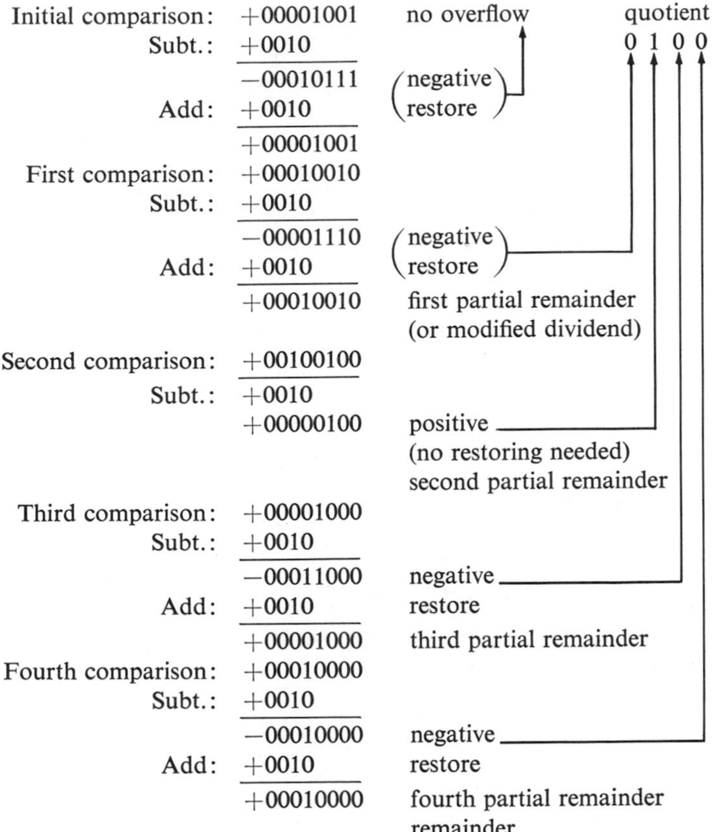

Initial comparison: +00001001 no overflow
quotient
0 1 0 0
Subt.: +0010
−00010111 negative restore
Add: +0010
+00001001
First comparison: +00010010
Subt.: +0010
−00001110 negative restore
Add: +0010
+00010010 first partial remainder (or modified dividend)
Second comparison: +00100100
Subt.: +0010
+00000100 positive (no restoring needed) second partial remainder
Third comparison: +00001000
Subt.: +0010
−00011000 negative
Add: +0010 restore
+00001000 third partial remainder
Fourth comparison: +00010000
Subt.: +0010
−00010000 negative
Add: +0010 restore
+00010000 fourth partial remainder
remainder

3.4.3 Nonrestoring Method

The nonrestoring division can be viewed as an improvement of the restoring division to speed up the division. If we inspect the restoring division, we find that each restore operation (addition of the divisor) is followed by a subtraction of the divisor after it is shifted one position to the right (or the dividend is shifted one position to the left). The subtraction is needed for the next comparison. These two operations, addition of the divisor followed by subtraction of one-half of the divisor (shifting the divisor to the right is equivalent to dividing by 2), can effectively be replaced by adding one-half of the divisor. In other words, we can combine the restoration addition with the following test subtraction. Therefore, if we start with a test subtraction and the result is negative, the divisor is shifted and then added. This addition

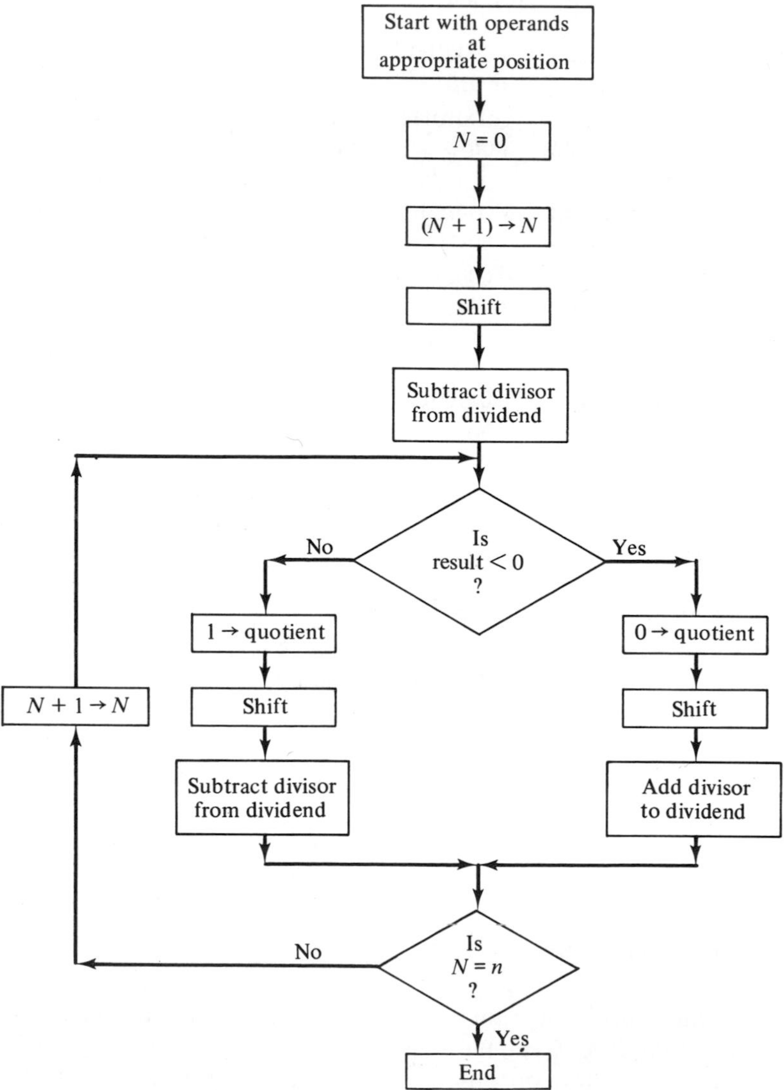

Figure 3.9 Non-restoring division.

combines the restoration and the next test subtraction, and depending on the sign of the result another cycle starts. If the remainder from the test subtraction is positive, no restoration is needed, and hence the next test cyle should subtract the shifted divisor. The flow chart for the nonrestoring division is given in Fig. 3.9. Example 3.8 illustrates the nonrestoring method described above.

Example 3.8

$$A = \text{Dividend} = 1011$$
$$B = \text{Divisor} = 0010$$

All numbers shown are in signed-magnitude notation.

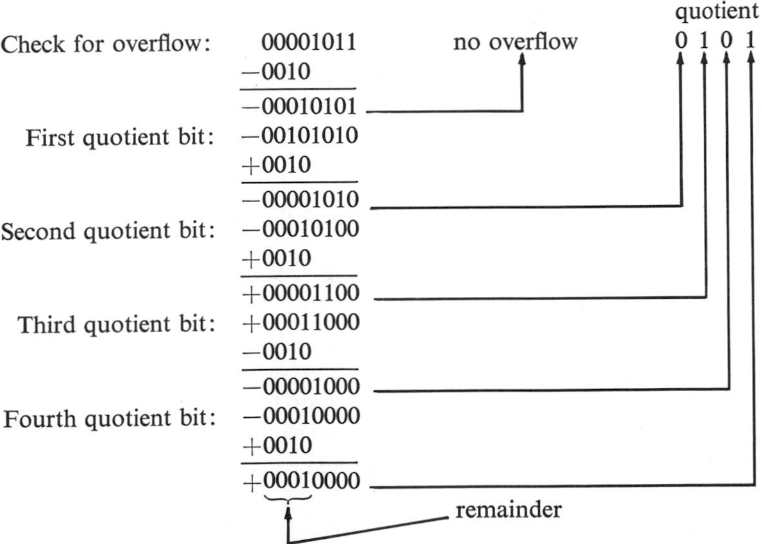

3.4.4 High-Speed Division Techniques

Division is the most time-consuming basic arithmetic operation in digital computers, and various techniques to speed up division exist. It is not the intention of this text to go through detailed† discussion of the various high-speed techniques. Instead, a brief discussion is presented below.

The basic methods for speeding up the division process are essentially extensions of the nonrestoring division technique with a normalized (the most significant bit is a 1) divisor and dividend. This framework of division requires no arithmetic test operation if the dividend is negative or positive and

†See Volume II of this text for details.

contains a high-order 0. Test addition or subtraction is required only if the high-order bit in the dividend is a 1. In the former case the result of any test addition or subtraction is determined by inspection since the divisor is normalized and its high-order bit is a 1. For instance, if the high-order bit of the modified dividend is a 0, we know that a test subtraction of the divisor will result in a negative remainder (partial dividend). Therefore, the test subtraction is not needed, and the corresponding quotient bit is a 0. This method is effectively shifting through 0s in the dividend. Similar methods for numbers in the 1's or 2's complement representation using speedup techniques by shifting across 0s or 1s of the dividend exist but will not be developed here.

Among other high-speed division techniques is the division using divisor multiples. This is a nonrestoring division in which the quotient bits are obtained in multiple-bit groups at a time. For instance, in a division by 2-bit multiples, the 3 high-order bits of the dividend are inspected against the high-order 2-bits of the divisor. A proper multiple (either positive or negative) of the divisor is selected from a wired-in table and the following operation is performed:

$$\begin{aligned} &\text{Dividend} + \text{Multiple} \times \text{Divisor} \\ &\quad = \text{Partial remainder} \qquad \text{(or modified dividend)} \end{aligned}$$

According to whether the partial remainder is positive or negative, a proper 2-bits quotient is again selected from another wired-in table. This process continues until all the bits of the quotient are obtained. This technique applies to numbers in 2's complement representation. The entries in the wired-in tables are predetermined according to the rules of nonrestoring division. In fact the nonrestoring division of the previous section is a special case of the above technique, with the selected divisor multiple always being equal to 1.

Another high-speed division technique where division is carried out by multiplication and complementation is known as *division by repeated multiplication.* It is also sometimes called Wilkes-Harvard iterative division and is briefly described below.

Let N, D, and Q refer to the dividend, divisor, and quotient, respectively. Assume that D is in normalized fractional form ($\frac{1}{2} \leq D < 1$). The quotient Q can be written as

$$Q = \frac{N}{D} = (N)\frac{1}{D}$$

The main idea here is to substitute the division process by repeated multiplication. To achieve this, first the value of $1/D$ is approximated by using a wired-in table whose entries are the high-order bits of the divisor (5 to 7 bits are enough). If the approximation is sufficiently accurate, then the division can be replaced by a multiplication process. Since getting the exact value of the divisor reciprocal from a table requires an enormous amount of hardware, an approximation to the reciprocal (say R) could be obtained such that

$$Q = \frac{N \times R}{D \times R} = \frac{N \times R}{1 \pm \epsilon} \qquad \text{where } D \times R = 1 \pm \epsilon, \epsilon \ll 1$$

or

$$Q = \frac{(N \times R)(1 \mp \epsilon)}{(1 \pm \epsilon)(1 \mp \epsilon)} = \frac{(N \times R)(1 \mp \epsilon)}{(1 - \epsilon^2)} \frac{(1 + \epsilon^2)}{(1 + \epsilon^2)}$$

This iterative multiplication can continue and in general Q can be written as

$$Q = \frac{(N \times R)(1 \mp \epsilon)(1 + \epsilon^2)(1 + \epsilon^4) \cdots (1 + \epsilon^{2^n})}{1 - \epsilon^{2^{n+1}}}$$

The number of iterations depend on the accuracy required. Since the 2's complement of $(1 \pm \epsilon)$ is $(1 \mp \epsilon)$ and the 2's complement of $(1 - \epsilon^2)$ is $(1 + \epsilon^2)$, the above process is a repeated complementation and multiplication.

Additional methods of high-speed arithmetic algorithms and the implementation of those methods and the ones discussed in this section can be found in Volume II of this text.

3.5 FLOATING-POINT ARITHMETIC

Most earlier machines were designed with only a fixed-point arithmetic unit and most often used the fractional representation. Therefore, it was the programmer's responsibility to scale his numbers to be in a fractional form. Since this is a time-consuming job, computer hardware has been developed to obviate this task. Many computers have been constructed with built-in floating-point arithmetic capabilities. In floating point, the radix point for each number varies automatically. This feature is considered to be one of the major developments in digital computers, since it allows great flexibility in handling numerical quantities. Built-in floating-point arithmetic capabilities have resulted in some added complexity in the design of the arithmetic unit. Computers which do not have a floating-point arithmetic unit usually perform floating-point arithmetic by programming at the expense of speed.

3.5.1 Representation of Floating-Point Numbers

Any numerical quantity can be expressed in the form $C \times B^e$. For example, the decimal number 1260 can be written as 1.26×10^3 and the decimal number .00512 can be written as $.512 \times 10^{-2}$ or as 5.12×10^{-3}. In this representation, a number consists of two portions, the coefficient part (mantissa) and the exponent part (or characteristic). For paper and pencil computations, the radix point can be put anywhere with no fixed position. For machine computations in floating point, to be compatible with that for fixed-point numbers, the mantissa is usually in a fractional form with the radix point to the left of the most significant digit of the number digits. This is called a normalized mantissa. For instance, the floating-point representation of the binary number +1101.01 has two parts—the mantissa (fractional part) is .110101, and the exponent part is the binary equivalent of +4. Usually a signed floating-point number is represented by three parts, as sign bit, mantissa bits, and exponent bits. For a fixed word length, the number of bits allocated to either the mantissa or the exponent depends on the accuracy and the range of the numerical quantities to be represented.

The size of the mantissa is limited to values in the range $\frac{1}{2} \leq$ mantissa < 1; that is, the most significant bit of the mantissa has a magnitude of 1. The only exception is the 0 quantity which is represented by all-0 digits. The size of the exponent varies according to the number of digits allocated for it. For instance, when 8 bits are used, there are 256 (2^8) different codes to represent both positive and negative exponents. An approximately equal range would be $-128 \leq$ exponent $\leq +127$. One way of coding for the exponent is to have the smallest 8-bit fixed-point number (eight 0s) represent the smallest exponent (-128) and the largest number (eight 1s) represent the largest exponent ($+127$). This is a biased exponent since any exponent (e) is represented by the binary equivalent of $|(e + 128)|$. In this representation, positive exponents are distinguished from negative exponents by simply inspecting the high significant bit. A 1 in the high significant bit of the exponent indicates a positive exponent. This choice of exponent representation makes the comparison between the exponents follow the same rules of comparison as unsigned fixed-point numbers.

According to the above rules, if the exponent of a number has a value of 0, its representation should be $(128)_2$. The only exception is again the 0 quantity. The number 0, when represented by all-0 digits, effectively represents 0×2^{-128}, which is still zero. This representation of the 0 quantity produces valid arithmetic results.

3.5.2 Addition and Subtraction

The addition and subtraction of floating-point numbers is more difficult than that of fixed-point numbers, because of the requirement for an alignment of the radix point. If two floating-point numbers do not have equal exponents, it means that the digits of equal weights in their fractional parts are not aligned. Hence, an initial alignment is necessary before addition or subtraction takes place. Once the alignment is performed, the addition or the subtraction is the same as that of fixed-point fractional numbers. The resulting number would have an exponent equal to that of any of the aligned operands and a fractional part equal to the sum or difference of their fractional parts. If a and b are two floating-point members such that

$$a = C_a \times B^{e_a} \quad \text{and} \quad b = C_b \times B^{e_b}$$

then

$$a + b = (C_a + C_b \times B^{e_b - e_a}) \times B^{e_a}$$

The resulting number might not have a normalized mantissa, and hence a normalization process might be necessary.

The initial alignment could be accomplished by first obtaining the difference of the two exponents and then shifting the mantissa of the number with the smaller exponent to the right a number of places equal to the magnitude of that difference. In this case, the exponent of the result is taken as the larger exponent.

For normalization of the result, there are two possible cases.

1. The resulting fractional part is too large (overflow in the mantissa): In this case normalization can be done by shifting the mantissa one place to the right and increasing the exponent by 1.
2. The resulting fractional part is too small (underflow): This can be detected by having the magnitude of the most significant digit not equal to 1, and consequently the number is not normalized. In such case, the result can be normalized by shifting to the left and decreasing the exponent.

The normalization process takes care of any overflow or underflow in the mantissa part. An overflow in the exponent part is treated just as an overflow in fixed-point arithmetic is handled, that is, by generating an overflow signal. Floating-point addition is illustrated by Example 3.9.

Example 3.9

In this example, the numbers A = +24 and B = −14 are to be added using floating-point, 2's complement representation. We shall assume a 5-bit exponent and a 7-bit plus sign for the fractional part. The exponents are denoted by A_e and B_e and the fractions by A_m and B_m. A = +24 = 0,11000, and B = −14 = 1,0010.

a. Normalization:

$$A_m = 0.1100000, \qquad A_e = 00101$$
$$B_m = 1.0010000, \qquad B_e = 00100$$

b. Alignment ($A_e - B_e$):

$$\begin{array}{r} 00101 \\ -00100 \\ \hline \end{array} = \begin{array}{r} 00101 \\ +11100 \\ \hline 1\,00001 \end{array} \qquad \therefore A_e - B_e = 1$$

↑
drop end carry

Shift B_m one position to the right so that both A and B have the same exponent (00101 in this case). The aligned fractional parts are

$$A_m = .1100000, \qquad B_m = 1.1001000$$

c. Perform addition ($A_m + B_m$):

$$\begin{array}{r} 0.1100000 \\ +1.1001000 \\ \hline 1\ 0.0101000 \end{array} \qquad \text{a positive number}$$

↑
drop end carry

d. Normalization: Since ($A_m + B_m$) is positive, the most significant digit must be a 1. Hence, shift one position to the left and decrease the exponent by a 1.

$$(A_m + B_m) = 0.1010000$$

and the correct exponent is

$$00101 - 1 = 00100 \qquad \text{(binary 4)}$$

$$\therefore A + B = 0.1010000 \times 2^4 = 1010 \qquad \text{(binary 10)}$$

3.5.3 Multiplication and Division

Multiplication of two floating-point numbers does not need an initial alignment of the radix point. It can be done by adding the two exponents while multiplying the two fractional parts. Hence, the product $a \times b$ can be written as

$$a \times b = C_a \times C_b \times B^{(e_a + e_b)}$$

Since the numbers are initially in normalized form, the multiplication of the two fractional parts will never result in an overflow. An underflow might occur, and if it does, normalization of the product is necessary. Underflow normalization is the same as described in Case 2 of the addition normalization.

Division of two floating-point numbers may require an initial alignment. The algorithm requires the division of the fractional parts and the subtraction of the exponents; thus,

$$a \div b = \frac{C_a}{C_b} \times B^{(e_a - e_b)}$$

However, before the division of C_a by C_b can be performed, we must be sure that $|C_a| < |C_b|$. This will guarantee that when the fixed-point division is performed it will not overflow. Since both numbers are initially normalized, we can guarantee that $|C_a| < |C_b|$ by shifting C_a 1 bit to the right and increasing its exponent by 1 before we start the division. After the initial alignment, the division process is no more than that of a fixed-point division of the aligned fractional numbers and a fixed-point subtraction of the integer exponents. An underflow of the quotient may occur, and if it does, it is handled by the same method described in addition and subtraction.

Example 3.10

In this example we shall multiply A = +24 and B = −14 using floating-point notation. We shall assume a 5-bit exponent and a 7-bit plus sign fractional part. The numbers are the same as in Example 3.9:

$$\text{A:}\quad A_m = 0.1100000, \qquad A_e = 00101$$
$$\text{B:}\quad B_m = 1.0010000, \qquad B_e = 00100$$

No alignment is required. We add the exponents and multiply the mantissas:

$$\begin{array}{rl} A_e = & 00101 \\ (+)B_e = & \underline{00100} \\ & 01001 \end{array}$$

Now we must multiply the mantissas. We shall do this in magnitude form:

$$\begin{array}{rl} |A_m| = & .1100000 \\ (\times)\ |B_m| = & \underline{.1110000} \\ & .10101000000000 \end{array}$$

Since we have only 7 bits for the fractional part, we can retain only 7 bits in the answer:

$$|A_m| \times |B_m| = .1010100$$

The answer is negative, and we must convert the magnitude of the answer to a negative number in 2's complement form:

$$1.0101100 \times 2^{01001} = -336$$

Example 3.11

Let us perform B divided by A using the numbers of Example 3.10:

$$\begin{aligned} &B\colon\quad B_m = 1.0010000, \qquad B_e = 00100 \\ &A\colon\quad A_m = 0.1100000, \qquad A_e = 00101 \end{aligned}$$

Before we begin, we shall convert the mantissas into their magnitude form:

$$\begin{aligned} &|B|\colon\quad |B_m| = 0.1110000, \qquad B_e = 00100 \\ &|A|\colon\quad |A_m| = 0.1100000, \qquad A_e = 00101 \end{aligned}$$

Before we can divide the mantissas we must be sure that $|B_m| < |A_m|$. In the example this is not so. Therefore, we must shift $|B_m|$ to the right and increase its exponent:

$$\begin{aligned} &|B|\colon\quad |B_m| = 0.0111000, \qquad B_e = 00101 \\ &|A|\colon\quad |A_m| = 0.1100000, \qquad A_e = 00101 \end{aligned}$$

Now we can subtract the exponents and divide the mantissas:

$$\begin{array}{rrcr} B_e - A_e = & 00101 & \longrightarrow & 00101 \\ & \underline{-00101} & \longrightarrow & \underline{(+)11011} \\ & & & 00000 \end{array}$$

$$\frac{|B_m|}{|A_m|} = \frac{0.0111000}{0.1100000} = 0.1001010 = .57812$$

The answer is $0.1001010 \times 2^{00000}$ in magnitude form. However, we must convert it to its negative representation since the answer is negative:

$$\frac{B}{A} = 1.0110110 \times 2^{00000}$$

A comment about the answer. The answer to our problem of $(-14)/24$ is $-.58333$. The answer we have obtained is $-.57812$. The reason for the difference is due to the truncation error which we obtained when we retained only a 7-bit quotient in the division of the mantissas. This is also the case in digital computers, which, because of a finite word length, may have truncation errors in their arithmetic operations.

REFERENCES

CHU, Y., *Digital Computer Design Fundamentals*. New York: McGraw-Hill, 1962.

FLORES, I., *The Logic of Computer Arithmetic*. Englewood Cliffs, N.J.: Prentice-Hall, 1962.

GSCHWIND, H. W., *Design of Digital Computers*. New York: Springer-Verlag New York, Inc., 1967.

MACSORLEY, O. L., "High-Speed Arithmetic in Binary Computers," *Proc. IRE*, 49, No. 1 (Jan. 1961), pp. 67–91.

RICHARDS, R. K., *Arithmetic Operation in Digital Computers*. New York: Van Nostrand Reinhold, 1955.

STEIN, M. L., and W. K. MUNRO, *Introduction to Machine Arithmetic*. Reading, Mass.: Addison-Wesley, 1971.

WILSON, J., and R. LEDLEY, "An Algorithm for Rapid Binary Division," *IRE Trans. Electronic Computers*, EC-10, No. 4 (Dec. 1961).

PROBLEMS

3.1. Assume that 1's complements are used and that numbers are 7 bits each, including sign. What will the results be if each of the following numbers is shifted right two places if we permit a 9-bit answer?
(a) 0,110000 (b) 0,110010 (c) 1,001100
(d) 1,110011 (e) 1,000111 (f) 0,000111

3.2. Repeat Problem 3.1 for shifting the numbers left two positions.

3.3. Repeat Problem 3.1 assuming that the numbers are in 2's complement form.

3.4. Repeat Problem 3.2 for 2's complement.

3.5. For Problem 3.1, assume that the size of the shifted number is limited to 7 bits. What would be the result if each number in Problem 3.1 is shifted two positions to the right if we retain only 7 bits?

3.6. Repeat Problem 3.5 for shifting to the left two positions.

3.7. Repeat Problem 3.5 if the numbers are in 2's complement.

3.8. Repeat Problem 3.6 if the numbers are in 2's complement.

3.9. Repeat Problem 3.5 if the numbers are in signed-magnitude.

3.10. Repeat Problem 3.8 if numbers are in signed-magnitude.

3.11. Perform the addition and subtraction of the following decimal numbers using (1) signed-magnitude, (2) 10's complement, and, (3) 9's complement:
(a) $5250 + (-321)$ (b) $(-864) + 753$
(c) $3570 - 2100$ (d) $(-45) + (-2300)$

3.12. Perform the addition and subtraction of the following binary numbers using 6-bit words and signed-magnitude representation:
(a) $11010 - 1101$ (b) $(-10011) + (10010)$
(c) $100 - 110000$ (d) $111000 + 110100$
(e) $-101010 + (-011010)$ (f) $(-101101) - (+110001)$

3.13. Repeat Problem 3.12 using 1's complement representation.

3.14. Repeat Problem 3.12 using 2's complement representation.

3.15. If 1's complement is used, for which of the following expressions will an overflow or end-around carry be generated? Why?
(a) $+7 + (-4)$ (b) $+5 + (-8)$
(c) $+13 + (-13)$ (d) $+12 + (+9)$

3.16. Show all the partial results during the multiplication of two binary numbers using the repeated addition and shift algorithms. Assume the multiplicand to be 0,10101 and the multiplier to be 1,01101. Both operands are in signed-magnitude form.

3.17. What is the rule for rounding a double-length product to a single-length number?

3.18. Additions can produce overflows. How are overflows handled during the addition steps of a multiplication?

3.19. Assume that one clock cycle is required for a "shift" and two clock cycles are required for an "add." Try to estimate the average time for the multiplication of two 36-bit binary numbers if
(a) A shift and an add is initiated for each bit in the multiplier.
(b) Adds are initiated only for 1s in the multiplier, but a shift is initiated for each bit in the multiplier.
(c) Adds are initiated only for 1s in the multiplier and shifts over an arbitrary number of binary positions are possible within one clock cycle.

3.20. Compare the multiplication time, for the uniform shifts by two position, with the results of Problem 3.19.

3.21. Using the division by comparison algorithm, perform the following division. Show the partial dividend for every step:
(a) 01101010 ÷ 1001 (b) 1010 ÷ 0010
(c) 1001 ÷ 0101

3.22. Repeat Problem 3.21 for the restoring method.

3.23. Repeat Problem 3.21 for the nonrestoring method.

3.24. Compare the average execution times for
(a) Division by comparison.
(b) Restoring division.
(c) Nonrestoring division.

3.25. What is the largest and smallest positive quantity which can be represented by 36-bit floating-point numbers? Assume a reasonable size of exponent.

3.26. Explain the reason for restricting the size of the binary coefficients of floating-point numbers by $\frac{1}{2} \leq C < 1$.

3.27. Represent the following quantities by 36-bit floating-point numbers. Assume a reasonable size of exponent.

(a) $.006542)_8$ (b) $7251.5)_8$
(c) $-.00572)_8$ (d) $-7536.5)_8$

3.28. The numbers $X = -1.213$ and $Y = 35.4$ are to be added in floating-point, binary arithmetic. Assume 6 binary digits for the exponent and 18 binary digits for the normalized fractional part. Show all the steps of the operation.

3.29. The numbers $X = (.56)_8 \times 2_8^4$ and $Y = (.43)_8 \times 2_8^6$ are to be multiplied in the same format as in Problem 3.28. Show all the steps and partial results.

3.30. Show all the steps and partial results in the floating-point division X/Y, where X and Y have the values given in Problem 3.29. Use the same format as in Problem 3.28.

4

Codes

In digital computers we shall represent quantities by electrical signals. These signals will be binary in form. The signals can represent numbers or instructions or operation signals. We must encode the information we wish to represent in a computer into the binary format if the electronic circuitry of the machine is to be able to interpret the information.

We are all familiar with the decimal system and the decimal characters used to represent a number. In Chapter 2, we learned how to represent a number in another system, namely binary. In this chapter we shall develop other systems to represent numbers and entities.

Let us assume that we have a set of objects which we wish to discuss with our friends located in another city but wish to keep secret from our enemies. We have a private telegraph system between ourselves and our friends but suspect that our enemies are tapping the line; i.e., it's bugged. We give each object a code name which only our friends can decode, i.e., understand. We create a code book which uniquely identifies the objects and personally bring a copy of the code to our friends. Now that they have the code we can send telegrams back and forth using the code name for an object and not worry about our enemies. How can we create a simple and efficient code if we have many objects to identify?

4.1 UNIQUENESS AND LENGTH OF A CODE

If we have N objects, how can we uniquely encode them using a minimum number of bits to send over the telegraph system? We can determine this number by assigning a one-to-one correspondence between objects and the binary numbers from zero to $N - 1$. That is, the number of bits, n, needed for a fixed-length code word to uniquely identify N objects is

$$2^n = N \quad \text{or} \quad n = \log_2 N$$

Since this can result in a fractional number of bits, we round it up to the next whole integer.

Example 4.1

Assume that we have 98 objects to encode. How many bits must be in each code word if each word is the same size?

$$2^n = 98 \quad \text{or} \quad n = \log_2 98$$

$$\begin{aligned} n &= \log_2 98 = \log_{10} 98 \cdot \log_2 10 \\ &= (1.99123)(3.32192) \\ &= 6.6052 \text{ bits} \end{aligned}$$

Therefore, we need 7 bits per code word to uniquely encode 98 objects. Since 7 bits can uniquely encode 128 objects, we have $128 - 98 = 30$ codes which we shall not be using. Which 30 code characters will not be used? Let us look at a smaller problem to get a better feeling about choosing the characters to be used in a code.

Given a three-place binary number, we can represent eight different things. Let us assume that we wish to represent the decimal numbers 0 through 7 with their 3-bit binary equivalents as shown in Fig. 4.1. We were

Binary	*Decimal*
000	0
001	1
010	2
011	3
100	4
101	5
110	6
111	7

Figure 4.1 Binary codes.

not forced to use this representation, i.e., 100 does not have to stand for a 4, but it is convenient. We know that in order to encode the decimal integers 0 through 9 into a binary code we shall need at least

$$n = \log_2 10 = 3.32 \text{ bits}$$

We must therefore use 4 bits. But 4 bits permits us to encode 16 things. Which of the possible 16 combinations of 4 bits should be selected to represent the 10 decimal integers? How many possible selections of 4 bits can we have to represent 10 items? The answer is the permutation of 16 things taken 10 at a time, or

$$\frac{16!}{(16-10)!} = 2.9 \times 10^{10}$$

possible 4-bit codes to represent the 10 decimal integers 0 through 9.

For our original problem of Example 4.1 we have

$$\frac{128!}{(128-98)!} = \frac{128!}{30!}$$

possible codes. This is a very large possible number of codes. We choose a code in order to facilitate some use of the code. Thus, we may choose the binary codes which we used to encode the operations of a computer so as to simplify the decoding and minimize the amount of electronic hardware needed to decode the operations. In the following sections we shall discuss some useful codes which have been used to represent numerical and alphabetical characters.

4.2 POSITIVE WEIGHTED CODES

Some of the 2.9×10^{10} possible 4-bit codes can be represented by weights assigned to each of the four bit positions. The decimal value of the code group is the algebraic sum of the weights of the bit positions which contain a 1. The weights may be either positive or negative. If we consider only the possibilities of positive weights, then there are 17 such combinations:

$$(3, 3, 2, 1), \quad (4, 2, 2, 1), \quad (4, 3, 1, 1), \quad (5, 2, 1, 1),$$

Decimal digit	*Codes*			
	NBCD			
d	8421	2421	4221	5421
0	0000	0000	0000	0000
1	0001	0001	0001	0001
2	0010	0010	0010	0010
3	0011	0011	0011	0011
4	0100	0100	0110	0100
5	0101	1011	1001	1000
6	0110	1100	1100	1001
7	0111	1101	1101	1010
8	1000	1110	1110	1011
9	1001	1111	1111	1100
Unused bit combination	1010	0101	0100	0101
	1011	0110	0101	0110
	1100	0111	0111	0111
	1101	1000	1000	1101
	1110	1001	1010	1110
	1111	1010	1011	1111

Figure 4.2 Some positive weighted codes.

(4, 3, 2, 1), (4, 4, 2, 1), (5, 2, 2, 1), (5, 3, 1, 1),
(5, 3, 2, 1), (5, 4, 2, 1), (6, 2, 2, 1), (6, 3, 1, 1),
(6, 3, 2, 1), (6, 4, 2, 1), (7, 3, 2, 1), (7, 4, 2, 1),
(8, 4, 2, 1)

The 8421 weighted code corresponds to the first 10 natural binary numbers, and it is called the natural binary coded decimal code (NBCD). The table in Fig. 4.2 shows some of the more frequently used 4-bit positive weighted codes which represent the decimal characters 0 through 9 and also the unused code combinations. These bit combinations are often called binary coded decimal numbers or BCD codes. Using the NBCD code as an example we would represent the decimal number 673 by the following 4-bit code groups:

$$673 = 0110 \quad 0111 \quad 0011$$

4.3 NEGATIVE WEIGHTED CODES

There are 71 weighted codes which have one or two negative weights. These codes have been found by a computer search. An example of a negative weighted code is the 8, 4, −2, −1 code shown in Fig. 4.3.

Decimal digit	*Weights*			
	8	4	−2	−1
0	0	0	0	0
1	0	1	1	1
2	0	1	1	0
3	0	1	0	1
4	0	1	0	0
5	1	0	1	1
6	1	0	1	0
7	1	0	0	1
8	1	0	0	0
9	1	1	1	1
Unused bit combinations	0	0	0	1
	0	0	1	0
	0	0	1	1
	1	1	0	0
	1	1	0	1
	1	1	1	0

Figure 4.3 Negative weighted code.

4.4 UNWEIGHTED CODES

There are many 4-bit unweighted codes. The use of one such code instead of another code must be because of some useful property of the code. One unweighted code is the excess-3 code shown in Fig. 4.4. It is obtained by adding a binary code for 3 to each of the code characters of the NBCD code. Another useful unweighted code is called *the Gray code*. It has special properties, which we shall discuss in Section 4.6.

Decimal digit	*Excess-3*	*Gray code*
0	0011	0000
1	0100	0001
2	0101	0011
3	0110	0010
4	0111	0110
5	1000	0111
6	1001	0101
7	1010	0100
8	1011	1100
9	1100	1101
Unused bit combinations	0000	1000
	0001	1001
	0010	1010
	1101	1011
	1110	1110
	1111	1111

Figure 4.4 Unweighted codes.

4.5 SELF-COMPLEMENTING CODES

It is often necessary to form the 9's complement of a number while doing arithmetic. If the number is coded with a 4-bit code, then it will be necessary to form the 9's complement of the number using the same 4-bit code. Consider the following example: Given a number coded in NBCD, determine its 9's complement:

$$6\ 4\ 3 \longrightarrow 0110\ 0100\ 0011$$

$$\text{9's complement:}\quad 3\ 5\ 6 \longrightarrow 0011\ 0101\ 0110$$

The above example shows that the NBCD code is not self-complementing; i.e., if we change the 1s to 0s and the 0s to 1s, we do not obtain the 9's complement of the coded decimal number. If we now consider the same numbers in terms of the 2421 code, we get

$$6\ 4\ 3 \longrightarrow 1100\ 0100\ 0011$$

$$\text{9's complement:}\quad 3\ 5\ 6 \longrightarrow 0011\ 1011\ 1100$$

This is a self-complementing code. Likewise, the 4221 code, the excess-3 code, and the 8, 4, -2, -1 code are self-complementing.

4.6 REFLECTED CODES

A special type of code in which successive coded symbols differ in only one bit position is called a reflected or cyclical code. These codes are used in analog-to-digital signal converters in order to reduce possible ambiguity in the converted numbers. One of the most used reflected codes is the Gray code. Figure 4.5 shows the corresponding 4-bit combinations for the binary numbers and the corresponding Gray code.

The Gray code is unweighted but is derivable directly from any binary number using the following algorithm:

Given a binary number expressed as $b_n b_{n-1} \ldots b_1$ we wish to determine the Gray code number to which it corresponds, expressed as $g_n g_{n-1} \ldots g_1$:

$$g_k = b_k + b_{k+1} \pmod 2, \qquad k = 1, 2, \ldots, n-1 \text{ and } g_n = b_n$$

Decimal number	*Binary number*	*Gray code*	
0	0000	0000	
1	0001	0001	
2	0010	0011	
3	0011	0010	
4	0100	0110	Reflected BCD code (0–9)
5	0101	0111	
6	0110	0101	
7	0111	0100	Reflected excess-3 code (3–12)
8	1000	1100	
9	1001	1101	
10	1010	1111	
11	1011	1110	
12	1100	1010	
13	1101	1011	
14	1110	1001	
15	1111	1000	

Figure 4.5 Gray code.

Thus, g_k is obtained by adding b_k and b_{k+1} and discarding the carry. The inverse problem of converting a Gray code number into a binary number can be performed by

$$b_n = g_n$$

$$b_k = \sum_{i=k}^{n} g_i \pmod 2, \qquad k = n-1, n-2, \ldots, 1$$

Thus, b_k is obtained by forming the sum of the bits from k through n, dividing by 2, and retaining the remainder as b_k.

Example 4.2

Given the binary number 101101, what is its reflected form (Gray code)?

$$\begin{aligned}
g_1 &= b_1 + b_2 = 1 + 0 = 1 \pmod 2) = 1 \\
g_2 &= b_2 + b_3 = 0 + 1 = 1 \pmod 2) = 1 \\
g_3 &= b_3 + b_4 = 1 + 1 = 2 \pmod 2) = 0 \\
g_4 &= b_4 + b_5 = 1 + 0 = 1 \pmod 2) = 1 \\
g_5 &= b_5 + b_6 = 0 + 1 = 1 \pmod 2) = 1 \\
g_6 &= b_6 \qquad\qquad\qquad\qquad\qquad = 1
\end{aligned}$$

Thus, the Gray code form is 111011. Let us convert the Gray code number back to a binary number using the algorithm

$$\begin{aligned}
b_6 &= g_6 = 1 \\
b_5 &= g_6 + g_5 \pmod 2 = (1 + 1) \pmod 2 = 0 \\
b_4 &= g_6 + g_5 + g_4 \pmod 2 = (1 + 1 + 1) \pmod 2 = 1 \\
b_3 &= g_6 + g_5 + g_4 + g_3 \pmod 2 = (1 + 1 + 1 + 0) \pmod 2 = 1 \\
b_2 &= g_6 + g_5 + g_4 + g_3 + g_2 \pmod 2 = (1 + 1 + 1 + 0 + 1) \pmod 2 = 0 \\
b_1 &= g_6 + g_5 + g_4 + g_3 + g_2 + g_1 \pmod 2 = (1 + 1 + 1 + 0 + 1 + 1) \\
&\quad \pmod 2 = 1
\end{aligned}$$

Therefore, the binary number is 101101.

If we wish to use 10 of the reflected code numbers to represent the 10 decimal digits 0 through 9, we have many choices. Two of the most common are shown in Fig. 4.5. One is to use the first 10 numbers of the Gray code; another is to use the fourth through thirteenth numbers, which corresponds to a reflected excess-3 code. The latter choice has the advantage that the 9's complement can be obtained by complementing the high-order bit of the group of 4 bits.

4.7 NUMERIC CODES WITH MORE THAN 4 BITS

If we use only 4 bits to represent the code for the 10 decimal characters and if one of these bits should change during the transmission of the code, we shall decode, at its reception, a wrong number or an invalid bit combination. Under either condition we shall not know what character was transmitted. To overcome this possibility, especially in noisy transmission channels, we could develop codes with greater than 4 bits. This redundancy may permit us to detect an error in transmission.

4.7.1 Parity

One method of increasing the size of the code character is to just add another bit. We can add this bit so that the total number of 1-bits in the new character is odd. Such a system is called *odd parity*. Likewise, we could make the system *even parity* by adding a 1-bit to make the total number of 1-bits an even number. Figure 4.6 shows the NBCD code with an odd parity bit added

Decimal digit	*NBCD odd parity*	
0	1	0000
1	0	0001
2	0	0010
3	1	0011
4	0	0100
5	1	0101
6	1	0110
7	0	0111
8	0	1000
9	1	1001

Figure 4.6 NBCD code with parity.

to the front of the code. Notice that each word now contains 5 bits with an odd number of unity bits. With this type of code, we can detect if any single bit should change, for if it does, it will present an even number of unity bits. However, if 2 bits should change, we may not be able to detect this. Also, even though we can detect a single-bit error, we cannot correct the error since we do not know which bit changed.

4.7.2 2-out-of-5 Code

Another code which provides a built-in error detection capability is the 2-out-of-5 code. It obtained this name because each character in the 5-bit code contains exactly two 1s.

The number of combinations of 5 bits with two 1s is the combination of five things taken two at a time:

$$_5C_2 = \frac{5!}{2!(5-2)!} = \frac{5 \cdot 4}{2 \cdot 1} = 10$$

Thus, we can exactly cover all of the 10 decimal characters. There are many possible permutations of the 10 characters in this type of code. One of the codes is shown in Fig. 4.7. It is a semiweighted code with weights 74210; however, the code for decimal 0 is not weighted but assigned the leftover 11000. Another 2-out-of-5 code with weights 63210 is often used. Again, the decimal 0 is not a weighted code but is assigned the remaining combination 01001.

Decimal digit	*2-out-of-5 code*
	74210
0	11000
1	00011
2	00101
3	00110
4	01001
5	01010
6	01100
7	10001
8	10010
9	10100

Figure 4.7 A 2-out-of-5 code.

This code is error detecting since we know that any change in a single-bit location will present either two few or too many 1s. Double errors involving two 1s or two 0s can also be detected. However, double errors involving a single 0 and a single 1 will not be detected. Thus, this code has more error-detecting capability than a single parity bit.

4.7.3 Biquinary Code

Another code which has been used is a 2-out-of-7 code, composed of a 1-out-of-2 group and a 1-out-of-5 group. Again, we have 10 combinations but many possible permutations. Figure 4.8 shows one possible weighted code.

Decimal digit	Biquinary code 05 01234
0	10 10000
1	10 01000
2	10 00100
3	10 00010
4	10 00001
5	01 10000
6	01 01000
7	01 00100
8	01 00010
9	01 00001

Figure 4.8 Biquinary code.

4.8 ERROR-DETECTING AND -CORRECTING CODES

The three codes studied in Section 4.7 were shown to have single-error-detecting capabilities. Let us see why this is so.

The *minimum distance* of a code is the minimum number of bits that must change in a code character so that any other valid character of the code will result. Thus, the minimum distance for the NBCD, excess -3, 2421, or Gray codes is 1, in each case. The minimum distance for the NBCD with parity, 2-out-of-5, or biquinary codes of Section 4.7 is 2. Thus, to get from one character in the code to any other character of the code we must change at least two of the bits. In fact, any binary coded word can be made single error detecting by using the concept of parity, i.e., adding a single bit which will make its minimum distance equal to 2.

Suppose that the minimum distance between any characters of a code is 3. What can we say about its error detection and/or correction properties? If a single bit were to change in a minimum distance 3 code, then this "wrong" code word would be detected since it does not correspond to any of the characters in the code. Also, this incorrect code word, when compared against all the legitimate code characters, is closer to one of them than to any other. That is, it has a distance of 1 from one of the code characters but a minimum distance of 2 to any other of the code characters. Thus, with the assumption of only a single error, we can determine which of the code characters was really transmitted, i.e., the one for which the distance is 1. Thus, we can correct the "wrong" character into the correct character. Thus, a minimum distance of 3 in a code is single error correcting or double error detecting. Of course, if 2 bits should change, we would obtain an incorrect code word which we could detect but not correct.

We can determine a relationship between the minimum distance of a code and its error-detecting and/or -correcting capabilities. Let

L = minimum distance of a code
D = number of bits which can change and still be detected as a wrong code character
C = number of bits which can be corrected to obtain a correct code character

Then $L - 1 = C + D$ and $D \geq C$. We can show this relationship in the following table:

L	D	C
1	0	0
2	1	0
3	2 1	0 1
4	3 2	0 1
5	4 3 2	0 1 2

Thus, with a minimum distance of 3, the code can detect two errors *or* it can detect and correct a single-bit error. With a minimum distance of 4, we can change 3 bits in a code word and still detect that we have an incorrect character. If we assume that only 2 bits have changed, then we have an incorrectly coded character but cannot correct the character since it is halfway between two legitimate characters and we are not close to either one. Hence, we cannot choose which one to move to. If we assume that only 1 bit is in error, we shall be close to only one character of the code, and we can correct the single-bit error to correspond to a correct code word. Thus, we buy error correction at the trade-off for error detection. A minimum distance 4 code can be used either as error detection of up to 3 bits in error or as double error detection with single error correction.

4.9 HAMMING CODES

4.9.1 Minimum Distance 3 Code

A class of codes which can be used as error-detecting and-correcting codes was developed by R. W. Hamming. We shall now use Hamming's technique to develop a minimum distance 3 code which has single-bit error correction capability.

Let us assume that we wish to transmit the NBCD code as information. To the four information bits of this code, we must add additional bits to make the total transmitted word have a minimum distance of 3. The total transmitted word will contain 7 bits. We shall label these bit positions, starting from the left, with the numbers 1 through 7:

Bit positions	1	2	3	4	5	6	7
Bit names	C_1	C_2	b_4	C_4	b_3	b_2	b_1

We shall use the bit positions corresponding to powers of 2 as check bits, i.e., our added bits, and call them C_1, C_2, and C_4, respectively. The other bit positions will contain our original NBCD coded character and will be called b_4, b_3, b_2, and b_1 as shown.

We shall select our check bits as follows:

C_1 is chosen such that positions 1, 3, 5, and 7 have even parity.
C_2 is chosen such that positions 2, 3, 6, and 7 have even parity.
C_4 is chosen such that positions 4, 5, 6, and 7 have even parity.

Example 4.3

Let us determine the transmitted character used to send the NBCD character 5:

Bit positions	1	2	3	4	5	6	7
Bit names	C_1	C_2	b_4	C_4	b_3	b_2	b_1
			0		1	0	1

Since we are going to transmit an NBCD 5, $b_1 = 1$, $b_2 = 0$, $b_3 = 1$, and $b_4 = 0$, i.e., 0101.

C_1 must be selected to have even parity on bit positions 1, 3, 5, and 7. Thus, $C_1 = 0$. C_2 establishes even parity on positions 2, 3, 6, and 7, and therefore $C_2 = 1$.

Finally, C_4 must establish even parity on positions 4, 5, 6, and 7, and therefore $C_4 = 0$. Thus, our transmitted character is

Bit positions	1	2	3	4	5	6	7
	0	1	0	0	1	0	1

How do we determine if the transmitted word is incorrect and which bit is in error if we assume that a single bit has been changed? The three parity check bits C_4, C_2, and C_1 are compared against those positions which they check. If the correct parity is determined, i.e., even parity, we place a 0 in the pointer word corresponding to that check bit position. If the incorrect parity is found, we place a 1.

Pointer word	p_4	p_2	p_1
Check bits	C_4	C_2	C_1

The binary number found in the pointer word after all the comparisons have been completed is the bit position which is in error and must be changed. If a binary 0 is found, then no bit is in error.

Example 4.4

Let us assume that we transmitted the Hamming code word for the NBCD character 5 which we found in the previous example. During transmission, bit position 6 was changed, and we received the following word:

0 1 0 0 1 1 1

Is it correct? If not, can we determine which bit is in error? Let us check the check bits:

Bit position	1	2	3	4	5	6	7
Word received	0	1	0	0	1	1	1
Bit name	C_1	C_2	b_4	C_4	b_3	b_2	b_1

Check C_4, which was received as a 0. It checks bit positions 4, 5, 6, and 7, which have three 1s, and therefore the even parity does not check. Hence, p_4 equals a 1.

Check C_2, which was received as a 1. It checks bit positions 2, 3, 6, and 7, which also have three 1s, and therefore the even parity does not check. Hence, p_2 equals a 1.

Check C_1, which was received as a 0. It checks bit positions 1, 3, 5, and 7, which have two 1s, and therefore the even parity checks. Hence, p_1 equals a 0.

Thus, the check word is $p_4 = 1, p_2 = 1, p_1 = 0$, or 110, the binary number for a 6. Thus, bit position 6 must be changed, and we correct the received word to 0100101. This indeed is the word which was transmitted. If we remove the check bits, we obtain the four information bits 0101 or an NBCD 5, which was what we wanted to send.

Only single-bit errors can be corrected by this Hamming code. If 2 bits are in error, the error correction method will detect a single error, but the single corrected bit will not correct the two errors, and hence the word will still be in error. Thus, this code can detect two errors but can be used as error correcting only if a single error has been made.

4.9.2 Minimum Distance 4 Code

A minimum distance 4 code is a single-error-correcting and double-error-detecting code. The single-error-correcting Hamming code developed above can be made into a minimum distance 4 code with the addition of another check bit at bit position 8.

Bit position	1	2	3	4	5	6	7	8
Bit name	C_1	C_2	b_4	C_4	b_3	b_2	b_1	C_8

This bit has even parity over the entire 8-bit word. If a single error occurs in any of the first seven bit positions, then two things will occur: (1) the parity over the entire word will be wrong, and (2) the check bits C_1, C_2, and C_4 will indicate the bit which is wrong. If the single error occurs in C_8, the parity over the entire word will be wrong, but check bits C_1, C_2, and C_4 will indicate a binary 0. This tells us that it is C_8 which is wrong. If a double error occurs, the parity over the entire word is correct, but the check bits C_1, C_2, and C_4 will indicate a bit in error. Under such conditions, no correction should be made, but a double error has been detected.

4.10 ALPHANUMERIC CODES

4.10.1 A 6-Bit Code

If in addition to the 10 decimal digits we wish to encode the 26 letters of the alphabet into a binary code, we shall need a total of 36 representations. This is greater than 5 bits since 2^5 equals 32, which is less than the required 36. Thus, we shall need 6 bits, which permits 2^6 or 64 possible representations.

Instead of using only 36 of the possible 64 combinations, we can encode punctuation symbols and other special symbols which are usually found on typewriters and key punches. A binary group of 6 bits encoded to represent a decimal digit, an alphabetic letter, a punctuation mark, or a special symbol is usually called a *character*. Figure 4.9 shows such a code.

Most significant three bits

Least significant bits

Char.	000 0	001 1	010 2	011 3	100 4	101 5	110 6	111 7
000 0	0	8	+	*H*	×	*Q*	blank	*Y*
001 1	1	9	*A*	*I*	*J*	*R*	/	*Z*
010 2	2	#	*B*	.	*K*	$	*S*	'
011 3	3	@	*C*	[	*L*	*	*T*	%
100 4	4	?	*D*	&	*M*	-	*U*	≠
101 5	5	:	*E*	(	*N*	)	*V*	=
110 6	6	>	*F*	<	0	;	*W*	]
111 7	7	≥	*G*	←	*P*	≤	*X*	▪

Figure 4.9 A six-bit alphanumeric code.

To refer to a specific character in the code, the usual practice is to group the 6 bits into two octal characters and refer to the character by the octal equivalent. The letter "L" would have the code 43, a "blank" would have the code 60, and a "W" would be 66. Notice that the decimal digits 0 through 9 are coded in octal as 00 through 11, which when converted to a 6-bit binary are the NBCD equivalents with two leading 0s. Also, the letters are coded such that "A" has the lowest octal value code, a 21, while a "Z" has the highest code value of 71. Thus, it is possible to sort names by examining the octal codes since the codes for the letters have an ascending value.

4.10.2 ASCII and EBCDIC Codes

Although the 6-bit alphanumeric codes have permitted the encoding of the 10 decimal numbers, the 26 letters of the alphabet, and a multitude of special characters and punctuation marks, it has not provided enough capability to include the lowercase letters and special control characters needed to print, i.e., carriage return, feed a line, backspace, tabulate, etc. Because of the need for these additional characters and in order to standardize codes among computer manufacturers, the American National Standards Institute developed a 7-bit code called the American Standard Code for Information Interchange (ASCII). This code is shown in Fig. 4.10.

With the introduction of its 360 line of computers, the IBM Corporation introduced an 8-bit-per-character code called the Extended Binary Coded Decimal Interchange Code (EBCDIC). This code is shown in Fig. 4.11. One

$b_7\, b_6\, b_5\, b_4\, b_3\, b_2\, b_1$

				$b_7\, b_6\, b_5$							
b_4	b_3	b_2	b_1	000	001	100	101	010	011	110	111
0	0	0	0	NUL	DLE	@	P	SP	0	`	p
0	0	0	1	SOH	DC1	A	Q	!	1	a	q
0	0	1	0	STX	DC2	B	R	"	2	b	r
0	0	1	1	ETX	DC3	C	S	#	3	c	s
0	1	0	0	EOT	DC4	D	T	$	4	d	t
0	1	0	1	ENQ	NAK	E	U	%	5	e	u
0	1	1	0	ACK	SYN	F	V	&	6	f	v
0	1	1	1	BEL	ETB	6	W	'	7	g	w
1	0	0	0	BS	CAN	H	X	(	8	h	x
1	0	0	1	HT	EM	I	Y	)	9	i	y
1	0	1	0	LF	SUB	J	Z	*	:	j	z
1	0	1	1	VT	ESC	K	[	+	;	k	{
1	1	0	0	FF	FS	L	\	,	<	l	:
1	1	0	1	CR	GS	M	]	—	=	m	}
1	1	1	0	SO	RS	N	^	.	>	n	~
1	1	1	1	SI	VS	0	—	/	?	o	DEL

Control Character Meanings

NUL	Null
SOH	Start of heading (cc)
STX	Start of text (cc)
ETX	End of text (cc)
EOT	End of transmission (cc)
ENQ	Enquiry (cc)
ACK	Acknowledge (cc)
BEL	Bell
BS	Backspace (FE)
HT	Horizontal tabulation (FE)
LF	Line feed (FE)
VT	Verticle tabulation (FE)
FF	Form feed (FE)
CR	Carriage return (FE)
SO	Shift out
SI	Shift in

Figure 4.10 ASCII 7-Level Code.

$b_0 b_1 b_2 b_3 b_4 b_5 b_6 b_7$

			00				01				10				11				Bit Positions 0,1
			00	01	10	11	00	01	10	11	00	01	10	11	00	01	10	11	Bit Positions 2,3
			0	1	2	3	4	5	6	7	8	9	A	B	C	D	E	F	First Hexadecimal Digit
			12				12	12		12	12	12		12	12				Zone Punches
				11				11	11	11		11	11	11		11			
					0		0		0	0	0		0	0			0		
Bit Positions 4, 5, 6, 7	Second Hexadecimal Digit	Digit Punches	9	9	9	9	9	9	9	9									Digit Punches
0000	0	8-1	(1) NUL	(2) DLE	(3) DS	(4)	(5) SP	(6) &	(7) -	(8)					(9) {	(10) }	(11) \	(12) 0	8-1
0001	1	1	SOH	DC1	SOS				(13) /		a	j	~		A	J	(14)	1	1
0010	2	2	STX	DC2	FS	SYN					b	k	s		B	K	S	2	2
0011	3	3	ETX	TM							c	l	t		C	L	T	3	3
0100	4	4	PF	RES	BYP	PN					d	m	u		D	M	U	4	4
0101	5	5	HT	NL	LF	RS					e	n	v		E	N	V	5	5
0110	6	6	LC	BS	ETB	UC					f	o	w		F	O	W	6	6
0111	7	7	DEL	IL	ESC	EOT					g	p	x		G	P	X	7	7
1000	8	8	GE	CAN							h	q	y		H	Q	Y	8	8
1001	9	8-1	RLF	EM						\	i	r	z		I	R	Z	9	9
1010	A	8-2	SMM	CC	SM		¢	!	(15) ¦	:								LVM	8-2

Figure 4.11 EBCDIC code. (Reprinted by permission of International Business Machines Corp.)

1011	B	8-3	VT	CU1	CU2	CU3	.	$	,	#									8-3
1100	C	8-4	FF	IFS		DC4	<	*	%	@					⑀		⑁		8-4
1101	D	8-5	CR	IGS	ENQ	NAK	(	)	_	'									8-5
1110	E	8-6	SO	IRS	ACK		+	;	>	=					⑂				8-6
1111	F	8-7	SI	IUS	BEL	SUB	|	¬	?	"								EO	8-7
			12				12				12	12		12	12	12		12	Zone Punches
				11				11				11	11	11		11	11	11	
					0				0		0		0	0	0		0	0	
			9	9	9	9									9	9	9	9	

Card Hole Patterns

(1) 12-0-9-8-1
(2) 12-11-9-8-1
(3) 11-0-9-8-1
(4) 12-11-0-9-8-1
(5) No Punches
(6) 12
(7) 11
(8) 12-11-0
(9) 12-0
(10) 11-0
(11) 0-8-2
(12) 0
(13) 0-1
(14) 11-0-9-1
(15) 12-11

Control Character Representations

ACK	Acknowledge
BEL	Bell
BS	Backspace
BYP	Bypass
CAN	Cancel
CC	Cursor Control
CR	Carriage Return
CU1	Customer Use 1
CU2	Customer Use 2
CU3	Customer Use 3
DC1	Device Control 1
DC2	Device Control 2
DC4	Device Control 4
DEL	Delete
DLE	Data Link Escape
DS	Digit Select
EM	End of Medium
ENQ	Enquiry
EO	Eight Ones
EOT	End of Transmission
ESC	Escape
ETB	End of Transmission Block
ETX	End of Text
FF	Form Feed
FS	Field Separator
GE	Graphic Escape
HT	Horizontal Tab
IFS	Interchange File Separator
IGS	Interchange Group Separator
IL	Idle
IRS	Interchange Record Separator
US	Interchange Unit Separator
LC	Lower Case
LF	Line Feed
NAK	Negative Acknowledge
NL	New Line
NUL	Null
PF	Punch Off
PN	Punch On
RES	Restore
RLF	Reverse Line Feed
RS	Reader Stop
SI	Shift In
SM	Set Mode
SMM	Start of Manual Message
SO	Shift Out
SOH	Start of Heading
SOS	Start of Significance
SP	Space
STX	Start of Text
SUB	Substitute
SYN	Synchronous Idle
TM	Tape Mark
UC	Upper Case
VT	Vertical Tab

Special Graphic Characters

¢	Cent Sign
.	Period, Decimal Point
<	Less-than Sign
(	Left Parenthesis
+	Plus Sign
|	Logical OR
&	Ampersand
!	Exclamation Point
$	Dollar Sign
*	Asterisk
)	Right Parenthesis
;	Semicolon
¬	Logical NOT
-	Minus Sign, Hyphen
/	Slash
¦	Vertical Line
,	Comma
%	Percent
_	Underscore
>	Greater-than Sign
?	Question Mark
`	Grave Accent
:	Colon
#	Number Sign
@	At Sign
'	Prime, Apostrophe
=	Equal Sign
"	Quotation Mark
~	Tilde
{	Opening Brace
⑀	Hook
⑂	Fork
}	Closing Brace
\	Reverse Slant
⑁	Chair
|	Long Vertical Mark

Figure 4.11 (continued).

of the advantages in the use of an 8-bit-per-character code is that instead of using the 8 bits to represent a single character, the machine can use these bits to pack two 4-bit NBCD coded characters. Thus, the 8 bits can be decoded as two NBCD digits or as a single alphanumeric character.

REFERENCES

GARNER, H. L., "Generalized Parity Checking," *IRE Trans. Electronic Computers*, EC-7, No. 3 (Sept. 1958), pp. 207–213.

GRAY, H., U.S. Patent No. 2632058, March 17, 1953.

HAMMING, R. W., "Error Detecting and Error Correcting Codes," *Bell System Tech. J.*, 29, No. 2 (April 1950), pp. 140–147.

LIN, S., *An Introduction to Error-Correcting Codes*. Englewood Cliffs, N.J.: Prentice-Hall, 1970.

MARCUS, M. P., *Switching Circuits for Engineers*, 2nd ed. Englewood Cliffs, N.J.: Prentice-Hall, 1967.

PETERSON, W. W., *Error-Correcting Codes*. Cambridge, Mass. M.I.T. Press, 1961.

SCOTT, N. R., *Electronic Computer Technology*. New York: McGraw-Hill, 1970.

SELLERS, F. F., M. Y. HSIAO, and L. W. BEARSNSON, *Error Detecting Logic for Digital Computers*. New York: McGraw-Hill, 1968.

WEEG, G. P., "Uniqueness of Weighted Code Representations," *IRE Trans. Electronic Computers*, EC-9, No. 4 (Dec. 1960), pp. 487–489.

WHITE, G. S., "Coded Decimal Number Systems for Digital Computers," *Proc. IRE*, 41, No. 10 (Oct. 1953), pp. 1450–1452.

PROBLEMS

4.1. Represent the decimal number 5382 in
(a) NBCD.
(b) Excess-3 BCD.
(c) 2, 4, 2, 1 BCD.
(d) As a binary number.

4.2. Represent the decimal number in Problem 4.1 in Gray code.

4.3. Obtain the weighted binary code for the base 12 digits using weights of 5, 4, 2, 1.

4.4. Determine the even parity bit generated when the message is the 10 decimal digits in the 8,4, −2, −1 code.

4.5. Consider the bit configuration 010110010111. Interpret its meaning if it represents
(a) Three decimal digits in NBCD.
(b) Three decimal digits in excess-3 BCD.

(c) Three decimal digits in 2, 4, 2, 1 BCD.
(d) Two characters in the 6-bit code of Fig. 4.9.

4.6. Encode the decimal digits from 0 to 9 using a 4-bit binary code with the following weights:
(a) 5, 3, 2, −1 (b) 7, −4, 2, 1 (c) 8, −4, 2, −1
Which of these codes is a self-complementing code?

4.7. Show the configuration for a 24-bit word when its content represents
(a) The number $(674)_{10}$ in binary.
(b) The decimal number 674 in NBCD.
(c) The character AM7 in EBCDIC.

4.8. Write your full name in an 8-bit code made up of the seven ASCII bits and odd parity bit in the most significant position. Include blanks between the first name, middle initial, and last name and a period after the middle initial.

4.9. Repeat Problem 4.8 using EBCDIC.

4.10. Write A + B = C using
(a) The ASCII code.
(b) The EBCDIC code.

4.11. Given the code 0000, 0010, 0101, 0111, 1000, 1010, 1101, 1111 for 0 to 7 inclusive, demonstrate that it is not a single-error-detecting code by calculating distances.

4.12. Given the 11-bit combination $1001101p_1p_2p_3p_4$, which is encoded with four redundant bits for single error correction, use odd parity to determine the values of the redundant bits. Demonstrate the single-error-correcting power of the resulting 11-bit combination by deliberately inserting a single error in the sixth position and performing a complete check.

4.13. If p_x and p_y are the even parity bits for the binary numbers $X = x_nx_{n-1} \ldots x_1x_0$ and $Y = y_ny_{n-1} \ldots y_1y_0$, respectively, then we may write

$$p_x = x_n \oplus x_{n-1} \oplus \cdots \oplus x_2 \oplus x_1 \oplus x_0$$
$$p_y = y_n \oplus y_{n-1} \oplus \cdots \oplus y_2 \oplus y_1 \oplus y_0$$

Given p_x and p_y for two numbers X and Y, prove that the parity bit p_s for their sum $S = X + Y$ is

$$p_s = p_x \oplus p_y \oplus c_n \oplus c_{n-1} \oplus \cdots \oplus c_1$$

where $c_n, c_{n-1}, \ldots, c_1$ are the carries; that is, $s_i = x_i \oplus y_i \oplus c_i$.

4.14. Consider the augmented single-error-correcting Hamming code

$$c_1c_2b_4c_4b_3b_2b_1c_0$$

where c_1, c_2, and c_4 are the Hamming parity check bits and c_0 is an even parity bit added after the Hamming code is completed. Form the error-checking algorithm for the above code to demonstrate that the addition of c_0 makes it possible to recognize error conditions with two or more errors which the Hamming check $(c_4c_2c_1)$ alone cannot locate.

5

Switching Algebra and Logic Gates

In this chapter, we shall review the fundamentals of logic design, logic building blocks, and the electronic implementation of logic circuits.

Although digital computers manipulate numerical quantities, they do so in an ordered or logical manner. Therefore, an understanding of mathematical logic helps one to understand the circuit design of the computer units. The logic circuits from which the computer units are built employ electronic switching circuits because these circuits have an excellent signal-to-noise ratio and operate with very high speed.

5.1 SWITCHING ALGEBRA

The mathematical structure which is used as a model for the design of switching circuits is called switching algebra. It is a special case of a more general class of algebra known as Boolean algebra. The original presentation of concepts of Boolean algebra was made by George Boole (1815–1864). Boolean algebra provides a method where complex logic statements can be manipulated in order to rigorously determine their validity. Later, Claude Shannon developed switching algebra as a vehicle for the analysis of switching circuits.

From a computer logic design point of view, the concern in this text is not with mathematical logic, but rather with the logic of switching circuits. Therefore, switching algebra rather than Boolean algebra is to be considered here. Many of the properties of switching algebra, however, do carry over to more general Boolean algebra.

5.1.1 Basic Operations

There are three fundamental operations of switching algebra. These operations are the AND operation, the OR operation, and the NOT operation. While the AND and the OR are binary operations (they require at least two operands), the NOT is a unary operation (it requires one operand). In each case the operands take values from the set {0, 1}, and the result is also assigned a value from the same set {0, 1}. Some of the notations used in referring to these operations are shown in Table 5.1.

Table 5.1 *Basic Operations*

Operation	*Notations used*
AND	$x \cdot y, xy, x \wedge y, x \cap y$
OR	$x + y, x \vee y, x \cup y$
NOT (or COMPLEMENT)	$x', \bar{x}, \sim x$

A *switching variable* is defined as an arbitrary element which can be assigned a value from the set {0, 1}. If x and y denote switching variables, then the definition of the three fundamental operations of switching algebra can be written as

$$xy = \begin{cases} 1 & \text{when both } x = 1 \text{ and } y = 1 \\ 0 & \text{otherwise} \end{cases}$$

$$x + y = \begin{cases} 1 & \text{when either } x = 1 \text{ or } y = 1 \text{ or both} \\ 0 & \text{when both } x = 0 \text{ and } y = 0 \end{cases}$$

$$x' = \begin{cases} 1 & \text{when } x = 0 \\ 0 & \text{when } x = 1 \end{cases}$$

The above definitions can be represented by a truth table, as shown in Table 5.2.

A *switching expression* consists of a finite number of switching variables combined by the use of a finite number of occurrences of switching operations. For instance, x, x', $x + y$, $x + y'$, and $x(y' + z)$ are switching expressions. Parentheses may be used in the same way as in other mathematical notations. If parentheses are not used, then the order of precedence of the three fundamental operations is NOT, AND, OR.

A *switching function* of n switching variables $x_1, x_2, \ldots, x_n$ is a rule

Table 5.2 Truth Table Definitions for the AND, OR, and NOT Operations

x	y	$x \cdot y$	$x + y$	x'	y'
0	0	0	0	1	1
0	1	0	1	1	0
1	0	0	1	0	1
1	1	1	1	0	0

which assigns to each of the 2^n possible combinations of values of those variables a value of 0 or 1. A switching function may be specified by either a truth table or a switching expression. The truth table for a switching function lists the value of the function corresponding to each of the possible 2^n combinations of values of the variables. For example, if $f(x, y, z)$ is expressed by

$$f = xy + x'z \tag{5.1}$$

then the truth table for f can be obtained by evaluating f for each of the possible 2^n combinations of values of the variables x, y, z. Thus, if $x = 0$, $y = 1$, $z = 1$, then

$$\begin{aligned} f &= xy + x'z \\ &= 0 \cdot 1 + 0' \cdot 1 \\ &= 0 \cdot 1 + 1 \cdot 1 \\ &= 0 + 1 = 1 \end{aligned}$$

Evaluating f for each of the eight possible cases, the results are given in Table 5.3. Let us assume that a switching function is defined by the truth table

Table 5.3 Truth Table for f = xy + x'z

x y z	x'	xy	$x'z$	f
0 0 0	1	0	0	0
0 0 1	1	0	1	1
0 1 0	1	0	0	0
0 1 1	1	0	1	1
1 0 0	0	0	0	0
1 0 1	0	0	0	0
1 1 0	0	1	0	1
1 1 1	0	1	0	1

(Table 5.3) and try to derive a switching expression for the function. The truth table indicates that $f = 1$ if and only if

$$x = 0 \quad \text{and} \quad y = 0 \quad \text{and} \quad z = 1, \qquad \text{which implies that } x'y'z = 1$$

or

$$x = 0 \quad \text{and} \quad y = 1 \quad \text{and} \quad z = 1, \qquad \text{which implies that } x'yz = 1$$

or

$$x = 1 \quad \text{and} \quad y = 1 \quad \text{and} \quad z = 0, \qquad \text{which implies that } xyz' = 1$$

or

$$x = 1 \quad \text{and} \quad y = 1 \quad \text{and} \quad z = 1, \qquad \text{which implies that } xyz = 1$$

Therefore, $f = 1$ if and only if

$$x'y'z = 1 \quad \text{or} \quad x'yz = 1 \quad \text{or} \quad xyz' = 1 \quad \text{or} \quad xyz = 1$$

Thus, we can write

$$f = x'y'z + x'yz + xyz' + xyz \tag{5.2}$$

The switching expression at the right-hand side of Eq. (5.2) equals 1 if any of the terms equals 1 and is 0 otherwise. This is exactly the specification of f given in Table 5.3.

Notice that the switching expression at the right-hand side of Eq. (5.2) is not the same as the one in Eq. (5.1). This illustrates that the specification of a switching function by a switching expression is not unique and that there might be many different switching expressions that specify the same function. For this reason, it is necessary to study the properties of switching algebra in order to be able to manipulate switching expressions so as to go from one form to another. This will also enable us to develop algorithms that transform expanded expressions such as Eq. (5.2) to a simpler expression such as Eq. (5.1).

Since all the functions, *expressions and variables*, in this chapter, are switching functions, the word switching will be dropped and the word function will be used to mean switching function. Similarly, variables and expressions will be used to mean switching variables and switching expressions, respectively, unless otherwise stated.

5.1.2 Properties of Switching Algebra

As a first step of developing switching algebra, a few basic rules or *postulates* are stated to define the algebra. The first postulate is that of *uniqueness.* It states that the switching variables are two-valued; that is, a variable may assume only one value, 0 or 1, at a given time. Denoting switching variables by lowercase letters, the first postulate can be written as

1. (a) $x = 0, \quad \text{if } x \neq 1$
   (b) $x = 1, \quad \text{if } x \neq 0$

The second postulate defines the AND and the OR operations. This definition is restated below:

2. (a) $xy = \begin{cases} 1, & \text{if } x = 1 \text{ and } y = 1 \\ 0, & \text{otherwise} \end{cases}$
   (b) $x + y = \begin{cases} 0, & \text{if } x = 0 \text{ and } y = 0 \\ 1, & \text{otherwise} \end{cases}$

The last postulate is the definition of the NOT (COMPLEMENTATION or NEGATION) operation and can be expressed as

3. (a) If $x = 0$, then $x' = 1$ or $0' = 1$
   (b) If $x = 1$, then $x' = 0$ or $1' = 0$

The above three postulates completely define switching algebra. From these postulates, it is possible to develop several other properties which will be used later for the manipulation of expressions that will be employed to describe switching circuit operation. The properties to be derived from the above three apply to all Boolean algebra. These properties will be stated below, without proofs, and the reader is referred to any book on switching theory if he is interested.

All the properties of switching algebra come in pairs, where one is the *dual* of the other. The dual of 0 is 1; the dual of AND is OR. That is, if in any property 0 and 1 are interchanged and OR and AND are interchanged (but the NOTs remain unchanged), we obtain a *dual* property that also holds true.

The next three properties involve the concept of the *zero element*, the *unit element*, and the *complement*.

4. (a) $x + 0 = x$
   (b) $x \cdot 1 = x$

5. (a) $x + 1 = 1$
   (b) $x \cdot 0 = 0$

6. (a) $x + x' = 1$
   (b) $x \cdot x' = 0$

The *involution* (double negation) and the *idempotency* properties are stated below. The involution defines the double negation, while the idempotency states that some of the variables are redundant.

7. Involution: (a) $(x')' = x$
(b) $(x')' = x$

8. Idempotency: (a) $x + x = x$
(b) $x \cdot x = x$

For the algebra to be flexible, it must, like the more familiar algebra, obey some manipulative rules. Switching algebra obeys the *distributive*, *commutative*, and *associative* rules. These rules are listed in the following three properties.

9. Distribution: (a) $x(y + z) = xy + xz$
(b) $x + yz = (x + y) \cdot (x + z)$

10. Commutation: (a) $x + y = y + x$
(b) $xy = yx$

11. Associativity: (a) $(x + y) + z = x + (y + z) = x + y + z$
(b) $(xy)z = x(yz) = xyz$

Another property which is very useful in the reduction of expanded expressions is called *absorption* and is given below.

12. Absorption: (a) $x + xy = x$
(b) $x(x + y) = x$

A useful property which does not have a meaningful generic title is stated below as property 13.

13. (a) $(x + x'y) = x + y$
(b) $x(x' + y) = xy$

Another very useful property which deals with the complement is known as De Morgan's theorem. The general form of this property is given below.

14. De Morgan's Theorem: (a) $(x_1 + x_2 + \cdots + x_n)' = x_1'x_2' \cdots x_n'$
(b) $(x_1x_2 \cdots x_n)' = x_1' + x_2' + \cdots + x_n'$

The negation of an expression is realized by interchanging all OR and AND operations and replacing all variables with their complemented (negated) counterparts.

There are many other properties (theorems) in the form of identities which could be postulated; however, we are going to present one more property. The last property to be presented is useful in the manipulation of expressions and is stated below.

15. (a) $xy + y'z + xz = xy + y'z$
(b) $(x + y)(y' + z)(x + z) = (x + y)\cdot(y' + z)$

5.1.3 Simplification of Expressions by Algebraic Manipulation

So far we have defined switching algebra and stated some of its properties. There are many other equalities of expressions which could be proved using the postulates and the properties already presented. As a matter of fact there are 2^{2^n} functions of n binary variables. These functions are irreducible; i.e., they are in their simplest form. But any of the functions can be expressed by other expressions which might be in an expanded form. The above basic properties could then be used to simplify an expanded expression into a shorter one.

Let us discuss some examples which employ the basic properties to simplify switching expressions.

Example 5.1

Simplify each of the following expressions using the basic properties and the postulates of switching algebra:

a. $f_1 = a + abc' + abc + bc + b$.
b. $f_2 = abc' + bc' + ac'$.
c. $f_3 = (b' + c')(a + b' + c)$.

The application of any property will be illustrated by indicating the number of the property by which it was previously identified.

Solution:

$$
\begin{aligned}
f_1 &= a + abc' + abc + bc + b \\
&= a(1 + bc') + b(ac + c + 1) && [4(b), 9(a)] \\
&= a\cdot 1 + b(ac + 1) && [5(a), 12(a)] \\
&= a\cdot 1 + b\cdot 1 && [5(a)] \\
&= a + b && [4(b)]
\end{aligned}
$$

$$
\begin{aligned}
f_2 &= abc' + bc' + ac' \\
&= bc'(a + 1) + ac' && [9(a)] \\
&= bc'(1) + ac' && [5(a)] \\
&= bc' + ac' && [4(b)]
\end{aligned}
$$

$$
\begin{aligned}
f_3 &= (b' + c')(b' + a + c) && [10(a)] \\
&= b' + c'(a + c) && [9(b)] \\
&= b' + c'a + c'c && [9(a)] \\
&= b' + c'a + 0 && [6(a)] \\
&= b' + c'a && [4(a)]
\end{aligned}
$$

Example 5.2

Prove the following identities using the postulates and the basic properties of switching algebra.

a. $x'y'z + x'yz + xy'z' + xy'z + xyz = xy' + z.$
b. $yz' + (x' + z')\cdot(x'y + x'z) = yz' + x'z + x'y.$

Solution: a. The left-hand side equals

$$
\begin{aligned}
&x'y'z + x'yz + xy'z' + xy'z + xyz \\
&\quad = xy'z' + z(x'y' + x'y + xy' + xy) && [9(a), 11(b), 10(a)] \\
&\quad = xy'z' + z(x'(y' + y) + x(y' + y)) && [9(a), 11(b)] \\
&\quad = xy'z' + z(x' + x) && [6(a), 9(a), 4(b)] \\
&\quad = xy'z' + z && [9(a), 4(b)] \\
&\quad = xy' + z && [13(a)]
\end{aligned}
$$

b. The left-hand side equals

$$
\begin{aligned}
yz' + (x' + z')\cdot(x'y + x'z) &= yz' + (x' + z')\cdot x'(y + z) && [9(a)] \\
&= yz' + x'(y + z) && [12(b)] \\
&= yz' + x'y + x'z && [9(a)] \\
&= yz' + x'z + x'y && [10(a)]
\end{aligned}
$$

The above two examples illustrate the simplification technique using the basic properties to manipulate switching expressions. However, this method becomes somewhat difficult as the number of variables increase and the expressions become more complex. Some simplification techniques based on the properties of switching algebra will be developed to allow simplification in a less tedious way.

5.1.4 Truth Tables and Venn Diagrams

Another method of verifying the equality of two switching expressions is by the use of truth tables. In this method each expression is evaluated for all the possible combinations of values of the variables and then the results are compared. As an example, consider the identity

$$a + a'b = a + b \tag{5.3}$$

The truth table for the two expressions is shown in Table 5.4. It is clear from the truth table that the two columns for $(a + a'b)$, $(a + b)$ are identical, and therefore the identity is true.

Table 5.4 *Truth Table for Eq. (5.3)*

a	b	a'	$a'b$	$a + a'b$	$a + b$
0	0	1	0	0	0
0	1	1	1	1	1
1	0	0	0	1	1
1	1	0	0	1	1

A graphical technique which could be used for verifying switching expression identities is the Venn diagram. A variable is considered an event and mapped into an event space. The space occupied by a variable A plus the space occupied by its complement A' are considered to be the universe and to be represented by the square in Fig. 5.1. The intersection between the space occupied by A and that occupied by A' is the null space (no space). Thus, the product $A \cdot A'$ corresponds to the null space in a Venn diagram. Figure 5.1 shows the Venn diagram for some functions of two variables A and B.

As an example of using Venn diagrams to verify switching expression identities, consider the identity

$$A + A'B = A + B$$

Figure 5.2 shows the Venn diagram for each side of the identity. The two shaded areas $A + A'B$ and $A + B$ are the same, and hence the identity is

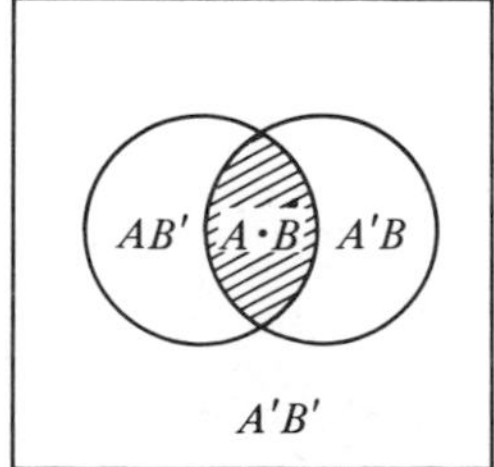

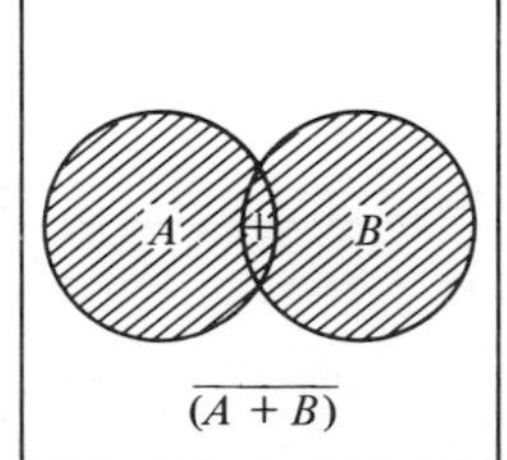

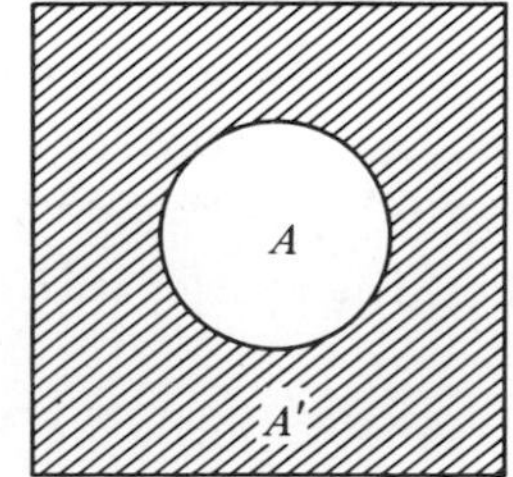

Figure 5.1 Venn diagram for some functions of A, B.

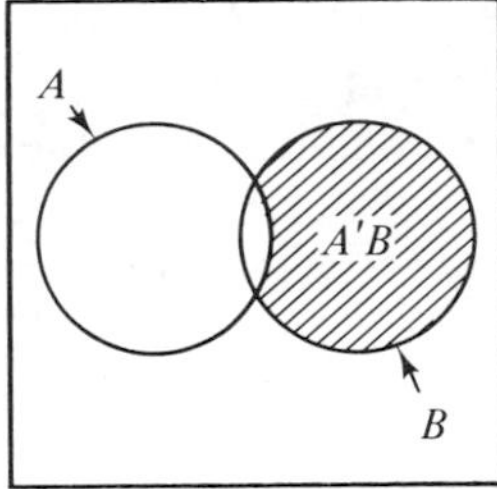

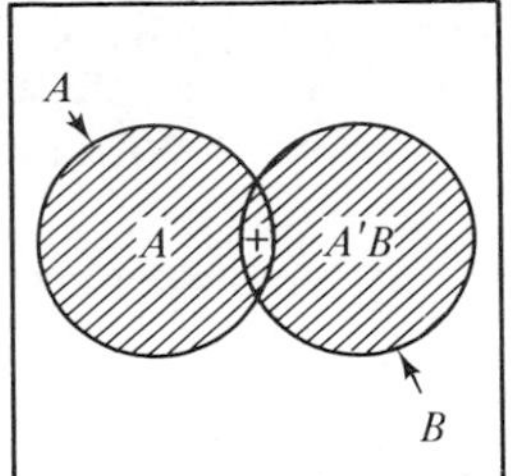

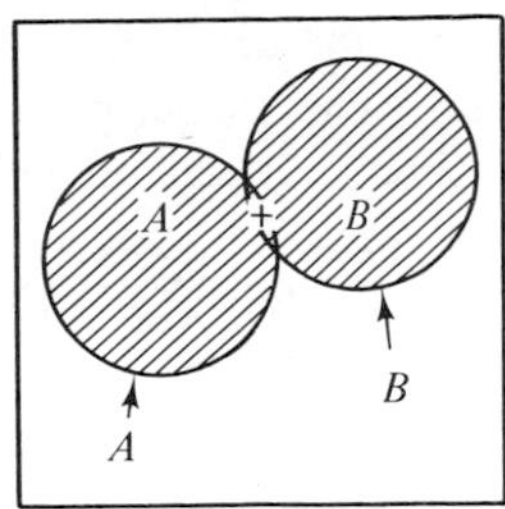

Figure 5.2 Venn diagrams for $A + A'B$, $A + B$.

true. Although this method is more illustrative, it becomes confusing when the number of variables increases.

5.1.5 Standard Forms

We stated earlier that the representation of a switching function by an expression is not unique. In this section, we shall develop a certain standard or *canonical* expressions for each function. As a matter of fact, any switching expression of more than two variables can be expressed as the sum of standard product terms (*minterms*) or as the product of standard sum terms (*maxterms*). A minterm (P term) of n variables is the ANDing of all the variables. A maxterm is the ORing of all the variables.† Table 5.5 shows all the minterms and maxterms for three variables. The columns labeled x, y, z list all the possible combinations of values for the variables x, y, z. The subscript of a P

Table 5.5 Minterms (P Terms) and Maxterms (S Terms) for Three Variables

x	y	z	*P Terms*	*S Terms*
0	0	0	$P_0 = x'y'z'$	$S_0 = x + y + z$
0	0	1	$P_1 = x'y'z$	$S_1 = x + y + z'$
0	1	0	$P_2 = x'yz'$	$S_2 = x + y' + z$
0	1	1	$P_3 = x'yz$	$S_3 = x + y' + z'$
1	0	0	$P_4 = xy'z'$	$S_4 = x' + y + z$
1	0	1	$P_5 = xy'z$	$S_5 = x' + y + z'$
1	1	0	$P_6 = xyz'$	$S_6 = x' + y' + z$
1	1	1	$P_7 = xyz$	$S_7 = x' + y' + z'$

†Here the term *variable* refers to either an unprimed variable or a primed variable. Sometimes the term *literal* is used to refer to the occurrence of a variable in either primed or unprimed form.

term is the decimal equivalent of the binary number given by the product terms. The subscript of an S term gives the complement of the decimal number given the value of the variables. With the above definitions of the P terms and S terms, some of the properties of switching functions can be shown in the form of theorems, but instead we shall state the important properties below.

5.1.5.1 Properties of the P Terms and S Terms

1. There are 2^n distinct P terms and 2^n distinct S terms for n variables. For example, for $n = 3$ the number of P terms equals the number of S terms equals 8.

2. A product term P_i is the complement of the corresponding sum term S_i; i.e.,

$$P_i = S_i', \qquad \forall_i$$

3. The logical product of any two distinct P terms is equal to zero; i.e.,

$$P_i \cdot P_j = 0 \qquad \text{for } i \neq j$$

This property can be stated in another way: For each combination of values of the variables, there is exactly one P term which is equal to 1; all other terms are equal to zero.

4. The logical sum of any two distinct S terms is equal to 1; i.e.,

$$S_i + S_j = 1 \qquad \text{for } i \neq j$$

This property also indicates that for every combination of values of the variables there is exactly one S term which is equal to zero; all other S terms are equal to 1.

5. There are 2^{2^n} distinct functions of n variables. These functions can be formed by taking 0, 1, 2, . . . , n terms at a time. Since there are 2^n terms,

$$\text{Number of distinct functions} = \sum_{j=0}^{n} \binom{2^n}{j}$$

$$= \sum_{j=0}^{n} \frac{2^n!}{j!(2^n - j)!} = 2^{2^n}$$

Therefore, there are 16 distinct functions of two variables and 256 distinct functions of three variables. Table 5.6 shows a list of the 16 functions of two variables.

6. Any switching function (f) of n variables can be expressed by one unique sum-of-products form,

Table 5.6 *The Sixteen Switching Functions of Two Variables A, B*

Function	*Terminology*
$f_0 = 0$	Zero
$f_1 = A$	Identity
$f_2 = B$	Identity
$f_3 = A'$	NOT
$f_4 = B'$	NOT
$f_5 = AB$	AND
$f_6 = A'B$	AND NOT A
$f_7 = AB'$	AND NOT B
$f_8 = A'B' = (A + B)'$	NOR
$f_9 = A + B$	OR
$f_{10} = A' + B$	OR NOT A
$f_{11} = A + B'$	OR NOT B
$f_{12} = A' + B' = (AB)'$	NAND
$f_{13} = AB' + A'B$	EXCLUSIVE-OR $\oplus$
$f_{14} = AB + A'B'$	COINCIDENCE $\oplus'$
$f_{15} = 1$	ONE

$$f = \sum_{i=0}^{2^n-1} \alpha_i \cdot P_i$$

where each α_i is either 0 or 1. For example,

$$f(A, B) = AB + A'B' = \sum_{i=0}^{3} \alpha_i \cdot P_i = 1 \cdot P_0 + 0 \cdot P_1 + 0 \cdot P_2 + 1 \cdot P_3$$

Therefore, $f(A, B) = AB + A'B' = P_0 + P_3$. Another common form of expressing the above function is

$$f(A, B) = \sum (0, 3)$$

7. Any switching function (f) of n variables can be expressed by one unique product-of-sums form,

$$f = \prod_{i=0}^{i=2^n-1} (\alpha_i + S_i)$$

where α_i takes a value of either 0 or 1; i.e.,

$$(x + y')(x' + y') = \prod_{i=0}^{2^n-1} (\alpha_i + S_i)$$

$$(x + y')(x' + y') = (1 + S_0)(0 + S_1)(1 + S_2)(0 + S_3)$$
$$(x + y')(x' + y') = S_1 \cdot S_3$$

Therefore,

$$f(x, y) = S_1 \cdot S_3$$

The above expression can be written in an equivalent form as

$$f(x, y) = \prod (1, 3)$$

8. The sum of all the P terms of n variables equals 1, while the product of all the S terms of n variables is equal to 0:

$$\sum_{i=0}^{2^n-1} P_i = 1, \qquad \prod_{i=0}^{2^n-1} S_i = 0$$

9. The complement of a given switching function of n variables can be expressed by a unique sum of P terms. It can also be expressed by a unique product of S terms; i.e.,

$$f' = \sum_{i=0}^{2^n-1} \alpha_i' \cdot P_i \quad \text{or} \quad f' = \prod_{i=0}^{2^n-1} (\alpha_i' + S_i)$$

where α_i' is the complement of α_i. That means that all the P terms which are not included in f are included in f', and similarly for the S terms. For example, if

$$f(x, y, z) = \sum (0, 1, 5, 7)$$

then

$$f'(x, y, z) = \sum (2, 3, 4, 6)$$

and if

$$f(x, y, z) = \prod (1, 2, 5, 6)$$

then

$$f'(x, y, z) = \prod (0, 3, 4, 7)$$

5.1.5.2 Formation of the P Terms

As stated earlier, corresponding to each combination of values of n variables there corresponds exactly one standard product term (P term) which is equal to 1. Therefore, when a switching function is given by a truth table, the corresponding canonical sum-of-products form is obtained by the sum of all the product terms for which the function is equal to 1.

Any sum-of-products form which is not a canonical (standard) form can be converted into a canonical one by repeated applications of the properties:

$$x \cdot 1 = x, \qquad x + x' = 1, \qquad x(y + z) = xy + xz$$

For example, the specification $f(x, y, z) = xy + x'yz$ gives f as a sum of products but not in canonical form. This is true, since the term xy does not contain the variable z and therefore it is not a standard product term (P term). The term xy can be expanded by using

$$xy = xy \cdot 1 = xy(z + z') = xyz + xyz'$$

and the canonical sum-of-products form for f is

$$f(x, y, z) = xyz + xyz' + x'yz$$

5.1.5.3 Formation of the S Terms

Expressing a function by a product of S terms (canonical product-of-sums form) follows steps dual to those used to obtain the canonical sum-of-products form. When a function is specified by a truth table, then the canonical product-of-sums form is obtained by the product of all the S terms for which the function is equal to 0.

Any product-of-sums form can be expanded into a canonical one by repeated application of the properties:

$$x + 0 = x, \qquad xx' = 0, \qquad x + yz = (x + y)(x + z)$$

For example, the specification $f(x, y, z) = x(y' + z)$, which is not a canonical form, can be expanded as follows:

$$\begin{aligned} x &= x + 0 = x + yy' = (x + y)(x + y') \\ &= (x + y + zz')(x + y' + zz') \\ &= (x + y + z)(x + y + z')(x + y' + z)(x + y' + z') \end{aligned}$$

Similarly,

$$(y' + z) = (y' + z) + xx' = (y' + z + x)(y' + z + x')$$

and therefore

$$f(x, y, z) = (x + y + z)(x + y + z')(x + y' + z)(x + y' + z')(x' + y' + z)$$

where the sum term $(x + y' + z)$ appeared only once in the final expression.

5.1.6 Minimization of Switching Expressions by Karnaugh Maps

The Karnaugh map is a simplified Venn diagram where the circular areas are represented by square areas. Figure 5.3 shows the layouts for Karnaugh maps of two, three, and four variables. The *P* terms (in two different forms) are placed in the squares of the maps to illustrate the similarity of the map to the Venn diagram. Notice that each two adjacent squares in the same row or column differ only in one variable, being complemented in one square and uncomplemented in the other. The two-variable map contains four squares, and then for every additional variable the size of the map doubles. In general there are 2^n squares in a Karnaugh map of n variables. These squares correspond to the 2^n distinct *P* terms or *S* terms of n variables.

The Karnaugh map is a very convenient tool for the simplification of switching expressions. Once an expression is mapped, then any two or more adjacent *P* terms can be grouped together and represented by one product

y \ x	0	1
0	$x'y'$	xy'
1	$x'y$	xy

y \ x	0	1
0	P_0	P_2
1	P_1	P_3

(a) Two-variable map

z \ xy	00	01	11	10
0	$x'y'z'$	$x'yz'$	xyz'	$xy'z'$
1	$x'y'z$	$x'yz$	xyz	$xy'z$

z \ xy	00	01	11	10
0	P_0	P_2	P_6	P_4
1	P_1	P_3	P_7	P_5

(b) Three-variable map

yz \ wx	00	01	11	10
00	$w'x'y'z'$	$w'xy'z'$	$wxy'z'$	$wx'y'z'$
01	$w'x'y'z$	$w'xy'z$	$wxy'z$	$wx'y'z$
11	$w'x'yz$	$w'xyz$	$wxyz$	$wx'yz$
10	$w'x'yz'$	$w'xyz'$	$wxyz'$	$wx'yz'$

yz \ wx	00	01	11	10
00	P_0	P_4	P_{12}	P_8
01	P_1	P_5	P_{13}	P_9
11	P_3	P_7	P_{15}	P_{11}
10	P_2	P_6	P_{14}	P_{10}

(c) Four-variable map

Figure 5.3 Karnaugh map for two, three, and four variables.

term with a fewer number of literals (variables in complemented or uncomplemented form). From Fig. 5.3 it is clear that the first row and the last row are adjacent. Similarly, the first and last columns are adjacent. Grouping two adjacent P terms eliminates one literal. In other words, two adjacent P terms can be represented by a product term with $n - 1$ literals in it. A group of four, eight, . . . eliminates 2, 3, . . . literals. Only binary-valued (2, 4, 8, . . ., 2^k) squares can be grouped together.

When using the Karnaugh map to simplify a given expression, we must identify the expression by its P terms (standard sum-of-products form) or its S terms. For each given P term (S term) a 1 (0) must be entered into the appropriate square in the map. The next step is to form as large a binary-valued (2, 4, 8, . . ., 2^k) group as possible and also a minimum number of these groups. The process of mapping expressions, grouping, and obtaining a minimum expression is illustrated in the following examples.

Example 5.3

$$f_1(x, y, z) = \sum (0, 1, 2, 3, 4, 5)$$

Here the function is already expressed as a standard sum-of-products form which is implied by the decimal notations. Figure 5.4(a) shows a reference map, while Fig. 5.4(b) shows the mapping and grouping of the function. There are two groups; each

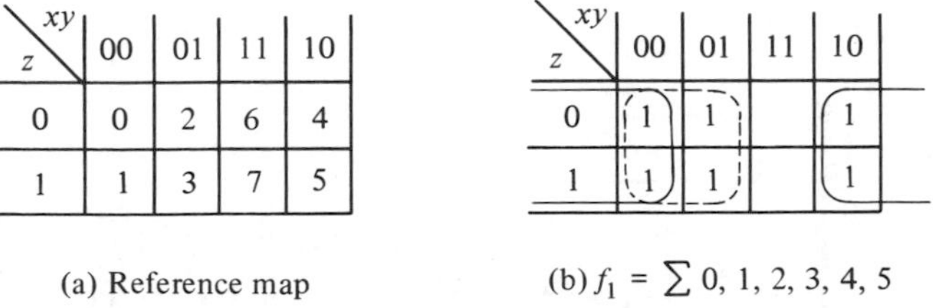

(a) Reference map (b) $f_1 = \sum 0, 1, 2, 3, 4, 5$

Figure 5.4 Karnaugh map for f_1.

contains four P terms. Using a specific P term in more than one group corresponds to saying $P_i + P_i = P_i$, which agrees with the rules of switching algebra. Also, whenever a group of, say, four squares are adjacent then grouping just two of them separately does not agree with the objective of finding the largest possible grouping. Therefore grouping (0, 1, 2, 3) together, in this example, leads to a simpler expression than if we grouped just (2, 3). To read out the result from the map, we look at any specific group and notice the variable which must be eliminated. The dashed group drops the variable y by extending into the first and second columns [equivalent to $(y + y')$], and similarly for z by extending over the first and second rows $(z + z')$. Therefore, the only term common to all four squares is x'. Similarly, the four adjacent edge squares have y' common to all four squares. Therefore,

$$f_1(x, y, z) = \sum (0, 1, 2, 3, 4, 5) = (x' + y')$$

Example 5.4

$$f_2(w, x, y, z) = w'x' + wy'z' + w'xy'z' + wy$$

In this example the function is represented by a sum-of-products form which is not canonical. To map the function, we must substitute for every product term its equivalent sum-of-standard-product terms (*P* terms).

In this case,

$$w'xy'z' = \sum (4)$$
$$wy'z' = xwy'z' + x'wy'z' = \sum (8, 12)$$
$$wy = \sum (10, 11, 14, 15)$$
$$w'x' = \sum (0, 1, 2, 3)$$

Therefore,

$$f_2(w, x, y, z) = \sum (0, 1, 2, 3, 4, 8, 10, 11, 12, 14, 15)$$

Figure 5.5 shows the Karnaugh map for f_2. The top row forms a group with $y'z'$ common to all its squares. The first column has $w'x'$ as the only common term to all of its squares. Similarly, the third grouping eliminates the variables z and x and has wy as the common term. Therefore,

$$f_2(w, x, y, z) = y'z' + w'x' + wy$$

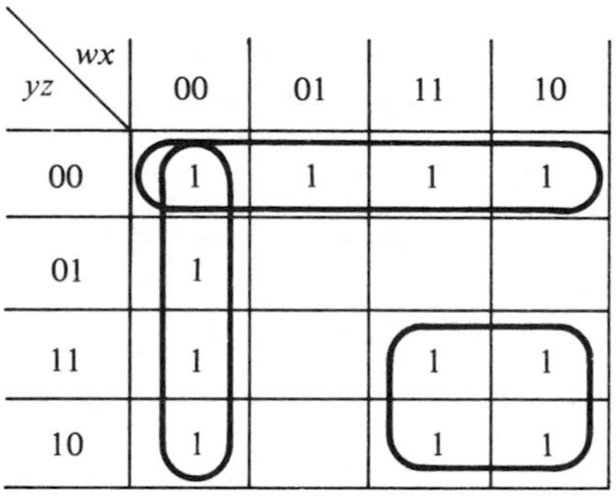

yz \ wx	00	01	11	10
00	1	1	1	1
01	1			
11	1		1	1
10	1		1	1

Figure 5.5 Karnaugh map for f_2.

Example 5.5

In this example, we shall consider a function which is not completely specified. That is, the function is not specified for some of the combinations of the values of the input variables. These are referred to as *don't care conditions* and may be utilized as an aid in simplifying the function. Corresponding to a don't care condition, the value of the function can be considered equal to a 1 if this helps in forming a larger grouping; otherwise it is considered to be 0. This process is illustrated by simplifying

$$f_3(w, x, y, z) = \sum (0, 1, 2, 3, 4, 8, 10, 11, 12, 14, 15), d(5, 6, 9, 13)$$

where $d(5, 6, 9, 13)$ constitutes the don't care *P* terms. Figure 5.6 shows the Kar-

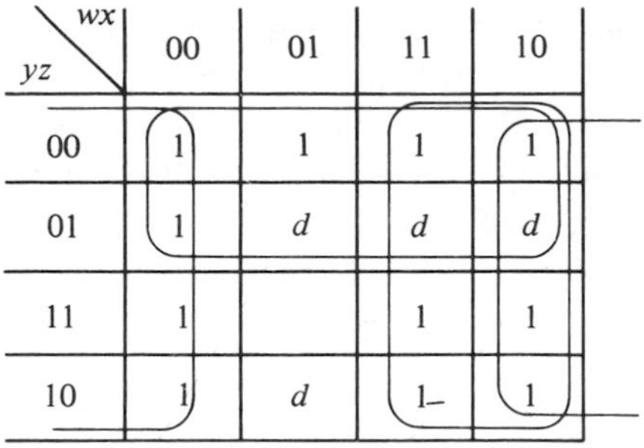

yz \ wx	00	01	11	10
00	1	1	1	1
01	1	d	d	d
11	1		1	1
10	1	d	1	1

Figure 5.6 Karnaugh map for f_3.

naugh map with the appropriate grouping. From the map the simplified expression for f_3 can be written as

$$f_3(w, x, y, z) = w + x' + y'$$

The Karnaugh map minimization technique can also be applied to the S terms of a given function in a manner similar to that for the P terms. Another method of obtaining a minimum product-of-sums form is to map and group the P terms of F' and then apply De Morgan's theorem to the result.

The minimization technique using the Karnaugh map does not guarantee an absolute minimal circuit. It yields to an optimal design only under certain minimality criteria such as using two levels of gates and using double rail logic and other restrictions which are well suited to diode-type logic. The use of the map becomes a difficult task when the number of variables increases. An extension to this technique is a method known as the Quine-McCluskey method, which extends easily to a large number of variables. The Quine-McCluskey method also extends to multiple output circuits where possible sharing of terms between more than one output is taken into consideration. It is not the objective of the authors to represent a detailed study of minimization schemes here. Practically, hardware minimization depends to a large extent on machine organization and the type of circuit elements employed in the design.

5.2 LOGIC GATES

To design a switching system, it is necessary to decide on the type of switching elements that will be used. The input/output behavior of a switching system can be described by switching functions, which usually use the basic logic operations AND, OR, and NOT. It is also possible to transform any switching expression into one which involves either NANDs or NORs or both as the basic operation. A logic element which performs a basic logic operation is referred to as a *gate*. Therefore, a knowledge of the logical and electrical behavior of the various available gates is of importance to a switching system designer. Due to the wide variations in the circuit design of gates from one manufacturer to another, the authors will concentrate more on the logical as opposed to the electrical behavior of gates. The electronic design is presented only as examples of how various gates might be constructed.

5.2.1 Diode Gates

Before introducing diode gates, we shall present the electrical characteristics of a diode. Figure 5.7 shows the symbolic representation for a diode.

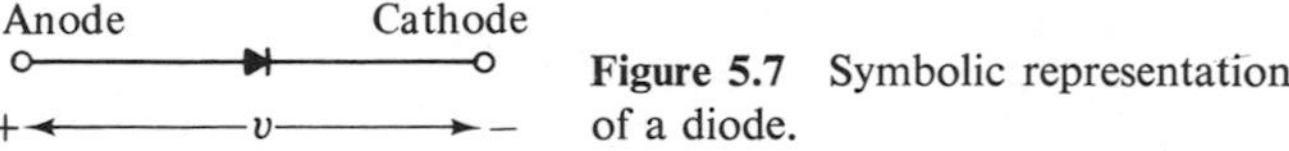

Figure 5.7 Symbolic representation of a diode.

When an electric voltage is applied to the diode such that the anode is positive with respect to the cathode, the diode is *forward-biased.* On the other hand, when the cathode is positive with respect to the anode, the diode is *reverse-biased.* In a forward-biased condition the diode offers an easy path to the current, while in the reverse-biased condition it presents a high resistance to the current flow (the current flows from high voltage to low voltage). Instead of presenting the exact voltage-current characteristics of the diode, the ideal diode characteristics are sufficient for our purpose. An ideal diode acts as a short circuit if it is forward-biased ($v > 0$) and as an open circuit when it is reverse-biased ($v < 0$).

Figure 5.8 shows two diode circuits; one is called an *up-follower*, and the other is called a *down-follower*. The input signals V_x and V_y take values of either +6 or 0 volts. It is easy to verify that for any input combination of V_x and V_y the output of the up-follower V_f is the same as the higher input. Also,

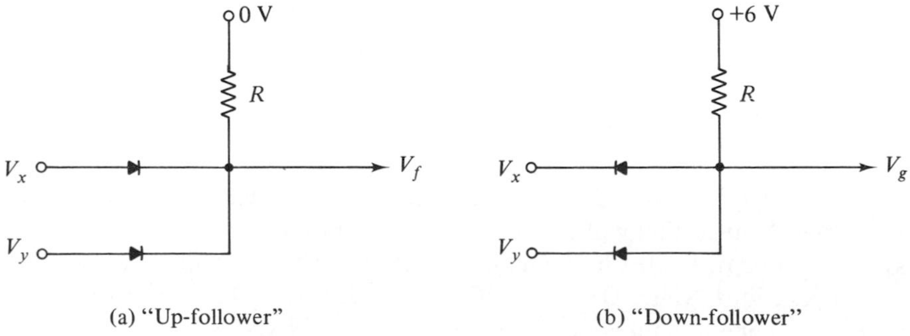

(a) "Up-follower"

(b) "Down-follower"

Figure 5.8 Diode gates.

(V_g) of the down-follower will be the same as the lower input voltage. If a higher voltage is assigned to the logical 1 relative to the voltage assigned to logical 0, then we speak of positive logic. In the circuit of Fig. 5.8, this means the assignment $1 \longleftrightarrow 6$ volts, $0 \longleftrightarrow 0$ volts is made. In this case the up-follower realizes the OR operation, while the down-follower realizes the AND operation. This type of assignment is referred to as positive logic and the up-follower and the down-follower are referred to as a positive OR gate and a positive AND gate, respectively. If the assignment is reversed such that $1 \longleftrightarrow 0$ volts, $0 \longleftrightarrow 6$ volts, then we speak of negative logic and the up-follower and down-follower realize negative AND and negative OR, respectively. The diode gates shown in Fig. 5.8 can be extended to more than two inputs by simply adding a diode for each additional input. There are some practical limitations on the number of diodes to be connected at the inputs of diode gates. This is due to the fact that actual diodes are not ideal devices with zero forward resistance and infinite reverse resistance. The finite forward and reverse resistance of the diodes puts limitations on the use of diode gates. If the number of inputs is large, then the finite resistance affects the output level and makes it vary depending on the state of the input. This problem is known as the *fan-in* problem. Another practical problem with diode gates is the fact that very few diode gates can be interconnected. This is due to the absence of any signal amplification and the decaying of the signal as it passes through several resistive elements. This is referred to as a *fan-out* problem. For the above reason other available gates are presented in subsequent paragraphs.

5.2.2 The Inverter (NOT) Circuit

Figure 5.9 shows a symbolic representation of a transistor. Figure 5.9(a) shows the *NPN* type, while Fig. 5.9(b) shows the *PNP* type. If we assume an ideal transistor, then the characteristics of a transistor can be approximated by a controlled switch. The switch is the current path between the collector and the emitter. This path is to be controlled by the voltage of the base relative to the emitter, V_{BE}. For an *NPN* transistor, the collector-to-emitter

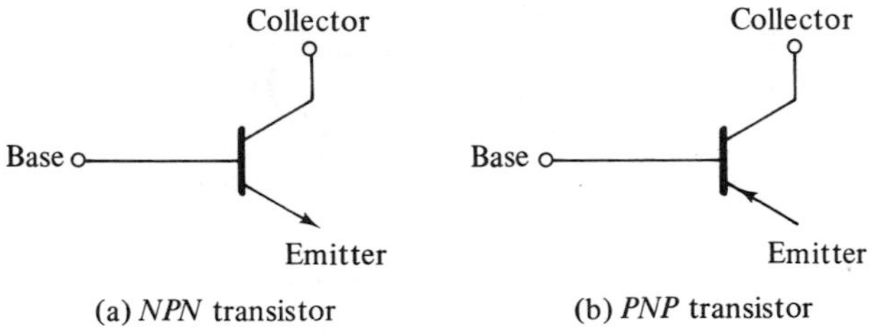

Figure 5.9 Symbolic representation of a transistor.

switch is considered as a short circuit if $V_{BE} > 0$; otherwise it is considered as an open circuit. The reverse statement is true for the *PNP* transistor. With these oversimplified characteristics of a transistor, Fig. 5.10 shows a possible realization of the NOT operation. The circuit is referred to as an inverter.

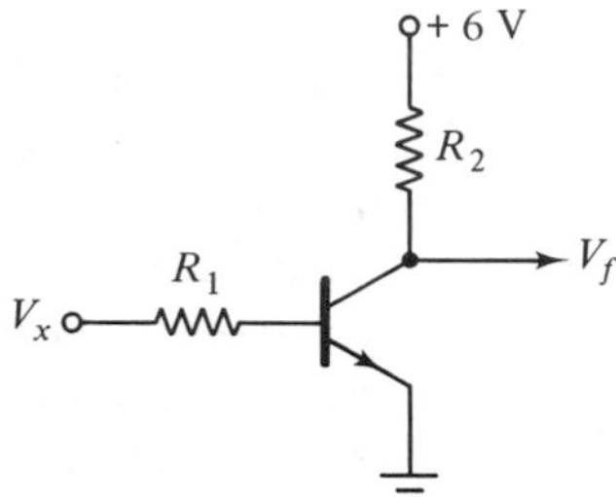

Figure 5.10 Inverter circuit.

Let us assume the logical 0 and 1 to be represented by the voltage levels 0 and +6 volts, in either a positive logic or negative logic representation. For the *NPN* transistor of Fig. 5.10, if $V_x = +6$ volts, the collector-emitter junction is a short circuit, and therefore $V_f = 0$. Similarly, if $V_x = 0$ volts, then the collector-emitter junction is an open circuit and $V_f = +6$ volts. Therefore, the circuit of Fig. 5.10 is an inverter for both positive and negative logic.

5.2.3 Transistor Gates

AND and OR gates can also be constructed using transistor circuits in the common-collector configuration. Figure 5.11(a) and (b) shows two such circuits for *NPN* and *PNP* transistors, respectively. Each of these circuits is referred to as an emitter follower. When an input, which does not exceed $+V$ or $-V$, is applied to the base, the output at the emitter will be an exact replica of the input. That is, the emitter follows the base. Practically, the

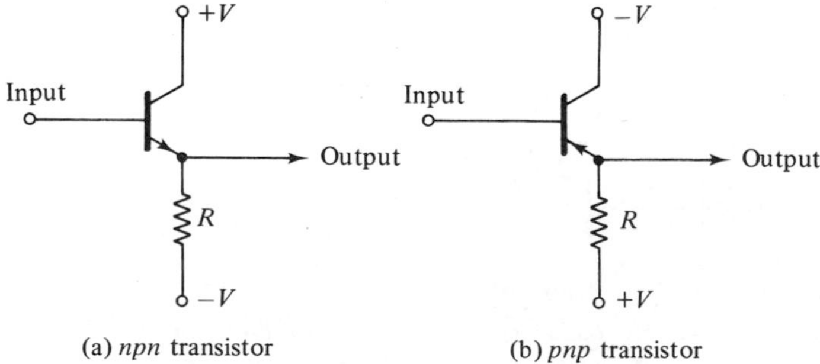

Figure 5.11 The emitter follower.

output will be slightly shifted in level from the input due to the drop in the junction. This shift in level is usually about .5 volt up for the *PNP* transistor and .5 volt down for the *NPN* transistor. The emitter follower as shown in Fig. 5.11 provides a current gain and consequently a power gain. The emitter follower is also characterized by a high input impedance and a low output impedance.

Figure 5.12 shows an AND gate which uses a transistor for each input variable. Let us use positive logic and assign the voltage levels $+5V$ and $0V$ to represent the logical 1 and 0, respectively. If any of the inputs are at 0 volts, then the emitter of this transistor follows its base and maintains the output of the whole circuit at 0 volts. Only if all the inputs are at +5 volts will the output of the circuit be held at 5 volts. Hence, the circuit realizes an AND gate.

A transistor OR circuit is shown in Fig. 5.13. Assume positive logic with the assignment $1 \leftrightarrow +5V, 0 \leftrightarrow 0V$. If any input is at +5 volts, the associated

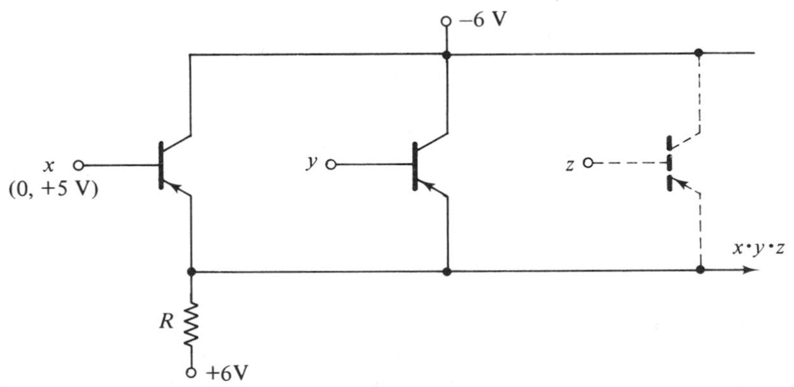

Figure 5.12 Transistor AND gate.

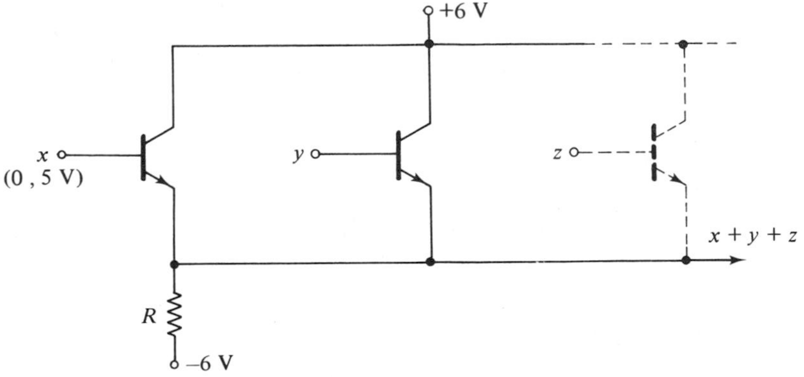

Figure 5.13 Transistor OR gate.

transistor is conducting and its emitter output follows the base and is held at $+5V$. This output will raise the voltage at all other emitters to $+5V$, reverse biasing all transistors from base to emitter by $+5V$, and will isolate the effect of other inputs. These are known as direct-coupled transistor logic (DCTL).

5.2.4 Diode-Transistor AND, OR Gates

The transistor gates discussed in the previous subsection require a transistor for every input. An AND gate which utilizes one transistor is shown in Fig. 5.14. The circuit consists of the diode AND circuit previously

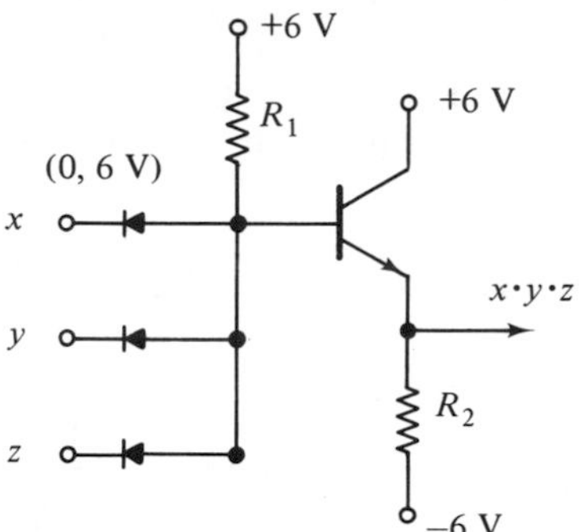

Figure 5.14 Diodes-transistor AND gate.

discussed followed by an emitter follower transistor. The diode's circuit realizes the AND operation, while the emitter follower provides for the current amplification. Thus, the transistor will enable the circuit to drive more circuits connected to its output than otherwise would have been possible.

A diode-transistor OR gate is shown in Fig. 5.15. Similar to the diode-transistor AND gate, the diodes are connected to perform the OR function,

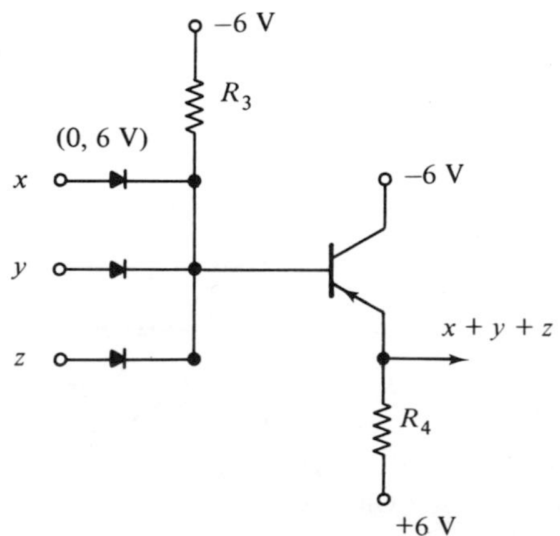

Figure 5.15 Diodes-transistor OR gate.

while the emitter follower is used for the current gain. In either circuit, the number of inputs per gate can be increased by adding a diode for each additional input.

5.2.5 NAND and NOR Gates

The NAND operation has been defined as an AND followed by a NOT. By combining the diode AND circuit (Fig. 5.8) and the inverter circuit (Fig. 5.10) previously discussed, it is possible to make a NAND gate. Figure 5.16 shows a diode-transistor NAND Gate. This line of logic circuits is called DTL, for diode-transistor logic.

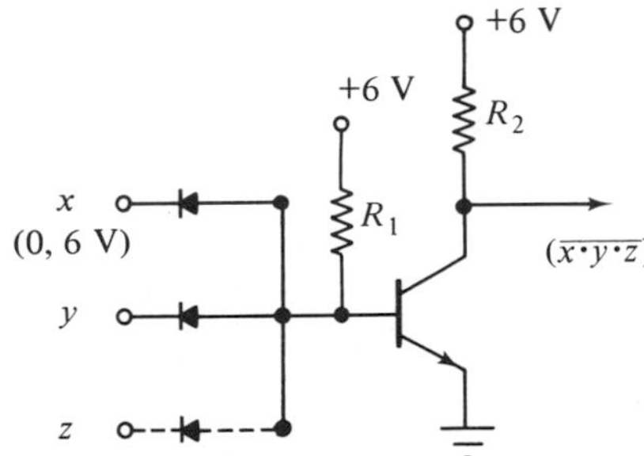

Figure 5.16 Diode-transistor NAND gate.

A NOR gate, which is defined as an OR followed by a NOT, could also be built by combining a diode OR circuit feeding a transistor inverter. A more common way of building NOR gates is to couple transistor inverter circuits together. A NOR circuit of this type is shown in Fig. 5.17. This line of circuits is known as RTL, for resistor-transistor logic. The standard RTL gate is the NOR gate, while that of the DTL gates is the NAND gate.

A high-speed line of circuits, called transistor-transistor logic (TTL), also exists and is widely used. The standard TTL gate is the NAND gate. TTL gates are faster and more complex than the DTL gates.

5.2.6 TTL and ECL Gates

The TTL gates are very widely used due to their high speed. Their speed is obtained at the expense of circuit complexity. The standard TTL gate is the NAND gate.

Figure 5.18 shows a TTL NAND gate in two different forms. The circuit shown in Fig. 5.18(a) is a discrete version of a TTL NAND gate. If any input, representing one of the three input variables x, y, z goes low (logical 0), the voltage at point A is pulled low, turning the output transistor off. This leads to a high voltage (logical 1) at the output. When all the inputs are high (1s), the

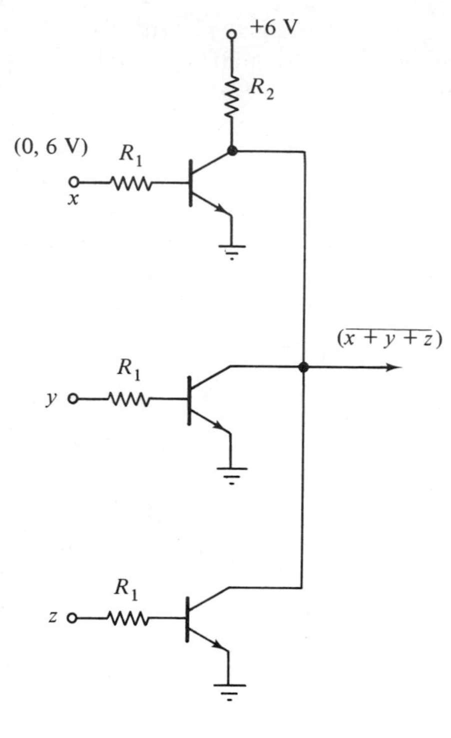

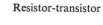

Figure 5.17 Resistor-transistor NOR gate.

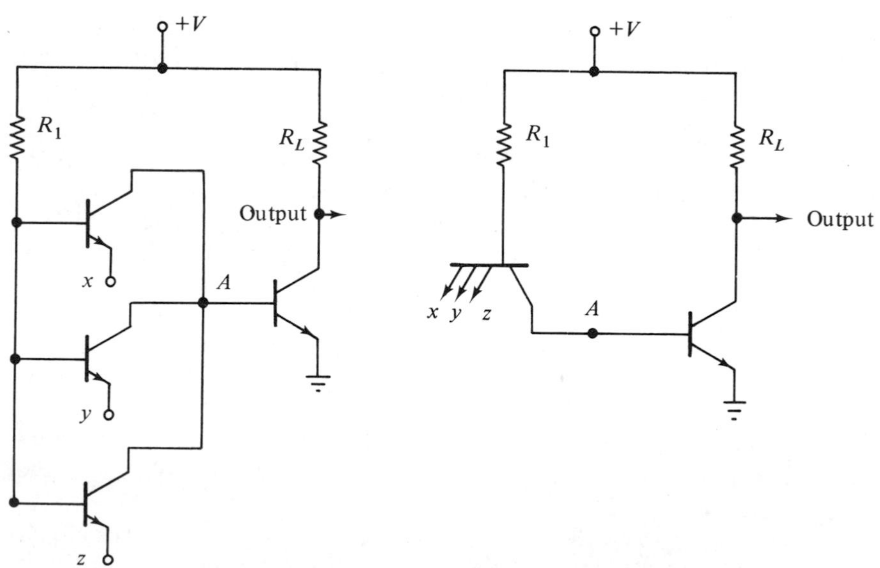

(a) Discrete TTL NAND

(b) Multiple-emitter TTL NAND

Figure 5.18 TTL gates.

voltage at point A is high and the output transistor is driven into saturation. In this case the output is pulled low (logical 0).

The TTL NAND gate shown in Fig. 5.18(b) is characterized by a multiple-emitter transistor at the input. This circuit configuration is available in integrated circuit form. Its behavior is the same as the discrete form shown in Fig. 5.18(a).

A faster but more complex configuration of TTL gates which is available in integrated circuit form is shown in Fig. 5.19. Other than the multiple-emitter transistor at the input, several transistors are used in the circuit, leading to a large gain. The two output transistors T_3 and T_4 are called a totem pole and provide for the circuit's speed and drive capabilities. A brief description of the circuit behavior is described in the next paragraph.

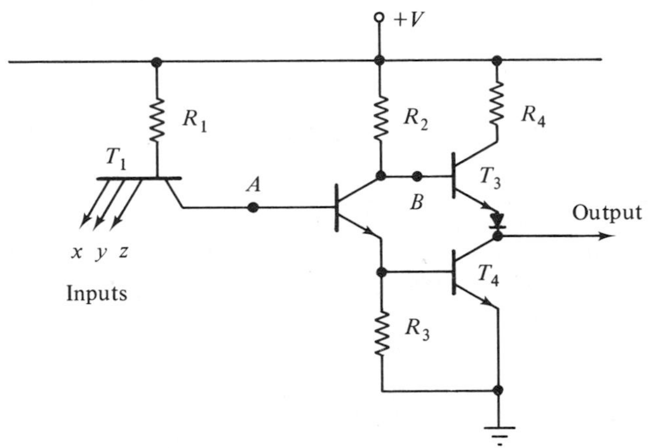

Figure 5.19 TTL NAND gate.

If any input goes low, point A is pulled low, turning off T_2 and hence T_4, while R_2 turns on T_3. Thus, the output goes high. On the other hand, with all inputs high, point A is pulled up, turning on T_2 and hence T_4. Point B is pulled low by T_2, and T_3 turns off. Thus, the output goes low.

ECL refers to emitter-coupled logic. This line of circuits is faster than the TTL circuits. High speed is obtained by operating the transistors either in the active or cut-off regions, instead of saturation and cut-off. Thus, the delay in *turnoff time* is eliminated. For this reason the ECL circuits are also referred to as *nonsaturating logic*. In the ECL circuits the voltage levels corresponding to the logical 0 and 1 have a small swing when compared to other types of circuits. Another common name for the ECL is current-mode logic or current-steering logic, since in these circuits the current is steered rather than the voltage levels passed around.

Figure 5.20 shows a dual OR, NOR ECL circuit. The circuit has three inputs and two outputs. The logic levels are -1.55 and $-.75$ volts, repre-

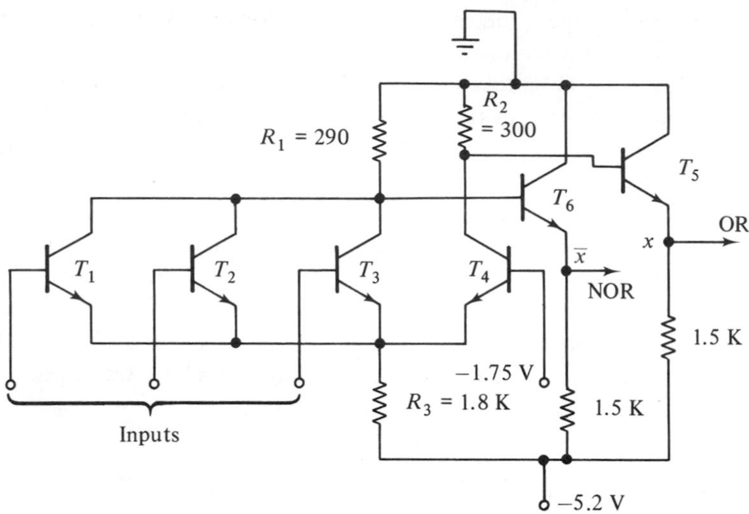

Figure 5.20 ECL three-input gate. (Courtesy Motorola Corp.)

senting logical 0 and 1, respectively. The basic circuit of ECL is an inverter, which is the circuit shown in Fig. 5.20 with T_1 and T_2 removed.

The inverter circuit's operation is based on the differential amplifier formed by T_3 and T_4. When the base of T_3 is at -1.55 volts (logical 0), T_3 will be off. The current through R_2 results in about an .8-volt drop across the resistor R_2, and thus the base of T_5 is held at $-.8$ volts. Assuming .75 volts across the base-emitter of T_5, the x output will be -1.55 volts, which is the same at the input. Since T_3 is cut off, very little current flows in R_1, and the base of T_6 is effectively at ground voltage. Thus, the x' output is maintained at $-.75$ volts, which is the base-emitter drop across T_6.

When the input to the base of T_3 is $-.75$ volts, T_3 will be on and T_4 will be off. Consequently, the output x will be at $-.75$ volts, while x' will be at -1.55 volts. Therefore, the x' is the complement of the input to T_3.

The analysis of the inverter circuits can be extended easily to show that the circuit of Fig. 5.20 is a dual OR, NOR gate. The x output is the OR output, while the x' output is the NOR output.

5.2.7 Integrated Circuits (IC, MSI, LSI)

All the circuits we have discussed so far have been presented as discrete components. In this case a gate can be constructed by assembling each part separately on a board, and then these parts are interconnected by either printed-circuit-plated connectors or by wires. The circuits used in computers are limited; i.e., they consist of a few basic types. Since these circuits are used in large repetition, it is more advantageous to fabricate and package them in groups rather than as discrete components. Today most circuits are fabricated using the technology of integrated circuitry. An integrated circuit package might contain a number of the basic circuits (NAND, NOR, OR, . . .) together on the same chip. If a few gates or a few flip-flops (see Chapter 6) are packages on the same chip, the process is known as MSI, for medium-scale integration. A MSI chip might contain a standard modulo-*n* counter or a decoder (see Chapter 6) as one package. When over 100 gates or flip-flops are fabricated in one package, we refer to large-scale integration (LSI). A typical example of LSI is the construction of a memory module (see Chapter 9) in one package.

5.3 GATE SYMBOLS AND LOGIC DIAGRAMS OF COMBINATIONAL CIRCUITS

Before implementing a switching function, it is helpful to represent the expression by a logic diagram. A switching expression describes the input/output relationship without specifying the logic blocks (gates) to be used. The designer must then select suitable gates to implement the given expression. Once the type of gates to be used is chosen, a logic diagram shows how these gates are connected in order to realize the given expression.

Table 5.7 *Logic Symbols*

Function	*Symbol*
AND	x, y → [·] → $x \cdot y$
OR	x, y → [+] → $x + y$
NOT	x → [I ▷] → x'
NAND	x, y → [NAND] → $x \cdot y = x' + y'$
NOR	x, y → [NOR] → $\overline{x + y} = x' \cdot y'$
EXCLUSIVE-OR	x, y → [⊕] → $x \oplus y = xy' + x'y$

5.3.1 Gate Symbols

There are various symbols which are used to represent each of the gates discussed in the previous section. Table 5.7 shows the symbols which will be used in this text. These symbols are not standard symbols, and the reader will find different symbols in different texts.

5.3.2 Logic Diagrams Using AND, OR

In this section, the construction of logic diagrams will be illustrated for some design examples.

Example 5.6

Given that

$$f_1(w, x, y, z) = \sum (1, 3, 4, 5, 6, 7, 9, 12, 13)$$

minimize and implement with AND-OR logic.

Using a Karnaugh map we can obtain a minimum sum of products as

$$f_1 = w'x + y'z + w'z + xy'$$

and a minimum product of sums as

$$f_1 = (x + z)(w' + y')$$

Figure 5.21(a) and (b) shows the block diagrams for the minimum sum of products and product of sums, respectively. This logic diagram assumes double rail logic,

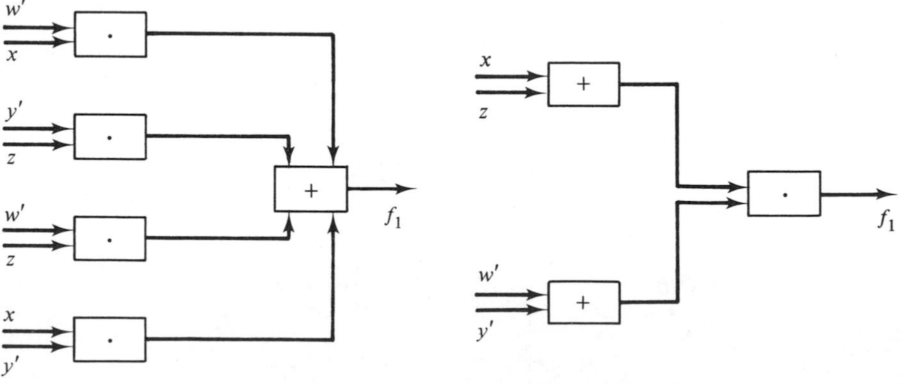

(a) Minimum sum-of-products (b) Minimum product-of-sums

Figure 5.21 Logic diagram for f_1.

where an input variable and its complement are available, and a maximum of two gate levels. If the variables are not available in complemented form, then inverters (NOTs) are needed to generate the complemented variables.

Example 5.7

Design and show a logic diagram for a code translation circuit. It receives an NBCD (8421) code and generates the corresponding excess-3 code.

The above information can be translated into the following truth table:

Input				*Output*			
w	x	y	z	g_1	g_2	g_3	g_4
0	0	0	0	0	0	1	1
0	0	0	1	0	1	0	0
0	0	1	0	0	1	0	1
0	0	1	1	0	1	1	0
0	1	0	0	0	1	1	1
0	1	0	1	1	0	0	0
0	1	1	0	1	0	0	1
0	1	1	1	1	0	1	0
1	0	0	0	1	0	1	1
1	0	0	1	1	1	0	0

The six input combinations (decimal 10 through 15) which are not shown in the truth table are not allowed to occur, and the corresponding outputs are considered as don't care conditions. A Karnaugh map corresponding to each output function is shown in Fig. 5.22.

The outputs g_1, g_2, g_3, g_4 can be written as

$$g_1 = w + xz + xy$$

$$g_2 = xy'z' + x'z + x'y$$

$$g_3 = yz + y'z'$$

$$g_4 = z'$$

The above expressions are minimum according to the map algorithm. The corresponding block diagram can be drawn similar to Example 5.6. The reader can verify that it requires 23 gate inputs. Further reduction in gate inputs might be obtained by manipulation:

$$g_1 = w + x(y + z)$$

$$g_2 = xy'z' + x'(y + z)$$

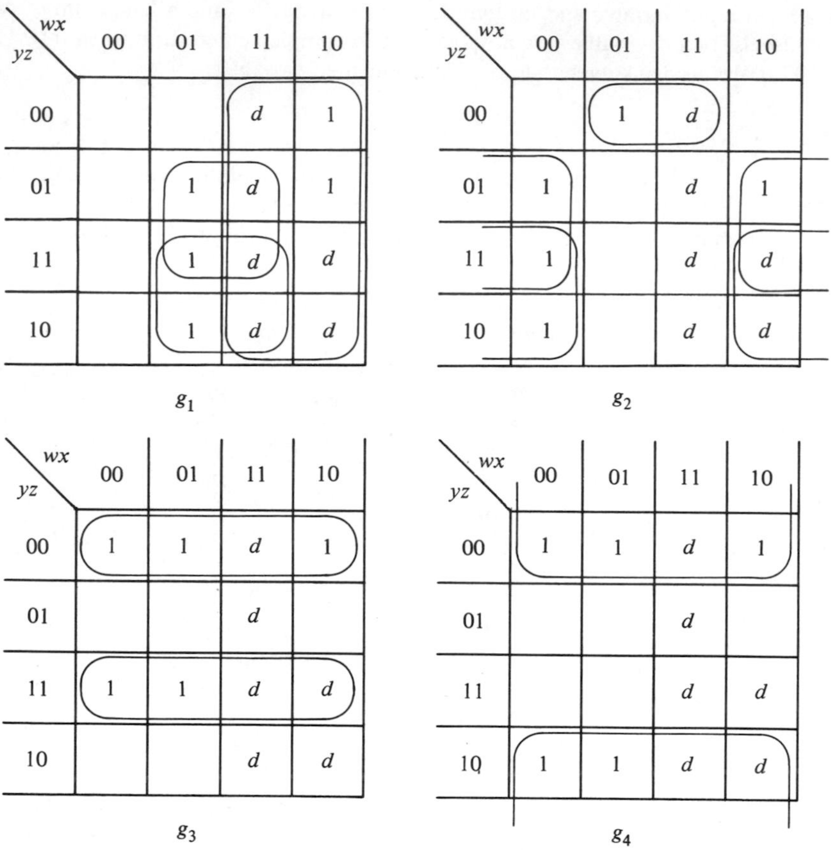

yz \ wx	00	01	11	10
00			d	1
01		1	d	1
11		1	d	d
10		1	d	d

g_1

yz \ wx	00	01	11	10
00		1	d	
01	1		d	1
11	1		d	d
10	1		d	d

g_2

yz \ wx	00	01	11	10
00	1	1	d	1
01			d	
11	1	1	d	d
10			d	d

g_3

yz \ wx	00	01	11	10
00	1	1	d	1
01			d	
11			d	d
10	1	1	d	d

g_4

Figure 5.22 Karnaugh maps for Example Two.

$$g_3 = yz + y'z'$$

$$g_4 = z'$$

The logic diagram for the above equations is shown in Fig. 5.23. The number of gate inputs is reduced to 18 at the expense of increasing the logic levels to 3.

5.3.3 Logic Diagrams Using NAND-NOR

The map minimization algorithm is based on the utilization of diode AND, OR logic with two gate levels. Due to their electronic properties, the NAND, NOR logic is more desirable and is very widely used.

Figure 5.24 shows the logical equivalence of the NAND and NOR gates in terms of AND and OR gates. Each of the NAND or NOR operations is not an associative operation. The absence of the associative property

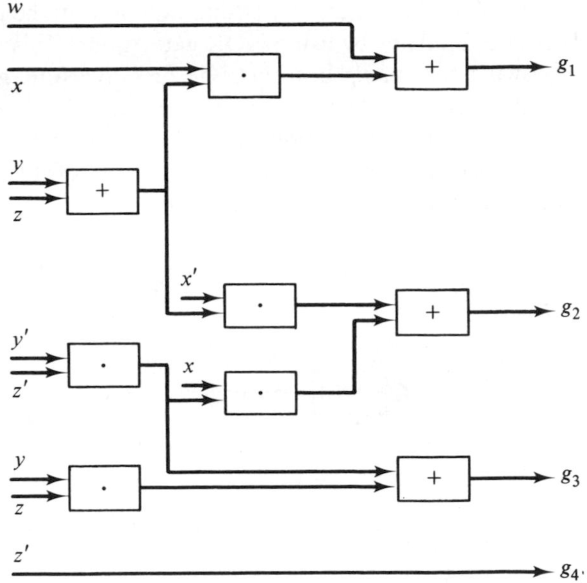

Figure 5.23 Logic diagram for Example Two.

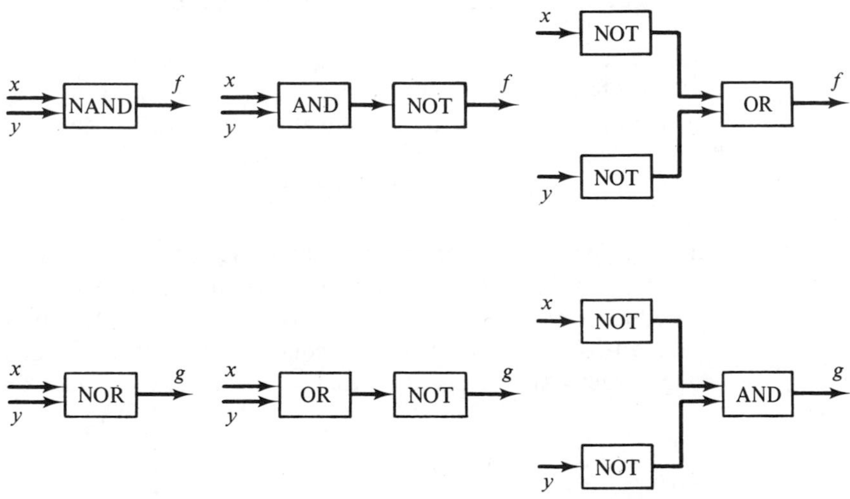

Figure 5.24 Logical equivalence for NAND and NOR gates.

makes it very difficult to manipulate a switching expression which involves NAND or NOR operations. Due to the above fact, the design of switching systems is usually done in terms of AND, OR, and NOT operations and is then translated into one with NAND or NOR or both NAND-NOR.

The above translation is possible because both the NAND and NOR operations are complete; that is, any switching function can be implemented by using NAND gates alone or by using NOR gates alone. This property can be proved by showing that using only NAND or only NOR gates we can perform each of the basic operations of switching algebra, i.e., AND, OR, NOT. Figure 5.25 shows the realization of NOT, AND, and OR operation using only NAND gates. The NOR implementation of the three basic operations can be shown in a similar manner.

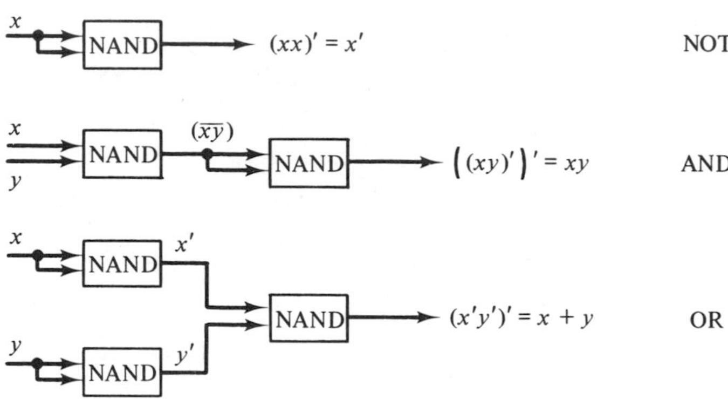

Figure 5.25 NAND realization of NOT, AND, OR.

The translation of a logic diagram using AND-OR into one which employs NAND and NOR gates can be accomplished by substituting each AND or OR gate by its equivalent set of NAND or NOR gates. This is an expensive way of realization because it requires a large number of gates. A simple conversion for a two-level network of AND gates followed by an OR gate (sum-of-products form) can be achieved by substituting every gate by a NAND gate. If every signal passes through exactly two levels (AND followed by OR), then the two-level NAND circuit is an exact equivalent. Similarly, any two-level product-of-sums realization (OR followed by AND) can be converted into an all-NOR circuit. The only condition is that each signal must pass through exactly two levels of gates. As an example, consider the function

$$f_1 = w'x + y'z + w'z$$

Figure 5.26(a) shows the AND-OR realization; the NAND equivalent is shown in Fig. 5.26(b). For the conversion of the two-level product-of-sums form, Fig. 5.27(a) shows the OR-AND realization for the function

$$f_2 = (x' + y)(x + y' + z')$$

The equivalent NOR gate is shown in Fig. 5.27(b).

The restriction of two levels can be relaxed and the conversion can be extended to multiple-level logic circuits. In general, any multiple-level . . .

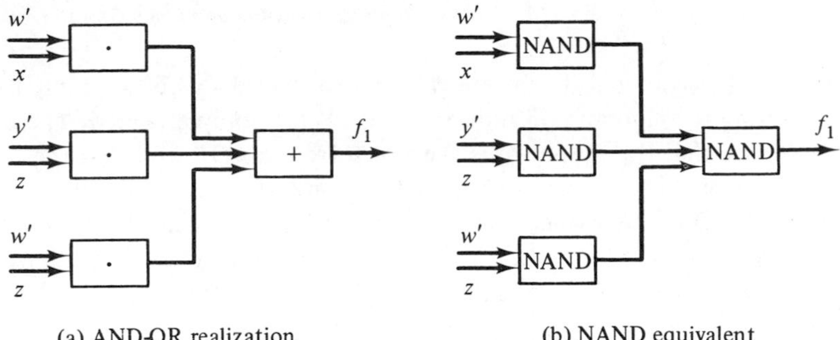

Figure 5.26 Logic diagrams for f_1.

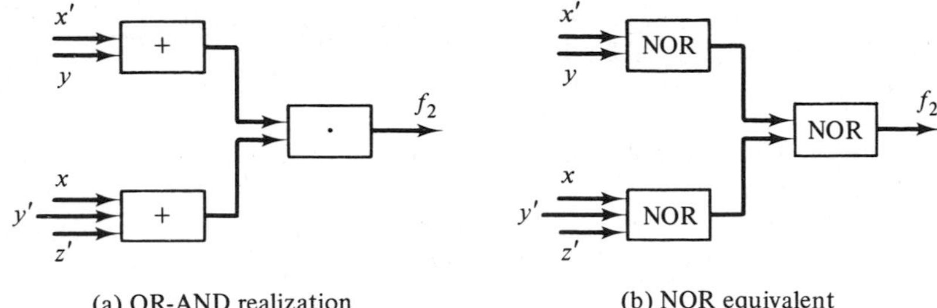

Figure 5.27 Logic diagrams for f_2.

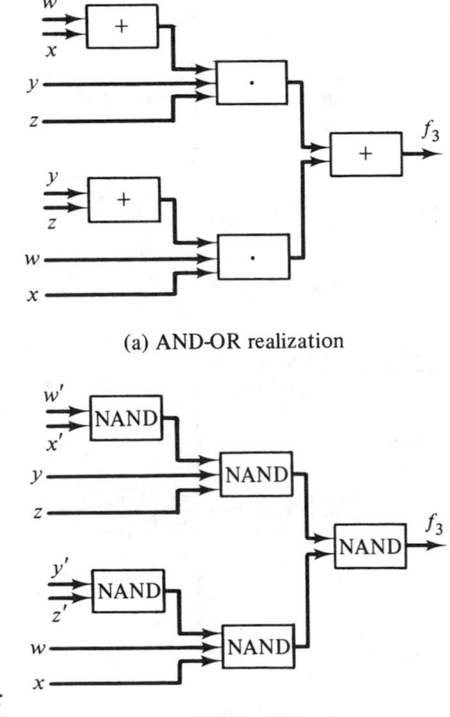

Figure 5.28 Logic diagrams for Example 5.8.

OR-AND-OR logic circuit may be replaced by an all-NAND circuit if all inputs passing through an odd number of levels are complemented. The only conditions are that the output must be taken from an OR and that the logic gates alternate (i.e., . . . OR-AND-OR). Similarly, any alternating . . . AND-OR-AND circuit (where the output is taken from an AND) may be replaced by an all-NOR circuit. Again we must complement all inputs which pass through an odd number of gates. The application of the above algorithm is illustrated by the following examples.

Example 5.8

Given $f_3 = wx(y + z) + yz(w + x)$, draw a NAND, NOR logic diagram.

A three-level AND-OR circuit which realizes f_3 is shown in Fig. 5.28(a). A three-level NAND equivalent is shown in Fig. 5.28(b).

Example 5.9

Given $f_4 = (wx + yz)(w + x)(y + z)$, draw a NAND, NOR logic diagram which realizes f_4.

Direct implementation of f_4 using AND-OR is shown in Fig. 5.29(a). The equivalent NOR circuit is shown in Fig. 5.29(b).

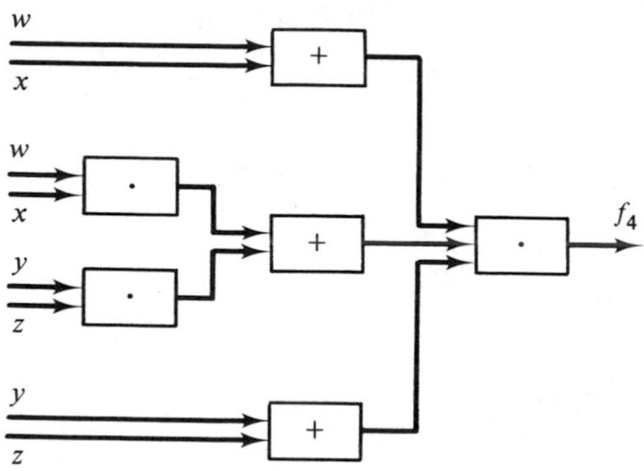

(a) AND-OR circuit

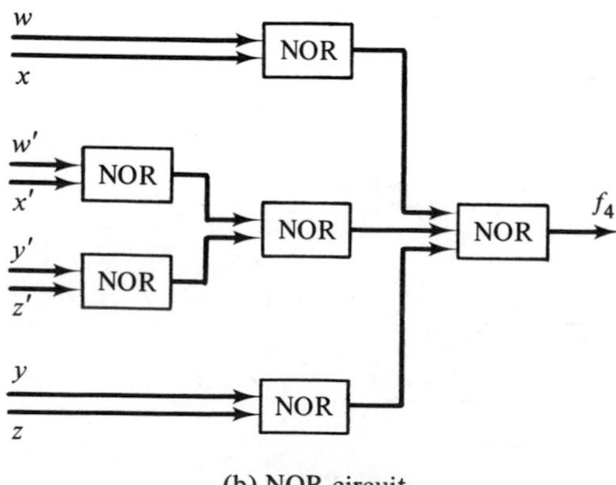

(b) NOR circuit

Figure 5.29 Logic diagrams for Example 5.9.

The only restriction in the multiple-level translation algorithm is the alternation of gates. It does not apply directly to logic circuits where the required alternation of gates does not exist. For such circuits some manipulations such as breaking the circuit into parts where the alternation exists might help in the translation. This point is illustrated in the following example.

Example 5.10

Consider the function given in Example 5.9 and draw a NAND, NOR logic diagram assuming that only two input gates are available.

Figure 5.29(a) is redrawn (with two input gates) in Fig. 5.30(a). The gates in Fig. 5.30(a) do not alternate. Notice that the outputs labeled f_1 and f_2 are generated

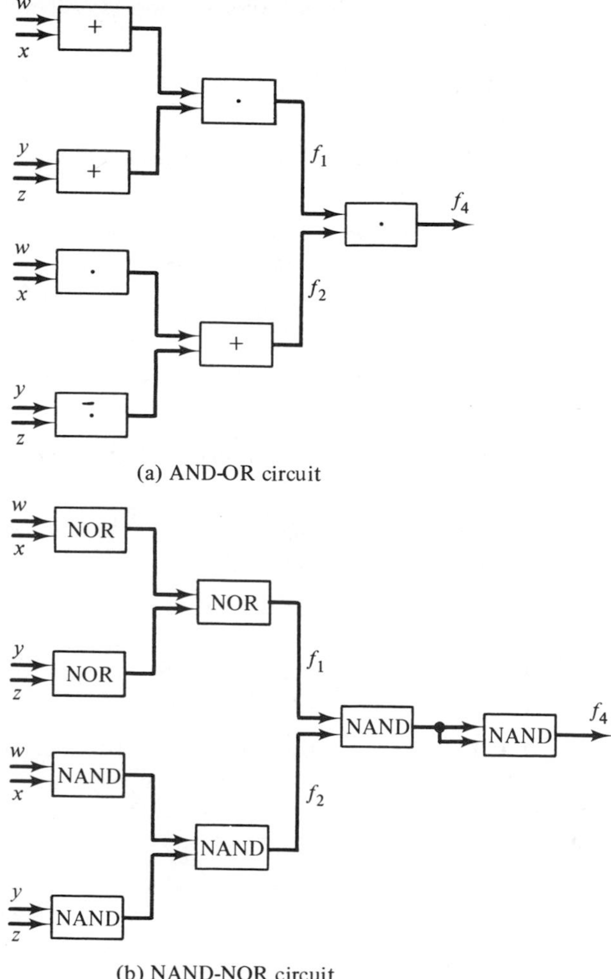

(a) AND-OR circuit

(b) NAND-NOR circuit

Figure 5.30 Logic diagram for Example 5.10.

by an alternating circuit. Thus, each part can be translated according to our algorithm. The AND circuit which generates f_4 can be replaced by two NAND gates. The equivalent NAND, NOR circuit is shown in Fig. 5.30(b).

REFERENCES

BARTEE, T. C., *Digital Computer Fundamentals*, 3rd ed. New York: McGraw-Hill, 1972.

DELHAM, L., *Design and Application of Transistor Switching Circuits.* New York: McGraw-Hill, 1968.

DIETMEYER, D. L., *Logical Design of Digital Systems.* Boston: Allyn and Bacon, 1971.

GIVONE, D. D., *Introduction to Switching Circuit Theory.* New York: McGraw-Hill, 1970.

HILL, F. J., and G. R. PETERSON, *Introduction to Switching Theory and Logical Design.* New York: Wiley, 1968.

MALEY, G. A., and J. EARLE, *The Logic Design of Transistor Digital Computers.* Englewood Cliffs, N.J.: Prentice-Hall, 1963.

MARCOVITZ, A. B., and J. H. PUGSLEY, *An Introduction to Switching System Design.* New York: Wiley, 1971.

MCCLUSKY, E. J., *Introduction to the Theory of Switching Circuits.* New York: McGraw-Hill, 1965.

MILES, T. E., "Schottky TTL vs. ECL for High Speed Logic," *Computer Design*, 11, No. 10 (Oct. 1972), pp. 79–86.

MORRIS, R. L., and J. R. MILLER, *Designing with TTL Integrated Circuits.* New York: McGraw-Hill, 1971.

PROBLEMS

5.1. If $x = 1, y = 0$, and $z = 0$, what is the value (0 or 1) of each of the following Boolean functions?

(a) $f_1 = x + y\bar{z}$.
(b) $f_2 = (x + y)\bar{z}$.
(c) $f_3 = xy\bar{z}$.
(d) $f_4 = x(y + \bar{z})$.
(e) $f_5 = \overline{\bar{y} + xz}$.
(f) $f_6 = \overline{\bar{x}y + x\bar{z}}$.

5.2. Construct the appropriate truth tables to verify the following rules for negating logical sums and products:
(a) $\overline{a + b} = \bar{a}\bar{b}$.
(b) $\overline{ab} = \bar{a} + \bar{b}$.

5.3. Justify the following relationships using truth tables:
(a) $a + ab = a$.
(b) $a(a + b) = a$.
(c) $a + \bar{a}b = a + b$.
(d) $a(\bar{a} + b) = ab$.

5.4. Draw a Venn diagram which will define the sample space designated by F given that $F' = (A' + B' + C')(A + B' + C')(A + B + C)(A' + B + C')$.

5.5. Show that the following identities are true using the algebraic method:
(a) $a + \bar{a}b = a + b$.
(b) $a(\bar{a} + b) = ab$.
(c) $ab + \bar{a}c + bc = ab + \bar{a}c$.

5.6. Given the truth table below, determine the two canonical forms of the function f.

x_1	x_2	x_3	f
0	0	0	1
0	0	1	1
0	1	0	1
0	1	1	1
1	0	0	1
1	0	1	0
1	1	0	0
1	1	1	1

5.7. For each of the switching functions (f, g, h, k, p) in the following truth table

x	y	z	f	g	h	k	p
0	0	0	0	1	0	1	0
0	0	1	1	0	1	0	1
0	1	0	0	0	0	0	0
0	1	1	1	1	0	1	0
1	0	0	0	0	0	1	0
1	0	1	1	0	1	0	0
1	1	0	0	1	1	1	1
1	1	1	1	1	1	1	0

(a) Express the function in canonical sum-of-products form.
(b) Express the function in canonical product-of-sums form.

5.8. Find the complement of each of the following expressions:
(a) $xy + y'z$.
(b) $a[b'c + d'(ef')]$.
(c) $ab' + c(d' + e) + fg'h$.

5.9. Prove each of the following identities by using the algebraic rules (no truth tables). Several steps may be combined, but make sure that each step is clear.
(a) $a'b + b'c + a'c = a'b + b'c$.
(b) $a'd + ac = ac + cd + a'c'd$.
(c) $xz' + x'y' + x'z + y'z = y' + x'z + xz'$.
(d) $ad' + a'b' + c'd + a'c' + b'd = ad' + (bc)'$.

5.10. Express $X \oplus Y$ in terms of only the NAND operation.

5.11. Use only NOR logic circuits to implement the Boolean equation $(AB' + A'B)C$. Use only NAND logic circuits to implement the Boolean equation $(A'B' + AB)C + (A'B + C')D$.

5.12. Draw the logic diagram to implement the Boolean equation $AB' + (AB) \cdot (A + B)$ using NAND, NOR logic.

5.13. Convert the following Boolean forms to standard sum-of-products forms:
(a) $f(w, x, y, z) = wy + \bar{x}(w + y\bar{z})$.
(b) $f(U, V, W, X, Y) = \bar{V}(\bar{W} + \bar{U})(X + \bar{Y}) + \bar{U}\bar{W}Y$.
(c) $f(V, W, X, Y, Z) = (X + \bar{Z})(\overline{Z + WY}) + (V\bar{Z} + W\bar{X})(\overline{Y + Z})$.
(d) $f(A, B, C, D) = (\bar{A} + \bar{C})D + \overline{(\bar{A} + B + C)}\bar{D}$.
(e) $f(A, B, C, D, E) = (\bar{C}\bar{E} + CE)(\bar{A} + B)D + \overline{(\bar{A} + B)}D\bar{C}E$.

5.14. Express the following functions in sum-of-products and product-of-sums standard forms:
(a) $f = x_1x_2 + x_1\bar{x}_3 + \bar{x}_1x_4 + x_2\bar{x}_4$.
(b) $f = (\bar{x}_1 + x_2x_3)(x_2 + \bar{x}_3)(\bar{x}_1 + \bar{x}_4)$.
(c) $f = x_1\bar{x}_2x_3 + \bar{x}_1x_2x_3 + \bar{x}_1x_3\bar{x}_4 + \bar{x}_2\bar{x}_4$.
(d) $f = x_1\bar{x}_2 + x_2\bar{x}_3(x_1 + \bar{x}_3)(x_1 + \bar{x}_4)$.
(e) $f = x_1x_2x_3 + x_1\bar{x}_3\bar{x}_4 + \bar{x}_1x_2x_3x_4 + \bar{x}_1\bar{x}_2\bar{x}_4 + x_2x_3x_4$.

5.15. Draw the circuit diagram of a positive logic OR gate using 0 and +10 volts as voltage levels. Prepare a voltage truth table and logic truth table of the circuit.

5.16. Prepare a voltage and logic truth table for a positive logic OR gate using 0 and −10 volts as voltage levels.

5.17. Draw the circuit diagram of a negative logic diode-transister NOR gate.

5.18. Determine the switching expression for the function *f* in Fig. P5.18 in terms of the input variables *a*, *b*, *c*.

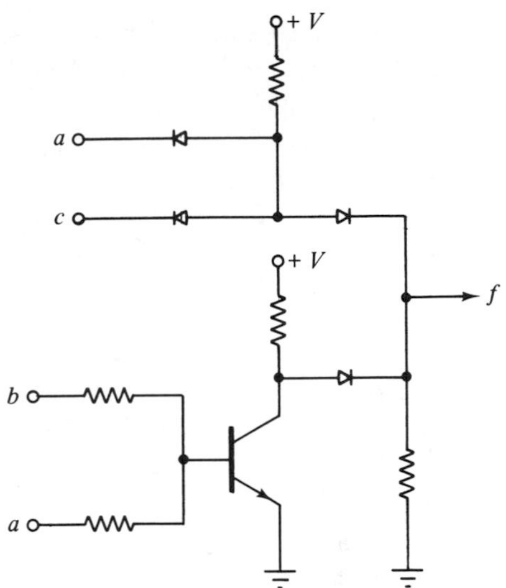

Problem 5.18

5.19. For the NAND gate shown in Fig. P5.19, obtain a voltage truth table and indicate whether positive or negative logic is used.

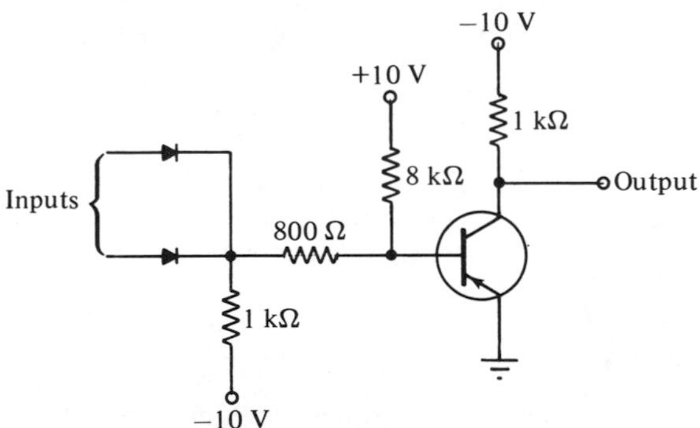

Problem 5.19

5.20. Draw a DCTL circuit to implement the Boolean function $A'B + A'BC$. See Figs. 5.12 and 5.13.

5.21. Draw a DCTL circuit to implement the Boolean function $(A + B)A'BC'$.

5.22. Draw the Karnaugh map for $ABC + BC' + AC + BC$.

5.23. Draw the Karnaugh map for $XY + Z + XYZ + Z'Y$. Using the map, read off a simplified expression.

5.24. For each of the following functions, find the minimum sum-of-products form:

(a) $f(a, b, c, d) = \sum (0, 1, 2, 3, 5, 7, 8, 10, 13, 15)$.

(b) $f(x_1, x_2, x_3, x_4) = \sum (0, 3, 5, 6, 9, 10, 12, 15)$.
(c) $f(w, x, y, z) = \sum (0, 6, 7, 13, 14, 15)$.

5.25. Simplify the following expressions (1) by the algebraic method (2) by the Karnaugh map method. Each function is a function of 4 variables.
(a) $F_1 = \sum (0, 1, 2, 4)$.
(b) $F_2 = \sum (2, 3, 6, 7, 10, 11, 13, 15)$.
(c) $F_3 = \sum (1, 4, 5, 7, 8, 13, 15)$.
(d) $F_4 = \sum (0, 1, 2, 3, 4, 5, 6, 7, 8, 9, 10, 11, 12, 14)$.

5.26. Find the simplest sum-of-products and product-of-sums expressions for
(a) $f_1(w, x, y, z) = \sum (1, 2, 3, 4, 5, 6, 7, 9, 10, 11, 12, 13, 14)$.
(b) $f_2(a, b, c, d) = \sum (2, 6, 7, 8, 9, 12)$.
(c) $f_3(p, q, r, s) = \sum (0, 1, 2, 5, 8, 9, 10, 11, 13, 14, 15)$.
(d) $f_4(x_1, x_2, x_3, x_4) = \sum (1, 2, 3, 4, 5, 9, 11, 12)\text{d}(10, 13, 15)$.

5.27. Find all minimum-input, two-level circuits using AND and OR gates:
(a) $f(a, b, c, d) = \sum (3, 6, 9, 11, 15)\text{d}(1, 4, 7, 13)$.
(b) $f(v, w, x, y, z) = \sum (1, 2, 3, 4, 5, 6, 7, 9, 12, 13, 17, 20, 21, 25, 28, 29)$.
(c) $h(a, b, c, d) = \sum (1, 2, 6, 7, 8, 12, 13)\text{d}(3, 5, 9, 15)$.
(d) $g(x_1, x_2, x_3, x_4) = \sum (0, 1, 2, 3, 4, 5, 6, 7, 9, 10, 11, 14)$.
(e) $h(w, x, y, z) = \sum (1, 3, 7, 11, 14)\text{d}(0, 5, 6, 15)$.

5.28. For the functions of Problem 5.27, find a minimum NAND, NOR circuit. Assume that complemented inputs are not available.

5.29. Find and draw the block diagram for the simplest two-level NAND circuit realization of $f(a, b, c, d) = \sum (0, 2, 6, 7, 8, 9, 13, 15)$.

5.30. Simplify $f = \overline{\bar{x}(y + xy\bar{z}) + z\bar{y}}$ using (1) De Morgan's theorem and algebraic manipulations and (2) the map method; then (3) determine the cost of the canonical and of the minimal realizations of this function using
(a) \$1 per input AND gate.
(b) \$1 per input OR gate.
(c) \$2 for a NOT circuit.

5.31. Two types of modules, each costing \$1.99, are available. Type A contains one NOT circuit. Type B contains three two-input AND gates and one three-input OR gate. There is no limit to the number of levels. All inputs are available both primed and unprimed. Design a minimum-cost circuit for $f(w, x, y, z) = \sum (5, 6, 9, 10, 13, 14)$.

5.32. A circuit receives two 3-bit binary numbers, $\text{A} = \text{A}_2\text{A}_1\text{A}_0$ and $\text{B} = \text{B}_2\text{B}_1\text{B}_0$. Design a minimal sum-of-products circuit to produce an output whenever A is greater than B.

5.33. It may be assumed that the four possible combinations of 4 bits not used in the following table will never occur. Find a minimal sum-of-products design for a circuit to produce an output only when the shaft is in the first quadrant (0–90°).

Shaft position	Encoder output			
	F_1	F_2	F_3	F_4
0–30°	0	0	1	1
30–60°	0	0	1	0
60–90°	0	1	1	0
90–120°	0	1	1	1
120–150°	0	1	0	1
150–180°	0	1	0	0
180–210°	1	1	0	0
210–240°	1	1	0	1
240–270°	1	1	1	1
270–300°	1	1	1	0
300–330°	1	0	1	0
330–360°	1	0	1	1

5.34. In a certain computer, two separate sections of the computer proceed independently through four phases of operation. For purposes of control, it is necessary to know when the two sections are in the same phase at the same time. Each section puts out a 2-bit signal (00, 01, 10, 11) called x_1x_2 on two separate lines. Design a circuit to put out a signal whenever it receives the same phase signal from the two sections.

6

Functional Logic Subunits

In this chapter we shall present the computer subunits from which the computer functional units are built. The computer functional units such as the arithmetic unit, memory unit, and control unit are presented in later chapters. The functional logic subunits used in the design of the computer units are discussed here, individually.

Any functional subunit consists, in general, of some storage elements and/or combinational logic interconnected so that they perform a particular logic or arithmetic operation.

The logic design of the combinational circuit was briefly discussed in Chapter 5, while the basic storage element is presented in Section 6.1. In the subsequent sections we shall present the various functional logic subunits as an interconnection of combinational logic gates and memory elements. The subunits to be discussed are registers, counters, decoders, encoders, pulse generators, adders, and subtractors.

6.1 FLIP-FLOPS

One class of storage elements which is capable of storing a single bit of information is the flip-flop. It has the capability of storing 1 bit because it has two stable states. If the flip-flop is storing a 1, it is said to be in the *one* (1) state, while if it is storing a 0, it is referred to as being in the *zero* (0) state. The flip-flop remains in a fixed state (stable) until a change in its input occurs which causes it to change its state. Every flip-flop must have at least one input line which determines what state it will assume next and at least one output line so that its present state can be determined. Most flip-flops, as we shall see, have two output lines and more than one input line. Flip-flops with two output lines have the value of the stored binary variable on one line and its complement on the other line. This is more common since most combination logic circuits require both the variable and its complement. The number of inputs in a flip-flop depends on the circuit design of the flip-flop.

6.1.1 Basic Logic of Flip-flops

The simplest flip-flop circuit consists of two gates tied back to back as shown in Fig. 6.1. This circuit is called a latch circuit, and its inputs *preset* and *clear* are only sensitive to dc signals. When NOR gates are used, the *preset* and *clear* inputs are normally equal to 0, since these 0 inputs do not affect the latch circuit output. In other words, whenever the *preset* and *clear* input are both 0, the latch circuit remains in either of the two stable states; the 1 state ($Q = 1$) or the 0 state ($Q = 0$).

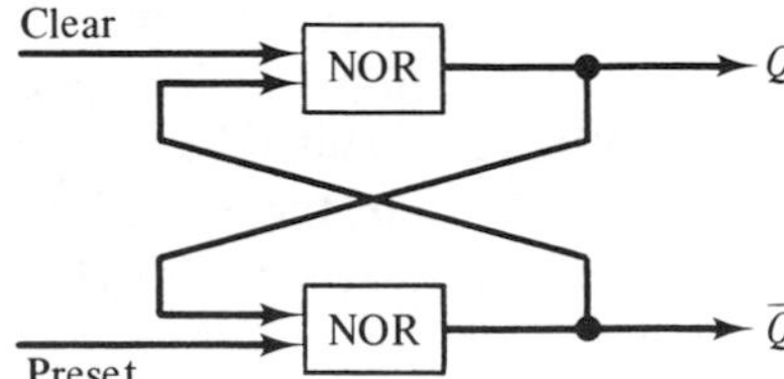

Figure 6.1 Latch circuit.

The latch can be driven into the 0 state (clear) by making the clear input go to 1. This will force the output of the upper gate to go to 0. Then, assuming that the *preset* input is 0, as it normally is, the lower gate will have its two inputs equal to 0, causing its output to go to 1. This 1 output will remain at the input of the upper gate after the clear input goes to 0 again. Consequently, the latch will remain in the 0 state even though it is no longer being cleared.

If the preset input goes to 1 while the clear input is kept at 0, the output of the lower gate goes to 0, which then forces the output of the upper gate to be equal to 1. If the latch circuit is preset and cleared simultaneously, then both outputs go to 0. The first input to return to 0 will determine the subsequent state of the latch. In case both inputs return to 0 simultaneously, the subsequent state of the latch will be undetermined. Therefore, the latch circuit, as described (also its equivalent using NAND gates), must not be driven into the undetermined state by not allowing the inputs preset = clear = 1 of the NOR gate latch to occur. In the case of the NAND gate latch, preset = clear = 0 must not be applied.

6.1.2 Flip-flop Structures and the Race Problem

The simple latch circuit described previously requires some time to change its state in response to the input signals. This switching time is limited by the inherent delays in the latch circuit gates and does not depend on the rise (or fall) time of the direct input.

The inputs to a flip-flop, in a digital system, are usually the outputs of other flip-flops in the system. For such a system to operate successfully, all flip-flops which are supposed to change state must do so simultaneously. If for some reason or another they do not change simultaneously, then a late-changing flip-flop may respond to the changed state of an early changing flip-flop rather than to its previous state. The variation in switching times of the flip-flops as described above is known as the *race* condition and might give rise to ambiguous performance of the system.

There are different ways to avoid the race problem. Each solution employs a clock input to control the time at which the flip-flop is allowed to change state. The pulse width must be extremely short, so that when the flip-flop changes state, the clock pulse disappears and the new state will have no effect on the input. On the other hand, the clock pulse must be wide enough to allow the flip-flop to react to a narrow pulse on either of the inputs (set or clear). A latch circuit with a clock pulse and the required gates is shown in Fig. 6.2. This arrangement is called a clocked flip-flop. When all the flip-flops in a system use the same clock, then we talk about synchronous memory elements.

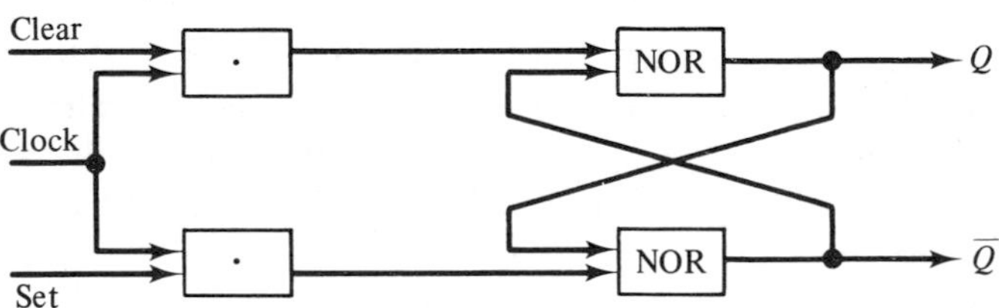

Figure 6.2 Clocked flip-flop.

Another practical approach, which is commonly used in integrated circuit flip-flops, is to build a flip-flop from a large number of gates. This approach is illustrated in Fig. 6.3 and the circuit is referred to as a *master-slave* flip-flop. This configuration is most commonly used in DTL (diode-transistor-logic) and TTL (transistor-transistor-logic) integrated circuits. The configuration contains two asynchronous flip-flops which together with the clock and the necessary gates form a synchronous, race-free flip-flop. When the clock signal equals 1, the master flip-flop responds to the set and clear inputs. The slave is disconnected from the master at this time by the second

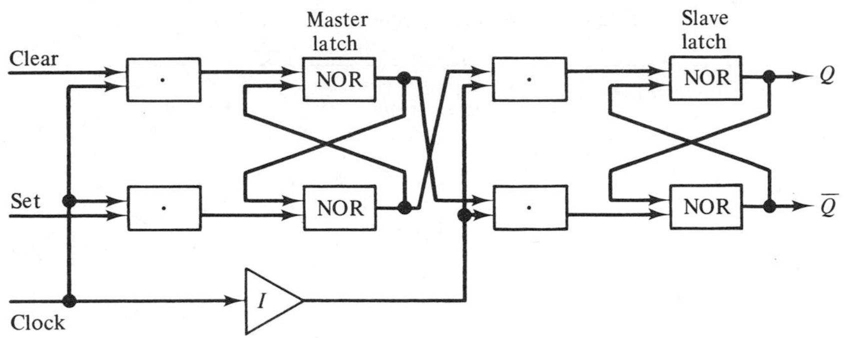

Figure 6.3 Master-slave flip-flop.

pair of AND gates and the inverted clock signal. Therefore, the slave flip-flop may not change while the master flip-flop is responding to the input signals.

There are different types of flip-flops, with different input/output characteristics. In this section no attempt is made to present a detailed analysis of how each of these flip-flop devices is constructed. Our main concern is to present the external properties of each type so that the reader will be able to use them in the design of the various parts of the digital computer.

6.1.3 The *S-R* Flip-flop

This type of flip-flop is called a set-reset flip-flop, or *S-R* flip-flop for short. Each of the logic diagrams shown in Figs. 6.1, 6.2, and 6.3 represents an *S-R* flip-flop. A symbolic representation of the *S-R* flip-flop (Q) is shown in Fig. 6.4. This flip-flop has two inputs. Input S, the set line, puts the flip-flop into the 1 state, and input R, the reset line, puts it into the 0 state. The input condition $S = 1$, $R = 1$ is an indeterminate condition and should never be allowed to occur. If the inputs are $S = R = 0$, the flip-flop will do nothing, and its state will remain unchanged.

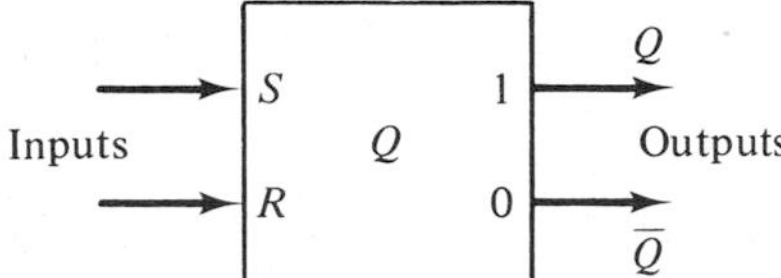

Figure 6.4 Symbolic representation of *S-R* flip-flop.

We shall consider the time delay between two successive clock pulses to be our time unit and look at the output of a flip-flop only at the discrete time intervals of the clock. Therefore, the input to the flip-flop at time t will have an effect on the output at time $t + 1$. If we denote the state of a flip-flop at time t by $Q(t)$, then its state $Q(t + 1)$ at time $t + 1$ is a function of $Q(t)$ and

Table 6.1 *Transition Table for S-R Flip-flop*

Input at time t		Present state, $Q(t)$	Next state, $Q(t+1)$
$S(t)$	$R(t)$		
0	0	0	0
0	0	1	1
0	1	0	0
0	1	1	0
1	0	0	1
1	0	1	1
1	1	0	Indeterminate
1	1	1	Indeterminate

the input at time t, that is, $R(t)$ and $S(t)$. The input/output relationship which is usually referred to as the transition table of an *S-R* flip-flop is shown in Table 6.1.

If we impose the restriction that either R or S must always be 0, we can express the information given in the transition table by the following two equations:

$$Q(t+1) = \bar{R}(t)\bar{S}(t)Q(t) + S(t)\bar{R}(t)$$
$$R(t)S(t) = 0$$

Changing our notation a little to avoid repeating the parameter t, we can write

$$Q(t+1) = \bar{R}\bar{S}Q + S\bar{R} \tag{6.1}$$
$$RS = 0 \tag{6.2}$$

Since $RS = 0$, Eq. (6.1) can be written as

$$\begin{aligned} Q(t+1) &= \bar{R}\bar{S}Q + S\bar{R} + RS \\ &= S(R + \bar{R}) + Q\bar{R}\bar{S} \\ &= S + Q\bar{R}\bar{S} \\ &= S + Q\bar{R} \end{aligned} \tag{6.3}$$

Equation (6.3) is equivalent to Eq. (6.1), and consequently it is known as the characteristic equation (or next-state equation) of the *S-R* flip-flop. It simply reads, "The next state will be the 1 state if the present state is the 1 state and we do not "reset" it or if the flip-flop is "set" regardless of its present state."

The condition given by Eq. (6.2) still holds, and the designer has to restrict both of the inputs to the *S-R* flip-flop from being equal to 1 at the same time.

When a flip-flop is used as a memory element, a logic designer is faced with the following problem: "If we know the present state of a flip-flop and also the state we want it to be in next, what are the appropriate inputs which are needed in order that the flip-flop changes to the required state?" The answer to the above question can be obtained from Table 6.1, but in order to have a quick answer to the above question, Table 6.1 can be written in another equivalent form. This form is known as the excitation table and is shown in Table 6.2 for the *S-R* flip-flop. The *d* in Table 6.2 indicates a don't

Table 6.2 *Excitation Table for S-R Flip-flop*

Present state, $Q(t)$	*Next state,* $Q(t+1)$	*Input at time t*	
		$S(t)$	$R(t)$
0	0	0	*d*
0	1	1	0
1	0	0	1
1	1	*d*	0

care condition, that is, either a 0 or a 1 will cause the flip-flop to change to the required next state. In general, the input lines for the flip-flop are obtained as an output of a combinational network which has the present state of the flip-flop as one of its inputs. Therefore, the inputs to the flip-flop are derived as a Boolean expression of the present state and possibly some external input variables. These expressions are called the input equations of the flip-flop.

6.1.4 The *J-K* Flip-flop

The *J-K* flip-flop has the properties of an *S-R* flip-flop except that $J = K = 1$ is allowed and causes the flip-flop to change state. Here the *J* acts as the *S* and sets the flip-flop to the 1 state, while the *K*, like the *R*, resets the flip-flop to the 0 state. The transition table for the *J-K* flip-flop is shown in Table 6.3. The characteristic equation can be derived from the transition table as

$$\begin{aligned} Q(t+1) &= Q\bar{K} + J\bar{K} + \bar{Q}J \\ &= \bar{K}Q + J\bar{Q} \end{aligned} \tag{6.4}$$

Equation (6.4) signifies that the next state of the *J-K* flip-flop will be the 1

Table 6.3 *Transition Table for the J-K Flip-flop*

$J(t)$	$K(t)$	$Q(t)$	$Q(t+1)$
0	0	0	0
0	0	1	1
0	1	0	0
0	1	1	0
1	0	0	1
1	0	1	1
1	1	0	1
1	1	1	0

state if its present state is 1 and it is not reset or if its present state is 0 and it is set. The excitation table for the *J-K* flip-flop is shown in Table 6.4. A realization of the *J-K* flip-flop using an *S-R* flip-flop is shown in Fig. 6.5(a).

Table 6.4 *Excitation Table for the J-K Flip-flop*

$Q(t)$	$Q(t+1)$	$J(t)$	$K(t)$
0	0	0	d
0	1	1	d
1	0	d	1
1	1	d	0

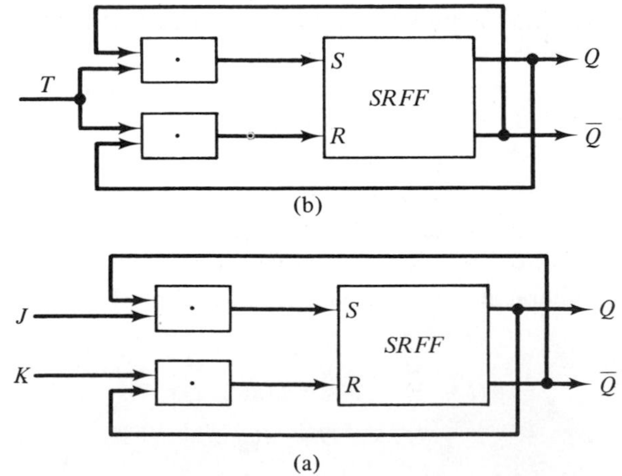

Figure 6.5 (a) Realization of JKFF from an SRFF; (b) realization of TFF from an SRFF.

6.1.5 The *T* Flip-flop

The *T* or trigger flip-flop has a single input line *T*. The flip-flop changes state only when $T = 1$; otherwise it remains in its former state. The *T* flip-flop is sometimes referred to as the complement flip-flop because when its input line is 1 it has the effect of complementing the content or (state) of the flip-flop. Table 6.5 gives the transition table for the *T* flip-flop. The characteristic equation for the *T* flip-flop can be written from the transition table as

Table 6.5 *Transition Table for the T Flip-flop*

$T(t)$	$Q(t)$	$Q(t+1)$
0	0	0
0	1	1
1	0	1
1	1	0

$$Q(t+1) = Q\bar{T} + \bar{Q}T \tag{6.5}$$

The excitation table is identical to the transition table. The *T* flip-flop can be constructed from an *S-R* flip-flop as shown in Fig. 6.5(b).

6.1.6 The *S-R-T* Flip-flop

The *S-R-T* flip-flop has three inputs. The *S* and *R* inputs have the same effect as in the *S-R* flip-flop; the *T* input line is a trigger input, and its effect is the same as the *T* input of a *T* flip-flop. The state of the flip-flop is again indeterminate if any two of its inputs are 1 simultaneously.

The above description of an *S-R-T* flip-flop is illustrated in Table 6.6, which is the transition table for this flip-flop. The *I* character in Table 6.6 indicates the indeterminate cases. These cases require that the inputs be restricted such that

$$SR = ST = TR = 0 \tag{6.6}$$

Eliminating these forbidden entries, the remaining part of the table can be described by

$$Q(t+1) = Q\bar{S}\bar{R}\bar{T} + \bar{Q}\bar{S}\bar{R}T + S\bar{R}\bar{T} \tag{6.7}$$

Equations (6.6) and (6.7) completely describe the behavior of the *S-R-T* flip-flop. Making use of the restrictions indicated by Eq. (6.6), we can sim-

Table 6.6 Transition Table for the S-R-T Flip-flop

$S(t)$	$R(t)$	$T(t)$	$Q(t)$	$Q(t+1)$
0	0	0	0	0
0	0	0	1	1
0	0	1	0	1
0	0	1	1	0
0	1	0	0	0
0	1	0	1	0
0	1	1	0	I
0	1	1	1	I
1	0	0	0	1
1	0	0	1	1
1	0	1	0	I
1	0	1	1	I
1	1	0	0	I
1	1	0	1	I
1	1	1	0	I
1	1	1	1	I

plify Eq. (6.7) to

$$Q(t+1) = S + T\bar{Q} + \bar{R}\bar{T}Q \tag{6.8}$$

Hence, Eqs. (6.6) and (6.8) are considered the characteristic equations for the *S-R-T* flip-flop.

The excitation table for the *S-R-T* flip-flop is given in Table 6.7. While the d in Table 6.7 indicates the usual don't care condition, the a and the b are special don't care conditions. We have to keep track of how we choose a or b so that the assignment for $\bar{a}$ or $\bar{b}$ will be the complement of a or b, respectively. That is, if $a = 1$, then $\bar{a}$ must be assigned a 0 and vice versa. The same holds for b. The arbitrary choice of what input lines we use to achieve a change of state, i.e., through the T or the *S-R*, makes the design using *S-R-T* flip-flop more involved when compared to other types of flip-

Table 6.7 Excitation Table for the S-R-T Flip-flop

$Q(t)$	$Q(t+1)$	$S(t)$	$R(t)$	$T(t)$
0	0	0	d	0
0	1	a	0	$\bar{a}$
1	0	0	b	$\bar{b}$
1	1	d	0	0

flops. This is true if we try to utilize the characteristics of the S-R-T flip-flop to its full extent.

6.1.7 The D Flip-flop

The D flip-flop usually refers to a storage cell with one input line and one output line. This type of flip-flop is also called a delay flip-flop. It is also possible to make a D flip-flop from an S-R flip-flop as shown in the accompanying figure. Such a flip-flop has the same characteristic as the delay

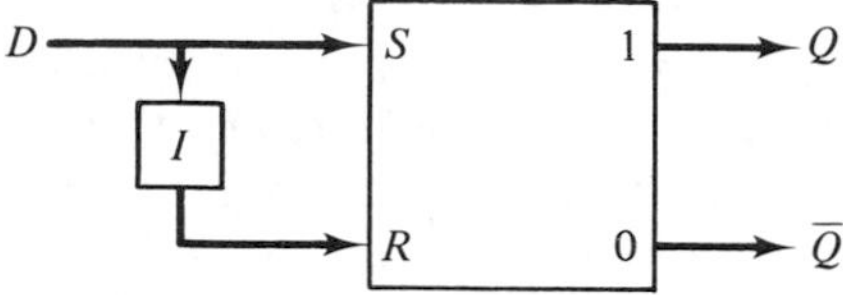

flip-flop, except that there will be two output lines, one representing the state and the other representing its complement. In this text, we shall consider the D flip-flop to be the delay flip-flop with one input line and one or two output lines. If we denote the input by D and the state by Q, then the characteristic equation of the D flip-flop can be written as

$$Q(t+1) = D(t) \tag{6.9}$$

That is, the input to the flip-flop at any clock interval t will appear at the output one clock pulse later. The excitation table for the D flip-flop is the same as the transition table. Table 6.8 gives the transition table for the D flip-flop.

Table 6.8 *Excitation Table for the D Flip-flop*

$Q(t)$	$Q(t+1)$	$D(t)$
0	0	0
0	1	1
1	0	0
1	1	1

6.2 REGISTERS

A storage cell is the basic storage device in a digital computer. It serves as a storage element for the smallest unit of information, that is, a *bit*. A register consists of one or more storage cells. The major use of a register is to hold (or store) information while it is being processed. Since information in computers usually consists of words of fixed length, the most common storage capacity of a register is one-word length.

The characteristics and capabilities of any specific register depend on the type of storage devices used to form the register and also on the arrangement of the logic circuits attached to the cells of the register.

6.2.1 Flip-flop Register

This is the simplest type of register. It consists of an ordered set of flip-flops. A register using *S-R* flip-flops is shown in Fig. 6.6. There are as many flip-flops as the number of bits to be stored in the register. Each flip-flop stores 1 bit of information. The discussion in this section applies to registers composed of any type of flip-flops as long as the flip-flop has two inputs, a clear input and a set input. For example, in a *J-K* flip-flop, the *J* input represents the set input, while the *K* input represents the clear input.

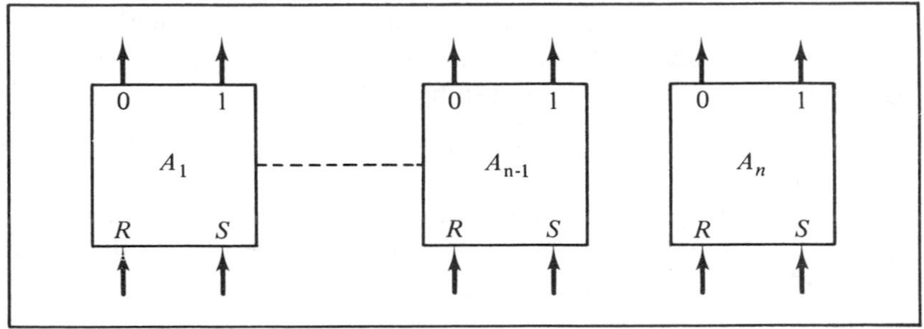

Figure 6.6 A register using *S-R* flip-flops.

To store some given binary information in a flip-flop register, simply energize the appropriate combination of inputs. For example, consider a register A with five cells and denote the cells A_1, A_2, A_3, A_4, and A_5. If the *set* inputs of A_1, A_2, and A_5 and the *clear* inputs of A_3 and A_4 are energized, then the register will contain the binary number 11001, regardless of its previous contents.

The register shown in Fig. 6.6 does not include any provision for transferring information into or out of the register. The contents of a register A can be transferred into another register B by providing a transfer path from the output of the A register to the input of the B register. This transfer path can be provided by attaching logic circuitry to both the output of register A and the input of register B. Figure 6.7 shows such a transfer path from register A to register B.

Assuming the flip-flops of registers A and B to be clocked flip-flops, the

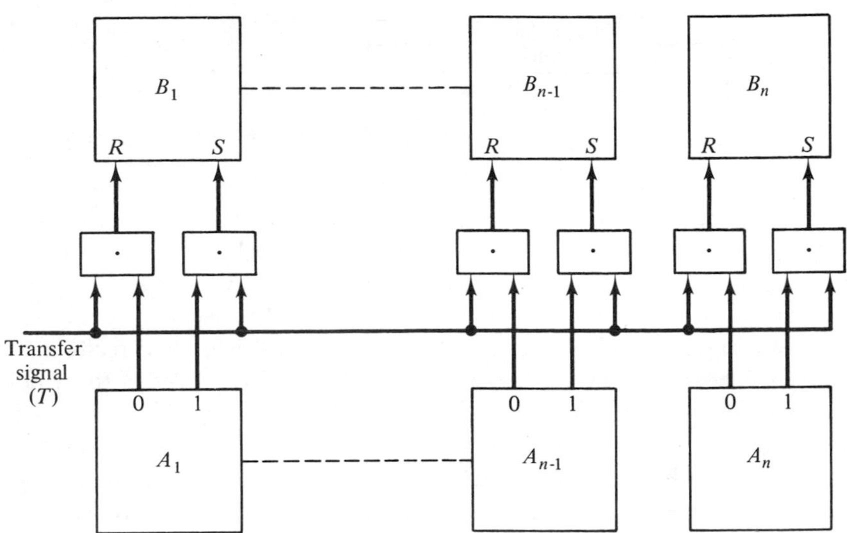

Figure 6.7 Transfer path between two registers.

state of register B can be described by the following state equations:

$$B_i(t+1) = A_i(t) \qquad \text{for } i = 1, 2, \ldots, n$$

The state equation indicates the relationship between register A and register B as a function of time. In other words, if the transfer pulse is applied at time t, then the contents of register A at this time will be available at the output of register B, one clock pulse later. So that the above transfer will occur, the inputs of the flip-flops of register B have to be energized properly. The logic relationship which describes the flip-flop inputs is called the *input equation*. In our case, the B register flip-flops' input equations are

$$S_{B_i} = A_i, \qquad R_{B_i} = \bar{A}_i \qquad \text{for } i = 1, 2, \ldots, n$$

Flip-flop registers are most often used in parallel computers where information transfers as well as arithmetic and logic operations are performed concurrently on all the bits of a word. For serial machines where the information is transferred and processed 1 bit at a time, other types of registers such as the recirculating registers are used.

6.2.2 Recirculating Registers

The recirculating register to be described here uses delay lines as storage devices. It is usually used in serial computers. A delay line recirculating register is shown in Fig. 6.8.

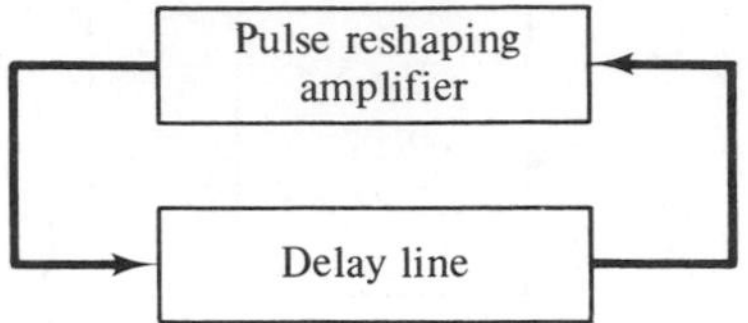

Figure 6.8 A delay-line recirculating register.

A binary bit is usually represented by a pulse if it is a 1 and by the absence of a pulse if it is a 0. Therefore, a combination of bits is stored in the register as a *pulse train* continuously circulating through the delay line. The delay line itself could be a coaxial cable or an LC network or any other wave-propagating media. The storage capacity of a delay line recirculating register depends on several factors. The first is the time needed for a pulse to travel through the delay line. The second factor is the pulse repetition frequency (or the time spacing between consecutive pulses). For example, if pulses are sent through the delay line with .4 microsecond (μs) between them and it takes a pulse 20 μs to cycle through the delay line, then the storage capacity of the register is 50 bits.

The amplifier shown in Fig. 6.8 is needed to amplify and reshape the attenuated and distorted pulses each time they propagate through the delay line media. Each time a bit of information enters the loop, it is AND-gated with the computer clock for synchronization. The clock of the computer is a pulse generator which synchronizes all operations in the computer. The spacing between the bits of a word stored in a recirculating register must be the same as the spacing between the clock pulses. Therefore, the timing for storing or reading out any bit in the register is synchronized with the clock pulses. This synchronization by itself is not sufficient since it cannot indicate the positional location of any bit within the word. An additional signal to indicate the beginning of a cycle (i.e., the first bit of a word) is necessary. A marker signal which signifies the beginning (or the end) of a recirculating cycle is usually used. This signal is generated by the computer periodically, once every cycle time, and any information transfer (input or output) must start with the marker signal.

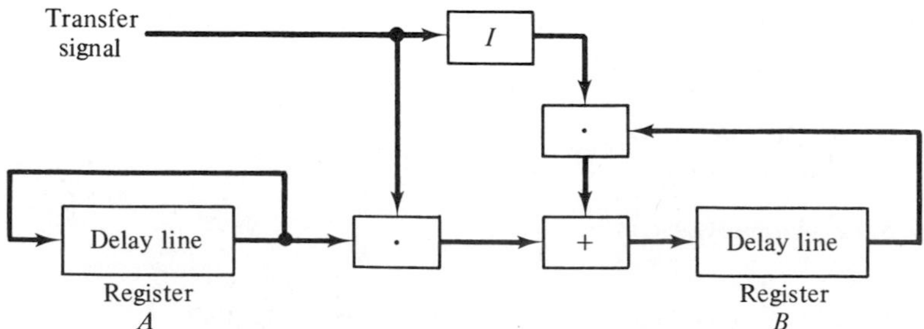

Figure 6.9 Transfer path between two recirculating registers.

Similar to the flip-flop register, additional logic circuitry has to be attached to a recirculating register for input/output capabilities. Figure 6.9 shows two recirculating registers *A* and *B* with the logic needed to establish a transfer path from *A* to *B*. The duration of the transfer signal is a complete cycle in order to be able to transfer the entire contents of *A* into *B* serially (1 bit every clock pulse).

The state equations for a recirculating register are the same as those for a flip-flop register.

6.2.3 Shift Registers

One of the most common operations in digital computer arithmetic is shifting. As we have seen in Chapter 3, many algorithms for performing arithmetic operations require shifting of operands either to the right or to the left. During the execution of any arithmetic operations, registers are the storage devices which hold the operands. Therefore, shifting of an operand can be accomplished if the registers have additional circuitry to perform this task. Registers which have built-in shifting capabilities are called *shift registers.*

6.2.3.1 Flip-flop Shift Registers

There are four basic shift operations: right shift, left shift, right shift circular, and left shift circular. For example, consider a symbolic representation of a register *A*, containing the binary number 11001010 as follows:

1	1	0	0	1	0	1	0

The four different shifts are shown in Fig. 6.10.

A right shift or a left shift one place moves every bit to the right or to the left, respectively, by one position, with the bit shifted out of one end

1	1	0	0	1	0	1	0	Originally
1	0	0	1	0	1	0	X	*A* after left shift of one position
X	1	1	0	0	1	0	1	*A* after right shift of one position
1	0	0	1	0	1	0	1	*A* after left shift circular of one position
0	1	1	0	0	1	0	1	*A* after right shift circular of one position

Figure 6.10 Various types of shifts.

being lost. In both cases, the contents of cells marked X are not defined by the shift. In this text we shall assume that 0 is entered into the X-marked cell unless otherwise specified. A circular shift is similar to a simple shift except that the bit shifted out of one end is circulated back to the other end cell of the register. Figure 6.11 shows a flip-flop register with a right shift capability. If a transfer path from the rightmost cell to the leftmost cell is added to a right shift register, it has a right shift circular capability. Figure 6.12 shows a four-cell register with a right shift circular capability using *S-R* flip-flops.

In both Figs. 6.11 and 6.12, we assumed that the circuit design of the flip-flops takes care of the race problem; therefore, the content of the flip-flop is transferred out at the same time its state is being changed according to the new inputs. If this is not the case, then shifting can be accomplished either by adding delay elements in the transfer path between the individual cells or by using an auxiliary register as an intermediate storage.

The state and input equations for an *S-R* flip-flop shift register $A = (A_1, \ldots, A_n)$ for the various shift operations (by one position) are shown in Table 6.9.

Table 6.9 *State and Input Equations for S-R Flip-flop Shift Register*

Type of shift	*State equations*	*Input equations*
Right shift	$A_i(t+1) = A_{i-1}(t),\ i = 2, \ldots, n$ $A_1(t+1) = d$†	$S_{A_i} = A_{i-1},\ R_{A_i} = \bar{A}_{i-1}$ $S_{A_1} = d,\ R_{A_1} = \bar{d}$
Right shift circular	$A_i(t+1) = A_{i-1}(t),\ i = 2, \ldots, n$ $A_1(t+1) = A_n(t)$	$S_{A_i} = A_{i-1},\ R_{A_i} = \bar{A}_{i-1}$ $S_{A_1} = A_n,\ R_{A_1} = \bar{A}_n$
Left shift	$A_i(t+1) = A_{i+1}(t),\ i = 1, 2, \ldots, n-1$ $A_n(t+1) = d$	$S_{A_i} = A_{i+1},\ R_{A_i} = \bar{A}_{i+1}$ $S_{A_n} = d,\ R_{A_n} = \bar{d}$
Left shift circular	$A_i(t+1) = A_{i+1}(t),\ i = 1, 2, \ldots, n-1$ $A_n(t+1) = A_1(t)$	$S_{A_i} = A_{i+1},\ R_{A_i} = \bar{A}_{i+1}$ $S_{A_n} = A_1,\ R_{A_n} = \bar{A}_1$

†*d* indicates an unspecified bit.

6.2.3.2 Shift Registers with Multiple Shifts

Some arithmetic techniques require shifting an operand by multiple positions at a time for speeding up an arithmetic operation. A shift register can provide such capability if a special arrangement of gates is added to it which establishes the required transfer path. For example, a shift left of four places of a register A can be implemented by providing a direct path from cell A_i to cell A_{i-4} for all possible cells, as shown symbolically in Fig. 6.13.

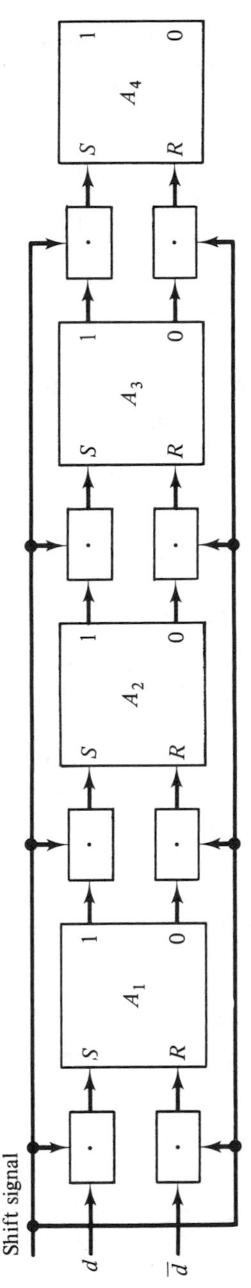

Figure 6.11 A flip-flop register with shift circuitry.

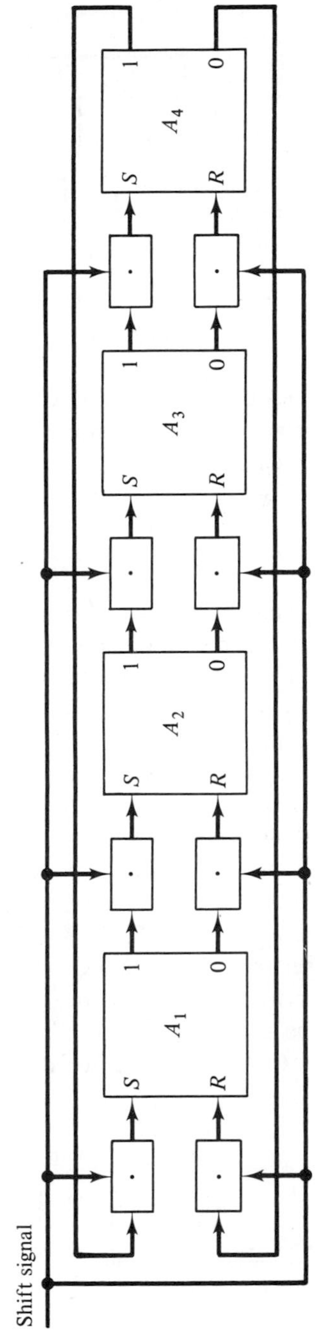

Figure 6.12 A flip-flop register with a right shift circular capability.

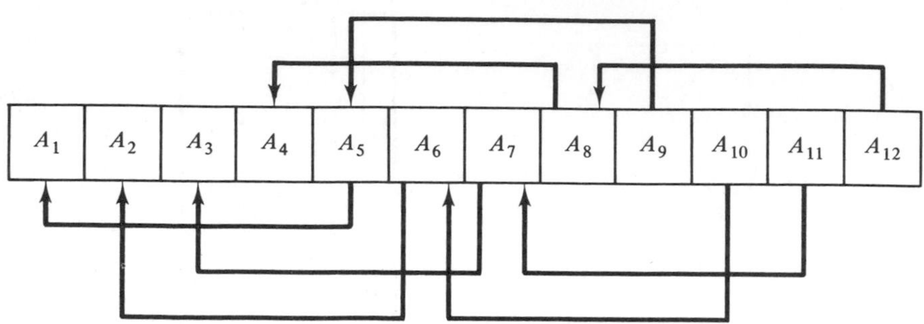

Figure 6.13 Shift left four positions.

In shift registers without a multiple shift provision, a four-position shift is performed (at the expense of speed) by four successive shifts of one position every clock pulse.

6.2.3.3 Use of Shift Registers as Parallel-to-Serial and Serial-to-Parallel Converters

Shift registers can be used as a means of converting information from serial to parallel form and vice versa. Parallel information means that it is available in parallel, that is, all the bits of a word can be transferred or operated upon at one time. If information is available in serial, then it can be transferred or processed only in serial fashion, that is, 1 bit at a time. The same definition also applies to registers. For example, the recirculating register shown in Fig. 6.8 is a serial register, while the flip-flop register of Fig. 6.6 is a parallel register. Digital computers are also classified as serial or parallel. Obviously, parallel computers are the most common due to their high speed when compared to serial machines. But even if a computer is exclusively parallel, conversion of information might be needed for communication between the computer and some input/output devices such as a magnetic tape where information is not stored in a parallel fashion. A shift register can be used to convert information from serial to parallel or vice versa. Figure 6.14 shows the serial-to-parallel conversion capability of a shift register.

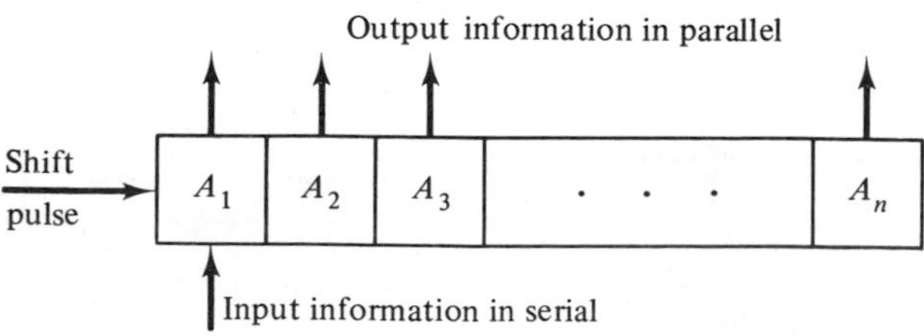

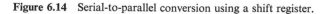

Figure 6.14 Serial-to-parallel conversion using a shift register.

First the A_1 flip-flop receives the first bit of the serial information to be converted. At the arrival of the second information bit, the shift pulse transfers the contents of A_1 into A_2 while A_1 stores the newly arrived bit. The process continues until all the bits have arrived and are stored in the register. Now all the bits are available at once in parallel but from different terminals.

The parallel-to-serial conversion capability of a shift register is shown in Fig. 6.15. In this case the register receives the information in parallel. The application of successive shift pulses will shift the information out of A_n, 1 bit at a time.

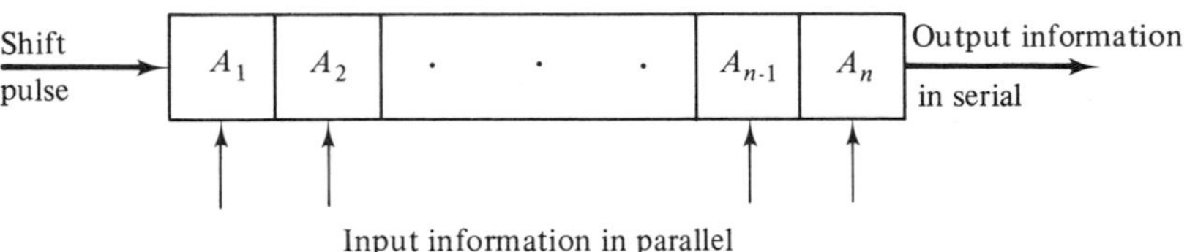

Figure 6.15 Parallel-to-serial conversion using shift register.

6.2.3.4 *Recirculating Registers as Shift Registers*

In recirculating registers, an information bit does not have a fixed location (i.e., a particular cell among an ordered set of cells), and its relative position is determined by synchronization with the clock signals and the mark signals. Therefore, shifting in a recirculating register is equivalent to a change in the time relation between the information and the mark signals. A mark pulse determines the beginning or the end of the pulse *train* representing the information stored in the register. Therefore, a delay of one pulse time in the recirculating loop of the register has the effect of shifting the information in the register by one position to the left. Assume that a shift signal is applied in synchronization with the mark signal and that a delay element of one clock pulse period is introduced in the recirculating loop as shown in Fig. 6.16. The first bit which enters the delay line of the recirculation

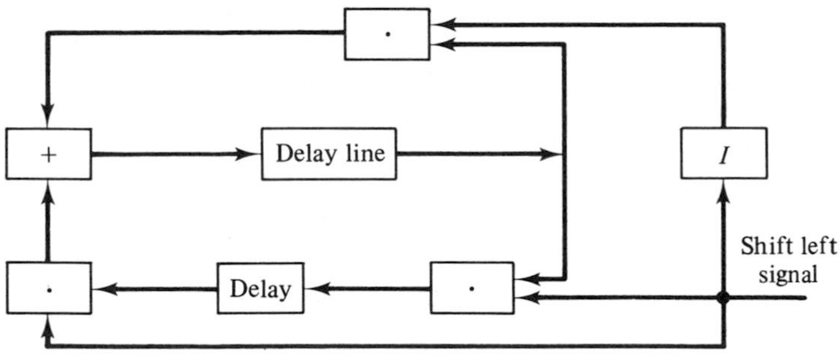

Figure 6.16 A left shift of a recirculating register.

loop after the application of the shift signal is the original output of the delay element, which is a 0. The following bits entering the delay line correspond to bits originally stored in the delay line but delayed one clock period. When the next mark pulse appears all the original information has entered the delay except the last bit (in this case the most significant bit). When the shift signal is removed in synchronization with the mark pulse, the most significant bit is lost.

Similar logic arrangements could be attached to a recirculating register in order to add other types of shift capabilities. For further details, see the book by Gschwind in the References.

6.3 COUNTERS

Counters are important functional subunits in digital computers. A counter is a device which counts the number of inputs it receives. It consists of a number of storage cells which store the current count and an arrangement of logic gates interconnected in a way which allows the count to change every time an input signal is received.

There are different kinds of counters depending on the applications for which they are used. The design of a particular counter depends on its function and the type of storage cells used. Some of the common counters are presented below.

6.3.1 Binary Counters

A binary counter is a counter which counts in the binary number system. It counts through a sequence of binary numbers, then resets itself to an initial value, and then repeats the counting process. A basic arrangement of a binary counter with three storage cells is shown in Fig. 6.17. This binary counter starts with all the storage cells in the 0 state; then upon receiving a count pulse, it goes through the sequence of states indicated by Table 6.10. A modulo-n binary counter has a maximum count of $n - 1$. It counts, in

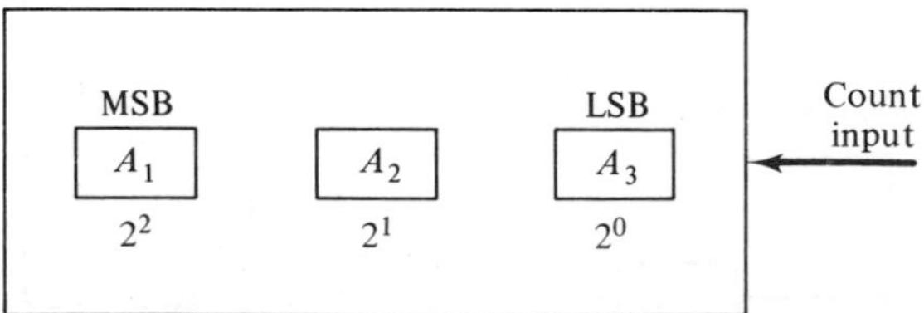

Figure 6.17 Three-cell binary counter.

Table 6.10 States of a Binary Counter

State			Remarks
A_1	A_2	A_3	
0	0	0	Initial state
0	0	1	After receiving the first count pulse
0	1	0	After receiving the second count pulse
0	1	1	After receiving the third count pulse
.	.	.	
.	.	.	
.	.	.	

binary, through the sequence 0, 1, 2, . . . , $n - 1$, and when it receives the nth count pulse it cycles back to 0 (the initial state).

The most common binary counters use flip-flops as their storage devices. Several standard circuits are available to realize counters, and many binary counters are manufactured using integrated circuits.

6.3.1.1 Design of Binary Counters Using Flip-flops

The design procedure of general sequential circuits can be applied to counters and is illustrated in Example 6.1.

Example 6.1

Design a modulo-8 binary counter using D flip-flops.

Table 6.11 shows the state table for the counter, and the state diagram is shown in Fig. 6.18. The counter advances its count only upon receiving a count pulse $C = 1$; otherwise (when $C = 0$) the counter remains in its current state. This property can be used to simplify the design by omitting the variable C and remembering to AND the expressions for the flip-flop inputs with the count pulse C. This is true for J-K or S-R flip-flops. It is not true for D flip-flops.

Using the characteristics (transition table) of the D flip-flops, the truth table for the counter is shown in Table 6.12. Using Karnaugh maps for individual minimiza-

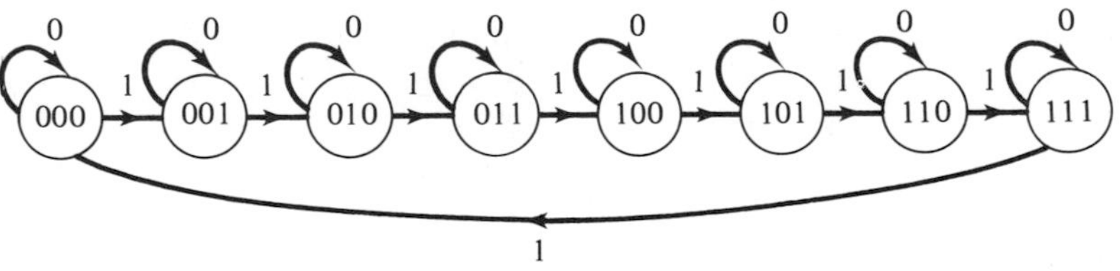

Figure 6.18 State diagram for modulo-8 binary counter.

Table 6.11 *State Table for Modulo-8 Binary Counter*

Present state			Next state: Count pulse $C = 0$			Next state: Count pulse $C = 1$		
A_1	A_2	A_3	A_1	A_2	A_3	A_1	A_2	A_3
0	0	0	0	0	0	0	0	1
0	0	1	0	0	1	0	1	0
0	1	0	0	1	0	0	1	1
0	1	1	0	1	1	1	0	0
1	0	0	1	0	0	1	0	1
1	0	1	1	0	1	1	1	0
1	1	0	1	1	0	1	1	1
1	1	1	1	1	1	0	0	0

Table 6.12 *Truth Table for a Modulo-8 Counter*

Input, C	Present state			Next state			Inputs to flip-flops		
	A_1	A_2	A_3	A_1	A_2	A_3	D_1	D_2	D_3
0	0	0	0	0	0	0	0	0	0
0	0	0	1	0	0	1	0	0	1
0	0	1	0	0	1	0	0	1	0
0	0	1	1	0	1	1	0	1	1
0	1	0	0	1	0	0	1	0	0
0	1	0	1	1	0	1	1	0	1
0	1	1	0	1	1	0	1	1	0
0	1	1	1	1	1	1	1	1	1
1	0	0	0	0	0	1	0	0	1
1	0	0	1	0	1	0	0	1	0
1	0	1	0	0	1	1	0	1	1
1	0	1	1	1	0	0	1	0	0
1	1	0	0	1	0	1	1	0	1
1	1	0	1	1	1	0	1	1	0
1	1	1	0	1	1	1	1	1	1
1	1	1	1	0	0	0	0	0	0

tion of D_1, D_2, and D_3 we obtain

$$D_1 = \bar{C}A_1 + A_1\bar{A}_2 + A_1\bar{A}_3 + C\bar{A}_1A_2A_3$$
$$D_2 = \bar{C}A_2 + A_2\bar{A}_3 + C\bar{A}_2A_3$$
$$D_3 = C\bar{A}_3 + \bar{C}A_3$$

A logic diagram which realizes the above equations is shown in Fig. 6.19.

The same design procedure could be used to obtain logic circuits using other types of flip-flops.

Binary counters using T flip-flops are very simple to design and can be obtained by inspection. Figure 6.20 shows a modulo-8 binary counter using T flip-flops. Figure 6.20 is derived by inspection of the state table of the binary counter (Table 6.11).

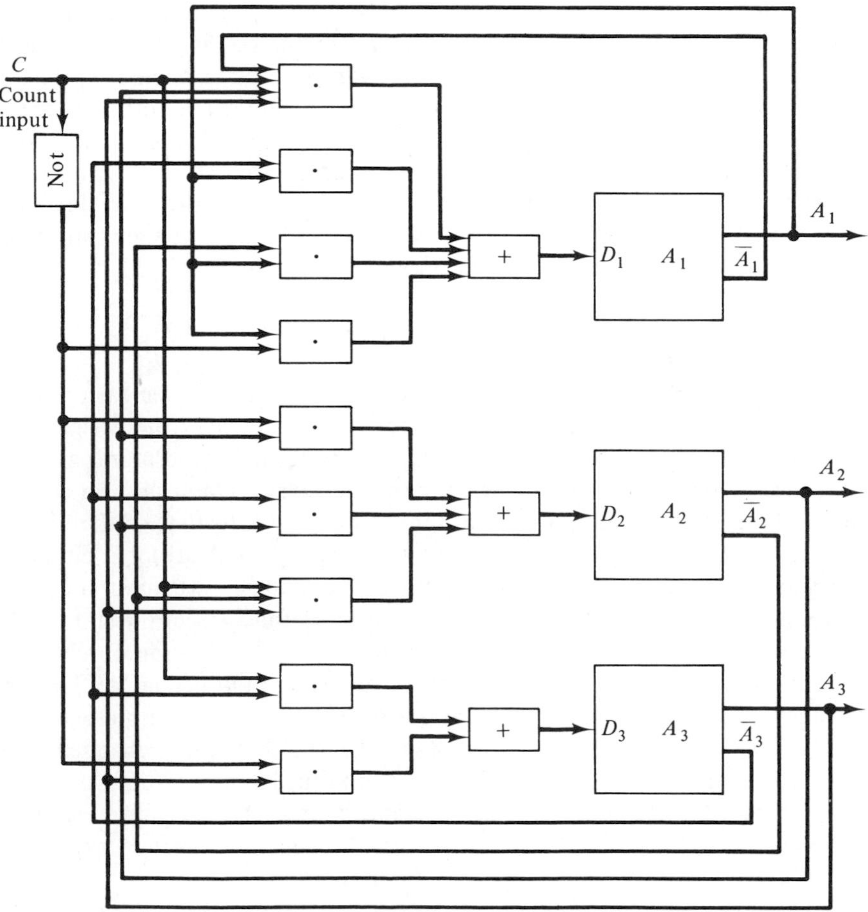

Figure 6.19 Modulo-8 binary counter using D flip-flops.

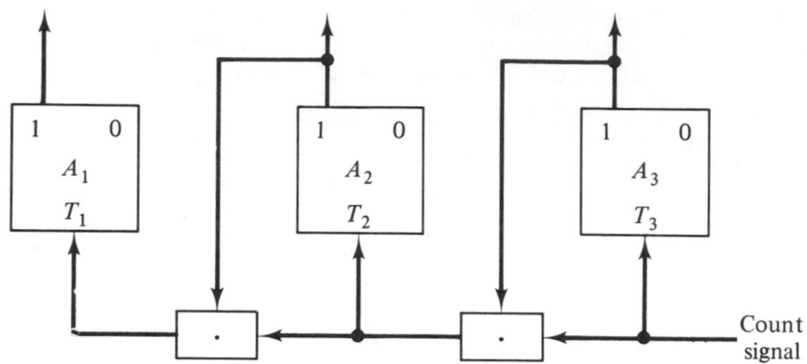

Figure 6.20 A modulo-8 binary counter using T flip-flops.

The A_3 flip-flop changes state with the arrival of every count pulse. Therefore, T_3 should be energized every time a count signal arrives. The A_2 flip-flop has to change state with the arrival of the count pulse only when the lower-order flip-flop (in this case, A_3) is in the 1 state. Similarly, the A_1 flip-flop has to change state with the count pulse only when both the lower-order bits (that is, A_2 and A_3) are in the 1 state. The design of a modulo-8 binary counter using T flip-flops, as described above, can be extended to binary counters with a higher number of cells in a straightforward fashion. To avoid the race problem in the above counter, the count signal must be very narrow pulses with short duration and low frequency.

6.3.1.2 Design of Binary Up-Down Counters

Counters which have the capability of increasing and decreasing their count are designated as up-down counters. Let us consider the design of a three-cell binary up-down counter using T flip-flops. Furthermore, assume that the counter, at any clock period, either receives a count-up or a count-down pulse. The count-up circuitry should be similar to that of the binary counter of the previous subsection, shown in Fig. 6.20. Additional circuitry is needed to enable the counter to decrease its count by 1 whenever a count-down pulse is received. To derive such circuitry, let us inspect the state table of the down binary counter shown in Table 6.13. The A_3 flip-flop changes its state whenever the counter receives a count-down pulse. The A_2 flip-flop changes state with the arrival of a count-down pulse only if the present state of the lower-order flip-flop A_3 is a 0. Similarly, the A_1 flip-flop changes state only when both A_2 and A_3 are in the 0 state and the counter receives a count-down signal. Combining both count-up and count-down circuitry results in an up-down counter. An up-down, three-cell binary counter is shown in Fig. 6.21.

Design of an up-down counter using any type of flip-flops can be done in three steps. First, derive the flip-flop's input equations for an up counter.

Table 6.13 State Table for a Down Counter

Present state			Next state		
A_1	A_2	A_3	A_1	A_2	A_3
0	0	1	0	0	0
0	1	0	0	0	1
0	1	1	0	1	0
1	0	0	0	1	1
1	0	1	1	0	0
1	1	0	1	0	1
1	1	1	1	1	0
0	0	0	1	1	1

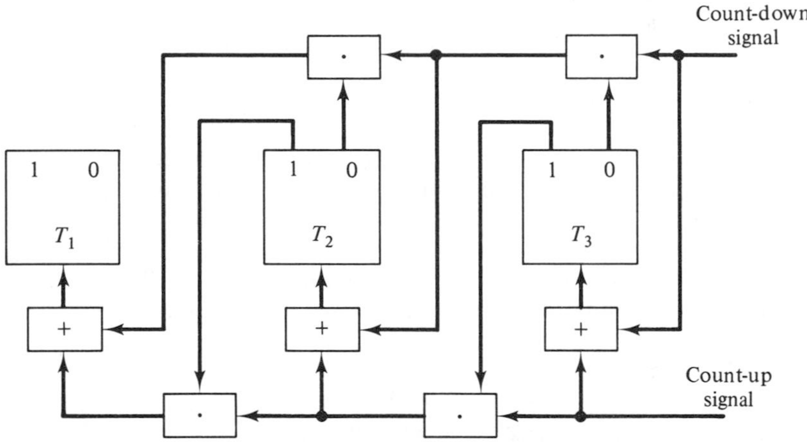

Figure 6.21 Three-cell up-down counter.

Then in a similar manner derive the input equations for count-down circuitry. The third step is to combine using logical-OR gates. This method is to be used whenever the count-up and count-down signals are received on separate input lines. If the count-up and count-down signals are both coded on one input line, for instance, say that a pulse represents a count-up and that no pulse represents a count-down, then the counter design could be achieved by following the design procedures of a general sequential machine.

6.3.2 Decade Counters

Decade counters are used for counting in a number system other than the binary system. For example, a decimal digit counter is called a *decade* of a decimal counter. This is a counter which counts from 0 up to 9, and after receiving the tenth count pulse, it cycles back to 0 and generates a carry to the next higher decade counter. The counting could be in any BCD (binary coded decimal) code desired. Let us design a decade of a decimal counter using an NBCD (8-4-2-1) code and T flip-flops. The design truth table for this counter is shown in Table 6.14. Considering the remaining six states (unused

Table 6.14 *Truth Table for a Decade Counter*

Present state				*Next state*				*Flip-flops input*				*Carry output*
A_0	A_1	A_2	A_3	A_0	A_1	A_2	A_3	T_0	T_1	T_2	T_3	
0	0	0	0	0	0	0	1	0	0	0	1	0
0	0	0	1	0	0	1	0	0	0	1	1	0
0	0	1	0	0	0	1	1	0	0	0	1	0
0	0	1	1	0	1	0	0	0	1	1	1	0
0	1	0	0	0	1	0	1	0	0	0	1	0
0	1	0	1	0	1	1	0	0	0	1	1	0
0	1	1	0	0	1	1	1	0	0	0	1	0
0	1	1	1	1	0	0	0	1	1	1	1	0
1	0	0	0	1	0	0	1	0	0	0	1	0
1	0	0	1	0	0	0	0	1	0	0	1	1

six codes) as don't care conditions and simplifying T_0, T_1, T_2, T_3, and the carry output of a multiple-output network, we obtain the following expressions:

$$T_3 = C = \text{Count signal}$$

$$T_2 = C\bar{A}_0A_3$$

$$T_1 = C\bar{A}_0A_2A_3 = T_2A_2$$

$$T_0 = C\bar{A}_0A_1A_2A_3 + CA_0A_3 = T_1A_1 + A_0A_3C$$

$$\text{Carry output} = CA_0A_3$$

A complete logic diagram of the decade counter is shown in Fig. 6.22.

6.3.3 Ring Counters

A class of counters which do not require much logic circuitry to construct is known as ring counters. A ring counter's circuitry is the same as the basic circuitry of a shift register. Figure 6.23 shows a ring counter which has a cycle length n and uses D flip-flops.

The counter starts in an initial state in which cell A_0 contains a 1 while all other cells contain a 0. The count pulse has the effect of a shift right cir-

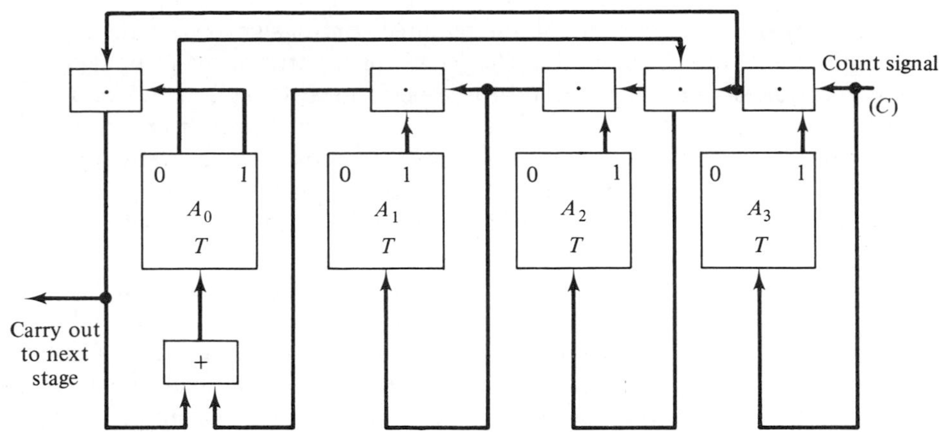

Figure 6.22 A decade counter using T flip-flops.

cular pulse. That is, upon the arrival of a count pulse, the content of any cell is shifted to the right one position, with the rightmost cell recirculating to the leftmost cell. The location of the cell which contains a 1 determines the count of the counter and therefore an n-cell ring counter counts modulo-n.

Let us compare a modulo-n binary counter with an n-cell ring counter. The current count of a modulo-n binary counter depends on the states of all the cells of the counter. Therefore, if every particular count is to be represented by a single output line, the states of the counter cells have to be decoded.† If an output line is connected to each cell of an n-cell ring counter, the resulting network is equivalent to the modulo-n binary counter plus the decoder. The ring counter requires more memory cells but less gates than the binary counter plus the decoder.

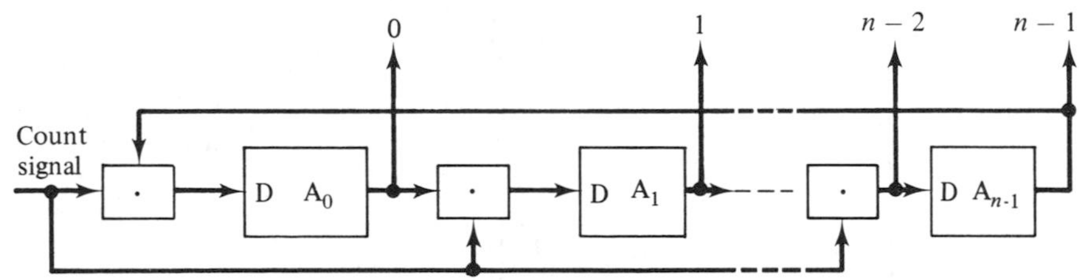

Figure 6.23 A ring counter.

6.3.4 Selectable Modulo Counters

An arithmetic or logical operation is usually performed, in digital computers, by breaking it into an ordered sequence of simple logic operations each of which can be done in a single clock pulse period. The control unit

†See Section 6.4 on decoders.

of a digital computer provides all the required control signals to the appropriate subunits of the computer. The length of a particular control signal sequence varies with the operation to be performed. Usually these control signal sequences can be generated as the output of a ring counter or a binary counter plus a decoder network. This implementation requires many counters. Each control sequence of length n requires a modulo-n counter.

A less expensive approach to the above problem is to use a selectable modulo counter. For example, a multiply operation requires a control signal sequence that is different from that for a divide operation. If these two different control signal sequences are to be generated by a selectable modulo counter, then the counter must follow different sequences of states depending on external signals, that is, whether the external signal signifies a multiply or a divide operation. Let us assume that multiplication and division require control sequences of five and seven pulses, respectively. To generate the necessary control signals, a selectable modulo counter, then, counts as modulo-5 or as modulo-7. The selectable modulo counter for this case is shown in Fig. 6.24. The reset signal clears all the cells of the counter.

When the counter receives a count signal, it behaves like an ordinary three-cell, modulo-8, binary counter. However, if there is a multiply signal and the counter state is 100, a clear signal is generated to reset the counter

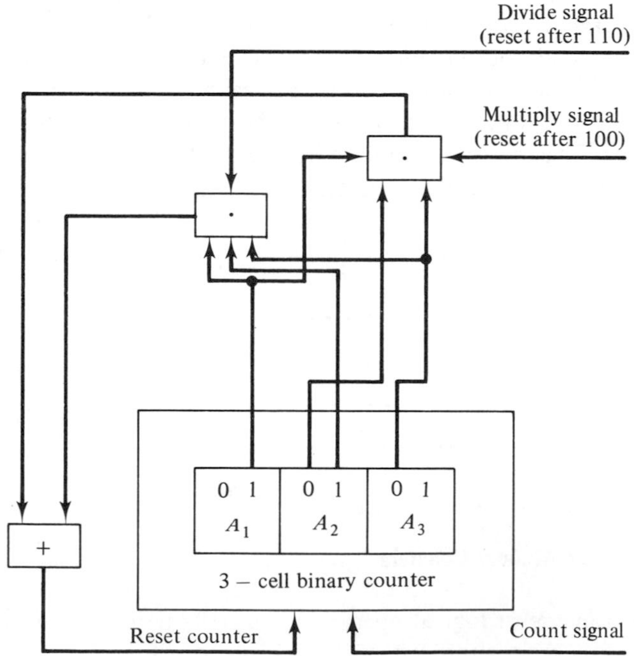

Figure 6.24 An example of a selectable modulo counter.

to 0. In effect the counter counts modulo-5. Similarly, a divide signal and state 110 generate a clear signal which causes the counter to count modulo-7.

6.4 ENCODERS AND DECODERS

An *encoder* refers to a multiple output combinational network whose function is to generate a binary code from nonbinary information. Every single item at the input of an encoder is mapped into an output of multiple lines. For example, consider an encoder to generate the NBCD code, discussed in Chapter 4; the 10 inputs to the encoder are the signals representing the decimal digits 0 through 9. The output is one of the 10 combinations of 4 variables corresponding to the specific input digit.

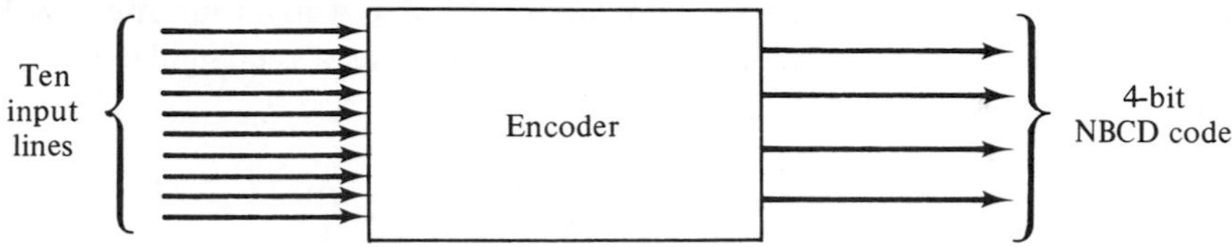

A *decoder* is also a combinational logic network, but its function is opposite to that of an encoder. A decoder maps a multiple-bit input back into a 1-bit output. In the case of the NBCD code, a decoder maps a 4-bit code into a 1-bit output, indicating the decimal number corresponding to the 4-bit combination. One of the most important uses of decoders in digital computers is the decoding of the operation code (op-code) for each instruction. As we shall see in later chapters, in any instruction the type of operation to be performed has to be interpreted first, before the instruction is executed. The only way to do so is by decoding the binary code associated with the instruction.

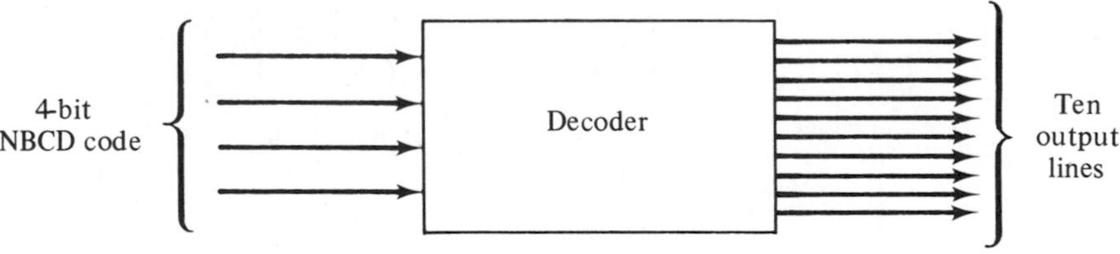

Encoders and decoders can be realized in different forms. Some of these forms are presented below.

6.4.1 Rectangular Matrix Encoders and Decoders

The design and structure of encoders and decoders is to be illustrated through an example. Let us consider the octal digits 0 through 7 and try to build an encoder which codes every octal digit to its equivalent 3 bits. Let us designate with the binary variables A, B, and C, respectively, the bits with weights 4, 2, and 1, and let $f_0, f_1, \ldots, f_7$ represent the digits 0 through 7. If the binary code generated by the encoder is to be stored in a flip-flop register, it is common to generate along with every bit its complement. Therefore, the encoder to be designed will have six output terminals, that is, A, $\bar{A}$, B, $\bar{B}$, C, and $\bar{C}$.

The Boolean expression for the output variables can be written as

$$A = f_4 + f_5 + f_6 + f_7, \qquad \bar{A} = f_0 + f_1 + f_2 + f_3$$
$$B = f_2 + f_3 + f_6 + f_7, \qquad \bar{B} = f_0 + f_1 + f_4 + f_5$$
$$C = f_1 + f_3 + f_5 + f_7, \qquad \bar{C} = f_0 + f_2 + f_4 + f_6$$

These Boolean expressions can be realized directly using four input *OR* gates and one-level logic. This realization is known as a rectangular matrix encoder, as shown in Fig. 6.25.

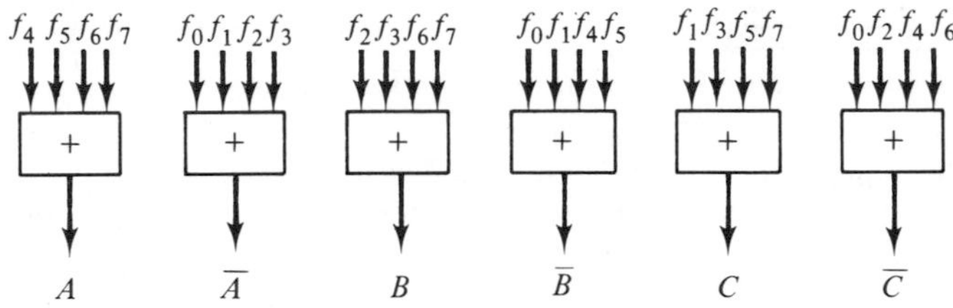

Figure 6.25 A rectangular matrix encoder for the octal digits.

To design a decoder, we have to express the outputs f_0 through f_7 in terms of the input variables A, $\bar{A}$, B, $\bar{B}$, C, and $\bar{C}$.

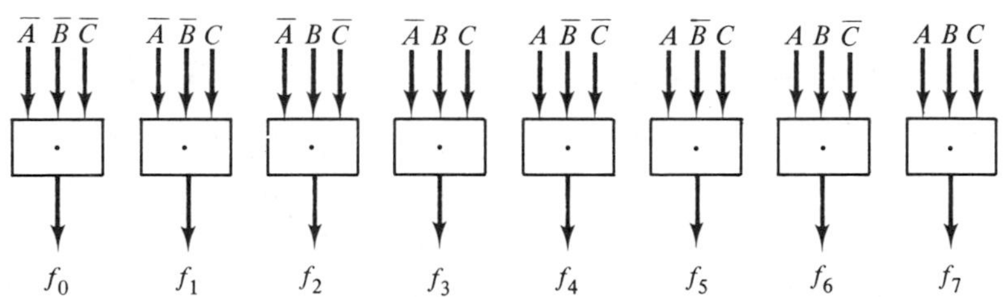

Figure 6.26 A rectangular matrix decoder for the octal digits.

These Boolean expressions are

$$f_0 = \bar{A}\bar{B}\bar{C} \qquad f_4 = A\bar{B}\bar{C}$$
$$f_1 = \bar{A}\bar{B}C \qquad f_5 = A\bar{B}C$$

$$f_2 = \bar{A}B\bar{C} \qquad f_6 = AB\bar{C}$$

$$f_3 = \bar{A}BC \qquad f_7 = ABC$$

A rectangular matrix decoder using AND gates and one-level logic is shown in Fig. 6.26. The use of AND gates is arbitrary, since if the output is expressed as standard sum terms, a similar network results with OR gates instead of the AND gates.

6.4.2 Diode Rectangular Matrices as Encoders and Decoders

Since a rectangular matrix decoder or encoder requires only one level of gates, the use of diode gates is very appropriate. Diode rectangular matrices are widely used for building decoders and encoders. If every AND gate block in Fig. 6.26 is replaced by its diode AND gate equivalent and the diodes are arranged to have a common voltage supply, the result will be a diode rectangular matrix decoder. Similar statements hold for the encoder of Fig. 6.25. An octal-binary diode rectangular matrix encoder is shown in Fig. 6.27. A binary-octal diode rectangular matrix decoder is shown in Fig. 6.28.

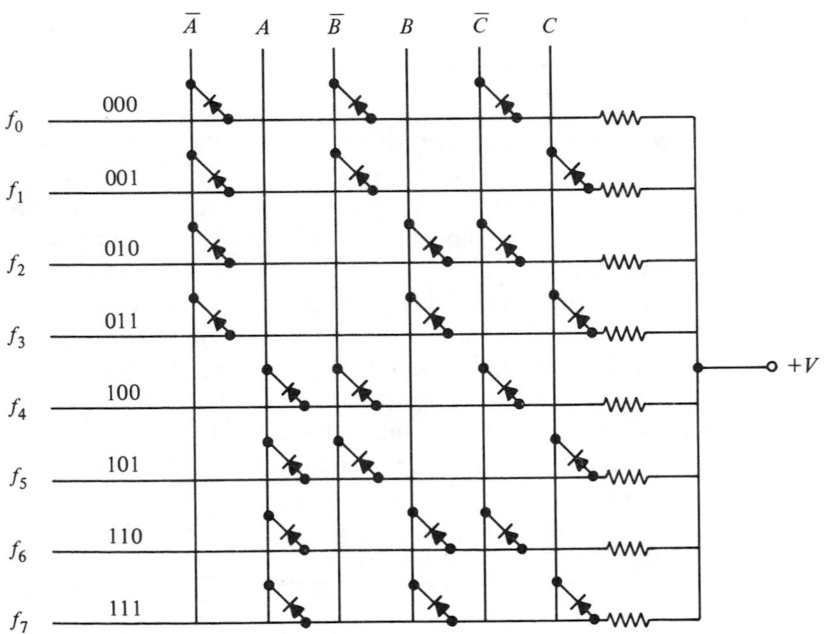

Figure 6.27 Diode rectangular matrix as octal-binary encoder.

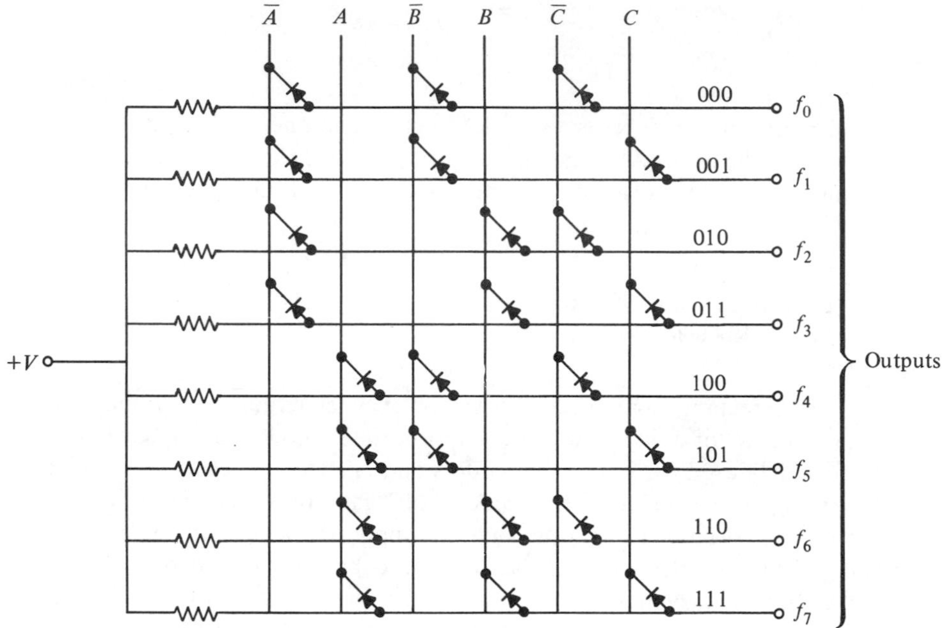

Figure 6.28 Diode rectangular matrix as binary-octal decoder.

6.4.3 Tree and Dual-Tree Matrices as Decoders

Another structure for a decoder which performs the same function as the rectangular structure but requires a fewer number of gate inputs is the *tree* (pyramid) structure. Figure 6.29 shows a tree structure decoder for 4 pairs of inputs and 16 outputs using AND gates. A similar tree can be constructed using OR gates.

A third structure which uses fewer gate inputs than the previous two is the *dual-tree*. Figure 6.30 shows a dual-tree decoder using AND gates with 4 pairs of inputs and 16 outputs. If the number of input Boolean variables is large, then the dual-tree structure is achieved by breaking the larger structure into smaller dual-trees. For example, consider a decoder with n input Boolean variables and 2^n outputs. Figure 6.31 shows a dual-tree decoder for $n = 8$ inputs and 256 outputs.

The number of gate inputs in a dual-tree decoder is less than that in the corresponding tree decoder. The gate inputs for the tree decoder are less than those for a rectangular decoder. There are features other than the number of inputs for which the three structures differ. The rectangular structure requires one level of gates with each gate having a number of inputs equal to the number of Boolean input variables (n). In the tree structure, if two-

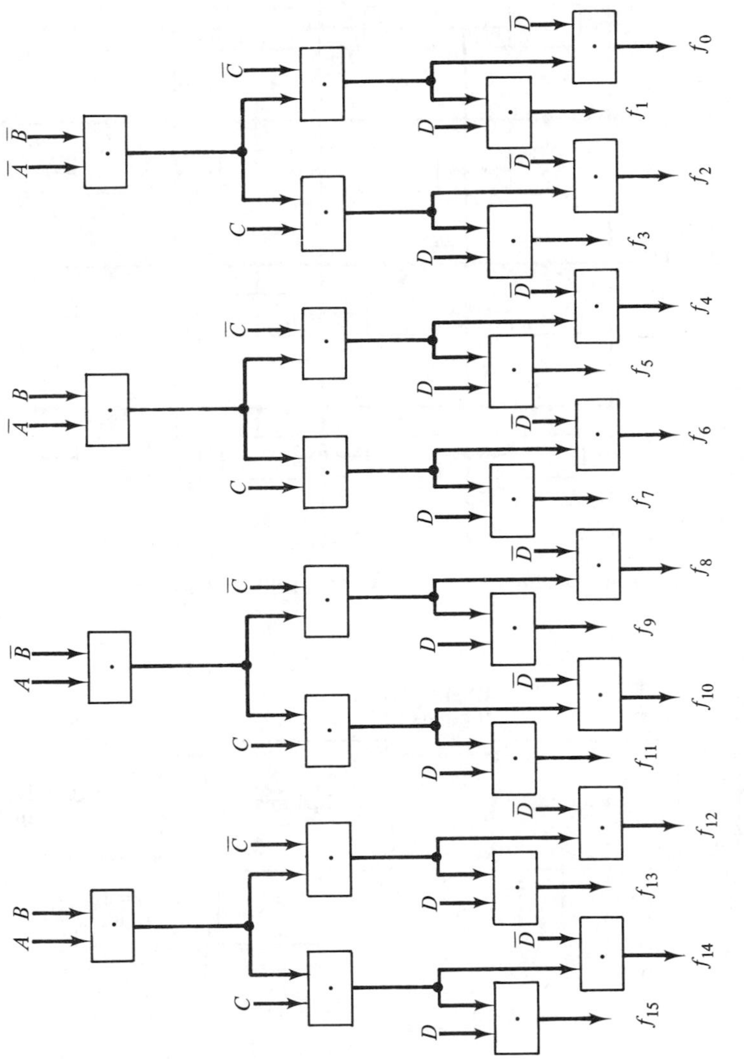

Figure 6.29 Tree decoder with four pairs of inputs and 16 outputs.

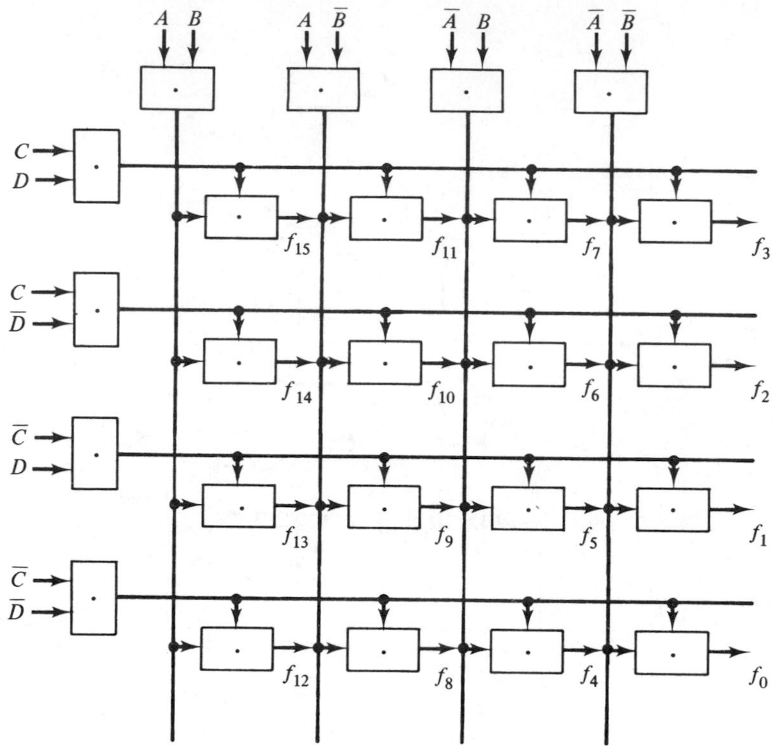

Figure 6.30 Dual-tree decoder with four pairs of inputs and 16 outputs.

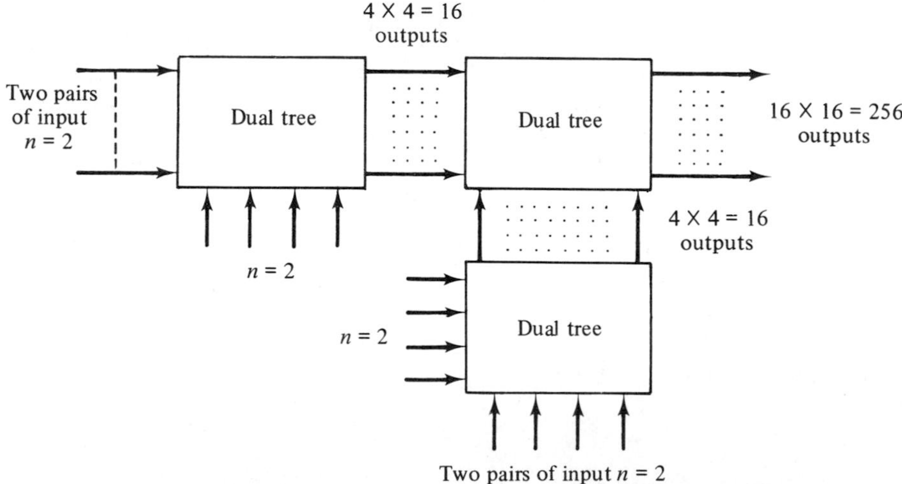

Figure 6.31 Dual-tree decoder for eight pairs of inputs and 256 outputs.

input gates are used, any signal goes through $n - 1$ gate levels between the input and the output. For the dual-tree, when two-input gates are used, the number of gate levels is one for $n = 2$, two for $n = 3$ and $n = 4$, three for $n = 5$ to $n = 8$, four for $n = 9$ to 16, and so on. A complete comparison of the three structures is given in Table 6.15.

Table 6.15 *Comparison of Rectangular, Tree, and Dual-Tree Decoders*

Input variables, n	Outputs, 2^n	Number of gate inputs		
		Rectangular	Tree	Dual-tree
2	4	8	8	8
3	8	24	24	24
4	16	64	56	48
5	32	160	120	96
6	64	384	248	176
7	128	896	504	328
8	256	2,048	1,016	608
9	512	4,608	2,040	1,168
10	1,024	10,240	4,088	2,240
11	2,048	22,528	8,184	4,368
12	4,096	49,152	16,376	8,544
13	8,192	106,496	32,760	16,788
14	16,384	229,376	65,528	33,414
15	32,768	491,520	131,064	65,472
16	65,536	1,048,576	262,136	132,288

6.5 TIMING PULSE GENERATORS

One of the basic functions of the control unit of a digital computer is to provide timing signals for the various units of the computer. These signals determine the sequence of the transfer and logic operations necessary to perform a specific task.

6.5.1 The Clock

The operations performed by a synchronous digital computer are sequenced by a *master clock*. The clock is a pulse generator whose output signal, most often, consists of evenly spaced pulses. The pulse repetition frequency of the clock signal is closely related to the speed of the computer.

This time interval must be long enough to allow for the transient response of the circuitry and for the signals to propagate through the gating circuitry of the computer. Other factors which usually determine the frequency of the clock pulses are the type and speed of internal memory used in the computer and the average time needed to execute an instruction.

Figure 6.32 shows a simple scheme for generating clock signals. Once a pulse is started around the loop, it will continue circulating indefinitely,

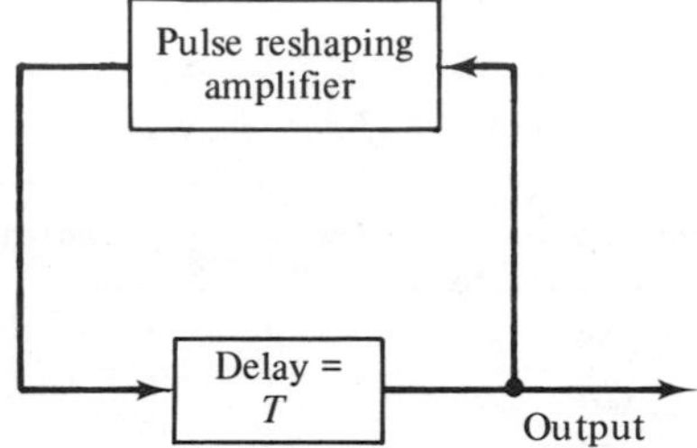

Figure 6.32 A clock pulse generator.

producing a sequence of evenly spaced pulses. Assuming that the length of the delay element is T seconds, the output of the clock pulse generator is a sequence of pulses with T seconds between two successive pulses. If the first pulse occurs at time t, then the successive pulses occur at $t + T$, $t + 2T$, . . . , etc. For convenience, these time intervals can be mapped into the positive integers 1, 2, 3, . . . , and p_1, p_2, p_3 . . . indicate the outputs of the clock pulses. Figure 6.33 shows the output of the clock versus time.

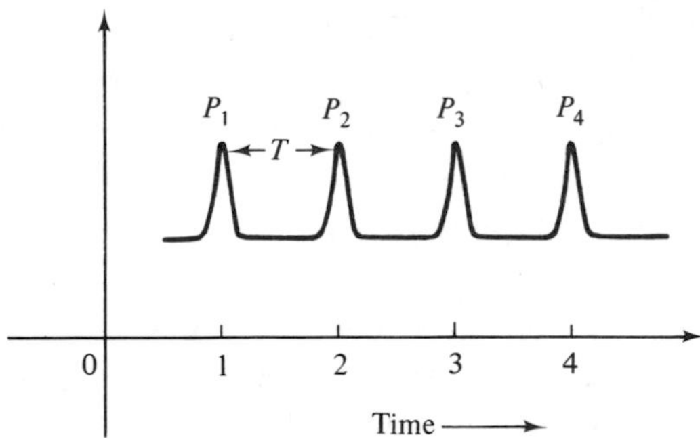

Figure 6.33 Clock output.

More than one pulse generator can be used. For example, if two pulse generators are used, then the first can generate the even pulses (p_2, p_4 . . .) while the other one generates the odd pulses (p_1, p_3, p_5 . . .). Figure 6.34 shows the time relation between these two sets of pulses. If pulse generators are used in the above fashion, then the resulting clock has a pulse frequency twice that of a single pulse generator. The above arrangement could be used

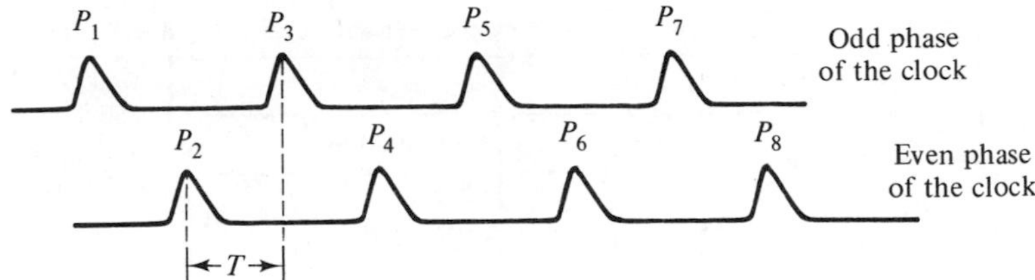

Figure 6.34 Two clocks.

in cases where the maximum frequency of a pulse generator is below the required clock frequency.

6.5.2 Pulse Distributors

In synchronous computers it is necessary to distribute clock pulses among a group of k control lines in cyclic form. Line 1 receives clock pulses p_1, p_{1+k}, p_{1+2k}, etc.; line 2 receives p_2, p_{2+k}, p_{2+2k}, etc.; and finally line k receives p_k, p_{2k}, p_{3k}, etc.

The logic circuitry which implements the distribution of pulses, as described above, is called a pulse distributor. One possible way of designing a pulse distributor is by the use of a ring counter. Figure 6.35 shows a modulo-6 ring counter as a pulse distributor.

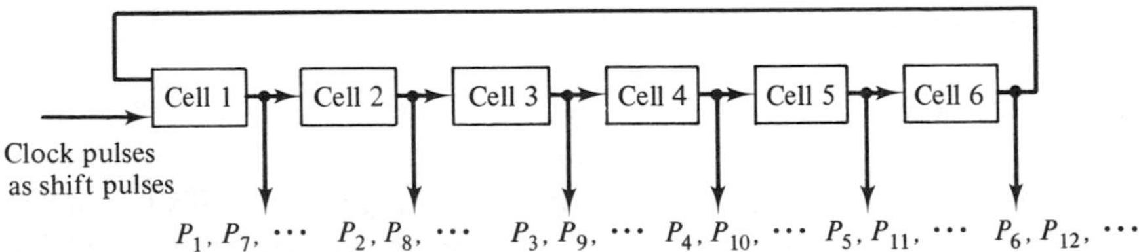

Figure 6.35 A pulse distributor.

The type of clock which is composed of two separate pulse generators is also a pulse distributor with $k = 2$. The output from each pulse generator represents a control line.

When the number of control lines (k) is large, the pulse distributor can be implemented more economically by using two ring counters each with a cycle length in the neighborhood of $\sqrt{k}$. For $k = 12$, two ring counters with cycle lengths 3 and 4 can be used. Such an arrangement is shown in Fig. 6.36. Each crosspoint corresponds to one of the 12 clock pulses. If an AND gate is driven by one vertical line and one horizontal line, the corresponding numbered pulse is derived.

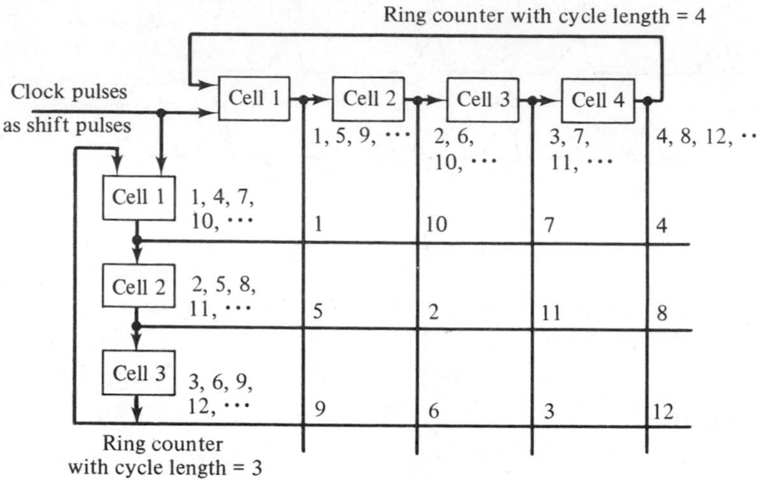

Figure 6.36 Dual ring counter distributor.

6.6 BINARY ADDERS AND SUBTRACTORS

Single-bit operand addition and subtraction tables were introduced in Chapter 2. In this section we shall present some logic circuitry which can realize the single-bit operand addition and subtraction. These logic circuits constitute the basic building blocks from which adders and subtractors, for full-length operands, are constructed.

6.6.1 Binary Half-Adder

A binary half-adder is a combinational logic network with two input variables and two outputs. The inputs represent a single-bit addend and a single-bit augend, while the outputs are the modulo-2 sum (s) and the associated carry (k) of the two operands. A symbolic (block) representation of a half-adder is shown in Fig. 6.37. Table 6.16 gives the truth table for the

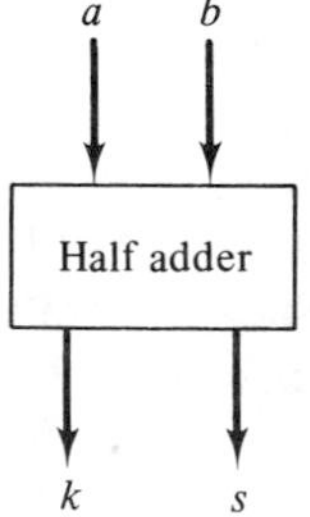

Figure 6.37 Symbolic representation of a half-adder.

Table 6.16 *Truth Table for Additon of 2 Binary Bits*

Inputs		*Outputs*	
Addend, a	*Augend, b*	*Sum, s*	*Carry, k*
0	0	0	0
0	1	1	0
1	0	1	0
1	1	0	1

addition of 2 binary bits. The sum-of-products form can be written as

$$s = a'b + ab'$$
$$k = ab$$

Another form which requires a fewer number of gate inputs can be written as

$$s = (a + b)\overline{ab}$$
$$k = ab$$

The corresponding realization of the above two terms is shown in Fig. 6.38.

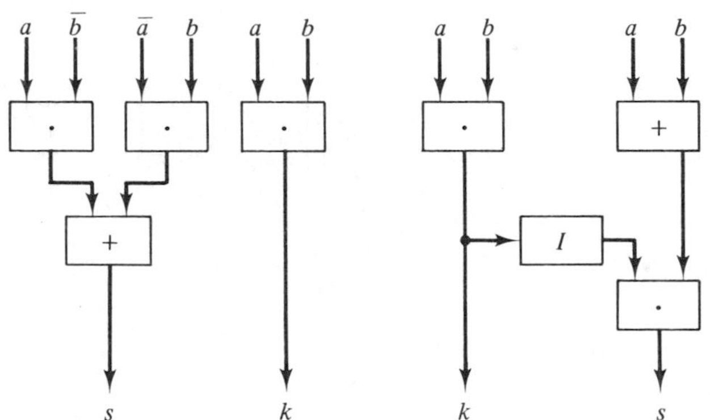

Figure 6.38 Some logic circuits for a half-adder.

6.6.2 Binary Full-Adder

A symbolic representation of a binary full-adder is shown in Fig. 6.39. It has three inputs and two outputs. The inputs represent the addend bit (a), the augend bit (b), and the carry (k^*) from a previous (lower-order) position.

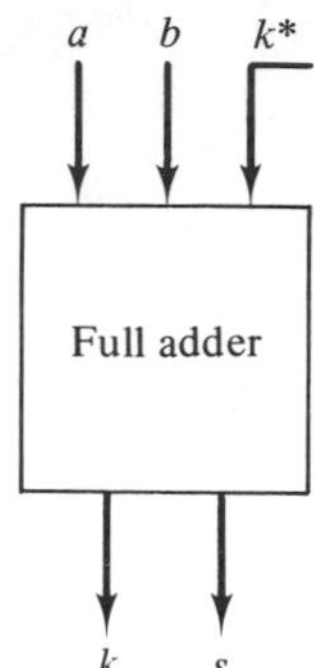

Figure 6.39 Symbolic representation of a binary full-adder.

Its outputs (s) and (k) represent the modulo-2 sum and the associated carry, respectively.

The input/output relationship of a binary full-adder is described in the truth table shown in Table 6.17. From the truth table, the sum and carry

Table 6.17 *Truth Table for Full-Adder*

Inputs			*Outputs*	
Addend, a	*Augend, b*	*Carry in, k**	*Sum, s*	*Carry out, k*
0	0	0	0	0
0	0	1	1	0
0	1	0	1	0
0	1	1	0	1
1	0	0	1	0
1	0	1	0	1
1	1	0	0	1
1	1	1	1	1

can be written as

$$s = a\bar{b}\bar{k}^* + \bar{a}b\bar{k}^* + \bar{a}\bar{b}k^* + abk^*$$

$$k = ab + ak^* + bk^*$$

Another expression for the sum s may be derived by inspection. The sum equals 1 if the three inputs are 1s or if there is exactly one 1. The first condition can be expressed by abk^*. The second condition can be expressed by $(a + b + k^*)\bar{k}$, since $\bar{k}$ indicates less than two 1s and $a + b + k^*$ indicates at least one 1. Therefore, the sum and the carry can be written as

$$k = ab + ak^* + bk^*$$
$$s = abk^* + (a + b + k^*)\bar{k}$$

A realization for the above expressions is shown in Fig. 6.40.

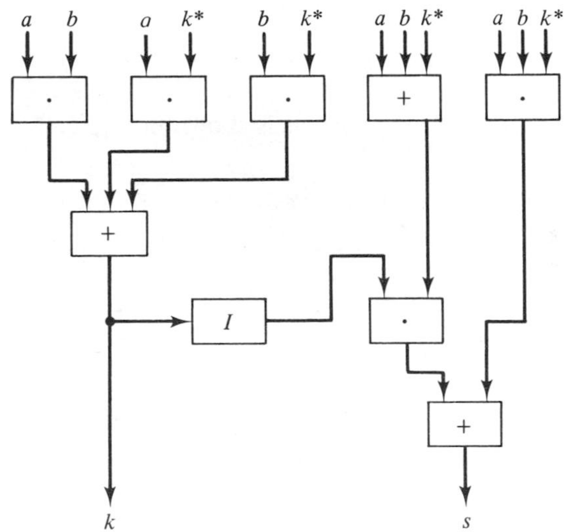

Figure 6.40 A full-adder.

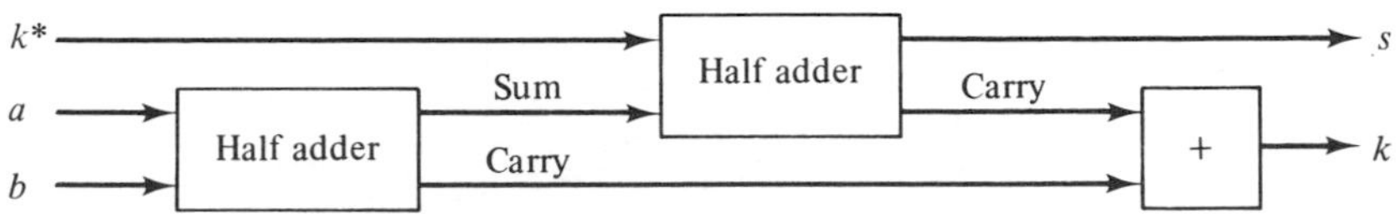

Figure 6.41 A full-adder composed of two half-adders.

A full-adder can be realized by using two half-adders. Figure 6.41 shows two half-adders combined to form a full-adder. The first half-adder adds the addend and the augend. The second half-adder adds the carry from the previous stage to the sum generated by the first half-adder. The carry out of the full-adder is the logical-OR of the carry generated by either half-adder.

The adder realizations given in this section are neither the only possible ones nor the simplest. Many variations in the design exist for various types of logic gates and also depend on the use of single rail or double rail logic.

6.6.3 Binary Half-Subtractor

A half-subtractor has two inputs; one represents the minuend bit (a), while the other represents the subtrahend bit (b). The outputs from a half-subtractor are the difference (d) and the borrow (k). The truth table for the half-subtractor is described in Table 6.18. From the truth table we obtain the following expressions:

$$d = a'b + ab'$$

$$k = a'b$$

A realization for the above expressions is shown in Fig. 6.42.

Table 6.18 *Truth Table for the Half-Subtractor*

a	b	d	k
0	0	0	0
0	1	1	1
1	0	1	0
1	1	0	0

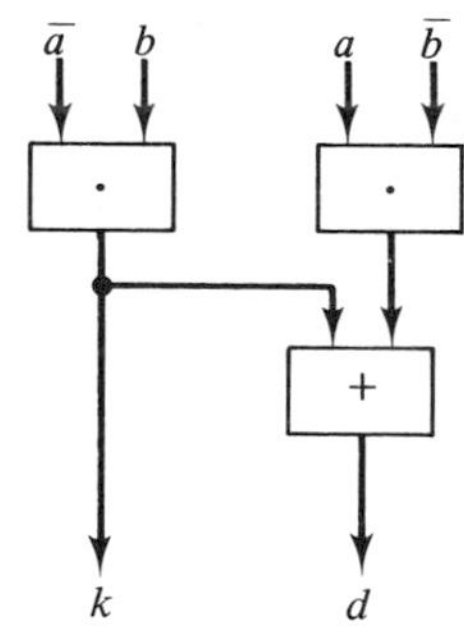

Figure 6.42 A half-subtractor.

6.6.4 Binary Full-Subtractor

Using k^* to signify a borrow from the previous stage, the truth table for the full-subtractor is shown in Table 6.19. The difference (d) and the borrow (k) can be written as

$$d = \bar{a}\bar{b}k^* + \bar{a}b\bar{k}^* + a\bar{b}\bar{k}^* + abk^* \qquad \text{[same as the sum } (s)\text{]}$$

$$k = \bar{a}\bar{b}k^* + \bar{a}b\bar{k}^* + \bar{a}bk^* + abk^*$$

Table 6.19 Truth Table for a Full-Subtractor

Inputs			*Outputs*	
a	*b*	*k**	*d*	*k*
0	0	0	0	0
0	0	1	1	1
0	1	0	1	1
0	1	1	0	1
1	0	0	1	0
1	0	1	0	0
1	1	0	0	0
1	1	1	1	1

These expressions are implemented in Fig. 6.43. It is a direct implementation of the sum-of-products form. Other implementations which might require a fewer number of gate inputs could be obtained by manipulating the standard form. A full-subtractor can be implemented by using two half-subtractors. Such an implementation is shown in Fig. 6.44. It is also possible to imple-

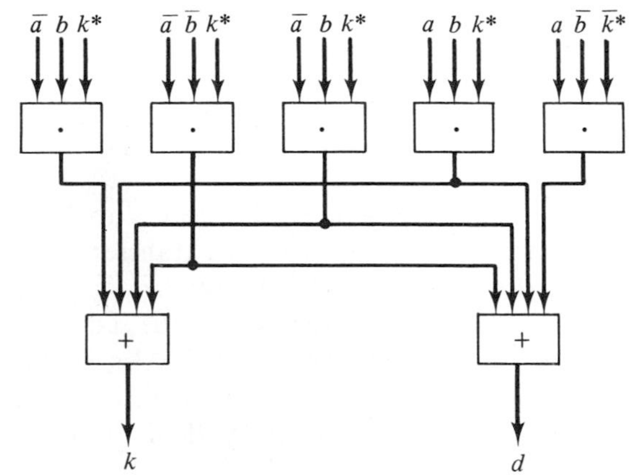

Figure 6.43 A full-subtractor.

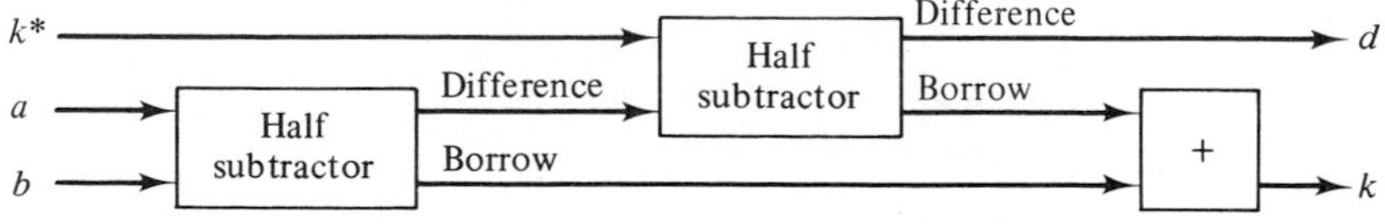

Figure 6.44 A full-subtractor composed of two half-subtractors.

ment a subtractor by using either two half-adders or a half-adder and a half-subtractor. In each case a special arrangement of additional logic gates is needed.

6.7 TRANSFER NOTATIONS

Most of the operations and the processing of information in a digital computer can be looked upon as a series of information transfers. Therefore, to simplify the development of a digital computer design, a shorthand notation for the various information transfers and processing is very helpful. This notation is to be developed through various chapters in the text whenever needed. In this section, a transfer notation for some register operations is introduced, to be referenced and used in later chapters.

6.7.1 Register-to-Register Transfer Notation

Let A and B be n-cell registers each having cells $A_1, A_2, \ldots, A_n$ and $B_1, B_2, \ldots, B_n$, respectively. The parallel transfer of the contents of A into the contents of B is denoted by

$$A \longrightarrow B$$

or

$$A_i \longrightarrow B_i \qquad \text{for } i = 1, 2, \ldots, n$$

We stated earlier that the information transfer from A to B does not destroy the contents of register A. Therefore, the above transfer leaves the contents of A unchanged. Also the above transfer destroys the previous contents of B.

The above transfer is defined for registers of the same storage capacity. This will always be the case throughout this text. Also we observe that the cells of the registers are labeled with identical subscripts $1, 2, \ldots, n$. This restriction can be removed, and a transfer with more general labeling is defined below.

If $A = (A_{i_1}, A_{i_2}, \ldots, A_{i_n})$ and $B = (B_{j_1}, B_{j_2}, \ldots, B_{j_n})$ such that each i_k and j_k is an integer with

$$i_{k-1} < i_k < i_{k+1} \quad \text{and} \quad j_{k-1} < j_k < j_{k+1}$$

then the transfer

$$A \longrightarrow B$$

means that

$$A_{ik} \longrightarrow B_{jk} \qquad \text{for } k = 1, 2, \ldots, n$$

This means that the transfer preserves the ordering of the positional location but without requiring identical labels. For example, if $A = (A_{11}, A_{12}, \ldots, A_{15})$ and $B = (B_1, B_2, \ldots, B_5)$, then

$$A \longrightarrow B$$

signifies that

$$A_{11} \longrightarrow B_1, A_{12} \longrightarrow B_2, \ldots, A_{15} \longrightarrow B_5$$

6.7.2 Definition and Transfer Notation for Subregisters

If A is an n-cell register $A_1, A_2, \ldots, A_n$, then any number of cells of A which are less than n can be defined as a subregister of A. For example, we can define the subregisters $F(A)$ and $G(A)$ as $F(A) = (A_1, A_2, \ldots, A_m)$ and $G(A) = (A_{m+1}, A_{m+2}, \ldots, A_n)$. The subregisters $F(A)$ and $G(A)$ can be looked upon, insofar as transfer, logic or arithmetic operation is concerned, as registers. The only thing to keep in mind is that, in this case, there exists only one physical register A.

The definition of transfer can now extend to subregisters. Let $A = (A_1, A_2, \ldots, A_n)$ and define the first m cells and the last $n - m$ cells of A as the subregisters $F(A)$ and $G(A)$, respectively. Let B be an m-cell register, $B = B_1, \ldots, B_m$, and C an $(n - m)$-cell register, $C_1, , \ldots C_2, C_{n-m}$. Then the transfer

$$B \longrightarrow F(A), \qquad G(A) \longrightarrow G(A)$$

puts new values in the first m cells of A, leaving the right $n - m$ cells unchanged. Similarly, the transfer

$$F(A) \longrightarrow F(A), \qquad C \longrightarrow G(A)$$

puts new values into the right $n - m$ bits of A, while leaving the left m cells unchanged.

The transfer

$$F(A) \longrightarrow F(A), \qquad G(A) \longrightarrow G(A)$$

leaves register A unchanged and is equivalent to

$$A \longrightarrow A$$

6.7.3 Transfer Notation for Shift Register

Let A be an n-cell register, $A_1, A_2, \ldots, A_n$, and D be a one-cell register. Define the subregisters $L(A)$ and $R(A)$ as

$$L(A) = (A_1, A_2, \ldots, A_{n-1})$$
$$R(A) = (A_2, A_3, \ldots, A_n)$$

That is, $L(A)$ is the leftmost $n - 1$ cells of A and $R(A)$ is the rightmost $n - 1$ cells of A.

A shift left one place can be described by the transfer

$$R(A) \longrightarrow L(A), \qquad D \longrightarrow A_n$$

These two transfers indicate the shift left by one place and that new datum is inserted into the A_n cell while the content of A_1 is lost. Similarly, a shift right by one place can be described by the transfer

$$L(A) \longrightarrow R(A), \qquad D \longrightarrow A_1$$

Consider again register A and define two operators ρ^1 and ρ^{-1} such that these operators when operating on A are defined as follows:

$$\rho^1(A) \quad = L(A) \longrightarrow R(A), \qquad A_n \longrightarrow A_1$$
$$\rho^{-1}(A) \quad = R(A) \longrightarrow L(A), \qquad A_1 \longrightarrow A_n$$

The operators $\rho^1(A)$, $\rho^{-1}(A)$ as defined represent a right shift circular and a left shift circular one place, respectively. Therefore, we can now use the transfer notation

$$\rho^1(A) \longrightarrow A$$
$$\rho^{-1}(A) \longrightarrow A$$

to signify the right shift circular and the left shift circular one place, respectively.

Circular shifts by more than one place can be derived by iteration. For example,

$$\rho^{-2}(A) = (A_3, A_4, \ldots, A_n, A_1, A_2)$$
$$\rho^2(A) = A_{n-1}, A_n, A_1, \ldots, A_{n-2})$$

Therefore, $\rho^2(A) \longrightarrow A$ and $\rho^{-2}(A) \longrightarrow A$ signify circular shifts by two places to the right or to the left, respectively. And obviously

$$\rho^k(A) = \rho^{-k}(A) = A \qquad \text{for } k = n, 2n, \ldots$$

Therefore,

$$\rho^n(A) \longrightarrow A \quad \text{or} \quad \rho^{-n}(A) \longrightarrow A$$

leaves A unchanged.

6.7.4 Conditional Transfer Notation

Sometimes information transfers between registers are subjected to the occurrence of some particular event. For example, some of the algorithms for arithmetic operations in Chapter 2 require the complement of a number if and only if its sign is negative. Let us consider a register A and assume that the operation of complementing A can be represented by

$$A_0\bar{A} + \bar{A}_0A \longrightarrow A$$

where A_0 is the sign bit of A. In this case, the above transfer signifies that either $\bar{A}$ or A is transferred to A depending on whether the value of A_0 is a 1 or a 0, respectively.

To develop conditional transfer notations, let us first consider a one-cell register b and an n-cell register $A = (A_1, A_2, \ldots, A_n)$ and define the logical multiplication bA as

$$bA = (bA_1, bA_2, \ldots, bA_n)$$

where bA_i is the usual logical product. Therefore, the product bA can be described as

$$bA = A \qquad \text{if } b = 1$$

$$bA = 0 \qquad \text{if } b = 0$$

This definition of multiplication of a one-cell register by an n-cell register can be used to conveniently describe a conditional transfer. Let A, B, and C be n-cell registers, and let a represent a one-cell register. Then the transfer

$$aA + \bar{a}B \longrightarrow C$$

implies that

$$A \longrightarrow C \qquad \text{if } a = 1$$

$$B \longrightarrow C \qquad \text{if } a = 0$$

The conditional transfer notation can be extended to more than two conditions. For example, the transfer

$$abA + \bar{a}bB + \bar{a}\bar{b}C + a\bar{b}D \longrightarrow D$$

can be used to describe the conditional transfer between the registers A, B, and C into D on the condition of the one-cell registers a and b as

$a = 1$	$b = 1$	$A \longrightarrow D$	
$a = 1$	$b = 0$	$D \longrightarrow D$	no change
$a = 0$	$b = 1$	$B \longrightarrow D$	
$a = 0$	$b = 0$	$C \longrightarrow D$	

REFERENCES

BARTEE, T. C., I. L. LEBOW, and I. S. REED, *Theory and Design of Digital Machines.* New York: McGraw-Hill, 1962.

BOOTH, T. L., *Digital Networks and Computer Systems.* New York: Wiley, 1971.

BREEDLOVE, P. S., "ECL/MOS for Optimum Minicomputer Systems," *Computer Design*, 11, No. 8 (Aug. 1972), pp. 61–66.

CHU, Y., *Digital Computer Design Fundamentals.* New York: McGraw-Hill, 1962.

GSCHWIND, H. W., *Design of Digital Computers.* New York: Springer-Verlag New York, Inc., 1967.

LO, A. W., *Introduction to Digital Electronics.* Reading, Mass.: Addison-Wesley, 1967.

PEATMAN, J. B., *The Design of Digital Systems.* New York: McGraw-Hill, 1972.

WICKES, W. E., *Logic Design with Integrated Circuits.* New York: Wiley, 1968.

PROBLEMS

6.1. Construct the latch circuit of Fig. 6.1 using NAND gates. Label each of the inputs and outputs and justify the operation of the circuit as an S-R flip-flop.

6.2. Derive an algebraic expression for the output of the circuit shown in Fig. P6.2 and analyze the operation of the circuit as compared to the operation of a standard S-R flip-flop.

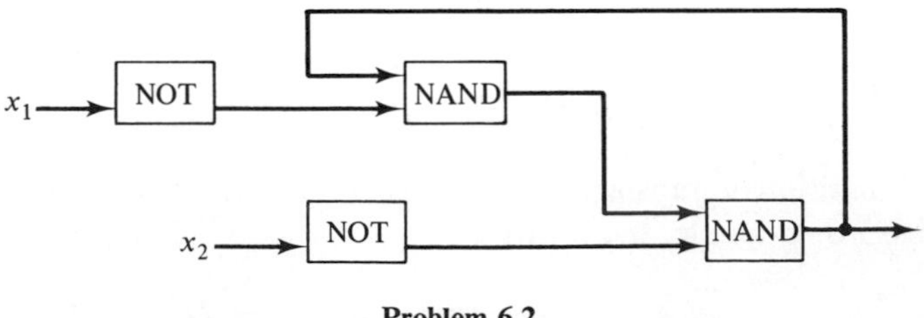

Problem 6.2

6.3. Using a *D* flip-flop and combination logic gates AND, OR, and NOT, show logic diagrams to realize
(a) An *S-R* flip-flop.
(b) A *J-K* flip-flop.
(c) A *T* flip-flop.
(d) An *S-R-T* flip-flop.

6.4. Draw a block diagram for a decade counter using a *T* flip-flop by modifying the binary counter in Fig. 6.20. Draw the block diagram for a five-stage count-down counter using
(a) *T* flip-flops.
(b) *S-R* flip-flops.

6.5. Assume that four flip-flops, A_1, A_2, A_3, and A_4, are to step through the following state combinations in the indicated sequence upon applications of clock pulses: 0000, 0011, 1101, 0001, 0010, 1010, 1011, 1000, 0111. After the last state combination the sequence is to repeat. Find suitable formulations for the input signals to each of the four flip-flops when the flip-flops are of the
(a) *T* type.
(b) *R-S* type.
(c) *J-K* type.
(d) *D* type.

6.6. Design a three-stage Gray code counter.

6.7. Design a Gray code up-down counter.

6.8. Design the complete logic diagram for a selectable modulo counter which counts either modulo-10 or modulo-8 depending on whether a control signal x equals 0 or 1, respectively.

6.9. Design two flip-flop registers of 4 bits each in such a way that the clock pulse will exchange their contents. Repeat the problem for recirculating shift registers. Design a controlled 3-bit shift register which has two control signals c_1 and c_2. It must perform the following operations:

c_1 c_2	*Operation*
0 0	No operation
0 1	Shift left one position
1 0	Complement the contents of the register
1 1	Set 1 into every cell of the register

Design this shift register using *D* flip-flops.

6.10. Repeat Problem 6.9 using an *S-R* flip-flop register.

6.11. Assume that two 3-bit registers, *A* and *B*, are available. Design the logic necessary to carry out the following parallel information transfer operations that are determined by the control signals c_1 and c_2. $[X] = [x_1, x_2, x_3]$ denotes a 3-bit external input signal.

c_1 c_2	*Operation*
0 0	No operation
0 1	The current value of X is loaded into A. The contents of B are unchanged.
1 0	The current contents of A are transferred to B. The current value of X is loaded into A.
1 1	The current contents of A are transferred to B. The contents of A are set to zero (i.e., $[A] = [0, 0, 0]$).

The A and B registers are to be constructed from S-R flip-flops.

6.12. Show the necessary gating for a delay line recirculating register with a right shift.

6.13. Design the logic diagram for a 6-bit shift register which can accept serial information, parallel information, and serial-parallel information 3 bits at a time. Show each of the three control signals which indicate the type of input.

6.14. Draw the circuit diagram for a diode matrix to decode the three counts of a modulo-3 counter.

6.15. Design an encoder to encode the 10 decimal digits into an excess-3 BCD code.

6.16. Using NAND gates, show a block diagram for a decoder of a two-stage binary counter.

6.17. Design a decimal-to-binary encoder for the decimal numbers up to 29 (13 inputs, 10 units, and 3 tens).

6.18. Design a decoder, binary-to-decimal, to match that of Problem 6.17.

6.19. Construct the block diagram for a rectangular decoding matrix with 4 pairs of inputs and 16 outputs
(a) Using NOR gates only.
(b) Using NAND gates only.

6.20. Repeat Problem 6.19 if the 16 outputs are sum terms.

6.21. Using two ring counters, show a block diagram for a pulse distributer which distributes 13 clock pulses on separate control lines.

6.22. Use three ring counters to construct a pulse distributer which distributes incoming clock pulses on 24 control lines. Choose the cycle length of each counter in a way to minimize the circuitry.

7

Computer Architecture and Programming

Although the authors assume that the reader has programmed a digital computer, they also assume that it probably has been done with a higher-level language such as BASIC, FORTRAN, COBOL, or PL/I. As such the reader may not be familiar with the structure of a computer and the programming of a computer at the machine language level. Of course, this entire text is devoted to the design of the computer and its structure. In this chapter we shall introduce the preliminary concepts of the design and structure. We shall also devlop here the elementary concepts of machine language programming. Without this concept it is not possible to adequately develop the design of the machine or possible alternatives in design.

There is no single digital computer organization. Each manufacturer has its own philosophy as to the "best" way to design the machine. Thus, there is no "ideal" computer or absolute basis of comparison. We can make relative comparisons about the structure of the subsystems which comprise the total computing system. The design of the system specifications at a general or subsystem level is called *computer architecture*. We shall investigate some system architectures in this chapter. In later chapters we shall be concerned with the implementation of different architectures, culminating in the implementation of a specific architecture in the design of a small computer.

Over the past 25 years the fraction of total system costs due to hardware has been decreasing due to the development of faster, cheaper, and smaller electronic components. Thus, the relative costs of programming are increasing. This presents the system architects with additional design trade-offs which must be considered in determining a final design. Should a function be implemented in hardware or developed in software (programming), it may be incorrect from the overall cost viewpoint to omit certain functions from hardware in order to try and economize on costs, since it may cost more to implement these functions in software.

In Section 1.2 we developed a simple block diagram of a digital computer with its five basic subsystems. We repeat that figure here as Fig. 7.1. We notice that the memory unit is central to all the other units of the computer. It, of course, holds the programs to be executed and the data which are used as operands during the execution of the program. In this chapter we shall investigate some additional aspects of each of the blocks and develop a more detailed block diagram of a general-purpose digital computer. We shall also develop the principles of programming a digital computer and some programming concepts. Finally, we shall discuss some of the components of programming systems. These studies may lead to a better understanding of the relationships of hardware and software to system architecture and overall system design trade-offs.

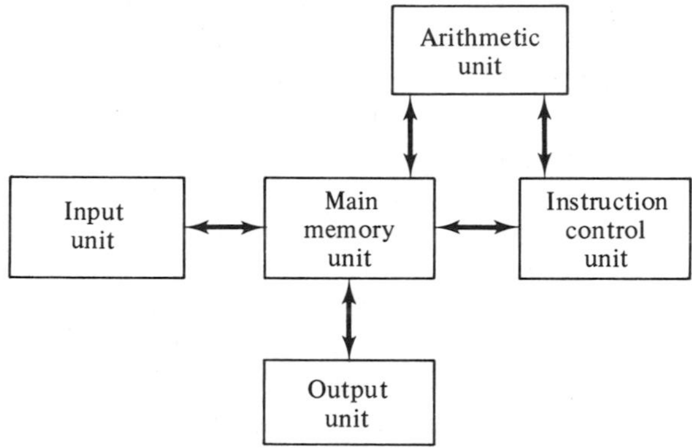

Figure 7.1 A simple block diagram of a digital computer.

7.1 ELEMENTARY COMPUTER ARCHITECTURE

7.1.1 Memory Addresses, Instructions, and Data

The storage of a computer is called its memory. The main memory consists of a large number of locations which can hold instructions, i.e., a program, or data. These locations are given numeric *addresses* just as houses on a street have numeric addresses. The addresses are invariant with a given location. Usually these addresses start with the number 0 and progress sequentially through the number of locations which the memory can contain.

At each address of the memory a group of bits is stored and is handled by the computer as a unit. The group of bits stored at each address is generally

Address	
0	1
1	0
2	2059
.	.
.	.
.	.
102	51
.	.
.	.
.	.
4095	3074613

Figure 7.2 A memory address layout.

referred to as a *word.* These words stored in memory can be either instructions or data. We shall not write the contents of an address as bits but will generally write decimal numbers to denote both data or addresses. Figure 7.2 depicts the memory and some numeric data therein.

Since the memory locations are built of electronic components, there is a maximum number which they can hold. Of course, the data are in the form of bits. Let us assume that each address of the memory can contain 32 bits of information. Thus, if we consider the data to be in positive integer form, the smallest number which we can store is a 0 and the largest magnitude will be $2^{32} - 1$. If there are 32 unity bits in the word, its decimal value is 4,294,967,295.

By using alphanumeric codes the contents of an address in memory can contain alphabetic information. If we assume 8 bits per character in the code, then the 32 bits of our memory word can contain 4 alphabetic code characters. If we assume a 4-bit NBCD code, then the 32 bits of our memory address can contain 8 NBCD-coded characters. Of course, if we group the 32 bits in groups of 4 bits, then we can represent the contents of an address as 8 hexadecimal characters immaterial of what the actual word contained. Figure 7.3 shows several possible representations of a word in our memory.

Binary	1100	0010	1101	0110	1110	0011	1100	1000
Hexadecimal	C	2	D	6	E	3	C	8
EBCDIC	B		0		T		H	

Figure 7.3 Representations of a word in memory.

The concept of the address is extremely important to the concept of the stored program electronic digital computer. One of its advantages is that the operands can be used in a random order rather than storing the operands in exactly the order they must be used. This permits the idea of a *random access* to operands instead of the *sequential access* of the early plugboard machines. Thus, the programmer does not have to order the operands.

Another important concept of the idea of *address* is that the same program can be used over again without modification, with different operands. One only has to load the new operands into the correct addresses in memory and then execute the program.

Since both instructions and operands can be obtained by addressing, it is possible to fetch an instruction and use it like an operand, that is, perform an operation on one instruction which we treat as an operand. This permits the computer to modify its instructions.

The ability of the instructions to use the address of the data rather than the data themselves permits the address to be used analogously to a variable in algebra. That is, we can manipulate the variable x without having to worry about the value of x at that specific instant. The variable x may stand for many different values in the course of describing an algorithm, and the same can be said for an address in memory.

7.1.2 Instruction Sequencing

The computer control unit controls the sequencing of information fetched from the memory unit. A typical sequencing of fetching and operation is

1. Fetch an instruction from memory.
2. Decode the instruction into
   a. Operation to be performed.
   b. Address of the data on which the operation is to be performed.
3. Fetch the data from the above address.
4. Perform the operation on that data.
5. Fetch the next sequential instruction and go to step 2 above.

The instruction of a computer program generally contains two parts: (1) the operation to be performed and (2) the address of an operand. The operation to be performed is given a code number, and the address of the operand is also a number. Thus, the instruction is just a bit pattern interpreted by the control unit as two different numerical values. Figure 7.4 depicts this.

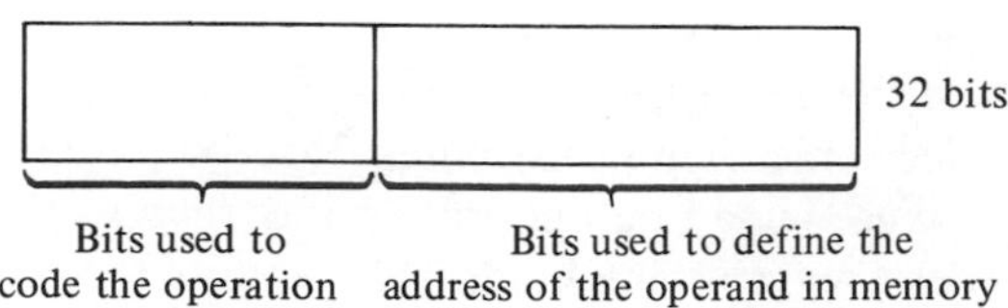

Figure 7.4 Instruction format.

From the above discussions we see that how the computer interprets the contents of a given memory address is a function of the time the control unit fetches the information from that location. Since all locations in memory contain bit patterns, one cannot determine by examining a given location whether the word is an instruction or a number.

The above sequence of events can be broken into two general phases. One is an *instruction phase, I* phase. During this part of the machine cycle, the control unit obtains an instruction from an address in memory, decodes the operation to be performed, and obtains the address of the operand in memory. The second phase is the *execution phase, E* phase, when the operand is obtained from the memory and the operation is performed. After the execution phase the control unit returns to the instruction phase and obtains a new instruction. The computer usually performs a single operation at a time. To perform these phases the control unit can be divided into two parts. One is the *instruction decoder*, and the other is the *control signal generator*. Figure 7.5 shows our new expanded block diagram for the digital computer.

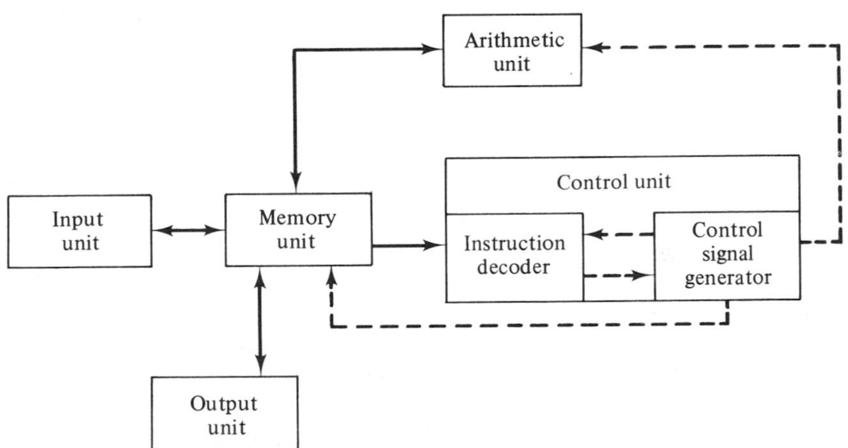

Figure 7.5 Subdivision of the control unit (solid lines indicate the flow of information; dashed lines indicate the flow of control signals).

7.1.3 Registers in the Computer

In addition to the locations in main memory which store instructions and data, there are special locations in the control unit and arthmetic units which also store instructions and data. These special locations are called *registers*. They are very fast electronic storage locations. Instructions or data generally move from slower main storage into these faster registers.

The computer manipulates this information and returns new information back into the main memory. When data are read into these registers from memory they replace the previous contents of the registers. However, when data are moved from these registers to memory, the data which were in the registers remain there and only a copy is stored in the memory.

What was just stated for the high-speed electronic registers is also true for any register in the computer, including all the locations in the main memory. When one *writes into* a location, the new data replace the old data that were there. When one *reads out* of a location, the data remain at that location and only a copy of the data is read out.

The arithmetic unit performs logical and arithmetic operations on data which are fetched from addresses in memory. To perform these mathematical operations, the arithmetic unit has registers which hold the data while the operations are being performed and a register to hold the results of the operation. The number of registers varies with the design of the machine. However, many machines have three registers which are used for fixed-point arithmetic and logical operations. As an example, consider the following arithmetic unit of a small digital computer, which is typical of many machines.

One register is called the *accumulator*, abbreviated as *ACC*. A second register is called the *multiplier-quotient*, *MQ*, and the third register is called the *X register*. During an addition operation the augend is in the *ACC*, and the addend is fetched from memory and placed in the *X* register. The contents of the *ACC* and *X* registers are added together, and the sum appears in the *ACC*, replacing the augend. For a subtraction operation, the minuend is in the *ACC*, the subtrahend is fetched from memory and placed in the *X* register, the contents of the two registers are subtracted, and the difference replaces the minuend in the *ACC*. For multiplication, the multiplier is fetched from memory and placed in the *MQ* register; the multiplicand is fetched from memory and placed in the *X* register; the contents of the registers are multiplied together. The most significant bits of the product are found in the *ACC*; the least significant bits are found in the *MQ* register. For division, the dividend is fetched from memory and placed in the *ACC*; the divisor is fetched from memory and placed in the *X* register. The operation is performed, and the quotient is found in the *MQ* register and the remainder in the *ACC*.

The above description can be abbreviated by the use of *register transfer equations*. Let *M* stand for any address in the memory. [*M*] stands for the contents of that address. *ACC* stands for the accumulator, and [*ACC*] stands for the contents of the accumulator. *MQ* stands for the multiplier-quotient register, and [*MQ*] stands for the contents of that register. *X* stands for the *X* register, and [*X*] means the contents of the *X* register. The operation of addition can be stated as

$$[M] \longrightarrow X$$
$$[ACC] + [X] \longrightarrow ACC$$

This is read as "The contents of an address is placed into the X register," followed by "The contents of the accumulator is added to the contents of the X register and the sum is placed back into the accumulator."

Likewise, subtraction is

$$[M] \longrightarrow X$$
$$[ACC] - [X] \longrightarrow ACC$$

Multiplication is

$$[M] \longrightarrow MQ$$
$$[ACC] \longrightarrow X$$
$$[MQ] \times [X] \longrightarrow ACC \,||\, MQ$$

This is read as "The contents from an address is placed in the MQ register; the contents of the ACC is placed in the X register; the product of the contents of the registers is placed into the accumulator and the MQ register." Division is

$$[M] \longrightarrow X$$
$$[ACC]/[X] \longrightarrow MQ \qquad \text{Remainder} \longrightarrow ACC$$

The control unit decodes the instruction to obtain the operation code (op-code) and the address of the operand on which to perform the operation. To do this the control unit must be able to store the instruction which it has fetched from main memory. It stores the instruction in a register called the *current instruction register* or *CIR*. Another register present in the control unit is called the *current address register* or *CAR*. This register contains the address from which the current instruction came. Thus, the control unit knows where in the main memory it obtained the current instruction.

A new block diagram of the digital computer now looks like Fig. 7.6.

7.1.4 Register Transfer Equations

A very useful notation for the design of a digital computer is the register transfer equations. Some examples of this notation were shown in Section 7.1.3. In this subsection we shall define this concept so that we may use it to design our computer in the following chapters. In Chapter 6 this notation was defined in general; here we wish to define it to specifically apply to its

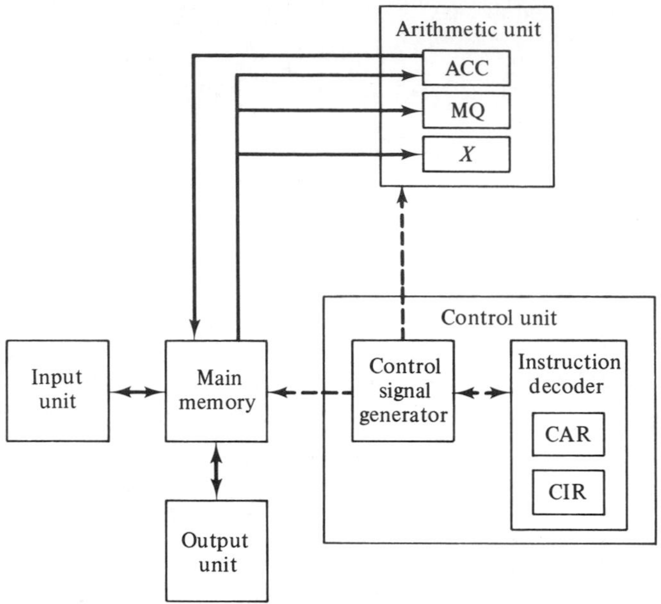

Figure 7.6 Block diagram of a digital computer showing some important registers.

use in digital computers. First let us define the registers of a computer.

ACC = accumulator of the arithmetic unit
MQ = multiplier-quotient register
X = register of the arithmetic unit
CAR = current address register
CIR = current instruction register
M = any memory address
R = any general-purpose register in the control unit

The brackets around a register means "the contents of." Sometimes we wish to specify particular bit positions of a register. Some special parts of a register are

1. *Op*(*R*) means the bit positions in register *R* corresponding to the op-code portion of the register.
2. *Ad*(*R*) means the address portion of register *R*.

Of course, *Op*(*CIR*) is just the op-code and *Ad*(*CIR*) is the address of the operand.

Sometimes it is necessary to specify the next sequential instruction. This is written as $[CAR] + 1 \longrightarrow CAR$. This shorthand notation saves many words and will be used to design a computer in the later chapters of this book.

7.1.5 Data Formats

The smallest unit of information in the computer is the binary integer or bit. Each register contained in the machine will store bits of information in them. However, it is generally inconvenient to specify the contents of a register in terms of the total bit pattern. For a 32-bit register this would require specifying all 32 bits. Instead of specifying the bits, they are usually grouped together in groups of 3 or 4 and the octal or hexadecimal equivalent character is specified.

Example 7.1

Consider a 24-bit register which contains the following data:

B		6		B		C		5		8		Hexadecimal
101	1	01	10	1	011	110	0	01	01	1	000	Binary
5	5		5		3	6	1		3		0	Octal

All three of the notations convey the same information.

If the bits in the register represent a fixed-point number, then they also convey a value to the bit pattern. That is, we can express the binary number in terms of another radix system if we so wish, such as a decimal number.

Example 7.2

Consider the decimal number for the 24-bit data word of Example 7.1. Assume that the binary point is at the left.

$$N = 1 \times 2^{-1} + 0 \times 2^{-2} + 1 \times 2^{-3} + \cdots + 0 \times 2^{-23} + 0 \times 2^{-24}$$

$$N \approx .722143_{10}$$

On the other hand, the register may contain encoded alphanumeric data.

Example 7.3

Assume that the data contained in the register of Example 7.1 contain alphanumeric data coded as shown in Fig. 4.9. Since that code is a 6-bit per character code we have four characters in the register.

55	53	61	30	Octal characters
)	*	/	H	Alphanumeric characters

Another representation for a bit pattern at an address of main memory is in binary scientific notation or binary floating point. For this notation, part of the data word is used for the exponent and part for the mantissa. A rule of thumb in computer design is that the exponent is about one-fourth of the word.

Example 7.4

Assume that the data contained in the register of Example 7.1 contain a floating-point number with the exponent being the left 6 bits and the mantissa being the right 18 bits. The leftmost bit of the word is the sign of the exponent which uses a 2's complement notation for a negative value. The exponent is an integer. The mantissa is a fraction, with bit 7 being the sign bit. The mantissa is also in 2's complement for a negative value.

$$N = 55 \quad 53 \quad 61 \quad 30 \qquad \text{Octal representation}$$

$$\text{Exponent} = (55)_8 \qquad \text{Mantissa} = (.536130)_8$$

$$(55)_8 = 101101 = (-19)_{10}$$

$$(.536130)_8 = 1{,}01011110001011000$$

$$= -.101\ 000\ 011\ 101\ 010\ 00 = (-.503520)_8$$

$$\approx (-.63210)_{10}$$

$$\therefore N \approx -.63210 \times 2^{-19} = -.63210 \times .000001907348$$

$$= -.120571 \times 10^{-5}$$

All the previous discussion has assumed that a data word and an instruction were the same size and occupied one location in main memory. Although this is true for many computing systems, it is not a requirement. Some computers have a *variable data-word length*. These machines permit words of different lengths to coexist in main memory. For this to be possible, the unit of addressing is not the word but a character. Thus, each character is addressable. A word comprises an integral number of characters. Sometimes a character is called a *byte*. This second name is usually used when the unit of addressing is a character and not a word. The word and the character or byte are essential data units for a computer. However, many machines have many more types of data structures; for example, the IBM 360 has eight different data structures.

7.1.6 Character-Oriented Computers

When the unit of addressing is the byte, that is, the address portion of the instruction specifies a location of a byte in memory, the computer is said to have a *variable field length* for its data. The length of the operand can be determined by three different methods. One method is implied by the operation to be performed by the instruction. Thus, a word may be specified as a fixed number of bytes, say 4. When an instruction which involves an operation to be performed on a *word* of data is used, the byte that is addressed is only the least significant 8 bits of the word and the control unit automatically fetches the next 3 bytes to obtain a full word. A second method is to specify the number of bytes on which the operation will be performed as part of the instruction. The third method is to keep fetching bytes from memory until a special character called a word mark is fetched. This delimits the word, and the control unit stops fetching additional bytes on which to operate.

Variable field length machines make efficient use of memory, especially when different precision data sizes may be appropriate for different applications or in using character strings. Character strings are used for compiling and editing. These types of computers have been used mostly for business applications.

7.1.7 Main Memory Subsystem

The function of the memory subsystem is to store programs, i.e., ordered sets of instructions, and data. In the course of executing the programs, new data are formed which must be stored for future reference in the memory. When information is transferred to the memory to be stored, we call this a *write operation*. When information is retrieved from the memory, it is called a *read operation*. Information is normally transferred from main memory one word at a time from or to a given address. To select the specific address and perform the transfer of information, the main memory subsystem has two special electronic registers associated with it. One register holds the specific address during the transfer of information and is called the *memory address register* or *MAR*. The other register holds the information which is to be written into the memory or receives the information which is read from the memory. It is called the *memory information register*, or *MIR*. Figure 7.7 shows a simplified block diagram of the main memory subsystem.

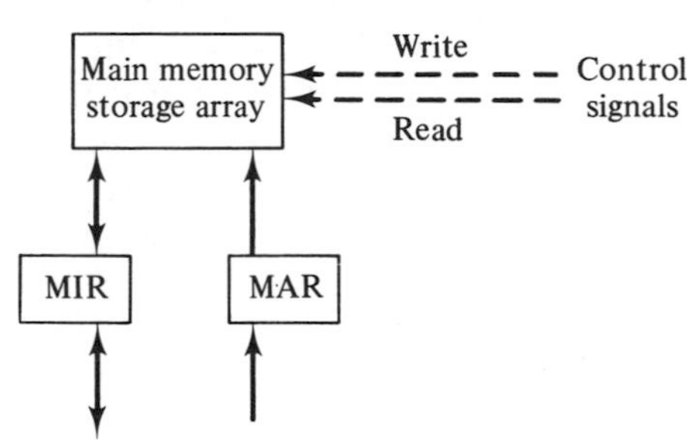

Figure 7.7 Simplified block diagram of main memory module.

The memory subsystem as shown in Fig. 7.7 is considered a *module* or *bank* of main memory storage. One method of increasing the effective speed of main memory is to utilize several such banks. This permits each bank to be accessed separately and permits an overlap on the accessing to memory. If we assign successive addresses to successive modules, we can achieve an increase in accessing speed. For example, suppose that we have two banks; one bank contains all the even addresses, while the other bank contains all the odd addresses. This structure is called an *interleaved* memory. The resolution of the accessed address among the number of banks is performed by the memory control unit (see Section 7.1.10).

7.1.8 Input/Output Device Characteristics

There are many different devices which are used to communicate with the computer. In general these devices require a form which is compatible with the binary nature of the computer when used for input and the alphanumeric nature of man when used for output.

To get information into the computer we might use two types of media, the punched card or the punched paper tape. Both of these media require a special device for preparation of the cards or tapes. The device usually has a typewriter-like machine for printing the information while it punches holes in the medium. The computer reads the presence or absence of holes. Both of these methods are performed prior to entry of the infromation into the computer. This is called an *off-line* method. It is possible to input information directly into the computer with an *on-line* device such as a typewriter. With this device, each time a key is depressed electronic signals corresponding to the alphanumeric code for the character are stored in a one-character location of a buffer register, and when the carriage return key is depressed the characters stored in the buffer register are read into the computer. All these input devices are character-oriented, and the computer usually reads one character at a time from the card, tape, or typewriter buffer register until it has exhausted all the characters from the medium. One card, the continuous paper tape, and the typewriter register are called a *unit record.*

The primary devices used for the output of information from the computer are the typewriter and the high-speed line printer. The computer fills the typewriter's buffer register with a line of characters coded as electrical binary signals. Then the typewriter empties the register one character at a time and prints each character of the line. The line printer prints an entire line at a time and therefore is a much faster output device. Table 7.1 shows the characteristics of these input and output devices. The words input and output are used so much in computing that they are usually abbreviated as I/O. In addition to the I/O devices some secondary storage devices such as magnetic drums and disks are also shown in Table 7.1.

***Table 7.1** I/O Device Data Rate Characteristics*

From	*To*	*Data rate*
Paper tape reader	Memory	60,000 ch./min†
Punched card reader	Memory	80,000 ch./min
Typewriter	Memory	Human typing speed
Memory	Typewriter	1800 ch./min
Memory	Line printer	200,000 ch./min
Memory	Cathode ray tube	500,000 ch./sec
Magnetic tape	Memory	300,000 ch./sec
Magnetic disk	Memory	1 million ch./sec
Magnetic drum	Memory	3 million ch./sec

†ch. = characters.

7.1.9 Channel Characteristics

Although some input and output devices have a high data rate, they are still much slower than the data rate of main memory. Thus, if an I/O operation is to be performed, the computer must slow down to the speed of the I/O device. This is very wasteful of expensive computer time. To overcome this slowdown requirement a small special-purpose computer called a *channel* is used between the I/O devices and the main memory. The channel has a limited number of registers used to store information which is passed between the main memory and the I/O devices. These registers are called a *buffer* memory. The buffer is filled at the slow data rate dictated by the I/O devices and empties at the high data rate of the main memory. The added cost of the channel is worth the price in the savings gained by not slowing down the entire computer to the speed of I/O devices. Another function performed by the channel is the selection of a specific I/O device requested by the control unit. The channel checks to see if the device is busy or free and whether an I/O operation can be performed on the device. The channel therefore performs two main functions: One is selection and status checking, and the other is a buffered data transfer to and from an I/O device and main memory. The latter function permits an *overlap* between the I/O operations and the computational operations of the computer.

The channel can be represented in a block diagram model as shown in Fig. 7.8. The *IOIB* is the input/output information buffer, which temporarily stores data on the way to and from memory. The *IOCR* is the input/output control register, which contains the address of the I/O device to or from which information is to be transferred as well as various other pieces of status in-

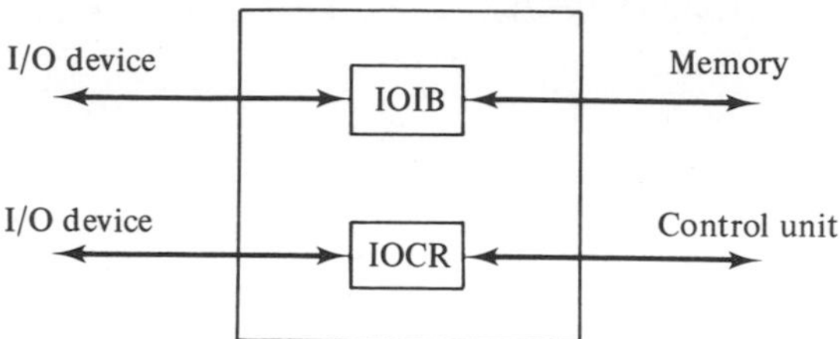

Figure 7.8 Block diagram of an I/O channel.

formation. A more detailed view and design of the channel will be presented in Chapter 10.

7.1.10 Memory Control Unit

In contemporary digital computing systems, several different subsystems can be competing for access to main memory. Several input/output devices can be transferring data to and from main memory via the channels to which they are connected. In addition, a program in process will be requesting instructions from main memory to be sent to the control unit and operands from main memory to the arithmetic unit. Generally, the latter two operations will not interfere with each other since one is requesting instructions during the *I* phase of machine operation, while the other is requesting operands during the *E* phase of operation. However, the I/O data transfers are performed at a rate controlled by the I/O devices and are asynchronous with the rest of the computer. Therefore, they may request a transfer to memory simultaneously with a request for an instruction or operand. Since the main memory of the computer can, generally, satisfy only one request at a time, provision must be made to "break the tie." The subsystem which performs this operation is called the *memory control unit*, which controls the memory bus. The word *bus* means a set of specific wires or connections over which information in the form of electrical signals passes. This subsystem performs a control or regulation of this information flow. In the case of simultaneous requests to the memory from the control unit or arithmetic unit and several channels, the memory control unit, by a priority method, allocates the order by which each requestor may access a word of main memory. Generally each channel has a preassigned priority, and the central processing unit has lowest priority. The memory control unit also determines which bank of main memory to access when a multiple bank memory system using interleaving is used.

7.1.11 A More Complex Block Diagram of a Digital Computer

In the previous parts of this section we developed the concept of address, instruction, and data. This leads to instruction sequencing and phases of operation of the computer. We investigated the arithmetic unit in greater depth and saw the need for registers to hold the operands during arithmetic operations. We also developed the idea of instruction decoding, the operation code and control signals which lead to the registers used to store the current instruction (*CIR*) and the address from which it was fetched (*CAR*). While considering the main memory, two additional registers were introduced which allow this subsystem to function. They are the memory information register (*MIR*) and the memory address register (*MAR*). Finally, while discussing input/output units, the concept of the channel was introduced in order to overlap I/O and computer operations within the computer. We can incorporate all these subsystems into a more complex block diagram of a typical digital computer. This is shown in Fig. 7.9.

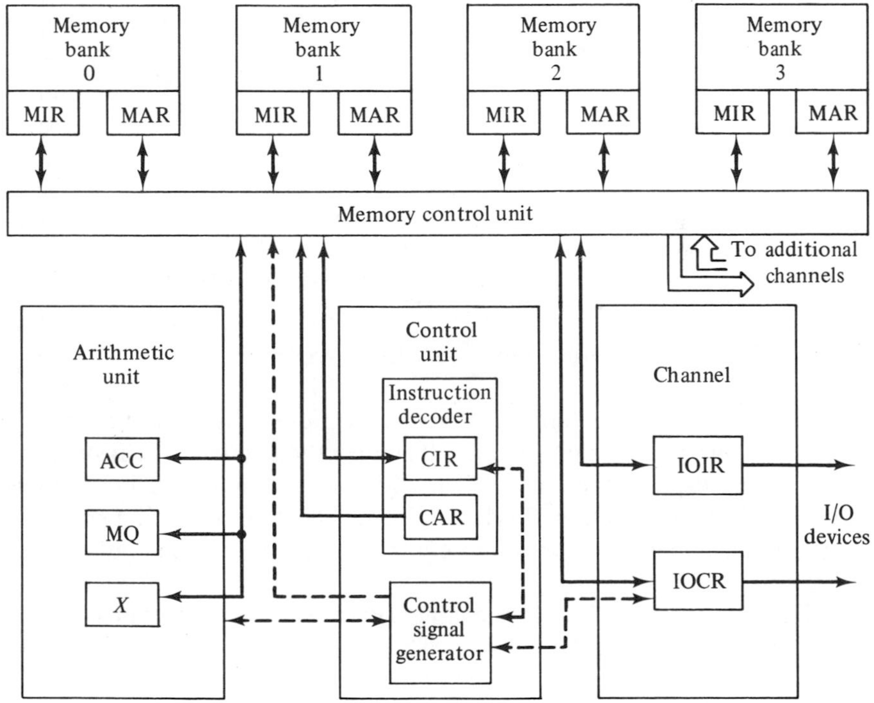

Figure 7.9 A more complex block diagram of a digital computer.

7.2 COMPARATIVE COMPUTER ARCHITECTURE

In this section we shall examine some different methods of designing the system architecture of a computer. We shall concentrate on the formats and types of instructions for various computers. This will imply certain hardware requirements in order to be able to implement these instruction types. We shall also introduce some fundamental concepts of machine language programming and the use of registers for various methods of gaining speed to obtain data from main memory. Once the basic principles of the format for instructions are understood, we shall examine the instruction formats and instruction repertoires of several computers designed over the past 25 years.

7.2.1 Instruction Address Formats

7.2.1.1 Four-Address Format

In one of the earliest digital computers, SEAC, each instruction comprised an operation code and four addresses. Figure 7.10 shows the format of such an instruction. Let us now program with this type of machine instruction. To do so we need a small programmer's manual for the machine.

Op-code	α	β	γ	δ

α = address of the first operand
β = address of the second operand
γ = address where the results of the operation are to be stored
δ = address of the next instruction

Figure 7.10 Four-address format.

Assume that the machine has a word length of 42 bits. It has a memory which contains 512 locations (words). The operation code of the instruction comprises 6 bits, and each of the four addresses is 9 bits. The instructions are coded in octal notation. Table 7.2 defines the operations and the octal code for each operation. An address in brackets means "the contents of" the address.

The problem that we wish to compute is to evaluate the formula

$$y(t) = v_{0y}t - \tfrac{1}{2}gt^2$$

where y = altitude of an artillery shell at any time
v_{0y} = the initial velocity in the vertical direction when the shell was fired

***Table 7.2** Definition of Operations for the Four-Address Machine*

Octal code	*Operation name*	*Definition*
77	STOP	The computer is ready but is sitting in neutral.
50	ADD	$[\alpha] + [\beta] \longrightarrow \gamma$; take the next instruction from address δ.
52	SUBTRACT	$[\alpha] - [\beta] \longrightarrow \gamma$ $[\delta]$ = next instruction
60	MULTIPLY	$[\alpha] \times [\beta] \longrightarrow \gamma$ $[\delta]$ = next instruction
63	DIVIDE	$[\alpha] / [\beta] \longrightarrow \gamma$ $[\delta]$ = next instruction
41	COMPARE	If $[\alpha] \geq [\beta]$, $[\gamma]$ = next instruction; otherwise $[\delta]$ = next instruction.
42	JUMP	$[\delta]$ = next instruction
30	READ	Read α number of words from the paper tape reader into memory. Place the first word at address β; $[\delta]$ = next instruction.
31	WRITE	Punch α number of words onto the paper tape punch from memory. The first word is located at address β; $[\delta]$ = next instruction.

g = the gravitational constant of 32 feet per second squared
t = time at which we compute $y(t)$

We shall compute the altitude 20 seconds after firing. Thus, $t = 20$ seconds. Suppose that $v_{0y} = 1600$ feet per second. First we must convert our decimal numbers into octal numbers:

$$(1600)_{10} = (3100)_8$$

$$(32)_{10} = (40)_8$$

$$(20)_{10} = (24)_8$$

$$(2)_{10} = (2)_8$$

We must store these data, which we shall use in our computation at addresses in our main memory. Our addresses will be given in octal code. The first address is $(000)_8$, and the last address is $(777)_8$. Let us store the numbers at the following addresses:

3100 is stored at address 020
40 is stored at address 021
24 is stored at address 022
2 is stored at address 023

We must also make provision for intermediate results used during the computation. Thus, we shall store these results as follows:

$v_{0y}\, t$ will be stored at 030
t^2 will be stored at 031
gt^2 will be stored at 032
$\frac{1}{2}gt^2$ will be stored at 033
$y(t) = v_0 t - \frac{1}{2}gt^2$ will be stored at 034

The program that we shall now write will have its first instruction stored at location 000. Figure 7.11 is a coding sheet for our problem. It shows the instructions used in the program and the data used by the instructions. It also shows where the intermediate results are stored. Remember that all the numbers shown on the coding sheet, except those in the column labeled "Remarks," are in octal coding.

Several observations can be made about this program. First, how does the program get into the computer to be executed? This is usually performed with the aid of another program called a *loader*. The loader is part of the operating system for the computer. We must, of course, first punch our code into paper tape, which is the input medium for this computer. Then we take the tape to the paper-tape reader and using procedures which are developed for this machine get the loader to load our program. If the loader is of the type called "a load and go," it will also start our program once it is loaded. If not, it will load the program beginning at a specified address, say 000, and when we push the start button, the computer will commence to fetch the first instruction from that prearranged address which will be the first instruction of our program. Another observation about the program in Fig. 7.11 is that it is wasteful of memory locations. It uses many more temporary storage locations, used for storing intermediate data, than is really necessary. In fact, only two temporary locations are needed for this program, and not five, which are used. Another observation is that we assumed integer numbers for the data constants which were supplied and for the computations. The computer is usually a fractional binary machine, and therefore the

Coding Sheet

Address	Instruction or data: Operation code	α	β	γ	δ	Remarks
000	60	020	022	030	001	$v_{0y} \times t \longrightarrow v_{0y}t$
001	60	022	022	031	002	$t \times t \longrightarrow t^2$
002	60	021	031	032	003	$g \times t^2 \longrightarrow gt^2$
003	63	032	023	033	004	$gt^2/2 \longrightarrow \frac{1}{2}gt^2$
004	52	030	033	034	005	$v_{0y}t - \frac{1}{2}gt^2 \longrightarrow y(t)$
005	31	001	034	000	006	Punch out data
006	77	000	000	000	000	STOP
007						
010						
011						
012						
013						
014						
015						
016						
017						
020	00	000	000	003	100	$v_{0y} = (1600)_{10}$
021	00	000	000	000	040	$g = (32)_{10}$
022	00	000	000	000	024	$t = (20)_{10}$
023	00	000	000	000	002	2
024						
025						
026						
027						
030						$v_{0y}t$
031						t^2
032						gt^2
033						$\frac{1}{2}gt^2$
034						$y(t) = v_0t - \frac{1}{2}gt^2$
035						
036						

Figure 7.11 Coding sheet.

programmer must keep track of the location of the binary point during the computations. This process is called *scaling*. This is a very bothersome process and prone to errors. It led to the development of floating-point arithmetic units for scientific-type computation. We have not included the concept of signed numbers in our example. Of course, this is an essential part of any computer. For our paper machine, let us assume a signed-magnitude notation with a 0 designating a positive number.

The last observation is about the four-address format of our instruction and the ability of the computer to perform instructions one at a time. Since the machine performs the instructions sequentially, why not code the algorithm in a sequential manner and therefore eliminate the need for the δ address? If we did that, we could use the extra 9 bits left in the instruction to be divided among the three remaining addresses and be able to utilize a larger memory without increasing the size of the memory word. The next section shows such an addressing format.

7.2.1.2 Three-Address Format

The three-address format has an operation code and three address fields. Figure 7.12 shows this format.

Op-code	α	β	γ

α = address of the first operand
β = address of the second operand
γ = address of the results

Figure 7.12 Three-address format.

If we use the machine word of the example machine in the previous subsection, then we have 42 bits in the word. Keeping the op-code the same at 6 bits leaves 36 bits for the three addresses. We have 12 bits for each address, and we can therefore address up to $2^{12} = 4096$ addresses of memory. Thus, by decreasing the number of addresses in our instruction format we can increase the amount of addressable memory we can use, if we keep the word length constant.

Without the fourth address, δ, how does the computer know where to obtain the next instruction? The computer will obtain the address of the next instruction by adding 1 to the contents of the current address register, *CAR*. Thus, except for special jump instructions, the next sequential instruction will be obtained automatically by the control unit after it completes the current instruction. Because of this automatic increase in the *CAR*, it is sometimes called the *current address counter* or *instruction counter*.

The instruction set of the three-address machine is similar to that of the four-address format of the previous subsection. Of course, the δ address is no longer used. The comparison instruction is somewhat different and is redefined as follows: If $[\alpha] \geq [\beta]$, $[\gamma]$ = next instruction; otherwise use the next sequential instruction. Thus, if $[\alpha] \geq [\beta]$, the current address register will be set to address γ; if $[\alpha] < [\beta]$, the CAR will be increased by 1; that is, we do not jump but take the next sequential instruction, $[CAR] + 1 \longrightarrow CAR$.

7.2.1.3 Two-Address Format

The two-address format usually eliminates both the γ and δ addresses. The format is

Op-code	α	β

α = address of the first operand

β = address of the second operand

The computer determines the address of the next instruction by the same method as in the three-address format. The results are now placed in the accumulator register of the arithmetic unit. The accumulator must therefore be addressable so that we can transfer data into and out of the accumulator.

If we again use a 42-bit word length with 6 bits used for the operation code, then we have 36 bits which can be used for the address portions. With a two-address format, there will be 18 bits per address, and we can address a memory of 2^{18} or 262,144 words. Some of the instructions for a two-address machine are shown in Table 7.3.

***Table 7.3** Two-Address Format Operation*

Op-code	*Operation name*	*Definition*
77	STOP	
50	ADD	$[\alpha] + [\beta] \longrightarrow ACC$
52	SUBTRACT	$[\alpha] - [\beta] \longrightarrow ACC$
60	MULTIPLY	$[\alpha] \times [\beta] \longrightarrow ACC$
63	DIVIDE	$[\alpha] / [\beta] \longrightarrow ACC$
41	COMPARE	If $[\alpha] \geq [ACC]$, $\beta \longrightarrow CAR$; otherwise $[CAR] + 1 \longrightarrow CAR$.
43	JUMP	$\alpha \longrightarrow CAR$
33	TRANSFER	$[ACC] \longrightarrow \alpha$, $\beta \longrightarrow CAR$
51	ADD and TRANSFER	$[\alpha] + [ACC] \longrightarrow \beta$
61	MULTIPLY and TRANSFER	$[\alpha] \times [ACC \longrightarrow] \beta$

7.2.1.4 One-Address Format

In the one-address format only the op-code and the α address remain. The accumulator performs a double function. It is usually the implied address of the second operand and the implied address of the results. The address of the next instruction is handled the same as the two- and three-address systems. If we again assume a 42-bit word length, with 6 bits reserved for the op-code portion of the instruction, we have 2^{36} possible addresses or 68,719,476,736 possible addressable words of memory. Figure 7.13 shows a comparison of the capability of the four address formats to address memory. We see that the one-address computer with a 42-bit word length and a 6-bit operation code can permit an addressing capability of over 68 billion words of memory. This is an extremely large addressing range and is not needed even by the largest of computers. Therefore, the word size can be reduced to, say, 30 bits. Again using 6 bits for the op-code permits an addressing range of $2^{24} = 16{,}777{,}216$ words of memory, which is still a very large memory size.

Number of addresses	*Bits in word length*	*Bits in op-code*	*Bits per address*	*Size of addressable memory*
Four	42	6	9	512
Three	42	6	12	4096
Two	42	6	18	262,144
One	42	6	36	68,719,476,736

Figure 7.13 Comparison of addressable memory size.

Although we may decrease the number of bits allocated to the address of an instruction, the one-address format computer generally requires many more instruction types to be able to accomplish its work. Typical one-address computers have over 200 op-codes as compared to 60 for a two-address format computer. Later in this chapter we shall discuss many types of instructions used in typical computers. For the present we shall limit our instructions to the list shown in Table 7.4.

Let us now assume a one-address-format digital computer with a word length of only 18 bits; 6 bits are used for the op-code, and 12 bits are used for the address. We shall again code the evaluation of

$$y(t) = v_{0y}t - \tfrac{1}{2}gt^2$$

using the same values as in the previous example. The program is shown in Fig. 7.14.

Table 7.4 *One-Address Format Operation*

Op-code	*Operation name*	*Definition*
77	STOP	Computer is in idle
50	ADD	$[\alpha] + [ACC] \longrightarrow ACC$
51	CLEAR and ADD	$0 \longrightarrow ACC$, $[\alpha] + [ACC] \longrightarrow ACC$
52	SUBTRACT	$[ACC] - [\alpha] \longrightarrow ACC$
60	MULTIPLY	$[ACC] \times [\alpha] \longrightarrow ACC$
63	DIVIDE	$[ACC] / [\alpha] \longrightarrow ACC$
41	COMPARE	If $[ACC] < 0$, $\alpha \longrightarrow CAR$; otherwise $[CAR] + 1 \longrightarrow CAR$.
43	JUMP	$\alpha \longrightarrow CAR$
33	TRANSFER	$[ACC] \longrightarrow \alpha$
53	REPLACE ADD	$[\alpha] + [ACC] \longrightarrow \alpha$
54	REPLACE SUBTRACT	$[ACC] - [\alpha] \longrightarrow \alpha$
61	REPLACE MULTIPLY	$[ACC] \times [\alpha] \longrightarrow \alpha$
76	NO OPERATION	Do nothing and take the next sequential instruction.

	Instruction or data		
Address	*Op-code*	α	*Remarks*
0000	51	0012	$0 \longrightarrow ACC; t \longrightarrow ACC$
0001	60	0012	$t^2 \longrightarrow ACC$
0002	60	0014	$gt^2 \longrightarrow ACC$
0003	63	0015	$\frac{1}{2}gt^2 \longrightarrow ACC$
0004	33	0011	$\frac{1}{2}gt^2 \longrightarrow 0011$
0005	51	0013	$0 \longrightarrow ACC; v_{0y} \longrightarrow ACC$
0006	60	0012	$v_{0y}t \longrightarrow ACC$
0007	54	0011	$v_{0y}t - \frac{1}{2}gt^2 \longrightarrow 0011$
0010	77	0000	STOP
0011	00	0000	y, answer also temporary
0012	00	0024	t, time
0013	00	3100	v_{0y} constant
0014	00	0040	g
0015	00	0002	2
0016			

Figure 7.14 One-address program.

7.2.1.5 One-Plus-One Address Format

The one-plus-one address format specifies the op-code; the address of one operand, α; and the address of the next instruction, δ. This type of address format was used on early computers, which used rotating magnetic drums for main memory. This system has the coding characteristics of a one-address-type instruction and uses the second address to allow minimum access coding for the drum. That is, the instructions do not have to be in sequential addresses around the drum and can be placed so that the next instruction will rotate under the read station of the drum just as the previous instruction is completed. Hence, instructions will be executed in an optimized time sequence without having to wait between instructions for the drum to rotate.

7.2.1.6 Symbolic Coding

One of the first concepts to be developed in programming a digital computer was the realization that the programming of an algorithm did not have to be performed in terms of absolute addresses of main memory but could be done in terms of symbolic names. After the programming in terms of symbols was completed, absolute machine addresses could be assigned to each symbol, and the machine language code would result. This can also be done for the operations. Instead of using the op-code to write an operation, a symbolic name can be used and the numeric code substituted after the program is completed. The symbolic names used during the programming are generally suggestive of the contents of the address and make the programming easier. Such symbolic names are called *mnemonics*. Let us use symbolic coding to program the problem

$$y(t) = v_{0y}t - \tfrac{1}{2}gt^2$$

for the one-address computer as described in Section 7.2.1.4. We shall assign mnemonics to the op-code and variables as follows:

Operation name	*Op-code*	*Mnemonic*
STOP	77	STP
ADD	50	ADD
CLEAR AND ADD	51	CLA
SUBTRACT	52	SUB
MULTIPLY	60	MUL
DIVIDE	63	DIV

Operation name	*Op-code*	*Mnemonic*
COMPARE	41	CMP
JUMP	43	JMP
TRANSFER	33	TRA
REPLACE ADD	53	RAD
REPLACE SUBTRACT	54	RSB
REPLACE MULTIPLY	61	RML
NO OPERATION	76	NOP

Variable name	*Mnemonic address*
Vertical height	Y
Time	T
Initial vertical velocity	VOY
Gravity	G
2	TWO

Symbolic Coding Sheet

Symbolic address	*Symbolic op-code*	*Symbolic* α	*Remarks*
A1	CLA	T	Add t to cleared ACC
A2	MUL	T	$t^2 \longrightarrow ACC$
A3	MUL	G	$gt^2 \longrightarrow ACC$
A4	DIV	TWO	$\frac{1}{2}gt^2 \longrightarrow ACC$
A5	TRA	Y	$\frac{1}{2}gt^2 \longrightarrow$ Y
A6	CLA	VOY	$v_{0y} \longrightarrow$ cleared ACC
A7	MUL	T	$v_{0y}t \longrightarrow ACC$
A8	RSB	Y	$v_{0y}t - \frac{1}{2}gt^2 \longrightarrow$ Y
A9	STOP		
Y			Answer
T		24	Time
VOY		3200	v_{0y} constant
G		40	g
TWO		2	2

Now we assign the op/codes numbers for the mnemonics and assign main memory addresses for the symbols:

Symbol	*Address*
A1	0000
A2	0001
A3	0002
A4	0003
A5	0004
A6	0005
A7	0006
A8	0007
A9	0010
Y	0011
T	0012
VOY	0013
G	0014
TWO	0015

The absolute machine language program now is

Absolute address	*Op-code*	α	*Remarks*
0000	51	0012	
0001	60	0012	
0002	60	0014	
0003	63	0015	
0004	33	0011	
0005	51	0013	
0006	60	0012	
0007	54	0011	
0010	77	0000	STOP
0011	00	0000	Y, answer
0012	00	0024	T, time
0013	00	3100	VOY
0014	00	0040	G, gravity
0015	00	0002	TWO, 2
0016			

The reader should compare this with the program in Section 7.2.1.4.

7.2.1.7 Use of Registers

In the two- and one-address computer we introduced the idea of the accumulator. This is a high-speed register within the arithmetic unit which was the implied third address (address of the results) and also the implied second address (address of an operand) for the one-address machine. In larger computers we may have several high-speed general-purpose registers which can each have multiple functions, among them being the accumulator. When we have such registers it is necessary to specify which one we wish to use with each instruction. Thus, instead of using an address of main memory, we can specify the location of an operand or the location to store the results by specifying one of the registers. Machines which utilize general-purpose registers must have instructions which permit operands to be loaded into the registers and also instructions to transfer the contents of registers into an address in memory.

Let us consider a computer which has 16 general-purpose registers. It can specify one address in memory but can also specify one of its registers. It may also specify register-to-register operations. Table 7.5 lists and defines some possible instruction types using mnemonics for the op-code, R_i and R_j to specify registers, and α to specify an address of main memory. Brackets around a register or α means "the contents of."

Since the flip-flop registers can be accessed much faster than main memory, the movement of an operand into a register from memory and its subsequent reuse, for instance, as a repeated multiplier, reduces the number of accesses to main memory and thereby increases the effective speed of the machine.

7.2.1.8 A Comparison of Instruction Address Formats

It is not possible to make a quantitative comparison of the various instruction address formats that have been described. A qualitative judgment can be made and is subject to opinion. The four-address format is redundant and has not been used since the very earliest machines. The three-address format does not provide for the use of faster accessing to registers. It essentially stores each result back to memory and does not use hardware to the best advantage. The push-down stack (see Section 7.2.2.10) forces a rigidity in the use of data since in order to obtain speed the stack must be used properly. The two-address format has sufficient speed but has a limited addressing capability unless a large-sized instruction word is used. It makes good use of both registers and memory. The single-address format provides sufficient bits for addressing purposes but requires many additional instructions to be executed in loading and storing registers. It is best for a highly

Table 7.5 *Some Instructions Using Registers*

Name	*Mnemonic*	*Definition*
Clear and add	CLA, R_i, α	$0 + [\alpha] \longrightarrow R_i$
Add	ADD, R_i, α	$[R_i] + [\alpha] \longrightarrow R_i$
Subtract	SUB, R_i, α	$[R_i] - [\alpha] \longrightarrow R_i$
Multiply	MUL, R_i, α	$[R_i] \times [\alpha] \longrightarrow R_i$
Divide	DIV, R_i, α	$[R_i] / [\alpha] \longrightarrow R_i$
Load	LOD, R_i, α	$[\alpha] \longrightarrow R_i$
Store	STO, R_i, α	$[R_i] \longrightarrow \alpha$
Replace multiply	RMP, R_i, α	$[R_i] \times [\alpha] \longrightarrow \alpha$
Replace divide	RDV, R_i, α	$[R_i] \div [\alpha] \longrightarrow \alpha$
Replace add	RAD, R_i, α	$[R_i] + [\alpha] \longrightarrow \alpha$
Replace subtract	RSB, R_i, α	$[R_i] - [\alpha] \longrightarrow \alpha$
Stop	STP, 0, 0	
Unconditional branch	UNB, 0, α	Address of next instruction $= \alpha$
Branch on zero	BRZ, R_i, α	If $[R_i] = 0$, NSI $= [\alpha]$.†
Branch on positive	BRP, R_i, α	If $[R_i] > 0$, NSI $= [\alpha]$.
Branch on negative	BRN, R_i, α	If $[R_i] < 0$, NSI $= [\alpha]$.
	Some register-to-register Instructions	
Sum registers	ADR, R_i, R_j	$[R_i] + [R_j] \longrightarrow R_i$
Subtract registers	SBR, R_i, R_j	$[R_i] - [R_j] \longrightarrow R_i$
Multiply registers	MLR, R_i, R_j	$[R_i] \times [R_j] \longrightarrow R_i$
Divide registers	DVR, R_i, R_j	$[R_i] / [R_j] \longrightarrow R_i$
Load register	LDR, R_i, R_j	$[R_j] \longrightarrow R_i$
Decrement register	DER, R_i, 0	$[R_i] - 1 \longrightarrow R_i$
Increment register	INR, R_i, 0	$[R_i] + 1 \longrightarrow R_i$

†NSI = next sequential instruction.

scientific parallel-flow structure using only a few registers as implicit operands and locations to store results. A good compromise is a two-address format where one of the addresses specifies one of a possible number of general-purpose central registers. This has the programming advantages of a two-address machine with the speed and addressing size of a single-address machine.

7.2.2 Methods of Addressing

In Section 7.2.1, we discussed several instruction address formats and showed some typical instructions for machines using those formats. The addresses of the operands were used to fetch an operand from the specified address of the main memory. In this subsection we shall discuss some variations in the method of determining an address from which to fetch an operand. These new methods increase the speed of the computer by decreasing the number of references to main memory and increasing the references to higher-speed registers.

7.2.2.1 Implied Addressing

Many types of instructions have the address of an operand implied in the operation and not given as an explicit address. For example, an add instruction in a single-address format implies the address of the second operand as the accumulator. The instruction states, "add the contents of an explicit address, α, to the contents of an implicit address, the accumulator, and place the results into the accumulator." A push-down stack machine implies the location of its operands as those at the top of the stack and therefore does not need an address (see Section 7.2.2.10).

7.2.2.2 Immediate Addressing

In this type of addressing we violate the concept that the data are found at an address. Instead the bit pattern in the address portion of an instruction is *not* an address but is the actual data. This type of addressing is used when the operand is short, for instance, to specify the number of places to shift a register in a shift instruction. With this type of addressing no memory access is needed to obtain an operand and hence the name *immediate addressing*.

7.2.2.3 Direct Addressing

This is the name that is given to the normal method of addressing first introduced as fundamental to the method of addressing. The address portion of the instruction specifies a location in memory from which an operand is fetched. When used to specify an address of an operand the address portion of the instruction is called the *direct address*.

7.2.2.4 Indirect Addressing

With *indirect addressing*, the direct address does not contain the operand but contains another address in which we can find the operand. Thus, it requires two fetches from the memory to obtain the operand. The first fetch obtains an address, and the second fetch obtains the data. Indirect addressing can go on for more than one level, as just described. Some machines allow an infinite level of indirect addressing. To utilize this type of addressing, a special bit in the instruction must be designated as the indirect address bit. When this bit is a 1, we interpret the contents of the address as another address. When this bit is a 0, we interpret the contents of the address as an operand, i.e., direct addressing. Indirect addressing is very useful for maintaining pointers to information. Thus, if we have a list of data which we wish to move to a new location in memory, we need only change the contents of the first-level indirect address to the new location address. Let us consider an example.

Example 7.5

Assume that we wish to add a list of numbers which may be in some consecutive locations in memory but that we do not know where they are or how many numbers are in the list. By a convention we shall place the location of the first number on the list as the contents of address 0509. The length of the list will be placed as the contents of address 0510. Thus, by accessing address 0509 indirectly we shall find the address of the first element on the list.

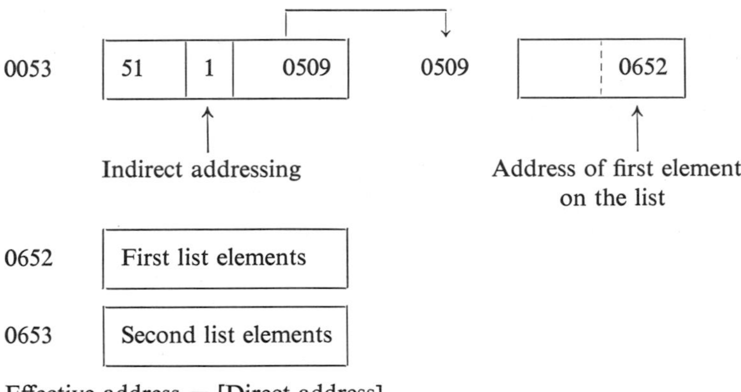

Effective address = [Direct address]

7.2.2.5 Base Addressing

This method of addressing utilizes a special register called a *base register*. It is located in the control unit of the computer. When base addressing is used, the direct address specified in the instruction is added to the contents of the base register to obtain the *effective address*. The effective address is used to obtain the operand from main memory. The direct address in the instruction is sometimes called the *displacement*, since it indicates the number of words displaced from the origin at which the operand is located. The base register contains the address which is the origin.

$$\text{Effective address} = \text{Direct address} + [\text{Base register}]$$

Base register addressing has many concepts embedded in it. One of these is the ability to address a large main memory utilizing only a small direct address in an instruction format. For example, assume that a computer has a 24-bit word length with a 12-bit direct address portion. Using direct addressing the computer can address $2^{12} = 4096$ words. To extend the addressing ability of this computer, a 24-bit base register is used with base addressing. By loading the base register with a proper number a memory with up to $2^{24} \approx$ 16 million locations can be addressed. Another concept embedded in base addressing is the ability to *relocate* the program anywhere in the main memory. If the programmer writes all the direct addresses relative to an origin of address 0, then we can displace the program by any amount by loading the base register with the new origin of the program, thereby shifting the program to another set of consecutive addresses in memory. Although this permits the instructions to be relocatable, it makes no provision for the data to be relocatable. If a second base register is used to address the data and all references to the data are written relative to an origin of address 0, we can shift a consecutive block of operands to any location in main memory by loading the origin of the block of addresses into the second base register. Since all references to the operands are written as if they started at address 0, the base register displaces the address referred to by the program to the proper location to which the operands have been relocated.

7.2.2.6 Indexing

The concept of *indexing* is similar to that of base addressing. The computer utilizes a special register called an *index register*. Some machines have many index registers, and therefore the particular one to be used with a given instruction must be specified as part of the instruction. The *effective address* is obtained by adding the contents of the specified index register to the direct address.

$$\text{Effective address} = [\text{Index register}] + \text{Direct address}$$

Although indexing looks like base addressing, it is used for a different purpose. The index register is usually loaded with a number which is modified during a *loop* of instructions. Thus, each time an instruction utilizing the index register is executed it will have a different effective address for the operands. The modification of the contents of the index register is done by special instructions which operate on the index registers. Some of these instructions are

1. Load index register with [α].
2. Load index register immediate
3. Store [index register] at α.
4. Decrement the index register by the amount α.

5. Increment the index register by the amount α.
6. Jump to address α if [index register] < 0.

The use of indexing can be explained best by an example.

Example 7.6

Assume that N numbers are to be added together to obtain the sum. The numbers are located in consecutive addresses of memory starting at address M. The number N is placed into an index register, and an add instruction is written using the contents of this index register to obtain the effective address for each number in the list. A special instruction which decrements the index register and tests for zero contents must also be written into this loop of instructions. Figure 7.15 shows the program for a one-address computer.

Symbolic coding sheet

	Instruction			
Symbolic address	*Op-code mnemonic*	*Index register*	α	*Remarks*
IN	LDII	5	N	N $\longrightarrow$ *I.R.*5
IN + 1	TDN	5	OUT	[*I.R.*5] − 1 $\longrightarrow$ *I.R.*5; if [*I.R.*5] < 0, $\longrightarrow$ OUT
	ADD	5	M	Eff. Add = M + [*I.R.*5]
	JUMP		IN + 1	IN + 1 $\longrightarrow$ *CAR*
OUT	TRA		ANS	[*ACC*] $\longrightarrow$ ANS

LDII = Load index register immediate $\alpha \longrightarrow$ I.R.

TDN = Tally down the index register

[*I.R.*i] − 1 $\longrightarrow$ *I.R.*i; if [*I.R.*i] < 0, $\alpha \longrightarrow$ *CAR*;
if [*I.R.*i] ≥ 0, [*CAR*] + 1 $\longrightarrow$ *CAR*

Figure 7.15 Example program using indexing.

The first instruction loads the specified index register with the number shown in the address portion of the instruction. This utilizes immediate addressing and saves a reference to memory. The second instruction is called a *tally-down* instruction. It is the special instruction which modifies the index register. First it subtracts a 1 from the index register, and then it checks to see if the contents of the index register is negative. If it is not negative, the contents of the current address register, *CAR*, is incremented by 1; i.e., the next sequential instruction is performed. If it is negative, the contents of the *CAR* is loaded with the address specified in the address portion of the tally-down instruction; i.e., a jump is performed. In the program shown we jump to the instruction located at address OUT. After the tally down we perform the addition using indexing. Each time through the loop we shall fetch a new number from the list of numbers to be added. Notice that we start with the last number of the list first, and on the final pass through the loop we shall add the first number on the list, i.e., the number at location M. Since the tally-down instruction will not see a negative number for N times, we must jump back and perform the loop again. After N times through the loop, the test on the index register will see a negative number, and we shall jump out of the loop to the transfer instruction at location OUT.

There are many different ways of performing the loop and test using indexing. The particular method is a function of the design of the instruction set which is used to modify the index register and the method of testing the contents of the index register. Some machines test for zero contents; others test for a positive contents. In the example we chose to test for a negative contents. The reader should consider the program needed to perform the algorithm of Example 7.6 using different tests. This is not so easy as it appears since you may go through the loop the wrong number of times or add the wrong effective address if you are not careful.

7.2.2.7 Self-relative Addressing

Self-relative addressing utilizes the contents of the current address register, *CAR*, as a base register. The effective address is the contents of the *CAR* added to the direct address of the instruction

$$\text{Effective address} = [CAR] + \text{Direct address}$$

This method of addressing specifies the address of operands as a displacement from the address of the current instruction which utilizes those operands. It permits the program and data to be relocated anywhere in memory without changing the addresses in the program. However, the programmer must know the relative position of the operands with respect to the instructions, and this position may not change if the program is to be relocatable.

7.2.2.8 Augmented Addressing

Augmented addressing is similar to base addressing, but instead of adding the contents of a special register to the direct address to obtain the effective address, this system concatenates the two addresses to obtain the effective address.

$$\text{Effective address} = [\text{Augmented address reg.}] \,||\, \text{Direct address}$$

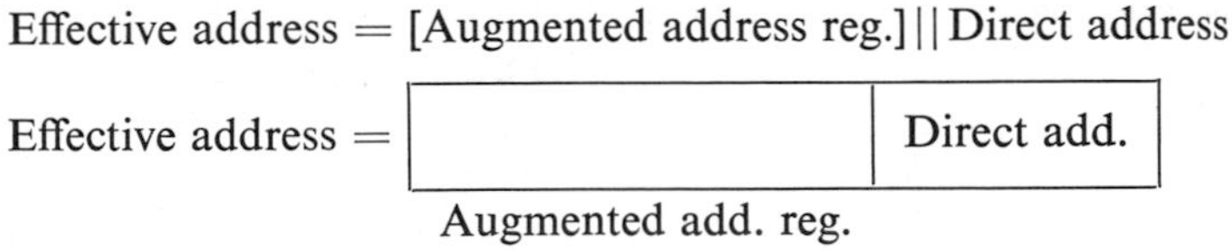

This also provides a means of relocating a block of instructions or data. The word in the augmented address register is sometimes called a *page* number, while the direct address is called the *word within a page*.

7.2.2.9 Block Addressing

Until now we have assumed that an address always refers to a single storage location. We could, however, use the address to specify the first word in a *block* of information. The blocks could be of variable length, with the length specified as part of the instruction. Or the instruction could indicate the length of the block by indicating the first and last address of the block of information. A third method of specifying the length of the block is to have a special character called an *end-of-block* character. The computer keeps fetching information starting with the first address until it reaches the end-of-block character. Of course the blocks could be of fixed length, and then only the address of the first word of the block needs to be specified. Block addressing is used extensively to address data stored in secondary storage devices such as magnetic drums, disks, and tapes. We can fetch blocks of data from such devices only in a *block access*. A whole block has to be read or recorded, even though only one word within the block is needed.

An important variation of block addressing is used in byte-addressed computers. In these machines each byte (character) is addressable. Each instruction specifies the address of the first byte of an operand. The length of the operand, i.e., the number of bytes fetched, is implied by the op-code in the instruction. Thus, if the operation is a fixed-point binary addition, 4 bytes are fetched even though the instruction specifies only the address of the first of the 4 bytes. If, on the other hand, the operation is a floating-point addition, 8 bytes are fetched, with the instruction again specifying only the address of the first of the 8 bytes. This type of addressing is sometimes called overaddressing since we have every character addressable even though we do not have to address each one to obtain them as words of data. The IBM Systems/360 and /370 use this type of addressing.

7.2.2.10 Push-down Stacks

A particular structure which utilizes implied addressing is known as the push-down stack. All operations have their implied operands as either the top element of the stack or the top two elements of the stack. The result of any operation is returned to the stack and becomes the top element. When an operand is required it is fetched from its memory address and placed on the top of the stack, thereby pushing down all other elements on the stack. Figure 7.16 shows a stack before and after the operation of addition is performed. The push-down stack is utilized in many types of compilers to generate a machine language equivalent program. The Burrough's Corpora-

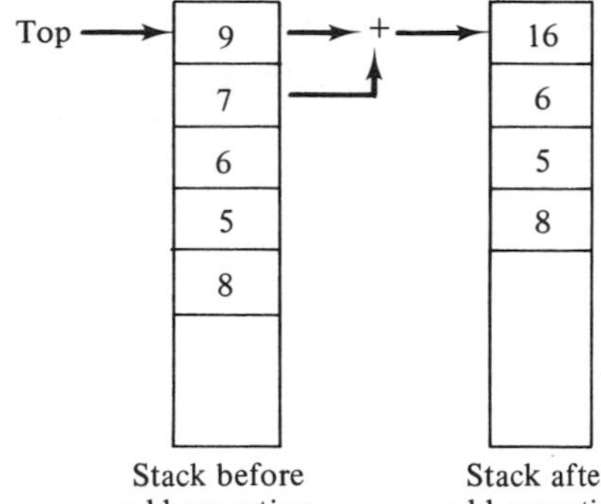

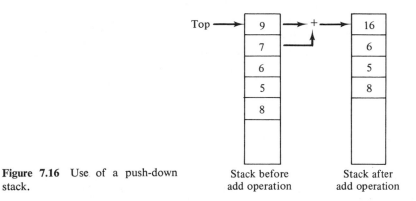

Figure 7.16 Use of a push-down stack.

tion B-5000 computer and several successor machines utilize the stack mechanism in hardware. These machines compile higher-level languages such as ALGOL and also execute the compiled program using the stack mechanism. Since the stack mechanism is constructed in hardware, it executes faster than a software implementation.

7.2.2.11 Combinations of the Previous Addressing Methods and Their Uses

Let us assume a digital computer which has 16 general-purpose registers. These registers can be used for any of the following purposes:

1. As accumulators.
2. To hold operands.
3. As index registers.
4. As base registers.

In addition, this computer has indirect addressing capabilities and self-relative addressing. The machine is a one-address word-oriented computer with a 32-bit word length. Its instruction format is

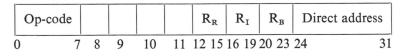

Op-code					R_R	R_I	R_B	Direct address
0 7	8	9	10	11	12 15	16 19	20 23	24 31

Bits 0–7 are the op-code.

Bits 8 and 9 specify indirect addressing, as shown below.

Bits 10 and 11 are used for self-relative addressing. If [bit 10] = 1, self-relative addressing applies:

$$\text{Eff. address} = [CAR] \pm \text{Direct address}$$

$$[\text{Bit } 11] = 0\ (+\ \text{Direct address})$$

$$[\text{Bit } 11] = 1\ (-\ \text{Direct address})$$

Bits 12–15 specify the results register R_R.
Bits 16–19 specify the index register R_I.
Bits 20–23 specify the base register R_B.
Bits 24–31 specify the direct address.

Bit 8	*Bit* 9	*Effective address*	*Comments*
0	0	$[R_B]$ + direct address	No indexing, no indirect
0	1	$[R_I] + [R_B]$ + direct address	Indexing, no indirect
1	0	$[[R_I] + [R_B]$ + direct address]	Indexing before indirect
1	1	$[R_I] + [[R_B]$ + direct address]	Indexing after indirect

Let us now consider an algorithm which will utilize the addressing capabilities of this machine. A list of numbers is to be added together. They are in consecutive memory locations; the first number is located at a location called LIST. There are N numbers in the list and N is stored at a location called NUMBER. The address LIST is found as the contents of a location called BEGIN. The symbolic coding of the problem is shown in Fig. 7.17.

Symbolic address	*Op-code mnemonic*	*Ind. add.*	*Results reg.*	*Index reg.*	*Base reg.*	*α*	*Remarks*
	LDI	00	0	5	0	NUMBER	N ⟶ R5
	TDN	00	0	5	0	* + 3	
	ADD	10	6	5	0	BEGIN	[[BEGIN] + [R5]] + [R6] ⟶ R6
	JMP	00	0		0	* − 2	
	TRA	00	6		0	ANS	[R6] ⟶ ANS
NUMBER						N	
BEGIN						LIST	
LIST							First number
LIST + 1							Second number

Figure 7.17 Segment of a program using indexing, indirect addressing, and self-relative addressing.

In this program we load register 5 with the number of elements on the list. Register 5 is used as an index register. The next instruction tallys down register 5 by 1 and tests to see if the contents is negative. If it is negative, the program jumps ahead three instructions and performs the transfer instruction. The symbol * is used to denote self-relative addressing. If register 5 is not negative, the next sequential instruction is performed. This is an addition of the numbers on the list to register 6, which is used as an accumulator. The addressing method is as follows: Index before indirect addressing. The effective address of the operand is

$$\text{Effective address} = [[\text{BEGIN}] + [\text{R5}]]$$

The next instruction jumps back two instructions using self-relative addressing, and therefore the loop is repeated unless the contents of register 5 is negative, in which case the transfer instruction is performed.

This program, although very efficiently written, does not execute with great efficiency since for each access to a number on the list an indirect address must be used and hence two references to main memory must be made for each operand. A more efficient method would be to use base addressing. This method is illustrated in Fig. 7.18. Notice that by placing the address list into register 4 and using it as a base register the number of references to main memory is reduced. However, this method requires the computation of an effective address by adding together the contents of registers 4 and 5.

Symbolic address	*Op-code mnemonic*	*Ind. add.*	*Results reg.*	*Index reg.*	*Base reg.*	*α*	*Remarks*
	LDR	00	5	0	0	NUMBER	N ⟶ R5
	LDR	00	4	0	0	BEGIN	LIST ⟶ R4
	TDN	00	0	5	0	∗ + 3	
	ADD	00	6	5	4	0	[LIST + [R5]] + [R6] ⟶ R6
	JMP	00	0	0	0	∗ − 2	
	TRA	00	6	0	0	ANS	[R6] ⟶ ANS
NUMBER						N	
BEGIN						LIST	

Figure 7.18 Segment of a program using base addressing.

7.2.2.12 Loops, Instruction Modification, and Pure Procedures

The concept of a loop of instructions is fundamental to the principles of digital computer programming. Without the ability of repeating an ordered sequence of instructions with a modification of the instruction stream on each repetition, it would be necessary to write an instruction for each step of a program. Programs would become unmanageable in length, and the amount of time required to write the program and prepare it for input to the computer would be as long as computing the algorithm using a desk calculator. The ability to modify the address portion of an instruction so as to obtain a different effective address for the operands within a loop can be performed by two methods. One method is instruction modification; the other is indexing. We discussed indexing in Section 7.2.2.6. Let us consider instruction modification.

Instruction modification uses an instruction as an operand during the execution of a loop of instructions and modifies the address portion of that instruction by adding or subtracting a number from it to produce a new direct address. The address portion of the instruction is changed each time the loop is executed. Figure 7.19 is an example of a program which modifies an instruction. It sums N numbers. The first number is located at LIST. The value of N is found at address NUMBER. The sum is placed into address ANS. The contents of location INDEX keeps track of the number of times the loop has been executed.

Symbolic address	*Op-code mnemonic*	*α*	*Remarks*	
MAIN	CLA	LIST	[LIST] ⟶ ACC	Add the list
	ADD	ANS	[LIST] + [ANS] ⟶ ACC	
	STO	ANS	[ACC] ⟶ ANS	
	CLA	ONE	[ONE] ⟶ ACC	Increment the index
	ADD	INDEX	[INDEX] + 1 ⟶ ACC	
	STO	INDEX	[ACC] ⟶ INDEX	
	CLA	NUMBER	[NUMBER] ⟶ ACC	Test
	SUB	INDEX	[NUMBER] − [INDEX] ⟶ ACC	
	JUZ	OUT	[ACC] : 0 ; = ⟶ OUT, ≠ next instruction	
	CLA	MAIN	[MAIN] ⟶ ACC	Modify the add instruction
	ADD	ONE	[MAIN] + 1 ⟶ ACC	
	STO	MAIN	[MAIN] + 1 ⟶ MAIN	
	UNC	MAIN	MAIN	
INDEX		0		
ONE		1		
ANS		0		
NUMBER		N		
OUT			Remainder of the program	

Definition of Operation Mnemonics:

CLA	Clear the ACC and add; [α] + [ACC] ⟶ ACC
ADD	ADD; [ACC] + [α] ⟶ ACC
STO	Store; [ACC] ⟶ α
SUB	Subtract; [ACC] − [α] ⟶ ACC
JUZ	Jump on zero; [ACC] : 0; ≐ ⟶ α; ≠ [CAR] + 1 ⟶ CAR
UNC	Unconditional Jump; [α] ⟶ CAR

Figure 7.19 Instruction modification.

We should make some comments about the loop concept and instruction modification. First, a straight-line program could be written which performs the same algorithm without a loop or any instruction modification. The straight-line program would execute faster since it would not have to perform any of the housekeeping instructions of the loop. However, it would require many more instructions if the loop were executed many times. Thus, looping saves on total written instructions and main storage at the expense of execution time. Also the number of instructions written is independent of the number of times the loop is executed. Only the parameter N needs to be changed in our example to change the number of times the loop is executed. One of the disadvantages of looping is that the programmer must be careful in writing the code. The tendency is to go through the loop either once too often or once too few times.

Let us now compare indexing and instruction modifications. Using indexing and the special tally-down instructions we can perform looping and save a considerable number of instructions as compared to instruction modification. Compare the program of Section 7.2.2.6 with that of Fig. 7.19. Also with instruction modification we must perform some initializing instructions which are not shown in Fig. 7.19. If the program is executed more than once, we must initialize location INDEX to 0 since it has been changed during a previous execution of the program. We must also initialize ANS to 0 and change the address portion of MAIN to the address of the first element on the new list. These extra instructions are needed if the program is to be executed again. By using an index register, we never changed the direct address but only computed an effective address. This saved many accesses to memory, and therefore the program utilizing an index register will execute faster than the one using instruction modification.

A pure procedure is a program in which none of the instructions are modified and which any program can therefore utilize as a subroutine. By loading an index register with the value of N and placing the address of the first element on a list in a base register it is possible to write a subroutine using indexing and base addressing which will add consecutive numbers on the list and not modify any of the instructions in the subroutine. The latter type of programming has become more important with the development of multiprogramming, multiprocessing, and time-sharing systems.

7.2.3 Instruction Formats and Repertoires

The instruction format defines the use of the bits within an instruction word. In Section 7.2.1 we devloped several different types of address formats. In Section 7.2.2 we developed modifications to the method of addressing. We must now combine these two sections to show how the modified methods of addressing are incorporated into the instruction format. Some typical instruction formats of computers will be discussed as well as some typical operations used in these machines.

The instruction word is divided into various parts called *fields*. We have discussed two of these fields, namely the op-code and the address or addresses. Two additional types of fields may be incorporated into an instruction. One is the flag field. This is specific bits which indicate whether certain addressing features are used with that particular instruction. An example would be a flag bit indicating whether indirect addressing applies to a given instruction. The other type of field is called the *tag* or *register* field. This is usually a group of bits which indicates which register is to be used for some purpose such as an index register for a particular instruction. The size of the tag field depends on the possible number of registers which could be used for some function. Thus, if there are seven registers which can be used as index registers, the number of bits in the tag field would be three.

Before the instruction format of a computer can be finalized a number of compromises must be made. Some of these are the methods of addressing to be used; the number of special registers to be used for base, index, and arithmetic operations; the size of addressable memory; the type of operations and the market for which the machine is primarily designed; the number of op-codes; and the compatability which may be needed because of previously designed machines.

7.2.3.1 Characters Per Word

One of the characteristics of digital systems which influences the size of the instruction word is the number of bits used to represent the alphanumeric character within the computer. For example, in second-generation machines, the alphanumeric character was usually 6 bits and there could be 64 possible alphanumeric characters. Because of this 6-bit structure, second-generation machines have word lengths which are generally multiples of 6 bits. Table 7.6 shows some second-generation machines and their word

Table 7.6 *Second-Generation Computer Word Length*

Bits (word length)	*Computer*
12	CDC 160A, PDP-8
18	PDP-9
24	SDS 900 series, G.E. 400 series
36	Univac 1108, IBM 7090, G.E. 600 series, PDP-10
48	Burroughs 5500 series, CDC 1604, Honeywell 1800
60	CDC 6600

sizes. These machines are generally coded in octal notation since 2 octal characters can represent all the 64 alphanumeric characters in their character set.

With the advent of the 7-character ASCII code, one would expect that word lengths would be multiples of 7 bits in length. However, 7 is not a multiple of 2 and is not easily represented by a subcode within the machine. Thus, for third-generation machines many manufacturers went to an 8-bit alphanumeric code and used 2 hexadecimal characters to represent the 8-bit alphanumeric character. These machines, therefore, code in hexadecimal and usually have word lengths which are multiples of 8 bits. Table 7.7 shows some of these machines.

Table 7.7 *Third-Generation Computer Word Length*

Bits (word length)	*Computer*
16	PDP-11, HP series, H-516, etc.
24	Datacraft
32	IBM 360/370 series, Xerox Sigma series
48	Burroughs 5700 series

An alternative to developing a hexadecimal machine to incorporate the newer expanded character set is to use 9 bits to represent an alphanumeric character and code the character using 3 octal characters. This allows an octal coded computer to accommodate an ASCII character set. Thus, a 36-bit word length computer could store 6 Hollerith code characters (6 bits per character) in a word, or 4 expanded code ASCII characters (9 bits per character) in a word.

To add to the confusion of the word length of a computer, there are machines which have variable word length. Some of these machines have a fixed size for their instruction and permit a variable word length for the data. Others permit a variable word length for both instructions and data. These machines are classified as character-oriented computers. Each character of the machine is addressable. For those computers with a fixed instruction size, the instruction is fetched from memory a character at a time, until the entire instruction is in the *CIR* (current instruction register). Then it is decoded. For computers with a variable instruction size, the operation code implies the length of the instruction. The operation code is the first character fetched from memory during the *I* cycle and it is decoded. This, then, determines the number of additional characters which must be fetched before the entire instruction is available.

The length of the data in a character-oriented machine can be determined by three different methods. One of these methods implies the length of the data in the operation code of the instruction. A second method uses a field within the instruction which specifies how many characters comprise the datum for that particular instruction. The third method uses special marks within the data word stored in memory. Characters are fetched starting with the beginning address specified in the instruction until one of the characters has the special mark attached to it (i.e., an extra bit set for that character). This special character or bit is called a *word mark* and delimits the lengths of the data words. Table 7.8 shows some variable word length computers and their characteristics.

Table 7.8 *Characteristics of Variable Word Length Computers*

IBM 1401	Variable instruction length and data length using word marks
IBM 1620	Fixed instruction length, variable data length
RCA 301	Fixed instruction length, variable data length specified by a field in instruction
Honeywell 200	Same as IBM 1401
IBM 7080	Fixed instruction length, variable data length using word marks
IBM S/360	Variable instruction length determined by op-code, variable data length specified in instruction

7.2.3.2 Arithmetic Operations and Word Length

Another characteristic of the digital computer which will influence the size of the word length is the type of arithmetic operations which will be performed within the machine. This is influenced by the market for which the computer is being designed. A large scientifically oriented computer will require high-speed floating-point arithmetic with a sufficient number of bits to ensure accuracy, while a computer oriented toward business applications may perform its floating-point instructions via software and have a high-speed decimal arithmetic unit which the scientific computer does not have. The amount of accuracy required and the algorithm to perform the arithmetic will influence the word length of the computer.

For a fixed word length binary arithmetic operation, the word length will depend on the maximum size of an integer which can be represented within the machine and, more importantly, the maximum size of memory which can be directly addressed. Thus, if we wish to be able to address over 1 million words of main memory, we shall need 20 or more bits in our word length since $2^{19} = 524{,}288$ but $2^{20} = 1{,}048{,}576$. However, if we are designing a small machine and only wish to address slightly over 64,000 words of storage, then a 16-bit word length will suffice, since $2^{16} = 65{,}536$.

If we choose a word length of 24 bits, because we can address over 16 million words of memory but now wish to consider a high-speed floating-point arithmetic capability, then we may need a larger word length, for with only 24 bits in a word we could represent a floating-point number with an 18-bit mantissa and a 6-bit exponent. This would yield a decimal equivalent of a mantissa of only $\pm 131{,}072$ and an exponent of $2^{\pm 32} \approx 10^{\pm 9{,}632}$. This gives an accuracy of about 5 decimal places and a magnitude of about 4 billion. Although this seems quite large, it is insufficient for computation in large scientific problems. Table 7.9 shows the significance and magnitude of floating-point numbers where the exponent is to the base 2 and both the exponent and mantissa bits include a sign. Thus, for a large scientific machine we may wish a word length of say 48 bits with a 36-bit mantissa and a 12-bit exponent. If our instruction word were only 24 bits long, it would be possible to place two instruction words in each location of memory and use them sequentially. This would speed up our machine since we would fetch instructions from memory only half as often. We could also pack two 24-bit fixed-point binary numbers in each 48-bit memory location. A floating-point number would take an entire memory word of 48 bits; an alternative method is to keep the word length of the computer at 24 bits but make the floating-point instructions fetch two words from memory. The first word would contain the exponent and least significant bits of the mantissa. The second word would contain the remaining bits of the mantissa. Which method is used would depend on how many floating-point operations were included in the program for which the machine was designed, i.e., its market.

Table 7.9 *Mantissa and Exponent*

Word length (bits)	*Mantissa (bits)*	*Exponent (bits)*	*Decimal places in mantissa*	*Exponent X* $(10^{\pm X})$
24	18	6	5.12	9.63
32	24	8	6.92	38.53
36	30	6	8.729	9.63
48	40	8	11.74	38.53
48	36	12	10.54	616.45
60	49	11	14.44	308.22
60	45	15	13.24	4929.58
64	58	8	17.16	38.53
64	48	16	14.15	9863.17

If the user were a business-oriented programming operation where a high-speed decimal capability were needed, then a word length of 24 bits could only contain 6 binary coded decimal numbers. This is certainly too small to represent today's business data. Thus, a 48-bit word length which could store 12 binary coded decimal numbers may be sufficient. This, again, may be too small. An alternative would be to use a variable word length computer which performed its decimal arithmetic on a serial-by-character basis with each character from the two operands being supplied from two strings of numbers in memory, one digit on each memory reference cycle. The sum of the two digits would be stored back into memory, and any carry produced would be retained in the arithmetic unit to be added into the next two digits. This process would continue until a word mark was detected in either of the operands. This method of decimal addition, although relatively slow, could supply all the word length that is necessary to add large numbers.

The word length to be chosen from an arithmetic data point of view is a trade-off between cost and speed. One can choose a large word length which yields both high cost in memory and a large parallel arithmetic unit for high speed but produces sufficient accuracy to perform the scientific algorithms. Another choice may be a slow variable word length computer which also produces sufficient accuracy to perform the problem but takes a long time with an inexpensive serial decimal arithmetic unit. Thus, high speed yields high costs.

7.2.3.3 Flags and Tags

The previous two parts of this section have dealt with the selection of a word length based on the characters to be represented in a word or the arithmetic operands which will be represented. In this section we wish to present the effect of the CPU architecture on the word length and the trade-offs that are possible.

The first consideration which must be given to the size of the instruction word length is the number of addresses which must be specified to achieve the results of the instruction. This need not be the same number of addresses for every instruction. Thus, a machine can have different instruction formats for different types of instructions. In fact, almost all computers have several different instruction formats. Let us consider some of the ways of specifying the address of the operands, and this will yield many different variations in instruction formats and word length.

Almost all the arithmetic instructions in a digital computer, whether they be fixed-point binary, floating-point binary, serial-decimal, or whatever, must specify three pieces of information. This information may be either explicit as part of the instruction format or implicit within the architecture of the computer or instruction. The three pieces of information are

1. Where to find the first operand.
2. Where to find the second operand.
3. Where to place the results of the arithmetic operation.

If we use a three-address instruction format, we can specify these three pieces of information as locations within memory. However, to achieve a sufficiently large address space will require a very large word length. We can reduce the need for a large word length by implying the location of the results as an accumulator in the arithmetic unit, and then we can specify only the two addresses of the operands. Still further, we can imply the accumulator as the location of one of the operands as well as the implied location of the results and then only explicitly write a single address of an operand. If we use an accumulator, we have placed several constraints on our architecture. One is that we must indeed have an accumulator in our arithmetic unit. A second is that since the accumulator can be only of a finite size, we must pick a word length for our arithmetic data which will fit within our accumulator.

There are times when a specified arithmetic data size will not meet our computing requirements and we would like to have a variable length data capability. We can achieve this by using a two-address format and specifying the length of the maximum operand within the instruction. This would be a tag field in the instruction whose format may be

Op-code	Tag	Address 1	Address 2

The instruction could read, "Add the two decimal numbers whose length is specified in the tag field and whose least significant digits are found starting at the addresses of the operands. Place the results back into the location of the first operand."

A typical one-address instruction would need only to specify the operation and the address of the operand. It could read

Op-code	Address

Add the binary number found at the specified location to the accumulator, leaving the result in the accumulator.

One of the ways to increase the size of addressable memory without having to increase the address portion of the instruction is to use base addressing. If we wish to be able to select when to add the base register to the direct address specified in the instruction to achieve an effective address, then it will be necessary to add a flag bit which will determine the condition. A 1 in the flag bit means use the base register, and a 0 means use the direct address.

$$\text{Effective address} = \text{Flag bit}\cdot[\text{Base register}] + \text{Direct address}$$

The instruction format could be

Op-code	1-Bit flag for base addressing	Direct address

Another useful concept which we developed was the use of index registers. In some computers there is only a single index register and it can be brought into play by the use of another flag bit. Thus, if we have both a base register and an index register, the instruction could be

Op-code	Base address flag bit	Index register flag bit	Direct address

The effective address could be computed as

$$\begin{aligned}\text{Effective address} = {} & (\text{Base address flag bit})\cdot[\text{Base register}] \\ & + (\text{Index register flag bit})\cdot[\text{Index register}] \\ & + \text{Direct address}\end{aligned}$$

In some machines there may be several registers which can be used for base registers and others which can be used as index registers. Some computers have a number of general-purpose registers which can be used as either base or index registers. Still other computers have general-purpose registers which can be used for base registers, index registers, or fixed-point binary accumulators. If there are multiple registers available, the instruction must specify which register is being used and also for what purpose. The location of the field bits in the word gives the purpose, while the bit code contained therein specifies the register.

Example 7.7

Consider a computer which has 16 general-purpose registers which can be used as accumulators, index registers, or base registers. An instruction which uses a single address can be in the form

Op-code	R	I	B	DA
0 7	8 11	12 15	16 19	20 31

Bits 0–7 specify the operation.

Bits 8–11 specify which of the 16 registers will be used as the accumulator.

Bits 12–15 specify which of the 15 registers is used as the index register. A 0 specifies no indexing.

Bits 16–19 specify which of the 16 registers is used as the base register.

Bits 20–31 comprise the direct address.

The effective address $= [R_I] + [R_B] + \text{DA} = \text{EA}$. If the operation is a fixed-point binary addition, then

$$[R_R] + [\text{EA}] \longrightarrow R_R$$

Once the concept of multiple registers used as accumulators is established in the design of a computer, a new type of instruction format becomes available. This is the register-to-register type of instruction. In this type, the operands are already loaded into the registers, and the results are stored in a register. This type of instruction has two different formats. One format is

Op-code	R_1	R_2

where R_1 is the register which contains the first operand and R_2 contains the second operand and is the register which stores the results. Another format is

Op-code	R_1	R_2	R_3

where R_1 is the register containing one operand, R_2 is the register containing the other operand, and R_3 is the register which stores the results.

The register-to-register instructions have a very fast execution time, since no data need to be fetched from memory. The high-speed accessing of the registers allows a high effective speed. Of course, results must eventually be stored into a memory address and new data must also be loaded into the registers.

7.2.3.4 Typical Instruction Formats

Let us now look at some typical instruction formats used in computers. The numbers above the fields indicate the bit positions in the word.

Standards Eastern Automatic Computer (SEAC) The SEAC, Standards Eastern Automatic Computer, was the first stored program electronic digital computer in operation in the United States. It was placed in operation in May 1950 at the Bureau of Standards in Washington, D.C. The machine was initially a four-address computer but was later modified to a three-address instruction format. It performed fractional fixed-point binary arithmetic with a data word of 45 bits including the sign bit. Its instruction format was

0	1 4	5			8	9 20	21 32	33 44
	Op-code	a	b	c	d	α	β	γ

Bit 0 is a special breakpoint bit. It is used for on-line debugging. If bit 0 = 1, the computer stops after executing the instruction if the breakpoint switch located on the control console is set on.

Bits 1–4 are the op-code. The machine codes in hexadecimal characters.

Bits 5–8 are the instruction modification flag bits used for relative addressing. They are labeled a, b, c and d, respectively.

Bits 9–20 are the 12-bit direct address of the first operand, designated as α.

Bits 21–32 are the 12-bit direct address of the second operand, designated as β.

Bits 33–44 are the 12-bit direct address of the result, designated as γ.

The machine has two special registers for address modification. One is called the *instruction counter*, which is the current address register and is 12 bits long. The second register is the *base counter*, which is used for indexing and base addressing. It is also 12 bits long. The machine could therefore produce an effective address which was only 12 bits in length and hence the maximum size of its main memory was 2^{12} or 4096 words.

Effective addressing in the machine is based on the flag bits 5–8. If bit 5, the a bit, is set to 1, then the α direct address would be modified to an effective address; if bit 6, the b bit, is set to 1, then the β direct address is modified; and if the c bit, bit 7, is set to a 1, then the γ address is modified to an effective address. Bit 8, the d bit, is used to specify which of the two counters is used to form the effective address. If $d = 1$, then the base counter would be added to the direct address to produce an effective address, while if d = 0, the instruction counter is used to produce an effective address. Thus, if d = 1, we have base addressing (or indexing), while if d = 0, we have self-relative addressing.

Let α', β', and γ' be the effective addresses. Let C_b represent the base counter and C_I represent the instruction counter. The effective addresses may be written as

$$\alpha' = \alpha + \mathrm{a}\cdot[\mathrm{d}\cdot C_b + \bar{\mathrm{d}}\cdot C_I]$$

$$\beta' = \beta + \mathrm{b}\cdot[\mathrm{d}\cdot C_b + \bar{\mathrm{d}}\cdot C_I]$$

$$\gamma' = \gamma + \mathrm{c}\cdot[\mathrm{d}\cdot C_b + \bar{\mathrm{d}}\cdot C_I]$$

where · signifies the logical product, + represents the logical sum, + means plus, and [ ] means "the contents of." The effective addresses may be relative to either the base or instruction counters but not both.

The first electronic stored program digital computer had the capability of base or indexing or self-relative addressing. However, because of its three-address structure and limited-size registers, it could address only 4096 words of main memory.

IBM 7090 The IBM 7090 is a one-address binary transistorized parallel second-generation scientific computer with a 36-bit word length. It was first installed in June 1960 and was designed as a transistorized successor to the IBM 709 computer, which was first installed in September 1958. The 709 was a successor to the IBM 704 computer, first installed in December 1955. These machines are all one-address computers. The direct address is 15 bits, and thus a memory of 32,768 words can be addressed.

The IBM 7090 has an accumulator, a multiplier-quotient register, and three index registers. It has fixed- and floating-point arithmetic hardware. There are two basic instruction formats in this machine. One format is

S 1 11	12	13	14 17	18 20	21 35
Op-code	F	F		I	DA

Bits S–11 are the op-code portion of the instruction.
Bits 12 and 13 are the flag bits used to designate indirect addressing.
Bits 14–17 are not used in this instruction format.
Bits 18–20 designate one of the three index registers, called 1, 2, and 4.
Bits 21–35 are the direct address.

Indexing is performed by subtracting the contents of an index register from the direct address to obtain the effective address:

$$\text{Effective address} = \text{Direct address} - [\text{Index register}]$$

The machine provides for indirect addressing if the flag bits 12 and 13 are both set to 1. The machine also permits indexing before and after indirect addressing.

$$\text{Indirect address} = \text{Direct address} - [\text{Index register i}]$$
$$\text{Effective address} = [\text{Indirect address}]_{21\text{-}35} - [\text{Index register j}]$$

That is, if the instruction specified indirect addressing, the indirect address is computed using the index register tag bits of the instruction which specifies index register i. The contents of the indirect address is inspected, and if the tag bits specify indexing to index register j, the effective address is computed by using bits 21–35 of the indirect address minus the contents of index register j.

Another instruction format used by the 7090 for comparing the contents of an index register with the contents of a memory address or to modify the contents of an index register is

S 1 2	3 17	18 19 20	21 35
Op-code	Decrement	I	DA

Bits S, 1, and 2 are the op-code, also called the prefix.
Bits 3–17 are the decrement field.
Bits 18–20 are a tag field which specifies one of the three index registers.
Bits 21–35 are the direct address, which is nonindexable.

The instruction transfers control to the direct address based on a comparison of a specified index register and the contents of the decrement field.

UNIVAC-1107 As a basis of comparison, let us look at another second-generation computer which uses a one-address format. The UNIVAC-1107 is a binary, transistorized, parallel computer which also has a 36-bit word length. It has fixed- and floating-point arithmetic. The fixed point may be fractional or integer. Its basic instruction format specifies a results register and the address of an operand. It provides for 16 results registers, 15 index registers, and indirect addressing. It uses a 16-bit direct address field and can address a maximum memory of 65,536 words. All its registers are contained in a high-speed thin-film magnetic memory which constitutes addresses $(000000 - 000177)_8$ of main memory.

The basic instruction format is

35 30	29 26	25 22	21 18	17	16	15 0
Op-code	j	a	b	h	i	u

Bits 35–30 are the op-code. The machine is coded in octal.

Bits 29–26, j, are the minor function code. In some operations these bits specify subcodes and therefore a 10-bit op-code is possible. For arithmetic operations, this field specifies the length and location of that part of the operand which is transferred to the arithmetic unit.

Bits 25–22, a, designate one of the 16 possible result registers which are used as accumulators for arithmetic operations. For some instructions, they designate 1 of the 15 possible index registers. For I/O operations, they designate the channel.

Bits 21–18, b, designate which of the 15 index registers is used to determine the effective address.

Bit 17, h, designates incrementation of the index register specified in b. If h $= 0$, no incrementation. If h $= 1$, $[B]_Q + [B]_\Delta \longrightarrow B_Q$, where B specifies the index register; Q specifies the modifier portion of the index register, bits 17–0; and Δ specifies the increment portion of an index register used to charge B_Q, bits 35–18.

Bit 16, i, designates indirect addressing. If i $= 0$, no indirect addressing. i $= 1$, effective address $=$ [direct address $+ [B]_Q$]. Indirect addressing can be to any depth since the right 22 bits of the indirect address are inspected if the instruction specifies indirect addressing. If the indirect address bit is a 1 at bit 16 of the indirect address, the indirect addressing is repeated, etc.

Bits 15–0, u, are the direct address. If j $= (16)_8$ or $(17)_8$, immediate addressing applies.

IBM 1401 The fourth computer whose instruction format we shall consider is the IBM 1401. This is a second-generation transistorized business-oriented computer which has both a variable instruction length and a variable operand length. The computer is character string oriented with a limited arithmetic capability. Because of the wide difference between this machine and the others we have been considering, we shall digress and explain more than just the instruction format for the machine.

A character in this machine was 8 bits in length. Six of the bits coded the 64 possible alphameric characters. The seventh bit was used for odd parity, and the eighth bit was a special bit called the *word mark bit*. This word mark bit is used to delimit the length of the instructions and the data. Each character of memory is addressable, with the maximum size of memory being 16,000 characters. The CPU has three 3-character registers. One is the pointer to the first character in the next sequential instruction and is called the *I* register. Another is a pointer to the address of the low-order character of an operand and is called the *A* address register, while the third is a pointer to the low-order address of a second operand and is called the *B* address register. The CPU has no accumulator, and all instructions are performed character by character and returned to a location in memory. The machine is basically a two-address computer using the *A* and *B* address registers to specify the starting address of operands in memory. Results are usually stored back into the memory locations of one of the operands. Besides specifying the starting address of the operands, the character in the *A* address register also determines the use of three index registers, which are located in specified memory addresses.

Data are stored as character strings with the low-order character stored in the highest address of core. The length of the data word is determined by the setting of the word mark bit in the high-order character.

Example 7.8

Store the number 942815 in the 1401 with the low-order character being stored at location 457.

3	4	$\bar{9}$	4	2	8	1	5	*Contents*
450	451	452	453	454	455	456	457	*Address*

The line above the 9 indicates that the word mark bit is set.

The instructions are also character strings which may be either 1, 2, 4, 5, 7, or 8 characters in length. The *I* register points to the address of the first character in the instruction, which is the op-code. The instructions are stored with the op-code in a low-order address and the remainder of the instruction characters in sequential higher addresses. The op-code character has its word mark bit set.

Example 7.9

$\bar{M}$	2	3	4	8	7	6	*Contents*
137	138	139	140	141	142	143	*Location*

This instruction shows a 7-character-length instruction which moves characters starting at address 234 to address 876. The character located at 233 would be moved to address 875 and so on until one of the characters of an operand had its word mark bit set. This would end the instruction and the movement of data within the memory.

The instruction formats for the six possible different instructions are shown in Table 7.10, where I_1 is always the op-code and contains a word mark bit, and I_2 is called a *d*-modifier character used much as a minor function field is used. $I_{3,4,5}$ specifies the address of the first operand, or a branch address, and $I_{6,7,8}$ specifies the address of the second operand.

The essence of this machine structure is that the CPU continues to fetch either instruction characters or data characters from sequential addresses until it detects a word mark bit. This special character delimits the words and instructions of the machine. Operations are performed on a serial-by-character basis, with the results replacing the characters of one of the operands in memory.

Table 7.10 *Instruction Formats for IBM 1401*

Length (char.)	*Location M_i*			
1	I_1			
2	I_1	I_2		
4	I_1	$I_{3,4,5}$		
5	I_1	$I_{3,4,5}$	I_2	
7	I_1	$I_{3,4,5}$	$I_{6,7,8}$	
8	I_1	$I_{3,4,5}$	$I_{6,7,8}$	I_2

Control Data Corporation 6600 The Control Data Corporation's CDC 6600 is a large computer which was first delivered in 1964. The machine is a transistorized, multiple-register computer with 10 peripheral computers used to handle the I/O for the large central computer. The central processing unit (CPU) communicates with a central magnetic core main memory of 131,072 words, as do the 10 peripheral processor units (PPU). Each PPU also has a separate memory of 4096 words. The word length of main core is 60 bits, while the word length of the PPU memory is only 12 bits. The machine has 12 I/O channels, and only the PPUs can gain access to them. All I/O activity is vested in the PPUs, which are in effect separate small computers. The CPU is a very fast computing machine with no I/O capability. When the CPU requires an I/O operation it places the request in main memory and signals one of the PPUs. The PPU accepts the signal and commences to perform the I/O operation; the CPU is multiprogrammed and proceeds with another computation. The PPU is similar to a data channel. The CPU has 10 independent processing units which can be overlapped in performing operations. Figure 7.20 shows a block diagram of the CDC 6600.

One of the PPUs is assigned the job of containing the supervisor for the entire system. It allocates jobs, memory, I/O requests, and resources to the PPUs. The machine operates in the multiprogrammed mode. The basic idea is to keep the high-speed CPU busy with centralized computing while the slower PPUs do the I/O work. The CPU attempts to overlap its operations to its 10 independent units whenever possible. Although many instructions require from 0.3 to 0.4 microseconds to complete, the overlapping of instructions can yield an effective time of 100 nanoseconds per instruction. To achieve this high effective instruction rate, the main memory of the 6600 may be interleaved up to 32 ways. Figure 7.21 shows the registers and instruction flow within the CPU.

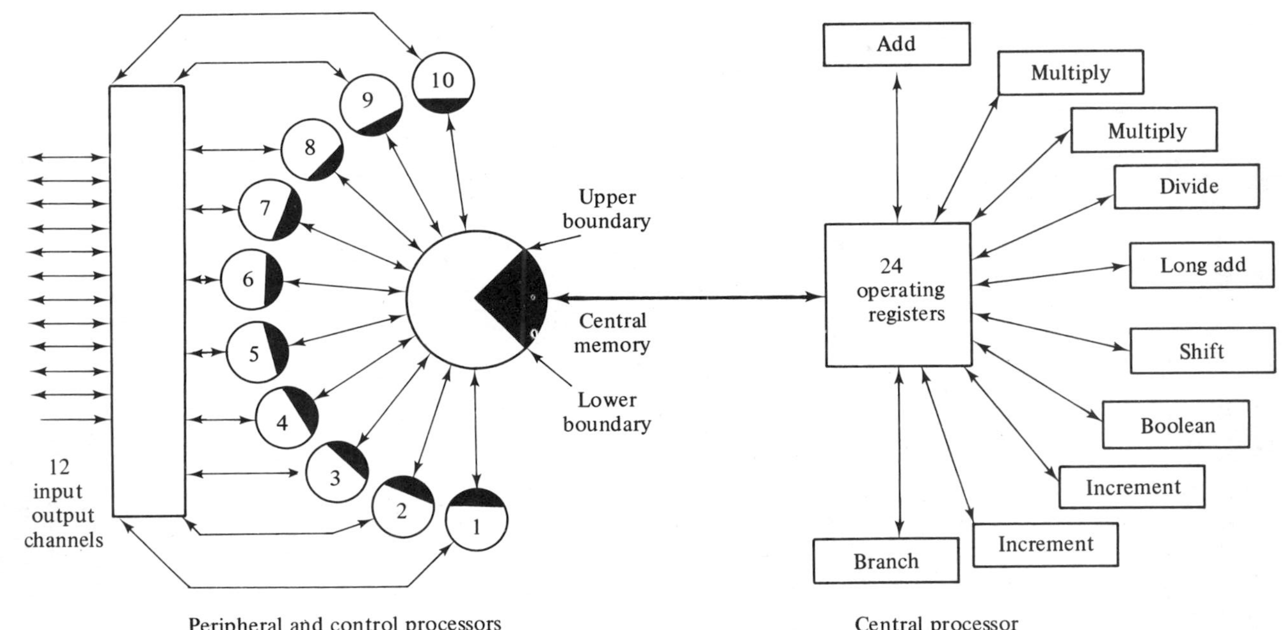

Figure 7.20 Block diagram of CDC 6600. (Courtesy of Control Data Corporation.)

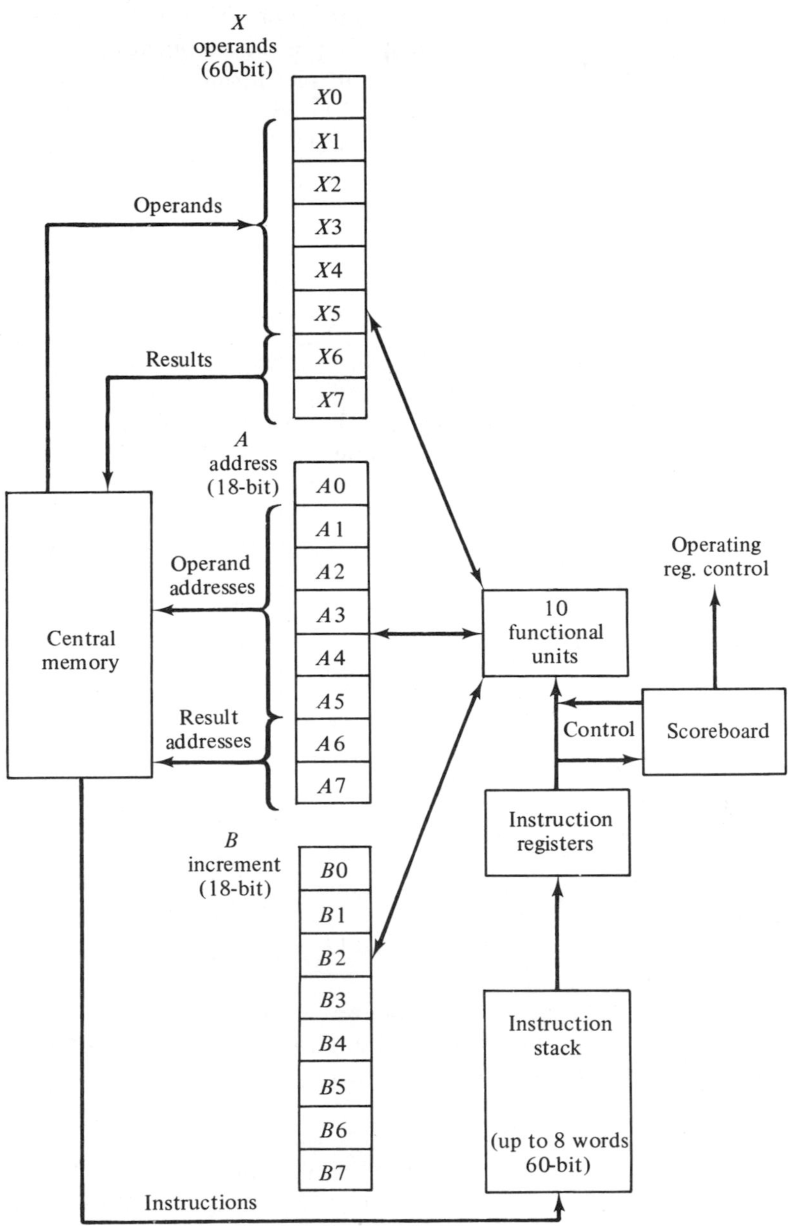

Figure 7.21 Central processor operating registers. (Courtesy of Control Data Corporation.)

The CPU can perform integer arithmetic on either 60-bit words or 18-bit words. Floating-point data have 11 bits in the exponent and 49 bits for the mantissa. The instruction format is of two forms. One form is 15 bits long and can be packed up to 4 to the 60-bit memory word. It is a register-to-register format and makes use of the eight 60-bit X operand registers (see Fig. 7.21).

0 5	6 8	9 11	12 14
Op-code	X_i	X_j	X_k

Bits 0–5 specify the op-code.
Bits 6–8 specify the results register, X_i.
Bits 9–11 specify the register containing the first operand, X_j.
Bits 12–14 specify the register containing the second operand, X_k.

The second instruction format is 30 bits in length and can be packed 2 to a memory word.

0 5	6 8	9 11	12 29
Op-code	i	j	k

Bits 0–5 specify the operation.
Bits 6–8 specify one of the 8 A address registers (18 bits) or one of the 8 B index registers (18) bits.
Bits 9–11 specify one of the 8 A address registers or one of the 8 B index registers.
Bits 12–29 specify an 18-bit direct address.

A word of memory can also contain one 30-bit instruction and two 15-bit instructions. Thus, the CDC 6600 makes very effective use of a large word length to pack many instructions, up to 4, into a word of memory.

IBM System 360/370 In 1964, the International Business Machine Corporation (IBM) announced a series of program-compatible digital computers utilizing hybrid integrated circuits which had a performance spread of 50 to 1. The IBM System/360 is the name given to this third-generation line of machines. These machines can provide fixed-point operations, floating-point operations, and variable word length operands with serial-by-character decimal operations. To some extent this family of machines combines aspects of both the IBM 7090 and IBM 1401 computers, which were the predecessors of the System/360. To enhance a conversion to the System/360 from predecessor IBM machines, the company provided through *emulation* the capability to execute programs written for the IBM 704, 709, 1401, 1410, 1440, 1460, 1620, 7010, 7040, 7044, 7070, 7074, 7080, 7090, and 7094.

The System/360 can address an 8-bit group called a byte. The byte permits up to 256 alphanumeric characters or 2 BCD characters to be represented. Two bytes is called a half-word, while 4 bytes constitute a word and 8 bytes is called a double word. There are eight basic data structures for these machines, and they are shown in Fig. 7.22.

To handle these various data formats, the CPU of the System/360 has 16 general-purpose 32-bit registers used for addressing, indexing, and as accumulators. Four 64-bit floating-point accumulators are available if the floating-point option is used.

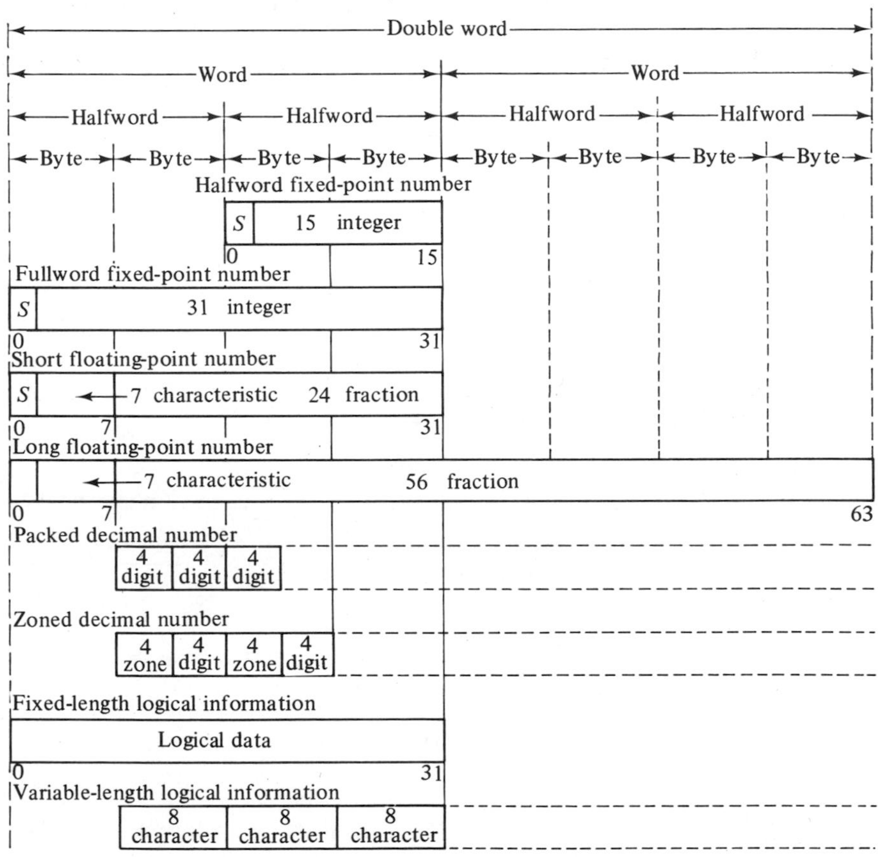

Figure 7.22 Data structures for System/360. (Reprinted by permission from the *IBM Systems Journal.* Copyright © 1964 by International Business Machines Corporation.)

The instruction formats of the System/360 have one, two, or three half-words depending on the data structure and the number of main storage addresses needed for an operation. If no main storage address is required for a particular instruction, one half-word is sufficient and the instruction uses a register-to-register format. A two-half-word instruction specifies one main memory address; a three-half-word instruction specifies two main memory addresses. There are five basic instruction formats, as shown in Fig. 7.23.

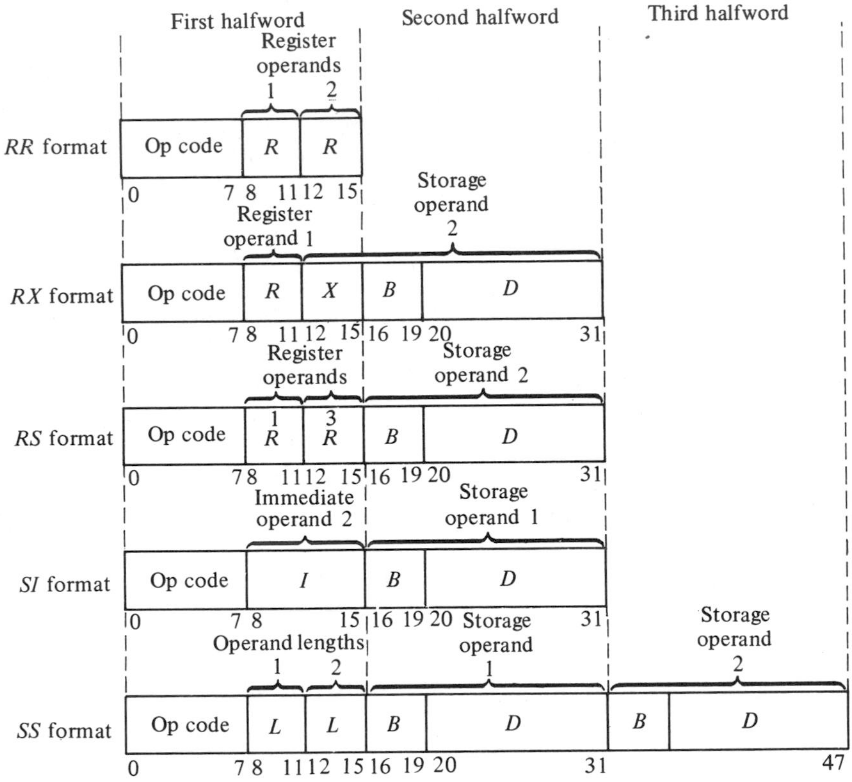

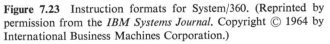

Figure 7.23 Instruction formats for System/360. (Reprinted by permission from the *IBM Systems Journal*. Copyright © 1964 by International Business Machines Corporation.)

The format mnemonics have the following meanings:

1. RR denotes a register-to-register operation.
2. RX denotes a register and indexed main storage address operation.
3. RS denotes registers and main storage address operation.
4. SI denotes a main storage address and an immediate operand operation.
5. SS denotes a main storage-to-main storage operation.

The first half-word of all instructions consists of two basic parts. The first byte contains the op-code. The length and format of the instruction is specified by the first 2 bits of the op-code byte. The second byte of the first half-word is used as either two 4-bit fields or one 8-bit field. This second byte is specified from the following types:

1. Four bits which designate a register which contains an operand.
2. Four bits which designate a general register which is used as an index register (X).
3. Four bits mask (M).
4. Four bits which designate a length (L).
5. Eight bits which designate a length.
6. Eight bits which are immediate date (I).

If the second and third half-words are present, they specify a base register by 4 bits (B) and a 12-bit address displacement (D).

The System/360 uses 24 bits to define a byte location in main memory. Thus, main memory can directly address 2^{24} (about 16 million) bytes. The effective address is computed by the use of base addressing as

$$\text{Effective address} = [\text{B}] + \text{D}$$

If the instruction is of the form RX, then the effective address is computed as

$$\text{Effective address} = [\text{X}] + [\text{B}] + \text{D}$$

7.2.3.5 Instruction Repertoire

Just as the instruction format varies among the computers of different manufacturers, so do the types of instructions which the computers execute. The number of instructions is, in general, a function of the word length of the computer. Computers with large word lengths generally execute more instructions per second and have a larger instruction repertoire than computers with small word lengths. The repertoire of contemporary computers is usually between 64 and 256 instructions. Of this number only about 8 are essential instructions. With this essential set of instructions any computable algorithm can be programmed on the computer. The remainder of the instructions provided by the manufacturer, although not essential, yield a higher effective speed than if all algorithms were done with only the essential set. That is, a single instruction may accomplish the same function as several instructions of the essential set. For example, a fixed-point binary multiplication need not be provided with a computer since it can be accomplished by programming using binary addition and shift instructions (i.e., add and shift algorithm). However, most computers provide fixed-point multiplication instructions since they are much faster than the software algorithm. Another example is hardware floating-point operations. Early computers, such as SEAC, and many contemporary minicomputers do not have floating-point hardware. Their floating-point operations are performed with software subroutines using fixed-point arithmetic and shifting instructions.

Some of the instructions which are part of the essential set for a computer are

1. An input instruction—necessary in order to load programs and data into the memory of the machine.
2. An output instruction—used to obtain results from the memory and to debug the programs.
3. Fixed-point binary addition.
4. Negation—this instruction and the addition instruction can be used to perform subtraction.
5. A shift instruction—used with the above two instructions to provide multiply and divide algorithms.
6. Logical operation, i.e., exclusive—OR—used to decode various data formats.
7. Conditional branch—needed to provide for loops and jumps in algorithms.
8. Store instruction—used to place temporary results into memory.

The repertoire of instructions which is shown below is not for any particular computer but is representative of instructions which can be found in existing digital computers. To list all the instructions available in the multitude of available machines would probably require a list of over 500 instructions. In general, the list is restricted to one-address machine types of instructions. More than one-address types are shown where they provide a unique and useful variation. Instructions are designed to manipulate various data formats available to a computer and to provide logical and decision capabilities within the machine. A listing of the subsets of instructions available for a large computer could be as follows:

1. Fixed-point binary arithmetic.
2. Floating-point binary arithmetic.
3. Variable length decimal arithmetic.
4. Word and byte transmission.
5. Control and branch.
6. Index and base register transmission.
7. Bit and logical operation.
8. Data conversion.
9. Sense indicators and status.
10. Input/output.

11. Data channel commands.
12. Compatibility with previous systems.

Fixed-point binary arithmetic usually provides for addition, subtraction, multiplication, and division. In some small minicomputers the operations of multiplication and division are not provided in hardware but are programmed. Many large computers also provide for double-precision fixed-point arithmetic operations where the operands can be two memory words long. In multiplication of two n-bit numbers, where the product may be $2n$ bits long, large computers provide for the storage of both or either of the n-bit parts of the answer. Another type of instruction may round up the most significant part of the answer and store the rounded product. The division of two n-bit numbers usually results in an n-bit quotient and an n-bit remainder. Either or both of these results may be stored or a rounded quotient may be developed.

Floating-point binary arithmetic is available in larger computers. They provide for addition, subtraction, multiplication, and division of numbers expressed in exponential form. Again, the machines may provide for double-precision data formats. Another variation may be provided for normalized or unnormalized multiplication and division. Normalization is the process by which the exponent is adjusted so that the most significant bit of the mantissa will be a 1 and not a 0.

Variable length decimal arithmetic is found in most large computers. It provides for arithmetic operations on binary coded decimal data words instead of binary data. The data format is usually not confined to be one word length but is expressed as a variable number of bytes. The number is supplied as part of the instruction. Addition, subtraction, multiplication, and division are provided and performed in a serial-by-character mode.

Word and byte transmission instructions are used to move various data formats among the registers and memory locations. For a single-address computer, the load and store instructions provide for the movement of words of data between memory and registers. Data can also be moved around in registers by the use of shift instructions such as shift left n bits, shift right, and circular shift a register n bits. Besides moving data on a word basis, most machines provide for the movement of data on a byte or character basis. This is done through a move instruction which provides for the movement of n bytes from one place in memory to another place in memory. Translation of bytes from one code to another and the editing of data strings of coded information can be provided by computers. These operations are useful to reformat input and output data.

Control and branch instructions are essential for the programming of algorithms. The branch instruction is also called a jump instruction. This class of instructions has two major forms. In one, a test is made, and if it is affirmative, the branch is taken to obtain an instruction from a new memory location. If the test fails, the next sequential instruction is executed. The other form provides for the setting of condition boxes based on the results of previously executed instructions. The branch instructions test the setting of the condition boxes to determine if control is to jump or remain sequential. Branch instructions come in many forms: branch on zero, branch on negative, branch on less than or equal to zero, branch on count, branch if the contents of one register is larger than the contents of a register, etc. A control instruction is the compare. This compares the contents of two registers and sets condition boxes or compares the contents of a register with a location in memory. The compare instruction usually provides for the comparison of binary digits or floating-point numbers.

Data conversion instructions are provided in larger computers which can have many different data formats. These instructions convert data from one of the formats to another of the formats. For example, data read from punched cards is usually placed in the computer memory as alphanumeric coded data such as the ASCII code. If the data are a decimal number, it must be converted from the ASCII code into a binary floating-point format before it can be used by a binary digital computer in arithmetic computations. Likewise, after the binary computations have been completed, the answers must be converted back to ASCII code (or EBCDIC) in order to be printed on a line printer. Data conversion instructions permit these conversions to be performed rapidly. If the computer does not have such instructions, it must use a software routine to achieve the same results but at a slower speed.

Sense indicators and status instructions are used to determine the state of the computer. Sense indicators are usually 1-bit registers which can be set by the computer upon the occurrence of a specific condition. Special instructions allow the programmer to sense the state of the registers in order to determine if the condition has occurred. Other instructions reset the registers back to a 0 bit. Status instructions are a more general type of sense instruction, where specific registers in the machine have fields which contain codes about the status or state of the computer. Some of these fields may be only 1 bit in length and are therefore just like sense indicators. Others may be several bits long, and their contents must be decoded in order to determine the specific status.

Input/output instructions are used to move information, programs and data, into the main memory and out of the main memory of the computer. These instructions probably have more variation than any other type of instruction as they depend on the I/O structure of the computer. The most simple types of I/O instructions are to read (input) one word of data in binary format from a given device into a specified location in memory or to write (output) one word from a given location in main memory to a specified output device. A large number of variations can exist within a computer and among computers. In Chapter 10 we shall deal specifically with I/O and shall detail the various types of instructions and architectures which exist in different computers.

Data channel commands are special types of instructions to parts of a large digital computer called the data channels. These instructions permit the channels to execute I/O operations independently of the central processing unit, which is executing programs. Thus, computers with data channels can have an overlapping of instruction execution and I/O operations. The specific structure of these channel commands is a function of the channel architecture. Again, in Chapter 10 we shall discuss channel commands in more detail.

An unusual set of instructions are those provided by the manufacturer of a computer in order to let a present computer execute programs designed for one of his previous computers. These compatibility instructions allow the user of the older machine to obtain the new machine without having to reprogram. If these compatibility instructions are provided such that the new machine attempts to look like the old machines for the old programs, it is called emulation. If, on the other hand, the manufacturer uses only the current instructions in special programs to be able to run the old programs without providing any hardware compatability instructions, it is called simulation.

In this section of the text we have attempted to introduce the student to the various instruction formats and data formats which computer manufacturers have developed over the years in their machines. The student should realize that this was only an introduction and not an exhaustive treatment. To become better acquainted with the computers mentioned in this section and many computers which we have not discussed, the student should obtain the literature about the machines from the manufacturer.

7.3 CONCEPTS OF PROGRAMMING

Almost with the inception of the stored program digital computer certain programming concepts have developed. These concepts have influenced the structure of digital computers through the ensuing years. Generally, concepts have first been implemented in software (programming) prior to being frozen in hardware. Thus, the concepts of indexing were first done through programs before index registers were built into computers and floating-point arithmetic was performed by subroutines before such arithmetic units were available.

This section of this chapter is not to be interpreted as an introduction to programming; there are many fine texts which perform this function. It is intended to be an introduction to certain concepts of programming which have affected the architecture of present computers or will affect the architecture in future machines, in the authors' opinion.

7.3.1 Assembling and Mnemonics

Once the concept of symbolic coding was developed, the realization that the computer could assign the absolute addresses to the symbols was inevitable. This process is called *assembling*. A set of standard mnemonics for the op-codes is supplied by the manufacturer and must be used by all the programmers who wish to use the assembler program also supplied by the manufacturers. The programmer is free to assign his own symbolic names for the address of the operands and locations of the instructions. The assembly program has a table which lists the mnemonics for the op-code and its equivalent machine language numeric code. It constructs several tables during the assembly process, which permits it to assign a machine address to each programmer-supplied symbol and to assemble all the instructions into a sequential machine language set which is the algorithmic equivalent of the symbolic program. The programmer has certain constraints imposed by the assembler. The language in which he programs is called the *source language*. The program he writes in the source language is called the *source program*, and the resultant machine language program produced by the assembler is called the *object program*.

The address of an operand is analogous to a symbol used as a variable in algebra. This was discussed in Section 7.1.1. The symbols used by the programmer in writing the source program are also the names of variables. It is the job of the assembly program to assign a one-to-one correspondence between source program names and machine addresses. The assembler does this by creating a symbol table, while it translates the source program to the object program. It lists all the symbols of the source program in this table and then assigns machine addresses in a one-to-one relationship. Figure 7.24 shows how a symbol table might appear.

Symbol	*Address*
Y	0011
T	0012
VOY	0013
G	0014
TWO	0015

Figure 7.24 Symbol table.

7.3.2 Subroutines and Linkages

During the course of programming a large algorithm one notices the existence of recurring sequences of subalgorithms within the program. It would be convenient if it were not necessary to repeat these subsequences. This would conserve storage and make the programming easier. Such subsequences of recurring instructions or subalgorithms are called *subroutines*. Many problems are constructed by dividing the problem into a set of smaller problems and evaluating the smaller problems. Likewise, many programs can be subdivided into smaller programs. Since a well-constructed subroutine can be used by other programmers, it is not necessary to reprogram this subunit. Thus, a library of subroutines can be utilized by everyone. Another advantage of programming using subroutines is that the main program can be constructed from a set of relatively independent subroutines, each of which has been *debugged* separately. This logical division of problems tends to reduce errors in programming their solution.

The ability to have the computer communicate between subroutines is called a *linkage*. The linkage between subprograms or subroutines is usually well defined for each computer system and is assisted by hardware instructions. The following concepts are common to most systems;

1. A *calling sequence* or *linkage*. This is the formal statements required by the calling program in order to initiate a subroutine. It consists of three parts. One part is the branch to the first address of the subroutine. Another part is a method to inform the subroutine where to jump back when it completes its algorithm, and the last part is a method to inform the subroutine of the location and length of the parameters which the calling program will transmit to the subroutine. In most high-level programming languages, such as FORTRAN, it is only necessary to specify the subroutine's name and parameter list. In machine language it is more complicated because the execution of the subroutine will require a change in the current address register (*CAR*). Thus, if the calling program does not "mark its place" before it calls the subroutine, there will not be a path back. This is usually done by a special jump instruction, sometimes called a *return jump*, which first saves the contents of the *CAR* at a specified location and then jumps to the location of the subroutine.

2. *Subroutine initialization*. Before the subroutine can start its algorithm, it must peform some "housekeeping." Since it will need to use the registers of the arithmetic unit, some index registers, some base registers, etc., it must first *save* the current contents of those registers. It can then use the registers for its own computation, and when completed it *restores* the registers to their prior values before jumping back to the calling program at the address after the branch to the subroutine. It must save the return address.

3. *Subroutine body.* After performing its housekeeping, the subroutine enters its algorithmic production. It may call another subroutine to perform this. Thus, it may become a calling program.

4. *Exit* from the subroutine. After the subroutine has completed its computation or services it must prepare to return control back to the calling program. First, it restores the registers to the values they had when the subroutine was called. Then it performs a special branch back to the place where it was called by the calling program. This return restores the address back into the *CAR*, which was saved in the return jump instruction.

The communication of parameters between calling program and subroutine is a matter of convention which is established by the manufacturer and used by all programmers. The parameters are usually placed into a list of consecutive addresses by the calling program prior to the linkage. As part of the calling sequence, the calling program sends the beginning address of the list to the subroutine. This can be accomplished by loading a preassigned register with this address just prior to executing the return jump instruction. The subroutine obtains the address from this register. The assigned register is established by the convention.

With the establishment of multiprogramming and multiprocessing (more than one CPU in the computer system) it has become necessary to develop subroutines which can have more than one call made to them. That is, one processor can call the subroutine before the other processor has completed using the subroutine. Subroutines which are designed for this type of operation are said to be *reentrant.* These subroutines are written so that they never modify their own instructions. They are called *pure procedures.* However, many subroutines require data parameters and indexing capabilities particular to each call made to the subroutine. There must be a programming convention established which permits a separate storage area for each calling sequence to the subroutine. There are two basic methods to establish this convention. In the first method, each calling program establishes a storage region for the subroutine including all space needed for the save requirement. This method is depicted in Fig. 7.25 and shows how an indefinite number of simultaneous users of the subroutine can be active without burdening the subroutine to make a storage allocation. The disadvantage is that each calling sequence must know the storage needs of the reentrant subroutine. The second method of controlling the storage space for calling parameters needed by a reentrant subroutine is to have the subroutine obtain space for each distinct call made to it. The method utilizes a fixed maximum amount of storage allocated to each reentrant subroutine. The subroutine then allocates some of this space to each call made to it. After a call is completed, the subroutine frees the space allocated to that particular call and can reallocate the space to a subsequent call. The mechanism used for space allocation by the subroutine

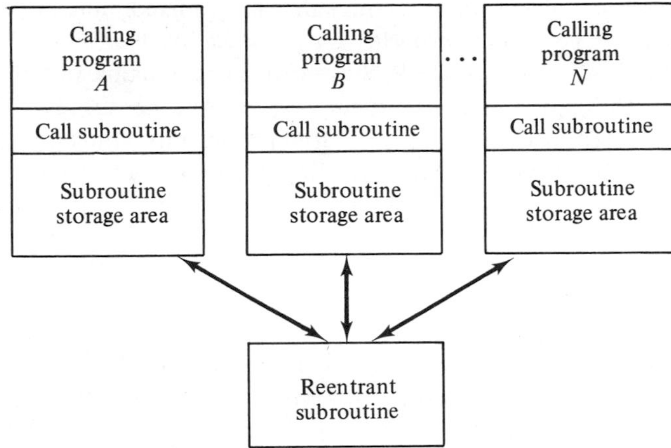

Figure 7.25 Storage method for reentrant subroutines.

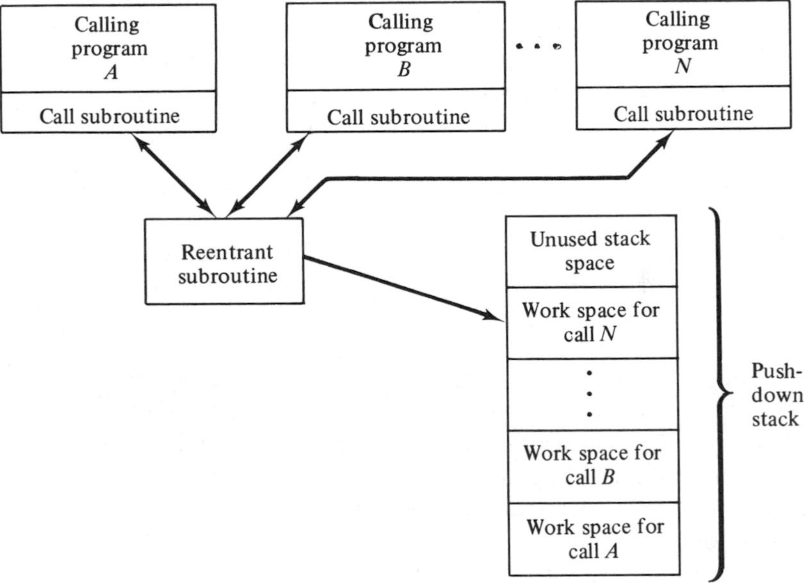

Figure 7.26 Another storage method using a pushdown stack.

is called a push-down stack. Figure 7.26 shows this method of handling reentrant subroutine storage. The subroutine releases space from the top of the stack as that call is completed. Thus, the calls are completed in the reverse order in which they are made; that is, the first call made to the subroutine is the last call to be completed.

One disadvantage of this method is that only a fixed number of calls can be executing the reentrant subroutine, since only a fixed amount of storage is allocated for the push-down stack. Since calls arise dynamically, the number may exceed storage and therefore go unserviced. Another disadvantage is the complexity of writing this type of subroutine, which must allocate and control the stack. One advantage is that the calling program does not need to know the storage requirements of the subroutine.

7.3.3 Locality of References

An examination of many machine language programs yields two general properties of such programs. These properties are dependent on the particular program and the particular computer on which the program is run but appear valid for all programs and systems. They are

1. Because of indexing, indirect addressing, and instruction loops, the number of effective addresses computed by the machine is far greater than the number of instructions written by the programmer in the program.
2. Most effective addresses tend to dwell within a few localized areas of main storage for some sufficiently long sequence of addressing.

An example of property 1 is the DO loop in a FORTRAN program, where a few instructions written by the programmer are executed repeatedly. An example of property 2 is the body of the DO loop itself, which exists in a small area of main core and is being referenced over and over again during the instruction cycle.

The design of modern computers utilizes these two general properties by attempting to capture the locality of the references in high-speed electronic registers or special high-speed local storage (sometimes called a cache). The high-speed registers provide a means to employ indexing and base addressing. Thus, a small direct address within the instruction can be used to address a large main storage using base addressing and can be reused in a loop by utilizing indexing. This is typically done when moving through sequential addresses of an array. An attempt to capture a loop can be done in either of two ways. One way is to retain instructions as they are used in a local store until a branch is made to one of the previously referenced instructions. Once the branch is made a loop of instructions will be in the fast local store. (This technique is called a *look-aside* memory.) The other technique is to first move several sequential instructions into the fast local store and hope that a loop will be captured. This is known as a *look-ahead*. By utilizing pointers and local store, the number of actual fetches from the slower main memory for both data and instructions can be significantly reduced. It is the responsibility of the programmer to modify the contents of the machine registers to properly point to his data and instructions so as to obtain a high degree of referencing to these registers and therefore reduce the references to main memory.

7.3.4 Relocatability

During the first and second generations of computers there was usually only a single problem program running in the main memory at any given time. It utilized the resources it needed, and when completed another program was loaded and executed. Since this program was loaded and generally executed to completion, it could be assembled using absolute addresses. The programmer could specify to the assembly program where he wanted the program to be placed in memory and controlled the input and output data into fixed locations of memory as he chose. With the introduction of the third generation of machines, the concept of multiprogramming was introduced where several problem programs could be in the main memory at the same time and they would be executed under the control of a resident supervisor program so as to optimize the utilization of the computer systems resources. Thus, at assembly time, or compile time, the amount of main storage being used or the configuration of programs in main memory when a program is being executed would not be known. The utilization of memory is dynamic and allocated to problem programs by the supervisor so as to improve system utilization. For this reason it is necessary that a program be able to run from any contiguous region of main memory. In addition, it may be necessary to move the program in main memory between the consecutive times it has gained access to the control unit to be executed. Thus, the *binding* time or the time at which the addresses are fixed should be when they are needed.

Two methods of relocation are in general use. One method uses a software technique and binds the addresses when the program is loaded into a region of main memory. It is therefore implemented within the loader utility program. The assembler or compiler assigns addresses relative to a zero address origin. To each instruction word, the loader adds a base address to every address in an instruction. The base address is supplied by the supervisor to the loader at the time the problem program is loaded into a region of memory. Thus, after the problem program has passed through the loader program every address will be an effective address or a physical address in memory. The second method is a hardware technique and binds the addresses when they are used with the use of a base register. The effective address is computed for each access to memory by adding the contents of the base register to the zero origin relative address. This method requires a separate adder to compute effective addresses.

The base register method has the advantage that a problem program can be interrupted by the supervisor and reloaded in another region just by changing the contents of the base register. If one attempts to do this using the loader technique, the problem program must be passed through the loader again in order to have all its addresses reassigned. This is much more time consuming than the base register method.

7.3.5 Initial Program Loading

When the machine arrives and is connected together by the field engineers, it has a blank memory, i.e., a newborn baby. It must have programs loaded into the memory. However, before a program is executed it must be in memory. How do we get the first program into memory? The first program to be loaded will be the loader. It is loaded by a hardware sequence of instructions which are built into the machine and executed when the *load* button is pushed on the machine console. This simple built-in program reads in a few instructions from a card in a card reader or some paper tape from a paper-tape reader. These few instructions are loaded in memory at locations starting with 0. Once they are loaded into memory the built-in sequence jumps to location 0 and commences to execute instructions from that location. The few instructions are sufficient to cause the remainder of the loader to be read into memory, and then the loader brings in the remaining programs and we are in business. Thus, the process is in three parts: The first is a built-in sequence of operations which causes a few instructions to be loaded at a precise location of memory. The second is the execution of these few instructions, which are called a *bootstrap loader*, and they load the relocatable loader. The third is the use of the relocatable loader to enter the supervisor and other routines which are needed to operate the computing system.

7.4 PROGRAMMING SYSTEMS

In today's large computing systems, it is not possible for a programmer to write all the programs required to make the system run efficiently and accomplish his work. A whole set of programs, written as subroutines, which link together are provided by the manufacturer and is called a *programming system*. These systems generally include an *assembler* program; several general high-level language *translators* such as FORTRAN, COBOL, and PL/I; several problem-oriented languages such as ECAP (Electronic Circuit Analysis Program), GPSS (General Purpose System Simulation), and CSMP (Continuous System Modelling Program); and a library of different programs some of which may be standard subroutines to be incorporated into user programs while others may be entire programs to be used by many people and thus not have to be rewritten each time they are used.

Another function of the programming system is to provide the efficient use of the computer systems hardware and software (programs). A special program, called by various names—among them a *control program*, a *monitor*, a *supervisor*, or an *operating system*—is supplied by the manufacturer and allocates the resources of the system among the problem programs which are running on the system. It allocates equipment such as main storage, secondary storage, input/output devices, and the CPU among the programs in a multiprogrammed environment. It attempts to resolve all conflicts for system resources by the problem programs and ensures privacy of each user who may be sharing the system.

It is not possible to perform the system design for a large computer system without considering the interplay between the programming system and the hardware system. The design of system software must be considered coincidentally with the design of system hardware if a viable, cost-effective computing system is to be built.

7.4.1 Assembler

An assembly program (assembler) is a translator supplied by the manufacturer which permits the programming of a problem in a source language using programmer-supplied mnemonics in place of machine language. In a simple assembler, each source statement is translated into a single machine language instruction. In a *macro* assembler an ability to specify rather long and complex sequences of instructions is supplied. These sequences are usually utility programs which are needed to perform functions within the system such as I/O operations or certain data management processes. Most assemblers also provide a method of assigning constants and reserving storage space for data and provide subroutine calling sequences.

The programmer also has the capability of giving instructions to the assembler. These are called pseudoinstructions, since they are not part of the source program but are used by the assembler to perform certain tasks which the programmer may specify. An example of a "pseudo" is the specification of the ORIGIN of the program. The assembler will utilize this information and also pass it on to the loader in the object program so that the program will be loaded where the programmer specifies.

The operation of the assembler is as follows: The assembler must be loaded into memory if it is not already there. It then reads in the entire source program. It performs a translation on the source program and in the process creates a symbol table which is a cross reference between the programmer-assigned mnemonics and the assembler-assigned absolute (or relative) machine addresses. The programmer must use the manufacturer's supplied mnemonics for the operations. The assembler already has these in a table, and when creating the object program instructions it supplies the correct op-code. After the completion of the assembly an output of both the source and object program is produced as well as a cross reference map, i.e., the symbol table. The assembler can also detect certain programming errors such as doubly defined symbols or undefined symbols. These errors are also part of the output.

Since almost all programming in a language which is close to machine language is performed in an assembly language, the designers of a computing system should produce instructions and software systems which facilitate the use of the assembler. Character-manipulating instructions should be available to make the translation process easier. Search instructions might be included to scan the symbol table for a previously defined mnemonic. The easier it is to assemble the source program in the source language, the closer the source language will be to the actual machine language.

7.4.2 Compilers and Interpreters

Compilers and interpreters are two types of a general class of programs called *translators*. A translator program has as its input a *source program* and delivers as its output an *object program*. The source program is treated as data by the translator, which transforms these data into new information, which is the object program. The use of translator programs permits the computer user to write his program in a high-level language, which facilitates the statement and solution of his problems. The high-level language, higher than assembly level, is called the *source language*. It has its own rules, symbols, and conventions which must be obeyed by the user. Violation of these rules, or syntax, results in a programming error, which is detected by the translator during its operations. In most cases, the object program is a machine language program and can be executed by the computer. Almost all programming performed is done in a source language such as FORTRAN, BASIC, COBOL, ALGOL, PL/I, GPSS, CSMP, ECAP, MIMIC, and WATFIV.

The advantages of using a translator instead of programming in machine language or assembly language are

1. Convenience in stating the problem in a more mathematical-like language.
2. Programming in a source language which is independent of the architecture of a machine.
3. Error detection by the translator and ease of making changes in the program.

Although a standard source language such as FORTRAN is suppose to be independent of the machine properties on which it is being executed, this is rarely the case, and some machine dependencies are introduced by the manufactuer with the translator he supplies. Since the programmer sees only the source language and its execution, he in effect sees a higher-level language machine. The machine designer should supply those features which will enable the translator to perform better and faster. There are two types of translators, which differ in their internal structure and operation. One is an *interpreter*, and the other is a *compiler*.

An interpreter is the type of translator which first translates a source program statement and then executes that statement before repeating these two steps for the next statement. The interpreter coexists in main memory with the source language program during its execution, and this execution tends to be slower than the compiler method since each statement gets translated each time it is executed. Thus, if a three-statement loop is executed five times, there would be a total of 15 translations of source instructions during this execution.

One can look at the execution of a machine language program as being executed by a translator. In this case, the translator is the hardware of the computer which interprets each instruction during its E cycle. A software interpreter is just another step in the hierarchy. First the source program written in the source language is interpreted into machine language statements, and then these statements are interpreted by hardware and executed. If the computer designer makes his machine language instructions closer to the source language statements of the interpreter, the machine could execute higher-level source language programs much faster since the amount of interpreting would be decreased.

A compiler is a type of translator that translates the entire source program into an object program and then executes the object program. Thus, the compiler is similar to an assembler. However, one source program statement in a high-level language will translate into many object-level statements if the object program is machine language. Since the compiler translates a source statement only once, it would translate the three-statement loop only once even though it is executed five times. The compiler therefore saves time when compared to an interpreter when there are many loops in the program or the program is executed many times with different sets of data. Thus, once the program is translated (compiled), the object program is executed just like any other machine language program, and no further translation is performed. If a change must be made in a source program, it must be completely recompiled. This is a disadvantage of a compiler-type translator.

A compiler operates in several distinct time and space phases. During the compilation time the entire source language program must be available to the compiler. However, the data need not be in memory. After the object program is produced, the compiler no longer needs to be in memory, and the object program can be executed whenever desired.

7.4.3 Service Programs

Service programs are a set of programs which are frequently used by programmers and are general enough so that once written they can be reused by different programmers. Thus, they form a library of housekeeping programs. The service programs consist of linkage editors, a loader, sort/merge programs, utility programs, and emulators.

7.4.3.1 Linkage Editors

A linkage editor provides a method of combining program segments which were individually compiled or assembled into a single program which can be executed. It is used with subroutine programs so that each part of a program, i.e., a subroutine, can be compiled separately. This enables changes to be made in the subroutine without recompiling or assembling the entire program, but only the part which has been changed.

7.4.3.2 Loader

The loader places object language programs which have been compiled or assembled or linked together into main memory. It provides for relocation of programs at load time.

7.4.3.3 Sort/Merge Programs

Sort/merge programs provide often-used algorithms to sort or merge records into ascending or descending order. Since these types of operations are performed quite often in business applications, an efficient program which utilizes magnetic tape or disks for input, output, and intermediate storage during the ordering is essential. It must be usable in a multiprogrammed environment and also be used as a subroutine by higher-level compilers.

7.4.3.4 Utility Programs

Utility programs are used by the system to perform operations on the sets of data stored in secondary storage. They therefore are analogous to the work performed by a librarian on books in a library. Some utility-type programs

1. Transfer, copy, or merge sets of data from one storage medium or device to another.
2. Edit, rearrange, and update data.
3. Change or extend the catalog of sets of data maintained by the programming system.

4. Print a catalog, for programmer use, of the names and locations of programs and data stored in secondary storage.

7.4.3.5 Emulator Programs

An emulator program is a special type of program which in conjunction with hardware in the computer allows object programs written for one computer system to be executed on another system without reprogramming.

7.4.4 Supervisor

The *supervisor* is that part of the programming system which provides a mechanism by which CPU time is allocated to various jobs running in a multiprogrammed computer. The jobs are stacked on a *job queue* and are presented to the programming system upon completion of a previous job. The jobs are initiated by the supervisor in accordance with specified conditions and priorities, and many jobs may be performed concurrently. Part of the supervisor is a *job scheduler*. It interprets the job control language and performs operations concerned with the allocation of resources to jobs and with scheduling and dispatching jobs. An important part of a job scheduler is the allocation of I/O devices and the communication with the operators to perform mounting and demounting of magnetic tape reels and disk packs.

In a multiprogrammed system, *foreground* or non-batch-processing jobs can be operating concurrently with *background* or batch-processing jobs. The supervisor switches the CPU between jobs on a priority basis, foreground having a higher priority than background. It can accomplish the switch by either *timing-slicing*, which is based on a time-out clock hardware, or by *event posting*, which is based on the posting of an event such as the arrival of data into main storage or an I/O interrupt.

The resource management function of the supervisor involves allocation, scheduling, and dispatching. It allocates regions of main memory to a job or an I/O device, etc. Scheduling is the ordering of future events, while dispatching is the activation of a scheduled event. The supervisor schedules and dispatches CPU time. An I/O supervisor, or a subset of the total supervisor function, coordinates the use of I/O resources by scheduling and dispatching all such devices.

The problem programs, object programs, compilers, interpreters, service programs, and all other programs communicate requests for service through the supervisor. They do not actually perform the service themselves, although it may appear as if they are performing such services. A call for a service such as I/O is really a call to the supervisor, which schedules and dispatches this call as the resources become available and as priorities permit. The supervisor is really the manager of all the hardware and software in the computing system.

7.4.5 Multiprogramming

Multiprogramming is a method of operating a computer system so that several independent programs reside in memory at the same time and time-share the CPU and the I/O resources. These programs are operating in the batch mode. That is, they were programmed and prepared off-line from the computer and entered into the computer through a card reader into an input job queue. The scheduler part of the supervisor schedules the order in which the jobs are executed. At any given time only one problem program or the supervisor itself has control of the CPU. All other jobs in memory are dormant, although the channels may be performing I/O operations that were requested by a job before it became dormant. When a problem requests an I/O operation, it releases the CPU to the supervisor, which schedules the I/O request, makes the requesting program dormant, and turns the CPU over to another dormant problem program, which becomes active. The first program will be eligible to become active once its I/O request has been satisfied and this information has been conveyed to the supervisor. The basic principle of multiprogramming is that the software system should optimize the use of the hardware. That is, I/O operations will be overlapped with computation, which yields a maximum utilization of the CPU. For this to occur the mix of jobs that are in memory must be such that one of them can compute while the others are awaiting the completion of their I/O.

One does not achieve the cost advantage of multiprogramming without some disadvantage. Since each program must be independent of any other program in memory, the programs must not interfere with each other. There must be adequate security provisions to ensure this type of operation. A simple method of protection is to allocate continuous memory locations to each program. This is called a *region*. The upper and lower addresses of the region are placed by the supervisor into special registers each time the program in that region becomes active. Any attempt by the program to access an address outside of the bounds of its region causes an interrupt and returns control to the supervisor.

Another disadvantage of multiprogramming is the competition for main memory. If several programs are to be in main memory simultaneously, then the system needs a larger main memory than if only one program were in main memory at any time. The solution is very expensive and is self-defeating since it increases the cost per program. One solution is to divide each program into several segments which are relatively independent and to overlay each segment into the region upon the completion of the work of the previous segment, that is, construct each program as a sequence of subroutines with a maximum size. This solution is also not advantageous. First, it requires a larger and more complicated supervisor which can provide automatic overlay features. Second, it requires the programmer to be cognizant of the multi-program environment and to divide his program appropriately. Since the concept of multiprogramming is supposed to be programmer-independent, this violates the premise.

A solution to the overlay problem and the proper allocation of main memory to programs is an automatic method called a paged and segmented virtual memory. Although this system is not perfect, it does promise certain advantages. This concept will be discussed in Volume II.

Another disadvantage of multiprogramming is the complicated schedules which must be designed and which must remain in main memory in the permanent region allocated to the supervisor. The design of this scheduler, which schedules jobs based on their I/O requirements and priority, is not easy and does not always maintain the system at full utilization. The alternative method is to just keep main memory full and hope that there exists sufficient diversity among the jobs to properly utilize resources.

Still another difficulty introduced by multiprogramming is the accounting procedures for sharing costs to each job. It is difficult to maintain the amount of time each job utilizes each of the system's resources and charge that job accordingly. Also, how do you allocate the cost of the supervisor's CPU time among the problem programs?

Despite all the above problems, multiprogramming still shows a cost/performance improvement over attempting to operate a large computing system with monoprogramming. All large computing systems have overcome these problems to some extent and have a multiprogrammed environment.

7.4.6 Time Sharing

Utilizing a batch-operated, multiprogrammed computer system, the programmer must submit his program in punched-card form and wait until he receives his output in the form of printed material. The *turnaround time*, i.e., the amount of time from submittal to reception of an answer, may be from several minutes to several hours. During this process the programmer cannot interact with the computer. He cannot see if there is an error in his program and correct it while the computer is running his program. He cannot change his program based on some intermediate results. He can only debug his program after the fact. During the days of monoprogramming on first-generation machines, the programmer had control of the entire computing system. He could perform on-line diagnosis and debugging of his program. The machine was his tool. Using batch multiprogramming, it may take many submittals to obtain a corrected running program. With a long turnaround time this could take days to get the program running. The programmer starts to feel as if he is the tool of the computer.

A solution to this turnaround time problem is to operate the system in a *time-shared* multiprogrammed mode with on-line input/output for each user. The I/O device is a typewriter terminal which can be used for input to the computer and output from the computer. The supervisor allocator *time slices* to each user. These time slices are quite short, usually about 50 milliseconds. If there are not too many users, the machine can return to the program in about 2 seconds. Hence, the user at a terminal submitting one line of a high-level source language program at a time thinks he has complete control of a computer.

To have conversational capability with the user and debug his program so as to give the illusion of having control of a computer, the software must be written to enhance the time-shared mode. The compiler must have syntax error-correcting capability on a line-by-line basis. Not all errors can be found before the entire program is entered; it must therefore present errors to the programmer after the completion of the compilation. It saves the entire source program in secondary storage and permits editing the source program to correct the errors. Upon successful compilation of the program, the programmer can request execution and present data when requested by the computer. In addition, the computer can be requested to save the program in secondary storage for reuse at a later time. This prevents the need to retype the entire program into the machine when it is to be reused.

The programming advantages are the ability to discover and correct trivial errors without having to go through a long turnaround time, to discover and correct run-time errors due to improved real-time diagnostics of a time-shared system, and to have man-machine interaction-type programs such as real-time control or computer-aided instruction.

The requirements of the hardware and software in the system are much more demanding. There must be a high resolution clock with a time-out resetable clock. This is used by the supervisor to time-slice. Each time a user receives control of the CPU the time-out clock is reset. When it reaches zero the supervisor switches to another user's program. The scheduler must be quite different from one for the batch mode of operation. It must maintain queues of activity associated with each of its on-line users. Also the computer cannot store the programs of all the users in main memory; there is not enough main memory. It must swap programs in and out of main memory with a secondary memory device, such as a drum. It overlaps this swapping with the execution of a time slice for a given program. Hence, besides its regular I/O operations with the terminals, the scheduler must provide for shifting the programs between main and secondary storage. This shifting must be transparent to, not seen by, the user. To overcome some of the problems in moving large programs in a short time slice, a paged and segmented virtual memory system can be used. This technique of memory management will be developed in Volume II.

REFERENCES

Bell, C. G., and A. Newell, *Computer Structures: Readings and Examples.* New York: McGraw-Hill, 1971.

Chu, Y., *Introduction to Computer Organization.* Englewood Cliffs, N.J.: Prentice-Hall, 1970.

Flores, I., *Computer Organization.* Englewood Cliffs, N.J.: Prentice-Hall, 1969.

———, *Computer Programming.* Englewood Cliffs, N.J.: Prentice-Hall, 1966.

Gschwind, H. W., *Design of Digital Computers.* New York: Springer-Verlag New York, Inc., 1967.

Hellerman, H., *Digital Computer System Principles.* New York: McGraw-Hill, 1967.

Hollander, G. L., "Architecture for Large Computer Systems," *Computer Design* (Dec. 1967), pp. 53–59.

Ledley, R. S., *Digital Computer and Control Engineering.* New York: McGraw-Hill, 1960.

Maley, G. A., and E. J. Skiko, *Modern Digital Computers.* Englewood Cliffs, N.J.: Prentice-Hall, 1964.

Schoeffler, J. D., and R. H. Temple, eds., *Minicomputers: Hardware, Software and Applications.* New York: IEEE Press, 1972.

Thornton, J. E., *Design of a Computer, The CDC 6600.* Glenview, Ill.: Scott, Foresman, 1970.

PROBLEMS

7.1. What is the largest decimal number which can be contained in a register with the following number of bits? Consider only unsigned positive numbers.
(a) 10 (b) 12 (c) 16 (d) 24 (e) 32
(f) 36 (g) 48 (h) 60 (i) 64

7.2. Write a program which computes e^x in which each operand and all intermediate results are stored in the sequential order in which they are used.

7.3. Show the register transfer equations for an arithmetic unit which uses four registers to perform the operations of multiplication and division.

7.4. Consider the following 24-bit patterns. Show their representation in octal, hexidecimal, and EBCDIC.
(a) 100 101 010 111 011 011 000 101.
(b) 001 110 111 000 101 100 011 010.
(c) 011 110 101 010 100 111 000 011.

7.5. For a computer which utilizes the byte as its unit of information, how will a word of 4 bytes be specified in order to perform arithmetic?

7.6. For an interleaved memory, what determines the bank in which an address is located?

7.7. Using the instructions as defined in Table 7.2, write a program for the four-address computer of Section 7.2.1.1 which will find the largest of the three numbers X, Y, and Z.

7.8. For the four-address computer of Section 7.2.2.1, write a program which will compute e^x with an accuracy of five decimal figures.

7.9. Repeat Problem 7.7 for the three-address computer defined in Section 7.2.1.2.

7.10. Repeat Problem 7.8 for the three-address computer defined in Section 7.2.1.2.

7.11. Repeat Problem 7.7 for the two-address computer defined in Section 7.2.1.3.

7.12. Repeat Problem 7.8 for the two-address computer defined in Section 7.2.1.3.

7.13. Repeat Problem 7.7 for the one-address computer defined in Section 7.2.1.4.

7.14. Repeat Problem 7.8 for the one-address computer defined in Section 7.2.1.4.

7.15. Repeat Problem 7.7 for a single-address computer with 16 general-purpose registers as defined in Section 7.2.1.7. Use the register-to-register operations as much as possible to obtain a program with fast execution.

7.16. Repeat Problem 7.8 for the single-address computer with 16 general-purpose registers as defined in Section 7.2.1.7.

7.17. Write a program to find the largest of three numbers X, Y, and Z using a computer with 16 general-purpose registers and a single-address field. Assume that the computer uses its registers as accumulators, index registers, and base registers. Use the instruction set of Table 7.5 and the instruction format of Section 7.2.2.11.

7.18. Using the computer as defined in Problem 7.17, write a program to determine e^x.

7.19. Consider the instruction format and data formats of the IBM System 360/370 as explained in the subsection beginning on p. 242. Using the list of instruction types as stated in Section 7.2.3.5, define a set of instructions for the System/360-type computer which you are about to design.

7.20. Repeat Problem 7.19, but use the structure and formats of the CDC 6600 as the bases for your proposed instruction set.

7.21. Do Problem 7.19 and then obtain the IBM System/360 instruction set. Compare your instruction set to the existing instruction set. Did you leave anything out?

7.22. Do Problem 7.20 and compare your instruction set with that of the CDC 6600. Where do they differ?

7.23. You wish to write a program which will perform a hardware checkout of a small computer. Develop the instruction set of the computer and then determine a minimum necessary subset of the instructions. Using this subset, write a program which will check the remaining instructions.

7.24. For the computer which you have developed for Problem 7.23, define the subroutine linkage connection.

7.25. Using FORTRAN, write a reentrant subroutine which you can call from three different main programs.

8

Arithmetic Units

In Chapter 7, digital computer architecture was presented, and the function of the various units has been defined. One of those units is the arithmetic unit. The function of the arithmetic unit is the performance of arithmetic operations such as addition, subtraction, multiplication, and division and also some logical operations. Various algorithms for performing arithmetic operations were discussed in Chapter 3. The design of arithmetic units which implement these algorithms is the purpose of this chapter. Just as the arithmetic operations were classified as fixed-point or floating-point, their implementation also results in either a fixed-point or a floating-point arithmetic unit, respectively. The design of an arithmetic unit depends, also, on whether the information is processed in parallel or in serial and on the representation of negative numbers. In the subsequent sections, arithmetic units of different configurations are presented. These include binary arithmetic units for both serial and parallel fixed-point operations. Floating-point arithmetic operations are also implemented in a parallel floating-point arithmetic unit. In the last section of this chapter we shall present a brief discussion of some logical operations.

8.1 SERIAL BINARY ARITHMETIC UNIT

In serial computers the bits of an operand are available one at a time, and therefore information is processed in serial, that is, bit by bit. In this section, the individual operations to be performed by a serial arithmetic unit are first implemented separately and then combined to form a complete unit.

8.1.1 Serial Addition and Subtraction

Serial addition of two binary numbers requires only one full-adder block, regardless of the number of bits in the operands. The hardware savings presented by serial adders is achieved at the expense of low speed of operation when compared to parallel adders. Figure 8.1 shows a block diagram of the basic hardware layout for serial addition and subtraction. In a single-address computer an add or subtract instruction is interpreted as follows: to add the contents of a memory location into the contents of the accumulator (a register). After the contents of the given memory location are transferred into X, the addend and the augend are now available in the two shift registers A and X, respectively, with each register feeding the adder with 1 bit at a time. Every time a register feeds a bit to the adder, all other bits in the register are shifted one place to the right, thus leaving the left end cells of the register vacant. Consequently, the sum of the two operands is returned back into A, while X is circulated back into X, 1 bit at a time. After processing all the bits of the operands, the result of the operation is held in A while X is left unchanged. The C flip-flop shown in the diagram is a carry flip-flop. It receives the carry output of the adder and stores it so that it can be added to the next higher significant bits.

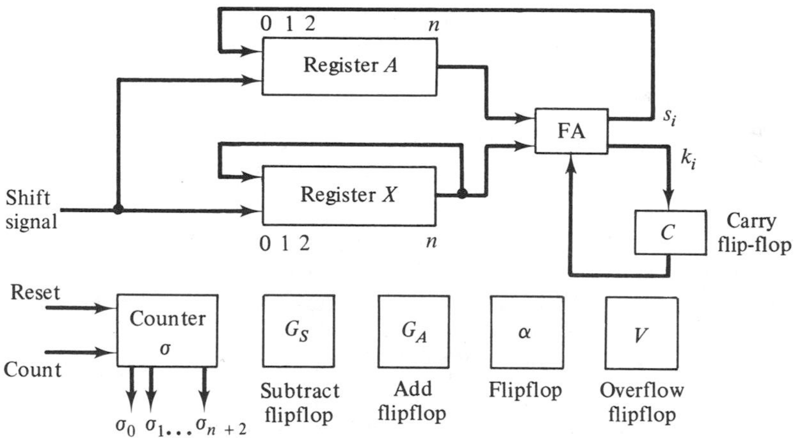

Figure 8.1 Basic layout for serial addition and subtraction.

Let us consider the operations of addition and subtraction in the 2's complement representation, which is described in Chapter 3. The following paragraphs describe the implementation of the 2's complement algorithm using the layout shown in Fig. 8.1.

The first step of the 2's complement algorithm is to convert the subtraction operation into addition by complementing (2's complement) the subtrahend. Let us assume two additional flip-flops, G_A and G_S, to store the type of operation; if G_A is in the 1 state, the operation is addition, and when G_S is in the 1 state, subtraction is to be performed. The G_A or G_S flip-flop is set when the operation starts and resets when it is finished. The 2's complement is equivalent to complementing the X register bit by bit and then adding 1 to it. To perform this operation in serial would require a long time. An alternative method, which saves time, is to complement the bits of X while they are being shifted into the adder (next step), while the addition of 1 could be done by storing 1 in the carry flip-flop, as indicated by the transfer

$$G_S \longrightarrow C$$

The second step is to add the two operands. In this case the addition is in serial and occurs on successive pulses. The necessary transfers are given by

$$\begin{gathered} \rho(X) \longrightarrow X, \quad L(A) \longrightarrow R(A), \\ G_A[\mathrm{S}^2(A_n, X_n, C)] + G_S[\mathrm{S}^2(A_n, \bar{X}_n, C)] \longrightarrow A_0, \\ G_A[\mathrm{K}^2(A_n, X_n, C)] + G_S[\mathrm{K}^2(A_n, \bar{X}_n, C)] \longrightarrow C \end{gathered}$$

repeated $n + 1$ times. The S^2 and K^2 indicate modulo-2 sum and carry, respectively.

The last step is to check for overflow. Let us assume an additional flip-flop V whose 1 state indicates an overflow. Therefore, the V flip-flop is to be set whenever an overflow condition exists. The overflow occurs if the two added numbers originally have the same sign and the addition results in the opposite sign. Since we are performing accumulative addition (i.e., the result is stored back in the A register), the original sign of the number in A is destroyed by the operation and an additional flip-flop to indicate the relationship between the sign of the operands is necessary. Let us denote the sign flip-flop by α. Before the addition takes place, α is to be set to 1 if the sign of A and X are the same; otherwise it is in the 0 state. This can be represented by

$$\overline{(A_0 \oplus (G_A X_0 + G_S X'_0))} \longrightarrow \alpha$$

After the addition takes place, the contents of the V flip-flop is adjusted according to the transfer

$$\alpha \cdot (A_0 \oplus (G_A X_0 + G_S X'_0)) \longrightarrow V$$

where α, equal to 1, indicates that the two operands have the same sign, and the exclusive-OR $(A_0 \oplus X_0)$ indicates that the sign of the result (A_0) is not the same as the original signs (X_0).

Let us assume that the addition or subtraction operation is indicated by either an add signal or a subtract signal. The implementation of the addition and subtraction is described below as a set of control (timing) signals and the transfers associated with each signal:

		Comments
Start signal	$1 \longrightarrow G_A$ for add,	Set either the add or subtract flip-flop
	$1 \longrightarrow G_S$ for subtract, $0 \longrightarrow \sigma$	Clear the counter σ
$(G_A + G_S)\sigma_0$	$G_S \longrightarrow C$, $\overline{(A_0 \oplus (G_A X_0 + G_S X'_0))} \longrightarrow \alpha$ $\sigma \boldsymbol{+} 1 \longrightarrow \sigma$	
$(G_A + G_S)\sigma_1$	$\rho(X) \longrightarrow X$, $L(A) \longrightarrow R(A)$ $G_A[S^2(A_n, X_n, C)] + G_S[S^2(A_n, \bar{X}_n, C)] \longrightarrow A_0$ $G_A[K^2(A_n, X_n, C)] + G_S[K^2(A_n, \bar{X}_n, C)] \longrightarrow C$ $\sigma \boldsymbol{+} 1 \longrightarrow \sigma$	Shift and add
$(G_A + G_S)(\sigma_2 + \sigma_3 + \cdots + \sigma_{n+1})$	Same as σ_1	Repeat n times
$(G_A + G_S)\sigma_{n+2}$	$\alpha \cdot (A_0 \oplus (G_A X_0 + G_S X'_0)) \longrightarrow V$, $0 \longrightarrow G_S$, $0 \longrightarrow G_A$	Check overflow and end the operation

The notation $(\sigma_i + \sigma_{i+1} + \cdots + \sigma_n)$ is used to indicate that the transfers associated with σ_i are also performed whenever any of the timing signals $\sigma_{i+1}, \ldots, \sigma_n$ occurs. In the above transfers, and also in the transfer notations in the remaining part of this text, the $+$ sign represents the logical-OR operation, while the boldface plus sign, $\boldsymbol{+}$, represents the arithmetic plus sign.

8.1.2 Serial Multiplication

In this section an implementation of the multiplication by repeated addition is presented using a serial multiplier. The operands are in fractional 2's complement representation. Each operand consists of n bits plus sign. A block diagram for the basic hardware layout is shown in Fig. 8.2. The A register, X register, full-adder, carry flip-flop, counter σ, and the sign flip-flop α are those used in the serial addition and subtraction of the previous subsection. The additional hardware and its function is described below:

G_m = multiply flip-flop. It is set to 1 at the start of executing the multiply instruction and reset when the multiplication is ended.

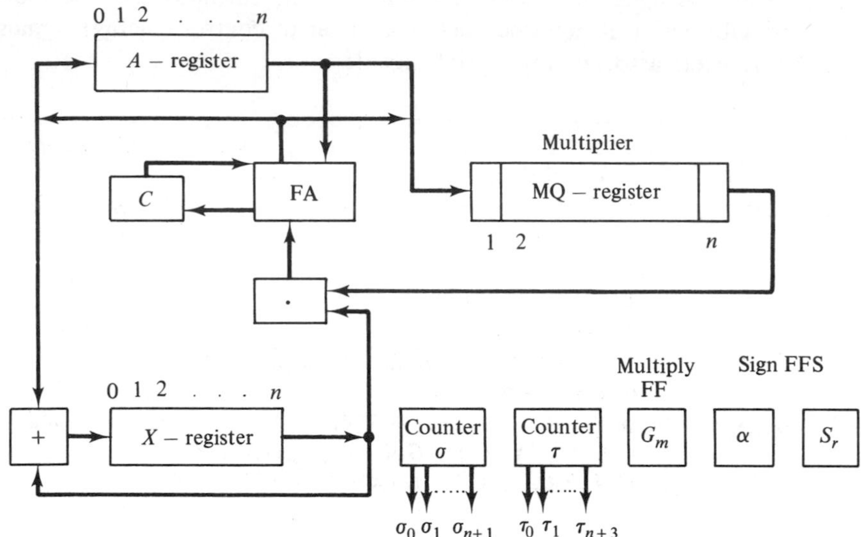

Figure 8.2 Layout for serial multiplier.

S_r = flip-flop to store the sign of the multiplier.

τ = modulo-$(n + 4)$ binary counter. It is used to control the major cycles of the multiplication process.

MQ = n-Bit register to hold the multiplier at the beginning of the operation. During the multiplication process MQ holds the remaining bits of the multiplier and the lower-order bits of the product.

A single-address multiply instruction is usually interpreted as performing:

$$[A] \times [M] \longrightarrow A$$

where A is the accumulator and M is a memory address specified by the instruction. When the operand is fetched from location M it will be available in the memory information register (MIR).

The actual multiplication by repeated addition starts with the magnitudes of the multiplier and the multiplicand stored in the MQ and X registers, respectively. The A register must be cleared and the product is derived in the combined A, MQ register with its sign and the most significant bits in A. Therefore, before implementing the repeated addition and shift process, the magnitude of each operand must be stored in the proper positions. This positioning of the operands is implemented by first transferring the magnitude of A into the X register, while the contents of the MIR are transferred into A by serial shifting. This requires one major cycle. In the second major cycle, the magnitude of A (the multiplier) is transferred into MQ. In both cycles the serial transfer and the complementation (if necessary) is done through the adder.

Once the magnitudes of the operands are in the proper positions, the repeated addition and subtraction process is straightforward. Depending on the multiplier bit being a 1 or a 0, the contents of the X register or 0 are added to the contents of the A register. This addition takes place serially, just as described in the previous subsection, and requires n consecutive control signals to add the $n + 1$ bits of the multiplicand. Let us refer to this process by an addition cycle and use the serial adder of the previous subsection.

After the first addition cycle, the (A, MQ) register must be shifted one position to the right, and then the next addition cycle starts. This process continues until all the multiplier bits are processed and requires n major cycles. In the next two major cycles, the sign of the product is derived and the product is transformed into 2's complement if necessary (if the result is negative). This requires two major cycles since the product is a double-length word. This step is performed in order to maintain the assumption of the number representation in the 2's complement form.

A sequence of transfers and the associated timing signals for this serial implementation of multiplication by repeated addition is listed below. For simplicity, the MQ register will be referred to as the Q register and the MIR register as the M register.

		Comments
START	$1 \longrightarrow G_m, 0 \longrightarrow \tau, 0 \longrightarrow \sigma$	
$G_m \tau_0 \cdot \sigma_0$	$A_0 \longrightarrow \alpha, A_0 \longrightarrow C, \sigma + 1 \longrightarrow \sigma$	
$G_m \cdot \tau_0 \cdot (\sigma_1 + \cdots + \sigma_n)$	$L(A) \longrightarrow R(A), L(X) \longrightarrow R(X)$ $L(M) \longrightarrow R(M)$ $M_n \longrightarrow A_0,$ $\bar{\alpha} \cdot A_n + \alpha \cdot S^2(\bar{A}_n, C, 0) \longrightarrow X_0$ $\alpha \cdot K^2(\bar{A}_n, C, 0) \longrightarrow C,$ $\sigma + 1 \longrightarrow \sigma$	Transfer $\lvert A \rvert \longrightarrow X$ using the adder and $M \longrightarrow A$
$G_m \cdot \tau_0 \cdot \sigma_{n+1}$	Same as $(\sigma_1 + \cdots + \sigma_n)$ plus $(\tau + 1) \longrightarrow \tau$	

By the end of this τ_0 cycle, the magnitude of the multiplicand is available in the X register, while the multiplier is stored in the A register as a signed number. In the next cycle (τ_1), the magnitude of A is transferred into the MQ register through the adder. By the end of this cycle, the magnitude of the multiplier is available in the MQ register.

$G_m \cdot \tau_1 \cdot \sigma_0$	$A_0 \longrightarrow S_r,\ A_0 \longrightarrow C,$ $\sigma + 1 \longrightarrow \sigma$
$G_m \cdot \tau_1 \cdot (\sigma_1 + \cdots + \sigma_n)$	$L(A) \longrightarrow R(A),\ L(Q) \longrightarrow R(Q),$ $0 \longrightarrow A_0,$ $\bar{S}_r \cdot \mathrm{A}_n + S_r \cdot \mathrm{S}^2(\bar{\mathrm{A}}_n, C, 0) \longrightarrow \mathrm{Q}_1$ $S_r \cdot \mathrm{K}^2(\bar{A}_n, C, 0) \longrightarrow C,$ $\sigma + 1 \longrightarrow \sigma$
$G_m \cdot \tau_1 \cdot \sigma_{n+1}$	$L(A) \longrightarrow R(A),$ $0 \longrightarrow A_0, 0 \longrightarrow C,$ $\sigma + 1 \longrightarrow \sigma, \tau + 1 \longrightarrow \tau$

The following n cycles ($\tau_2, \ldots, \tau_{n+1}$) are used for the repeated add/shift operation:

$G_m \cdot \tau_2 \cdot (\sigma_0 + \cdots + \sigma_n)$	$\rho(X) \longrightarrow X,\ L(A) \longrightarrow R(A)$ $\mathrm{S}^2(A_n, Q_n \cdot \mathrm{X}_n, C) \longrightarrow A_0$ $\mathrm{K}^2(A_n, Q_n \cdot \mathrm{X}_n, C) \longrightarrow C$	Add cycle
$G_m \cdot \tau_2 \cdot \sigma_{n+1}$	$L(A, Q) \longrightarrow R(A, Q)$ $0 \longrightarrow A_0$ $\sigma + 1 \longrightarrow \sigma, \tau + 1 \longrightarrow \tau$	Shift cycle
$G_m \cdot (\tau_3 + \cdots + \tau_{n+1})$ $\cdot (\sigma_0 + \cdots + \sigma_n)$	Same as $\tau_2 \cdot \sigma_0$	Add cycles
$G_m \cdot (\tau_3 + \cdots + \tau_{n+1}) \cdot \sigma_{n+1}$	Same as $\tau_2 \cdot \sigma_{n+1}$	Shift cycles

By the end of the above cycle (τ_{n+1}), the magnitude of the product is derived in the combined (A, Q) register. The next two cycles are used to derive the sign of the product, and if the result is negative, the 2's complement of the double-length product is derived:

$G_m \cdot \tau_{n+2} \cdot \sigma_0$	$\alpha \oplus S_r \longrightarrow \alpha,\ \alpha \oplus S_r \longrightarrow C,$ $\sigma + 1 \longrightarrow \sigma$
$G_m \cdot \tau_{n+2} \cdot (\sigma_1 + \cdots + \sigma_n)$	$L(A, Q) \longrightarrow R(A, Q),$ $\bar{\alpha} Q_n + \alpha \cdot \mathrm{S}^2(\bar{Q}_n, C, 0) \longrightarrow A_0,$ $\alpha \cdot \mathrm{K}^2(\bar{Q}_n, C, 0) \longrightarrow C,$ $\sigma + 1 \longrightarrow \sigma$
$G_m \cdot \tau_{n+2} \cdot \sigma_{n+1}$	Same as $(\sigma_1 + \cdots + \sigma_n)$ plus $\tau + 1 \longrightarrow \tau$
$G_m \cdot \tau_{n+3} \cdot (\sigma_0 + \cdots + \sigma_n)$	Same as $\tau_{n+2} \cdot (\sigma_1 + \cdots + \sigma_n)$
$G_m \cdot \tau_{n+3} \cdot \sigma_{n+1}$	$\alpha \longrightarrow A_0, 0 \longrightarrow G_m, 0 \longrightarrow \sigma, 0 \longrightarrow \tau$

8.1.3 Division

In this section, we shall implement the division of binary numbers using the nonrestoring division method. Consider fractional numbers in 2's complement representation. The layout of the arithmetic unit for performing division is shown in Fig. 8.3. The functional logic subunits are the same as

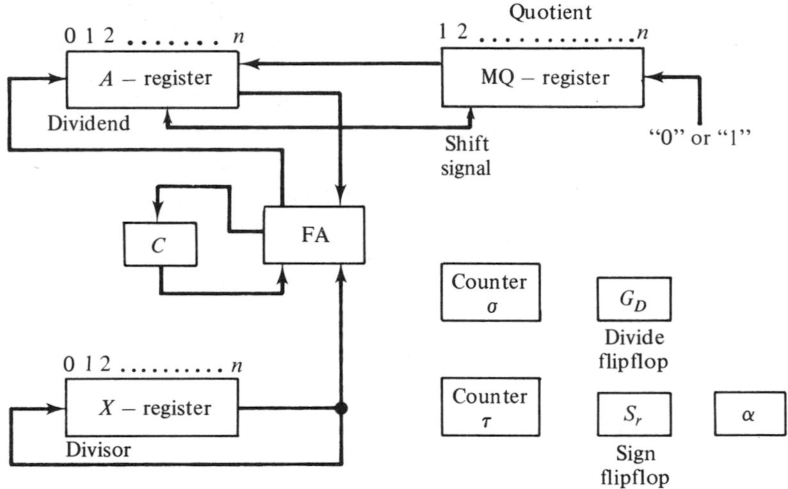

Figure 8.3 Layout for serial binary division.

in multiplication, except for two differences. The first is the use of the G_D flip-flop for division. The second is adding an additional shift left capability to both the A and MQ registers. The actual division starts with the dividend and the divisor in the A and X registers, respectively. The quotient is to be derived in the MQ register. The algorithm for the nonrestoring division is implemented on the magnitudes of the operands; the sign of the quotient is derived by inspection of the signs of the operands. Therefore, before the actual division starts, the magnitudes of the operands must be stored in proper locations of A and X. Similar to implementing the serial multiplication, this positioning process is performed in two major cycles. In the first cycle, the contents of the memory location M(the divisor) is transferred into X, while the magnitude of the dividend in the A register is formed through the adder. During the second cycle the magnitude of the divisor (X register) is derived. Once the magnitudes of both the dividend and the divisor are stored in the A register and X register, respectively, the actual division is straightforward. For simplicity, the MQ register will be denoted by Q.

The nonrestoring division algorithm, discussed in Chapter 3, involves repeated addition or subtraction and shifts. For serial division each addition or subtraction is performed serially as described in Section 8.1.1. A first test subtraction of the divisor from the dividend checks if there is any overflow condition. If the result of this subtraction is negative, it indicates no overflow; otherwise an overflow condition exists and the operation stops. Let us assume that the operands are properly scaled and therefore overflow conditions do not exist. The first quotient bit is derived by shifting the combined (A, Q) register one place to the left and subtracting X from A. A 0 or a 1 quotient bit is entered into Q_n depending on whether the subtraction result is negative or nonnegative, respectively. The second quotient bit is derived by first shifting the (A, Q) register to the left one position and then either adding or subtracting X according to the previous quotient bit being a 0 or a 1, respectively. The remaining quotient bits are obtained by repeating the process described for the second quotient bit. This repeated process requires n major τ cycles. By the end of the division process, the magnitude of the quotient is derived in the Q register, while the sign of the product is stored in the sign flip-flop (S).

Assuming subtraction is performed by the addition of the 2's complement, the timing signals and the associated transfers for the nonrestoring division is as described below:

START	$1 \longrightarrow G_D,\ 0 \longrightarrow \sigma,\ 0 \longrightarrow \tau$
$G_D \cdot \tau_0 \cdot \sigma_0$	$A_0 \longrightarrow \alpha,\ A_0 \longrightarrow C,$ $\sigma + 1 \longrightarrow \sigma$
$G_D \cdot \tau_0 \cdot (\sigma_1 + \cdots + \sigma_n)$	$L(A) \longrightarrow R(A),\ L(X) \longrightarrow R(X),$ $L(M) \longrightarrow R(M),\ L(Q) \longrightarrow R(Q)$ $M_n \longrightarrow X_0,\ 0 \longrightarrow Q_1$ $\bar{\alpha} \cdot A_n + \alpha \cdot S^2(\bar{A}_n, C, 0) \longrightarrow A_0$ $\alpha \cdot K^2(\bar{A}_n, C, 0) \longrightarrow C,$ $\sigma + 1 \longrightarrow \sigma$
$G_D \cdot \tau_0 \cdot \sigma_{n+1}$	Same as in $\tau_0 \cdot (\sigma_1 + \cdots + \sigma_n)$ plus $\tau + 1 \longrightarrow \tau$

By the end of the above cycle (τ_0), the magnitude of the dividend is derived in the accumulator, while the divisor is transferred from the *MIR* (denoted by M) into the X register. Also, the sign of the dividend is stored in the flip-flop α to be used later to derive the sign of the product. The Q register has been cleared to 0. In the next cycle, τ_1, the sign of the divisor is stored in the flip-flop S_r and the magnitude of the divisor is derived in the X register using the adder:

$G_D \cdot \tau_1 \cdot \sigma_0$	$X_0 \longrightarrow S_r,\ X_0 \longrightarrow C,$ $\sigma + 1 \longrightarrow \sigma$
$G_D \cdot \tau_1 \cdot (\sigma_1 + \cdots + \sigma_n)$	$L(X) \longrightarrow R(X),$ $\bar{S}_r X_n + S_r \cdot S^2(\bar{X}_n, C, 0) \longrightarrow X_0$ $S_r \cdot K^2(\bar{X}_n, C, 0) \longrightarrow C,$ $\sigma + 1 \longrightarrow \sigma$
$G_D \cdot \tau_1 \cdot \sigma_{n+1}$	Same as in $\tau_1(\sigma_1 + \cdots + \sigma_n)$ plus $\tau + 1 \longrightarrow \tau,\ 1 \longrightarrow C$

Now that the magnitudes of both the dividend and the divisor are derived in the A and the X registers, respectively, the first subtraction in the nonrestoring division algorithm starts and is described by transfers occurring at τ_2. The subtraction takes place during the first $n + 1$ minor cycles of τ_2. In the last minor cycle, σ_{n+1}, the combined (A, Q) register is shifted to the left and the sign of the subtraction result is used to derive the quotient bit. If the subtraction result is negative, a 1 is transferred into C since the next cycle has to be a subtraction cycle.

$G_D \cdot \tau_2 \cdot (\sigma_0 + \sigma_1 + \cdots + \sigma_n)$	$L(A) \longrightarrow R(A),\ \rho(X) \longrightarrow X$ $S^2(A_n, \bar{X}_n, C) \longrightarrow A_0, K^2(A_n, \bar{X}_n, C) \longrightarrow C$
$G_D \cdot \tau_2 \cdot \sigma_{n+1}$	$R(A, Q) \longrightarrow L(A, Q),\ A'_0 \longrightarrow Q_n,\ A'_0 \longrightarrow C$ $\sigma + 1 \longrightarrow \sigma, \tau + 1 \longrightarrow \tau$
$G_D \cdot \tau_3 \cdot (\sigma_0 + \cdots + \sigma_n)$	$L(A) \longrightarrow R(A),\ \rho(X) \longrightarrow X,$ $\bar{Q}_n \cdot S^2(A_n, X_n, C) + Q_n \cdot S^2(A_n, \bar{X}_n, C) \longrightarrow A_0,$ $\bar{Q}_n \cdot K^2(A_n, X_n, C) + Q_n \cdot K^2(A_n, \bar{X}_n, C) \longrightarrow C,$ $\sigma + 1 \longrightarrow \sigma$
$G_D \cdot \tau_3 \cdot \sigma_{n+1}$	$R(A, Q) \longrightarrow L(A, Q),\ \bar{A}_0 \longrightarrow Q_n,\ \bar{A}_0 \longrightarrow C,$ $(\sigma + 1) \longrightarrow \sigma, (\tau + 1) \longrightarrow \tau$

During the next $n - 1$ major cycles, we perform either a subtraction followed by a shift or an addition followed by a shift. The addition or the subtraction is performed depending on whether the sign from the previous operation is negative or positive, respectively, which is equivalent to the previous quotient bit being a 0 or a 1, respectively.

$G_D \cdot (\tau_4 + \cdots + \tau_{n+1}) \cdot (\sigma_0 + \cdots + \sigma_n)$	Same as in $\tau_3(\sigma_0 + \cdots + \sigma_n)$
$G_D \cdot (\tau_4 + \cdots + \tau_{n+1}) \cdot (\sigma_{n+1})$	Same as in $\tau_3 \cdot \sigma_{n+1}$

By the end of the τ_{n+1} cycle, the magnitude of the quotient is derived in the Q register. The next major cycle is the last cycle and is used to derive the sign of the quotient and convert the quotient into 2's complement form if it is negative:

$G_D \tau_{n+2} \cdot \sigma_0$	$\alpha \oplus S_r \longrightarrow \alpha,\ \alpha \oplus S_r \longrightarrow C,$ $(\sigma + 1) \longrightarrow \sigma$
$G_D \cdot \tau_{n+2} \cdot (\sigma_1 + \cdots + \sigma_n)$	$L(Q) \longrightarrow R(Q),$ $\bar{\alpha} Q_n + \alpha \cdot S^2(\bar{Q}_n, C, 0) \longrightarrow Q_1$ $\alpha \cdot K^2(\bar{Q}_n, C, 0) \longrightarrow C,$ $(\sigma + 1) \longrightarrow \sigma$
$G_D \cdot \tau_{n+2} \cdot \sigma_{n+1}$	$0 \longrightarrow G_D,\ 0 \longrightarrow \tau,$ $(\sigma + 1) \longrightarrow \sigma$

In the last major cycle the quotient is transformed into the 2's complement representation, while the remainder is left in the *A* register. The sign of the quotient is derived in α. It is also possible to adjust the last cycle such that the quotient with its sign is stored in register *A*.

8.1.4 A Serial Binary Arithmetic Unit

A serial binary arithmetic unit capable of performing addition, subtraction, multiplication, and division of binary numbers can be implemented by combining the various logic circuits used in implementing the individual operations. Let us assume fractional binary numbers in the 2's complement representation. The registers used in the various operations are the *A* register (accumulator), the *X* register, and the *MQ* register. Assuming that each

Table 8.1 *Registers of a Serial Binary Arithmetic Unit*

	Operand which it holds				
Register	*Addition*	*Subtraction*	*Multiplication*	*Division*	*Shift capabilities*
A	Addend	Minuend	Product, MS half	Dividend remainder	All four shifts
X	Augend	Subtrahend	Multiplicand	Divisor	Right circular
Q	—	—	Multiplier, LS half of product	Quotient	Right, left

operand consists of n bits as the number digits and 1 bit for the sign, then the A and the X registers consist of $n + 1$ cells each, while the MQ register consists of n cells. A summary of the shifting capabilities built into and the operand held by each register is shown in Table 8.1. Other than the shifting capabilities shown in the table, the contents of each register can be complemented bit by bit (1's complement) upon receiving a complement signal.

Figure 8.4 shows a block diagram for a serial binary arithmetic unit.

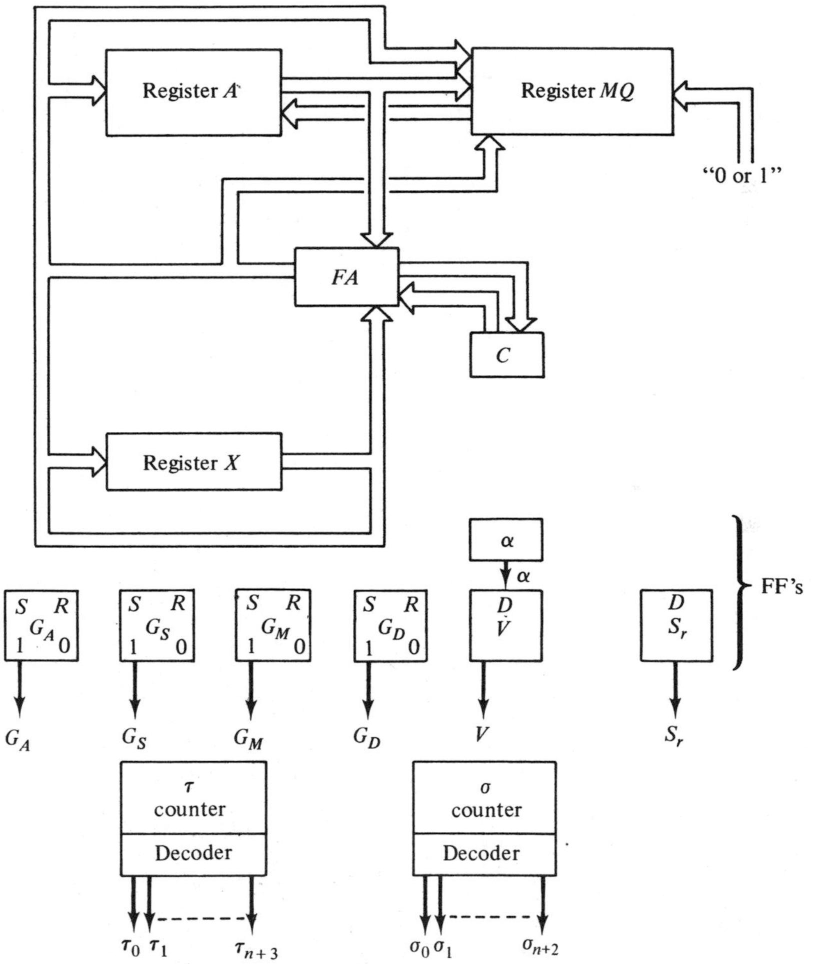

Figure 8.4 A serial binary arithmetic unit. This figure illustrates the layout of the basic hardware. The control signals are shown in Fig. 8.5.

The diagram shows the transfer paths for the combined transfer equations of addition, subtraction, multiplication, and division. The double-line arrows indicate the information bits, while the single-line arrows indicate the control signals. The timing control signals used are generated by the control logic shown in Fig. 8.5. In our previous discussion on multiplication and division, the algorithms implemented are based on multiplication or division of the operands after they are forced to be positive. In other words, the actual multiplication or division is performed on the magnitudes of the operands. Since we are assuming 2's complement representation, for all numbers in the

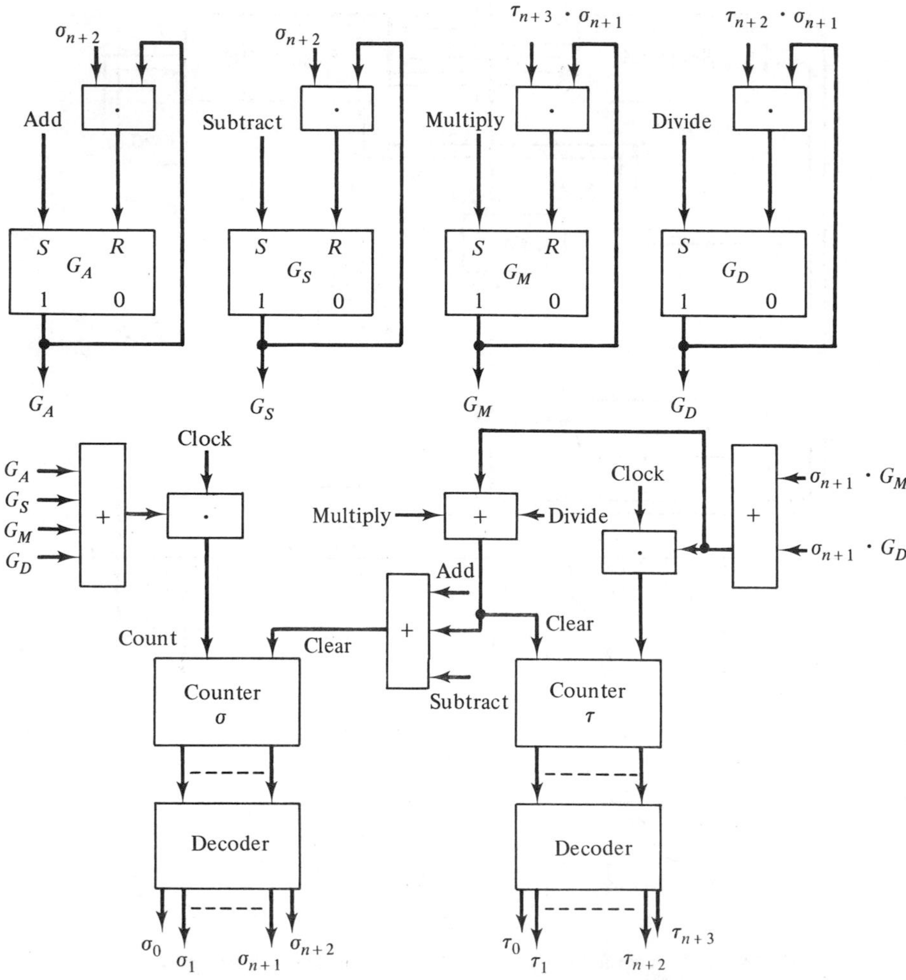

Figure 8.5 Control signals for the serial arithmetic unit.

machine multiplication and division will require complementation of negative numbers at the beginning of the operation. Similarly, if the result of the operation is negative, the 2's complement form has to be obtained in order to preserve the numbers representation within the machine. These two steps in the implementation of multiplication and division are implemented by passing the number to be complemented through the adder. An alternative implementation could have additional hardware to serially derive the 2's complement of a number. The cost of a separate 2's complement would be compensated by the gain in speed when compared to using the adder.

All the flip-flops used here, including the cells of any particular register, are assumed to be clocked flip-flops. Therefore synchronization of the occurrence of any transfer with the clock pulse is provided by the flip-flops themselves. The information transfers into any register shown in Fig. 8.4 take place only when a timing or control signal occurs; otherwise the contents of the register remains unchanged. For example, consider the transfer

$$\alpha X + \bar{\alpha} Y \longrightarrow Y$$

where α is a control signal and X and Y represent two different registers. Transfers of this form are represented by showing the transfer $\alpha X \longrightarrow Y$. This must be interpreted as follows: Whenever the conditions for transferring new information into Y do not exist, the register remains unchanged rather than transferring 0 into it. If the register is to be cleared, then a transfer of 0s into it is indicated explicitly.

Serial arithmetic requires less hardware to implement than parallel arithmetic. This saving in hardware is achieved at the expense of a slower speed than that for parallel operations.

The implementation of addition and subtraction, presented in Section 8.1.1, requires $n + 3$ time slices for the addition or subtraction of $(n + 1)$-bit numbers. Therefore, for $(n + 1)$-bit numbers, the addition time is $n + 3$ times that of a single-bit addition time. The single-bit addition time (clock period) is determined by the logic of the full-adder and the carry flip-flop. A single-bit addition requires, at most, two gate levels of delay plus the delay in the carry flip-flop.

The serial multiplication requires $n + 4$ major cycles, $\tau_0, \tau_1, \ldots, \tau_{n+3}$. Each major cycle requires $n + 2$ clock pulses. Thus, the multiplication can be performed in $(n + 4)(n + 2)$ times the single-bit addition time. Similarly, serial division requires $(n + 3)(n + 2)$ single-bit addition time. If we denote the switching time for the carry flip-flop as T_2, the single-bit addition time is equal to $2T_1 + T_2$, where T_1 is the switching time for a logic gate. The timing for the various serial operations is shown in Table 8.2 as a function of T_1 and T_2.

Table 8.2 Timing for Serial Operations

Operation	*Timing*
Addition or subtraction	$(n + 3)(2T_1 + T_2)$
Multiplication	$(n + 4)(n + 2)(2T_1 + T_2)$
Division	$(n + 3)(n + 2)(2T_1 + T_2)$

8.2 PARALLEL BINARY ARITHMETIC UNIT

In parallel machines all the bits of an operand are available at once, and therefore information is processed in parallel. As we shall see by the end of this section, parallel implementation of the various arithmetic operations is straightforward and simple when compared to serial implementation. It is much faster but requires more hardware.

8.2.1 Parallel Addition and Subtraction

In Chapter 5, we have already seen the basic adder or subtractor stage. Parallel addition and subtraction requires as many such adders (or subtractors) as the number of bits in an operand. An n-stage parallel adder is shown in Fig. 8.6. If the half-adder and the full-adders shown in Fig. 8.6 are replaced by a half-subtractor and full-subtractors, respectively, we obtain a parallel binary subtractor.

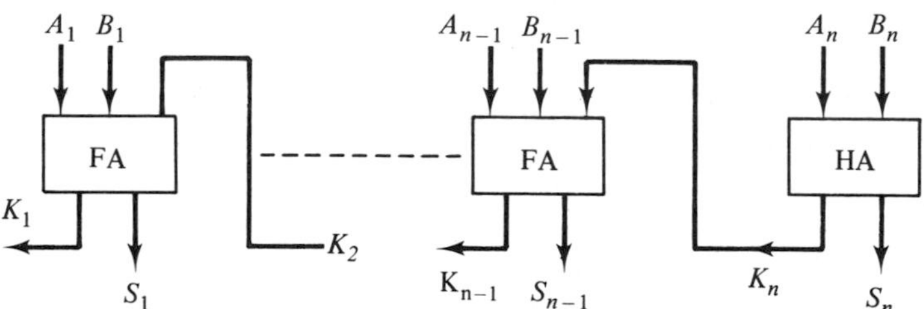

Figure 8.6 Parallel adder.

Consider the addition and subtraction algorithm for numbers in the 2's complement representation. Figure 8.7 shows the basic layout to implement the algorithm in a parallel arithmetic unit. During the operation register A (the accumulator) holds the addend (or the minuend), while register X holds the augend (or the subtrahend). The result (the sum or the difference) is stored in the A register. The G flip-flop indicates whether the operation is

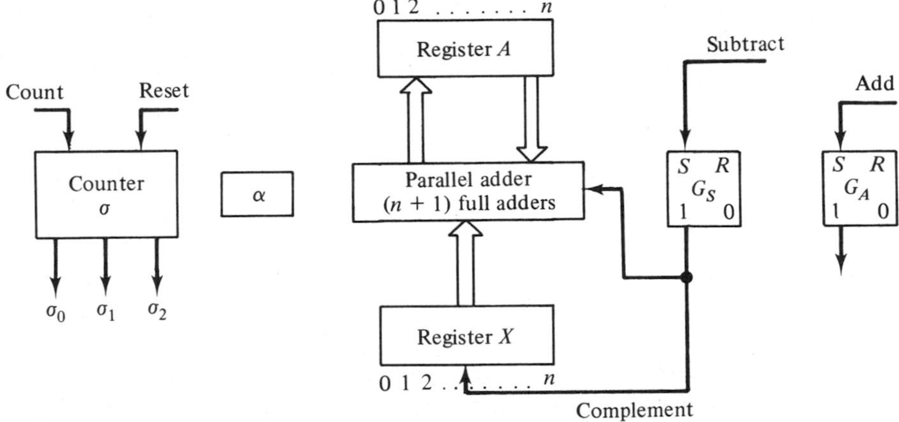

Figure 8.7 Basic layout for addition and subtraction in an arithmetic unit.

addition or subtraction. In the case of subtraction the output of the G flip-flop is used to generate the complement signal for the register X and add 1 to the least significant stage of the adder in order to form the 2's complement. The sequence of timing signals and the corresponding transfers are described below:

START	$1 \longrightarrow G_A$ for add, $1 \longrightarrow G_S$ for subtract, $0 \longrightarrow \sigma$
$(G_S + G_A)\cdot\sigma_0$	$G_S\bar{X} + G_AX \longrightarrow X, \overline{A_0 \oplus [G_AX_0 + G_SX_0']} \longrightarrow \alpha, (\sigma + 1) \longrightarrow \sigma$
σ_1	$S(A, X, G_S) \longrightarrow A, (\sigma + 1) \longrightarrow \sigma$
σ_2	$\alpha\cdot(X_0 \oplus A_0) \longrightarrow V, 0 \longrightarrow G_S, 0 \longrightarrow G_A$

The overflow flip-flop V and the α flip-flop are described in the earlier section on serial addition and subtraction.

The addition or subtraction operation requires three cycles (time slice) to perform. The σ_1 cycle must be large enough to allow the propagation of the carry through the ripple adder. The worst condition would result if a carry were generated in the least significant adder and then it propagated to the most significant one. Assuming two gate levels to generate a carry from a full adder, the carry propagation might require up to $2n$ gate levels for an n-bit adder. If it takes two gate levels to generate the sum of a full-adder, then the addition cycle σ_1 must last for $2n + 2$ gate-switching time. Assuming all the cycles $\sigma_0, \sigma_1, \ldots,$ have the same duration, then the addition or subtraction requires $3(2n + 2)$ gate-switching time, i.e., $(6n + 6)T_1$. This addition time is definitely smaller than but not so small as $1/n$ of that for serial addition (see Table 8.2). The ratio is less than $1/n$ due to delay in the carry propagation.

8.2.2 Multiplication

Consider the binary multiplication by repeated addition, discussed in Chapter 3. An implementation of the multiplication of binary numbers in the 2's complement representation is presented in this subsection. The multiplication operation employs three registers, A, X, and Q. The multiplication starts with the multiplicand and the multiplier occupying registers X and Q, respectively. The product is derived in the combined register (A, Q). The basic layout and the necessary paths are shown in Fig. 8.8. If we assume that the multiplier and the multiplicand are brought from memory into registers A and X, respectively, then, just as in the serial multiplication, a few preparatory steps are necessary before the repeated add/shift operations start.

First, the sign of the product is derived in the sign flip-flop α, while the

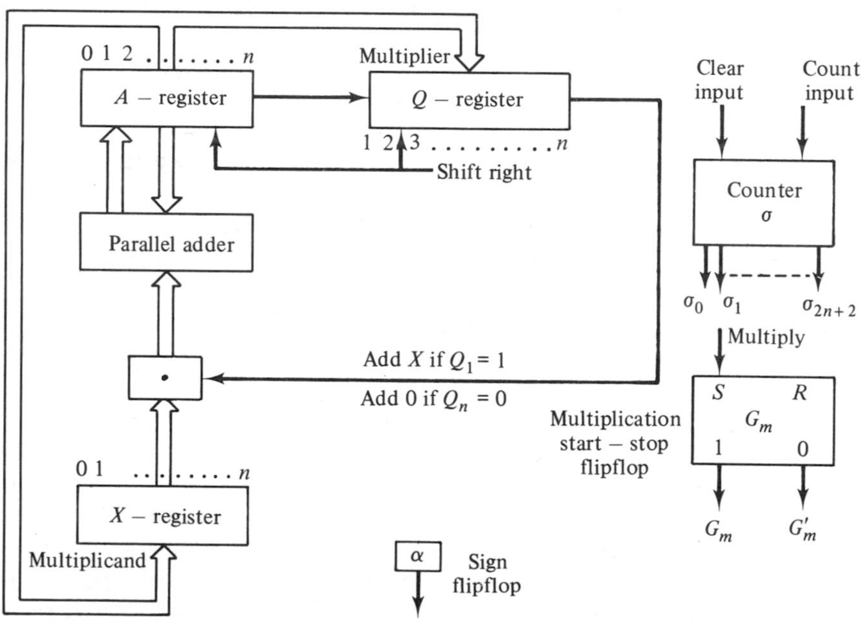

Figure 8.8 Basic layout for multiplication in a parallel arithmetic unit.

magnitude of the multiplicand (the X register) is derived in the X register. Also, the magnitude of the multiplier (the A register) is transferred into the MQ register and the A register is cleared to 0. These steps are described by the transfer equations during σ_0 through σ_2. The repeated additions and shifts are then performed in the following $2n$ cycles. A final cycle is then used to complement the product if it is negative. The timing signals and the corresponding transfers for the multiplication operation are described below:

		Comments
START	$1 \longrightarrow G_m,\ 0 \longrightarrow \sigma$	
$G_m\sigma_0$	$A_0 \oplus X_0 \longrightarrow \alpha,$	Store sign of product
	$A_0 \cdot S(\bar{A}, 1) + \bar{A}_0 \cdot A \longrightarrow A,$ $(\sigma + 1) \longrightarrow \sigma$	Makes A positive
$G_m\sigma_1$	$R(A) \longrightarrow Q,$	Transfer A into Q
	$X_0 \cdot S(\bar{X}, 1) + \bar{X}_0 X \longrightarrow A,$ $(\sigma + 1) \longrightarrow \sigma$	Make X positive in A
$G_m\sigma_2$	$A \longrightarrow X,\ 0 \longrightarrow A,$ $(\sigma + 1) \longrightarrow \sigma$	Transfer A into X and clear A
$G_m(\sigma_3 + \sigma_5 + \sigma_7 + \cdots + \sigma_{2n+1})$	$S(A, Q_n \cdot X) \longrightarrow A,$ $(\sigma + 1) \longrightarrow \sigma$	Repeated add and shift process (both rows)
$G_m(\sigma_4 + \sigma_6 + \cdots + \sigma_{2n+2})$	$L(A, Q) \longrightarrow R(A, Q),$ $0 \longrightarrow A_0,\ (\sigma + 1) \longrightarrow \sigma$	
$G_m\sigma_{2n+3}$	$\alpha \cdot S(\bar{A}, 1) + \bar{\alpha} A \longrightarrow A,$ $0 \longrightarrow G_m$	Put the product in a signed 2's complement form

8.2.3 Division

Assume fractional numbers with negative numbers represented in the 2's complement form. The division algorithm to be implemented is the non-restoring division which is described in Chapter 3. Binary division usually employs the same registers which are used in the multiplication operation. Figure 8.9 shows the basic hardware with the required transfer paths. The operation starts after the dividend and the divisor are brought (from the memory) into the A and X registers. Before the actual division takes place,

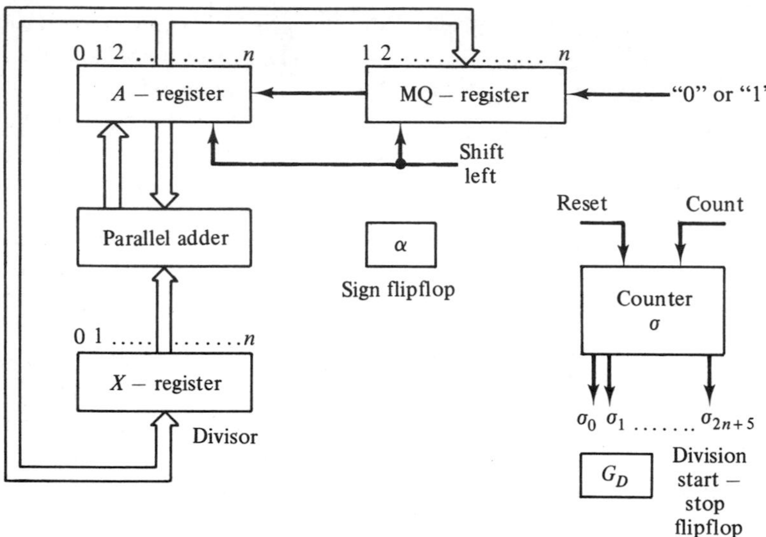

Figure 8.9 Basic layout for binary division in a parallel arithmetic unit.

each dividend and divisor is forced into a positive number equivalent to its magnitude. Thus, the division is to be performed on positive numbers represented by their magnitudes. The quotient is derived in MQ, while the remainder will be left in A. The sign of the quotient is derived in the sign flip-flop α.

With the magnitudes of the dividend and the divisor in A and X, respectively, a test subtraction of X from A determines the overflow condition. If the result is ≥ 0, this means that the dividend $\geq$ divisor, and an overflow signal is generated. If the result is < 0, this indicates no overflow, and the division continues after restoring the A register by adding X to it. We shall assume, in this section, that the operands are properly scaled and hence that no overflow check is necessary. The first quotient bit is obtained by shifting the combined (A, Q) register one place to the left and subtracting X from A. The quotient bit is a 0 if the result of the subtraction is less than zero, and a 1 if the result is greater than or equal to zero. The second quotient bit is obtained by first shifting the (A, Q) register one place to the left and then either adding X or subtracting X, depending on whether the previous quotient bit is a 1 or a 0, respectively. This process is to be repeated until the last bit of the quotient is derived.

Let us assume that the subtraction is performed by the addition of the 2's complement and therefore the adder shown in Fig. 8.9 performs addition and subtraction. The timing signals and the associated transfers for the nonrestoring division are shown:

START	$1 \longrightarrow G_D, 0 \longrightarrow \sigma$
$G_D\sigma_0$	$A_0 \oplus X_0 \longrightarrow \alpha,$ $A_0 S(\bar{A}, 1) + \bar{A}_0 A \longrightarrow A,$ $(\sigma + 1) \longrightarrow \sigma$
$G_D\sigma_1$	$A \longrightarrow Q,$ $X_0 S(\bar{X}, 1) + \bar{X}_0 X \longrightarrow A,$ $(\sigma + 1) \longrightarrow \sigma$
$G_D\sigma_2$	$A \longrightarrow X, Q \longrightarrow A, 0 \longrightarrow Q,$ $(\sigma + 1) \longrightarrow \sigma$
$G_D\sigma_3$	$R(A, Q) \longrightarrow L(A, Q), (\sigma + 1) \longrightarrow \sigma$
$G_D\sigma_4$	$S(A, \bar{X}, 1) \longrightarrow A, (\sigma + 1) \longrightarrow \sigma$
$G_D(\sigma_5 + \sigma_7 + \cdots + \sigma_{2n+3})$	$R(A, Q) \longrightarrow L(A, Q),$ $\bar{A}_0 \longrightarrow Q_n, (\sigma + 1) \longrightarrow \sigma$
$G_D(\sigma_6 + \sigma_8 + \cdots + \sigma_{2n+4})$	$Q_n S(A, \bar{X}, 1) + \bar{Q}_n S(A, X) \longrightarrow A,$ $(\sigma + 1) \longrightarrow \sigma$
$G_D\sigma_{2n+5}$	$\alpha \cdot Q + \bar{\alpha} Q \longrightarrow Q,$ $0 \longrightarrow G_D$

The first three cycles (σ_0 through σ_2) are used to derive the sign of the quotient and the magnitude of each of the operands. The first subtraction is performed at σ_4 after the shifting of the combined (A, Q) register to the left one position. The first quotient bit which is equal to $\bar{A}_0$ (sign of the subtraction result) is shifted into the Q register at σ_5, while the (A, Q) register is being shifted in preparation for the derivation of the next quotient bit. At σ_6 the divisor is added or subtracted from the dividend depending on whether the result of the previous operation is negative or positive, respectively. The shift and add (or subtract) operations are repeated until all the quotient bits are derived. The last cycle is used to form the 2's complement of the quotient if it happens to be negative..

8.2.4 A Parallel Binary Arithmetic Unit

A parallel arithmetic unit employs three basic registers: the A register (accumulator), the X register, and the Q register. The capacity of each register and its function while performing an arithmetic operation is the same as described in the serial unit. The only shift capability required is a right and a left shift for both registers A and Q. Figure 8.10 shows a block diagram of a parallel binary arithmetic unit. Similar to the serial arithmetic unit, the operands are assumed to be represented by their magnitudes in the cases of both multiplication and division. The various timing control signals shown are generated in Fig. 8.11. The counter σ is a binary counter with a maximum count of $2n + 5$, where $n + 1$ is the length of the operands. The counter is

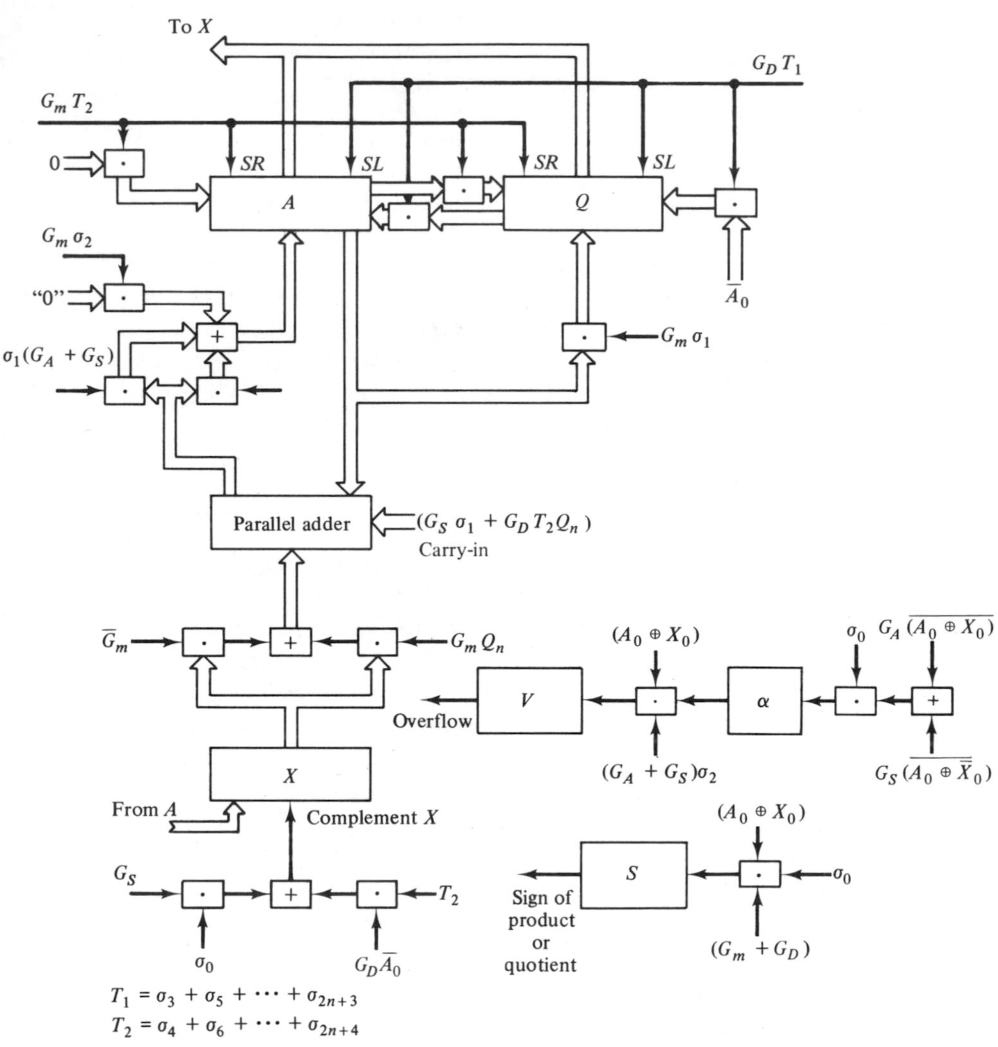

Figure 8.10 A parallel arithmetic unit.

reset to 0 at the beginning of each operation. The addition or subtraction operation uses the first three outputs only (σ_0, σ_1, σ_2), while the multiplication and division uses up the $2n + 3$ and $2n + 5$ outputs, respectively. In the case of addition and subtraction the corresponding start-stop flip-flops G_A and G_S are reset after the count σ_2, and consequently the operation ends at this time.

In Fig. 8.10, the double arrows into or from a register indicate a parallel transfer of information, and therefore the logic circuitry must be repeated

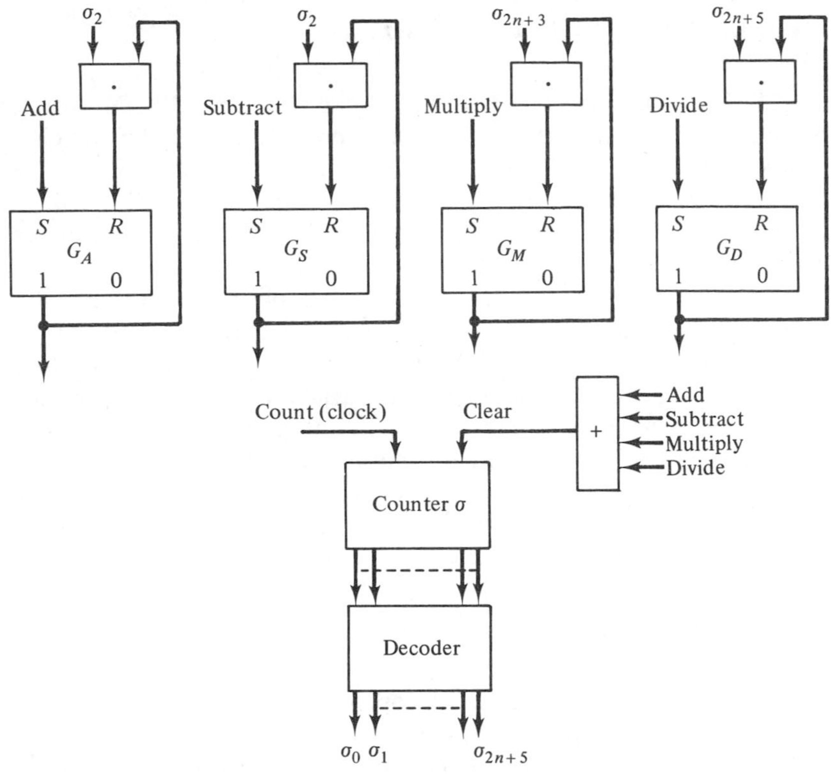

Figure 8.11 Control signals for a parallel arithmetic unit.

for every cell of the register. The only exception is where the input is shown at the right or the left side of the register; in such cases it indicates a single-bit transfer. The same rule applies to the control signals, which are indicated by single arrows.

Recall that previously we have denoted gate-switching time by T_1 and that both the sum and the carry of an adder stage can be generated by two gate levels. With these assumptions the timing for the parallel implementation of the various arithmetic operations can be derived as functions of T_1. Table 8.3 gives a summary of these timings for $(n + 1)$-bit operands.

Table 8.3 *Timing for Parallel Implementation*

Operation	*Timing*
Addition or subtraction	$3(n + 1)(2T_1)$
Multiplication	$(2n + 4)(n + 1)2T_1$
Division	$(2n + 6)(n + 1)2T_1$

8.3 DECIMAL ARITHMETIC UNIT

A decimal adder receives two binary coded decimal (BCD) operands and produces their sum in the same code. Therefore, a decimal adder circuit depends on the code used to represent the decimal digits. It is not our intention, in this section, to present all the possible decimal adders. The (8, 4, 2, 1) and the excess-3 BCD adders are presented to illustrate the problems associated with the design of decimal adders.

A typical stage of a decimal adder receives two 4-bit operands and a carry from the previous stage as its inputs, while its outputs consist of a 4-bit sum and a carry to the next higher stage. This is a network, similar to the parallel binary adder, which has nine input variables and five outputs. Therefore, deriving minimum expressions for the outputs would be an almost impossible task. One possible way to achieve decimal addition is to add the coded digits as if they represent binary numbers and then use additional logic to convert the result into the decimal system. This approach is used for the (8, 4, 2, 1) and the excess-3 adders presented below.

8.3.1 The (8, 4, 2, 1) Decimal Adder

Let us denote the 4-bit code for the operands A and B by (A_1, A_2, A_3, A_4) and (B_1, B_2, B_3, B_4), respectively. The bits A_1, A_2, A_3, and A_4 have the weights 8, 4, 2, and 1, respectively—and similarly for the bits of B. Using a four-stage binary adder to add A and B results in their sum base-16.

To have a base-10 sum, some corrections are needed. These corrections are determined by examining the following three different cases:

Case 1: If the sum of A and B is less than 10, then the sum base-10 is the same as the sum base-16, and no correction is needed.

Case 2: When the sum of A and B is greater than 9 but less than 16, the binary addition results in one of the illegal code words. We must add 6 in order to change the base and consequently skip over the illegal code words. In this case the addition of 6 will not only correct the sum but will also generate a carry out to the next higher decimal position.

Case 3: When the sum of A and B is greater than 15, the binary addition results in sum base-16 and a carry out which weighs 16. The addition of 6 is needed to correct the sum, while the carry out is now considered to be weight 10.

The addition algorithm can be stated as follows:

1. Add A and B as if they represent binary numbers.
2. If the sum of A and B is greater than 9, then add 6 in order to correct the sum.

3. The carry out (to next higher stage) is the logical-OR of the carry generated from the first step binary addition or the carry generated after correction (this is equivalent to generating a carry whenever correction is needed).

The addition algorithm is illustrated by Example 8.1.

Example 8.1

This example illustrates the three different cases which could arise in the addition of decimal digits in the (8, 4, 2, 1) code.

Case 1: $A = 0011, \quad B = 0101$

$$\begin{aligned} A &= 0011 = (3)_{10} \\ B &= \underline{0101} = (5)_{10} \\ A + B &= 1000 = (8)_{10}, \end{aligned}$$

which is the correct result

Case 2: $A = 0101, \quad B = 0110$

$$\begin{aligned} A &= 0101 = (5)_{10} \\ B &= \underline{0110} = (6)_{10} \\ A + B &= 1011 \\ +6 \quad & \underline{0110} \\ & 1\ 0001 = (1)_{10} \end{aligned}$$

and a carry out, which is the correct result (the leading 1 is a carry out)

Case 3: $A = 1000, \quad B = 1001$

$$\begin{aligned} A &= 1000 = (8)_{10} \\ B &= \underline{1001} = (9)_{10} \\ A + B &= 1\ 0001 \\ +6 \quad & \underline{0110} \\ & 0111 = (7)_{10} \end{aligned}$$

(the leading 1 of $A + B$ is a carry out) and a carry out, which is the correct result

To implement the above algorithm, we have to derive a boolean expression for a function (f), which represents the conditions for the correction. The correction function (f) is equal to 1 if the sum (before correction) is greater than 9 (that is, 10, 11, 12, 13, 14, 15, 16, 17, 18). The last three (16, 17, 18) are indicated by a carry out from the first binary addition. Therefore, if we denote the sum (before correction) by $C = (C_1, C_2, C_3, C_4)$ and the carry by K_1, then f can be written as

$$f = K_1 + \sum 10, 11, 12, 13, 14, 15$$

which, when minimized, can be written as

$$f = K_1 + C_1C_2 + C_1C_3$$

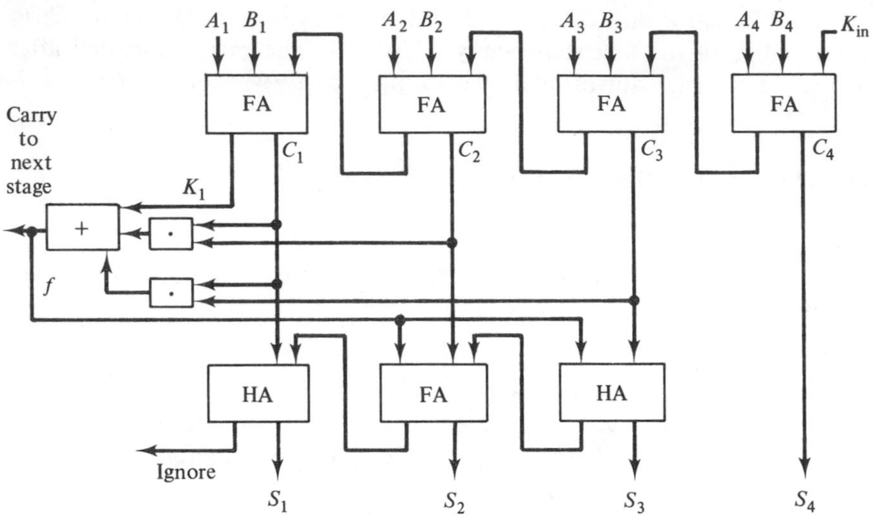

Figure 8.12 (8-4-2-1) BCD adder.

Figure 8.12 shows a complete block diagram for one stage of an (8, 4, 2, 1) BCD adder.

8.3.2 Excess-3 Adder

Addition of two decimal digits in the excess-3 code can be achieved in two steps. First perform binary addition on the coded digits; then the output of this adder is corrected to obtain the sum in excess-3 logical code word. The rules for the addition are shown below:

1. Add the coded digits as if they represent binary numbers. Depending on whether a carry is generated (from the last stage) or not, the sum has to be corrected according to the following rules:
   a. If a carry is generated, add binary 3 (011) to the sum.
   b. If no carry is generated, subtract 3 (or add 13, which is 1101, to the sum).
2. The carry generated from the first step is the correct carry for the decimal addition.

Figure 8.13 shows a block diagram of an excess-3 decimal adder stage. The correction always requires a 1 to be added to the least significant bit. This is accomplished in the diagram by adding the least significant bit to the

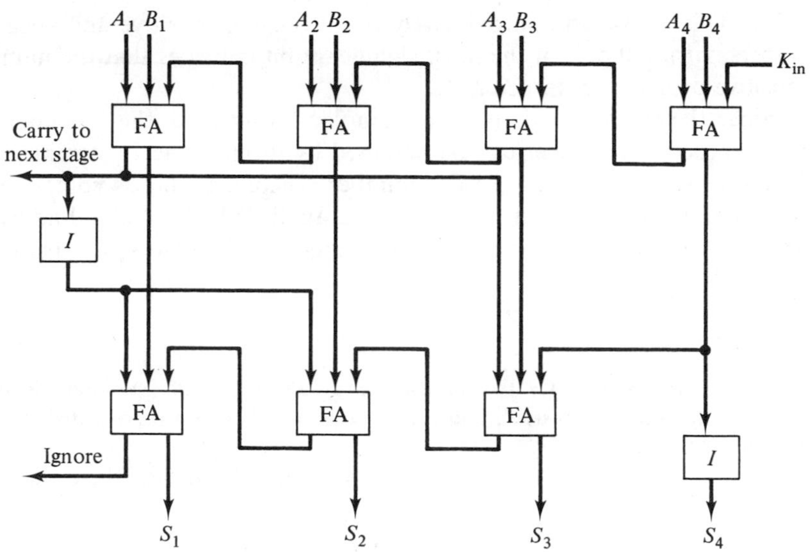

Figure 8.13 Excess three-decimal adder.

next higher position while this bit is complemented. It is easy to verify the equivalence of the two methods.

8.3.3 Decimal Addition and Subtraction

A decimal adder, as described above, is capable of adding two unsigned decimal digits (in excess-3 code) and produce their coded sum and a possible decimal carry to a higher-order stage. The bits of the coded digit are assumed to be available in parallel. The addition of binary coded decimal numbers could be performed in serial, parallel, or parallel-serial form. In the serial case, only 1 bit of the coded operand is available at one time. In the parallel case, all the coded digits are available at once. The third case assumes that the code for a decimal digit is available in parallel, while the decimal digits are available in serial; that is, a BCD number is available parallel by bit, serial by digit. Before we present any particular implementation, the representation of decimal numbers is first discussed. This number representation is then used for the serial, parallel-serial, and parallel implementations.

We shall consider that any operand consists of four decimal digits and that each digit is coded in the excess-3 code. Therefore, each operand is represented by four groups of 4 bits each, that is, a total of 16 bits. Negative numbers are represented in the 10's complement form. The range of numbers which can be represented by the four decimal digits is from -5000 up to

+4999. This allows an approximately equal range of positive and negative numbers. Also a 0 or 1 in the most significant bit indicates that the number is positive or negative, respectively.

Since the excess-3 code is a self-complementing code, the 10's complement of a decimal number can be obtained by forming the 2's complement of the corresponding excess-3 BCD number. Therefore, the 2's complement addition and subtraction algorithm can be applied to the decimal addition with negative numbers in the 10's complement form. This is illustrated by Example 8.2.

Example 8.2

This example illustrates the addition and subtraction of four-digit decimal numbers. Excess-3 code is used, and the negative numbers are represented in 10's complement form.

Let $A = 2631$, $B = -428$, $C = 4187$, and $D = -4700$; then these numbers will be stored as

$$A = 2631 = 0101\ 1001\ 0110\ 0100$$
$$B = 9572 = 1100\ 1000\ 1010\ 0101$$
$$C = 4187 = 0111\ 0100\ 1011\ 1010$$
$$D = 5300 = 1000\ 0110\ 0011\ 0011$$

1.

```
                A + B = 0101 1001 0110 0100
                      +
                        1100 1000 1010 0101
Carry                  1 0010 0010 0000 1001
(drop)                +
Indicates               0011 0011 0011 1101    (correction)
positive                0101 0101 0011 0110
result

            ∴ A + B =  2    2    0    3
```

2.

```
            C + D =     0111 0100 1011 1010
                      +
                        1000 0110 0011 0011
                        1111 1010 1110 1101
                      +
                        1101 1101 1101 1101    (correction)
Carry (drop)          1 1100 0111 1011 1010
Negative result
          ∴ C + D =     9    4    8    7       (in 10's complement)
                  =    −0    5    1    3       (in signed-magnitude)
```

3.

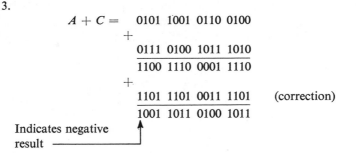

$$
\begin{array}{rl}
A + C = & 0101\ 1001\ 0110\ 0100 \\
+ & \\
& \underline{0111\ 0100\ 1011\ 1010} \\
& 1100\ 1110\ 0001\ 1110 \\
+ & \\
& \underline{1101\ 1101\ 0011\ 1101} \quad \text{(correction)} \\
& 1001\ 1011\ 0100\ 1011
\end{array}
$$

Indicates negative result (arrow pointing to the leading bit)

But since both operands are positive, a negative sum indicates an overflow.

8.3.3.1 Serial Implementation

There is more than one possible implementation for serial decimal addition and subtraction. One simple implementation which is compatible with the serial binary addition and subtraction is described below. The arrangement of the basic hardware for this implementation of decimal addition and subtraction is shown in Fig. 8.14. The numbers are represented in excess-3 code.

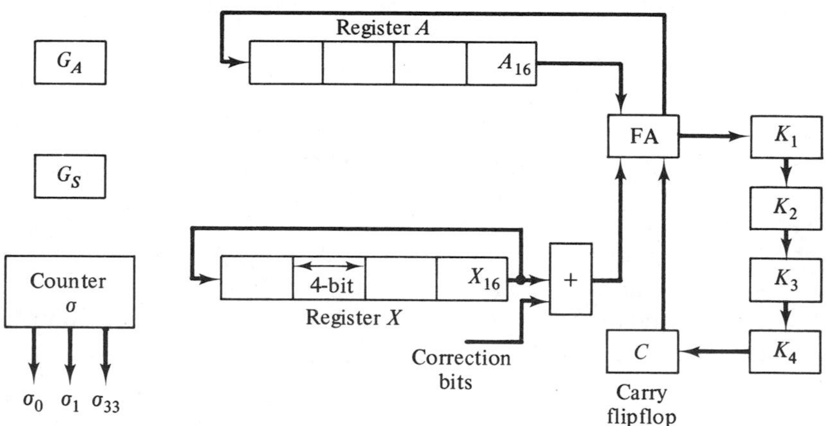

Figure 8.14 BCD addition and subtraction (serial).

First the numbers are added as if they are binary numbers; then the resulting sum is corrected. Therefore, the first part of the operation is similar to the binary addition and subtraction discussed in the previous subsection. The only difference is that we have to store the carry out from each decimal stage (group of 4 binary bits) to be used later for the correction. The register K shown in the diagram is used to store the carry out from the most significant stage. When the binary addition is completed, a correction operation starts in which the necessary correction is performed. The transfers for the binary addition, and storing of the carries between decimal stages are described below.

START	$1 \longrightarrow G_A$ for add, $1 \longrightarrow G_S$ for subtract, $0 \longrightarrow \sigma$
σ_0	$G_S \longrightarrow C$, $\overline{(A_1 \oplus (G_A X_1 + G_S X'_1))} \longrightarrow \alpha$, $G_S \bar{X} + G_A X \longrightarrow X$, $(\sigma + 1) \longrightarrow \sigma$
$\sigma_1 + \sigma_2 + \sigma_3 + \sigma_4$ $+ \sigma_5 + \cdots + \sigma^8 + \cdots$ $+ \sigma_{13} + \cdots + \sigma_{16}$	$\rho(X) \longrightarrow X$, $L(A) \longrightarrow R(A)$, $\mathrm{S}^2(A_{16}, X_{16}, C) \longrightarrow A_1$, $\mathrm{K}^2(A_{16}, X_{16}, C) \longrightarrow C$ $(\sigma + 1) \longrightarrow \sigma$
$\sigma_4 + \sigma_8 + \sigma_{12} + \sigma_{16}$	$\mathrm{K}^2(A_{16}, X_{16}, C) \longrightarrow K_1$, $L(K) \longrightarrow R(K)$

By the end of the timing period σ_{16}, the 16-bit numbers are added serially as if they represent binary numbers. The sum is available in the A register, while the K register contains the carries resulting from the addition of each of the 4-bit groups. The value of each carry bit determines the correction bits to be added to the corresponding sum. That is, the value of K_4 determines the correction factor for the first (least significant) decimal stage (A_{13}–A_{16}). Therefore, this particular bit of the K register must remain in position until all the correction bits are added to the contents of register A. The correction of the sum of a decimal stage is performed on four successive clock pulses according to the algorithm given in Section 8.3.2. After the correction of the sum of the first decimal stage is performed the K register is shifted one place and the correction of the second decimal digit is performed. This process continues until all the digits are corrected. Notice that during the correction no carries are propagated from one decimal stage to another. This requirement is satisfied since no addition takes place at $\sigma_{17,21,25,29}$, and if any carry is transferred to the carry flip-flop C at the previous clock period, it is not utilized. The correction ends by the end of the clock period σ_{32}, and the following clock period is used to detect any overflow and to terminate the operation.

$\sigma_{17} + \sigma_{21} + \sigma_{25} + \sigma_{29}$	$L(A) \longrightarrow R(A)$, $(\sigma + 1) \longrightarrow \sigma$, $\bar{A}_{16} \longrightarrow A_1$, $A_{16} \longrightarrow C$
$\sigma_{18} + \sigma_{22} + \sigma_{26} + \sigma_{30}$	$L(A) \longrightarrow R(A)$, $(\sigma + 1) \longrightarrow \sigma$, $\mathrm{S}^2(A_{16}, C, K_4) \longrightarrow A_1$, $\mathrm{K}^2(A_{16}, C, K_4) \longrightarrow C$
$\sigma_{19} + \sigma_{23} + \sigma_{27} + \sigma_{31}$	$L(A) \longrightarrow R(A)$, $(\sigma + 1) \longrightarrow \sigma$ $\mathrm{S}^2(A_{16}, C, \bar{K}_4) \longrightarrow A_1$, $\mathrm{K}^2(A_{16}, C, \bar{K}_4) \longrightarrow C$
$\sigma_{20} + \sigma_{24} + \sigma_{28} + \sigma_{32}$	The same transfers as in σ_{19}, $L(K) \longrightarrow R(K)$
σ_{33}	$\alpha(A_1 \oplus X_1) \longrightarrow V$, $0 \longrightarrow G_S$, $0 \longrightarrow G_A$

The transfers and the associated timing signals described above do not necessarily represent the fastest implementations. Some modifications of this particular implementation or maybe other implementations could result in higher speed. Once the above scheme is understood, the reader can easily find such modifications.

8.3.3.2 Parallel-by-Bit, Serial-by-Digit Implementation

In this implementation, the decimal digits are processed in serial, but the 4-bit group of a particular digit is processed in parallel. Therefore, the addition and subtraction operations require a complete decimal adder stage. The basic layout for performing addition and subtraction is shown in Fig. 8.15.

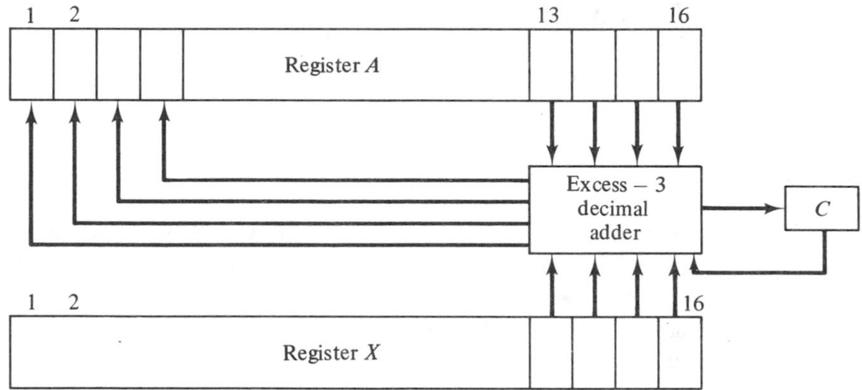

Figure 8.15 Addition and subtration.

The decimal adder above represents the excess-3 decimal adder shown in Fig. 8.13. The input information is transferred into the A and X registers from the memory (usually in serial). The least significant digits of the addend and augend are added first, and their sum is stored in $(A_{1,1} \ldots A_{1,4})$. At the same time a shift right and a right shift circular by four positions are performed on registers A and X, respectively. The addition and shift operations are then repeated three more times. Each time the decimal carry is stored in C and added to the next higher significant digits. After the most significant digits are processed, the decimal sum (in excess-3 BCD) is left in register A. The timing signals and the corresponding transfers for the decimal addition and subtraction are shown below. Subtraction is assumed to be performed by the addition of the 10's complement (which in this case is equivalent to the 2's complement).

START	$1 \longrightarrow G_A$ for addition, $1 \longrightarrow G_S$ for subtraction, $0 \longrightarrow \sigma$
σ_0	$G_S\bar{X} + G_AX \longrightarrow X$, $G_S \longrightarrow C$, $\overline{(A_{1,1} \oplus (G_AX_{1,1} + G_SX'_{1,1}))} \longrightarrow \alpha$, $(\sigma + 1) \longrightarrow \sigma$
$\sigma_1 + \sigma_2 + \sigma_3 + \sigma_4$	$\rho^4(X) \longrightarrow X$, $(L(A) \longrightarrow R(A))^4$, $S^3(A_4, X_4, C) \longrightarrow A_1$, $K^3(A_4, X_4, C) \longrightarrow C$, $(\sigma + 1) \longrightarrow \sigma$
σ_5	$\alpha \cdot (X_{1,1} \oplus A_{1,1}) \longrightarrow V$, $0 \longrightarrow G_S$, $0 \longrightarrow G_A$

In the above transfers $(\ )^4$ refers to a shift by four places in one step. The S^3, K^3 notations refer to the excess-3 decimal sum and carry, respectively. They represent the sum and the carry from the decimal adder stage shown in Fig. 8.13.

The parallel-by-bit, serial-by-digit implementation requires more hardware, but it is faster than the serial implementation.

8.3.3.3 Parallel Implementation

A parallel implementation of decimal addition and subtraction requires a number of decimal adders equal to the number of decimal digits in each operand. Using a block representation for the excess-3 decimal adder shown in Fig. 8.13, a basic layout for the parallel implementation of decimal addition and subtraction is shown in Fig. 8.16.

This layout is similar to that for parallel binary addition and subtraction. The least significant adder stage receives a 4-bit code from each of the A and X registers representing the least significant digit of the addend number and augend number, respectively. The same holds for higher-order decimal adder stages. The 4-bit sum from each adder stage is stored in A, while the carry output propagates to the next higher adder stage.

The algorithm for addition and subtraction in 10's complement is identical to the 2's complement algorithm, discussed in Chapter 3. Therefore, the timing signals and associated transfers are similar to those given in Section 8.3.1 for parallel binary addition and subtraction. The only difference is that the sum is decimal and not binary.

Since the excess-3 is a self-complementing code, the 9's complement of a decimal digit corresponds to the 1's complement of its excess-3 binary code word. Therefore, when 10's complement of the decimal number is required (as in the case of subtraction), it can easily be obtained by complementing the individual bits of the coded number and then adding 1 to the least signifi-

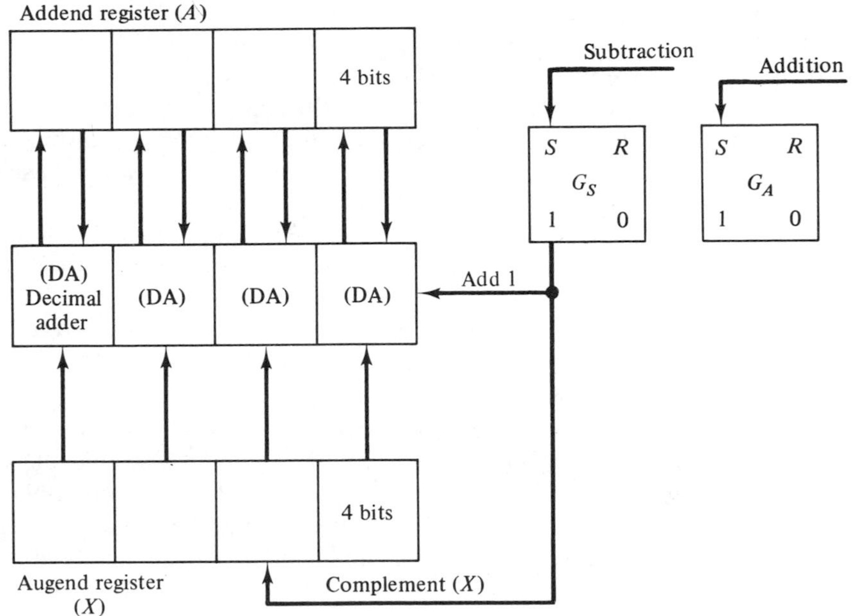

Figure 8.16 Layout for parallel implementation for addition and subtraction.

cant bit. This self-complementing property of the excess-3 code makes the implementation of addition and subtraction much easier, compared to the (8, 4, 2, 1) code or any other non-self-complementing code. In the latter case a special complementing logic is necessary to obtain the complement. Assuming that subtraction is performed by the addition of the 10's complement, the transfer equations are

START	$1 \longrightarrow G_A$ for addition, $1 \longrightarrow G_S$ for subtraction, $0 \longrightarrow \sigma$
σ_0	$\overline{A_1 \oplus (G_A X_1 + G_S X'_1)} \longrightarrow \alpha$, $G_S \bar{X} + G_A X \longrightarrow X$, $(\sigma + 1) \longrightarrow \sigma$
σ_1	$S^3(A, X, G_S) \longrightarrow A$, $(\sigma + 1) \longrightarrow \sigma$
σ_2	$\alpha(A_1 \oplus X_1) \longrightarrow V$, $0 \longrightarrow G_S$, $0 \longrightarrow G_A$

8.3.4 Decimal Multiplication

Decimal multiplication is more complicated than binary multiplication. In binary multiplication, the multiplier digit is either a 0 or a 1; therefore, the partial product is either 0 or the multiplicand. For decimal multiplication a multiplier digit takes values anywhere from 0 to 9; therefore, corresponding to each multiplier digit the derivation of the partial product is more complicated than the binary case. This complexity in decimal multiplication has led to a variety of possible solutions and algorithms. Once a method is chosen to derive the partial products, the addition of these partial products and the associated shifting could follow the same procedure as in the binary multiplication, with the only difference being the use of decimal adders instead of binary adders.

It is not the intention of the authors to present, in this text, the implementation of the various possible schemes, but rather a brief discussion of some possible methods to derive the partial products is presented.

The simplest, although the slowest, method is that which uses repeated addition. This method is similar to paper and pencil multiplication. For each multiplier digit, the multiplicand is added successively a number of times equal to the value of the multiplier digit. This requires a maximum of 9 additions, or an average of $4\frac{1}{2}$ additions per digit of the multiplier. One variation of this method, which conserves time, is to add only for multiplier digits which are equal to or smaller than 5. If the multiplier digit is larger than 5, then add 10 times the multiplicand and subtract the multiplicand an appropriate number of times. Knowing that the addition of 10 times the multiplicand corresponds to adding the multiplicand after it is shifted, this method results in a reduction in the average number of additions corresponding to any of the multiplier digits.

A method which achieves high speed, at the expense of cost, is shown in Fig. 8.17. In this method a logic network derives the multiple of the multiplicand at once, and therefore one addition is required for every digit of the multiplicand.

Other methods which compromise between the above two methods require a maximum of four additions for every multiplier digit. These methods inspect the binary code for a multiplier digit, 1 bit at a time. Suppose that the decimal numbers are represented by 2421 code. If for every bit in

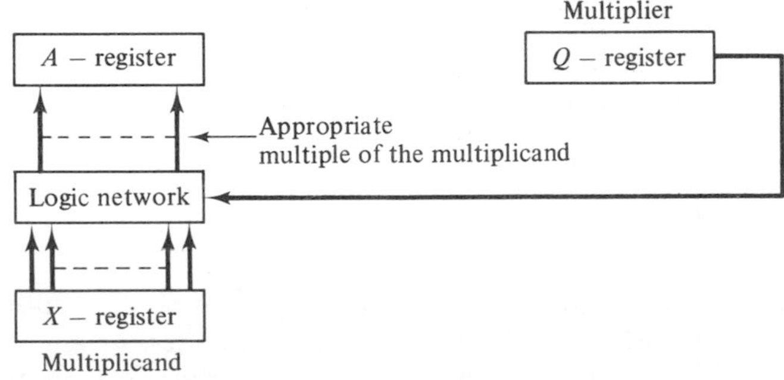

Figure 8.17 Immediate decimal multiplication.

the code only one addition is to be performed, then only 0, 1, 2, or 4 multiples of the multiplicand are to be added at one time. Such a method applies only to weighted codes, and the particular code used plays an important role in the complexity of implementation. For example, codes whose weights are powers of 2 are simpler to implement since some multiples of the multiplicand can be obtained by shifting.

High-speed decimal multiplication can also be realized by shifting across 0s and 1s, similar to binary multiplication schemes. Also, obtaining a multiple of the mutiplicand can be obtained through subtraction techniques whenever this leads to a higher speed of performance. For example, adding three times the multiplicand can be realized faster if it is broken into addition of four times the multiplicand (which corresponds to the multiplicand being shifted two positions) and subtraction of the multiplicand.

8.3.5 Decimal Division

Although the methods used in binary division can be applied to decimal division, the latter case is more complicated to perform. The complexity comes from the fact that in decimal division a quotient bit can take values from 0 through 9, which is more complicated than the case of binary division in which the quotient bit is either a 0 or a 1. Also associated with the choice of the quotient bit is the subtraction or addition of the appropriate multiple of the divisor, which is again more complicated than the binary case. Other than the above problems, decimal division can be performed according to one of the three basic schemes, that is, the division by comparison, the restoring division, or the nonrestoring division, as described for binary division. Therefore, instead of presenting the entire algorithm and implementation of each scheme, we shall concentrate here on obtaining a single quotient digit in each of the three basic schemes.

Consider first the comparison method. The quotient digit can be derived by subtracting the divisor from the dividend repeatedly as long as the divisor "goes" into the dividend. After each subtraction the comparator decides whether or not another subtraction is necessary. The number of subtractions we have to perform determines the value of the quotient digit.

In restoring division the comparison is replaced by a subtraction followed by addition (to restore). Therefore, in the restoring division, the divisor is subtracted from the dividend until a negative remainder (in the case of integers) results. Then the divisor is added to restore the partial dividend, and the quotient digit is equal to the number of subtractions with a positive remainder.

In the nonrestoring division, the restoring operation is eliminated. Therefore, when the partial dividend is negative, instead of restoring the partial dividend and then subtracting repeatedly one-tenth of the divisor (subtraction after shifting), the restoration is omitted and one-tenth of the divisor is added repeatedly until the sign of the partial dividend changes to positive. The quotient bit is equal to the complement (10's complement) of the number of additions required to obtain a positive partial dividend.

8.4 FLOATING-POINT PARALLEL ARITHMETIC UNIT

Floating-point numbers are represented by an exponent part and a fractional part. Therefore, the cells of a register which holds a floating-point number must be divided into an exponent part and a fractional part. A layout of a basic register for a floating-point number is shown in Fig. 8.18.

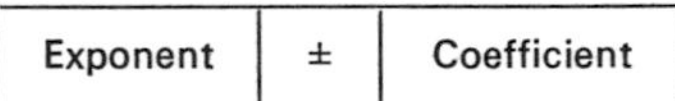

Figure 8.18 Basic register for floating-point number.

One cell holds the sign of the number, while the remaining bits are divided between the exponent part and the coefficient (fractional part). The most common exponent size for computer words of length 48, 36, and 32 is 11, 8, and 7, respectively. This size provides a good compromise between the accuracy indicated by the coefficient size and the range of the numbers reflected by the exponent size.

Before the design of a complete floating-point parallel arithmetic unit is discussed, the implementation of each individual floating-point operation is presented below. Without loss of generality, we shall assume a 36-bit machine with the exponent and the coefficients represented in the 2's complement form.

8.4.1 Floating-Point Additions and Subtractions

Recall, from Chapter 3, that floating-point addition (or subtraction) is performed by the addition (or subtraction) of aligned operands. Once the operands are aligned, the addition operation is simply that of normalized fractional operands.

To simplify the control logic in performing floating-point addition or subtraction, let us assume the availability of two auxiliary registers C and D to hold the exponent parts of the operands and an additional adder/subtractor (eight stages in our case). One possible layout of the hardware is shown in

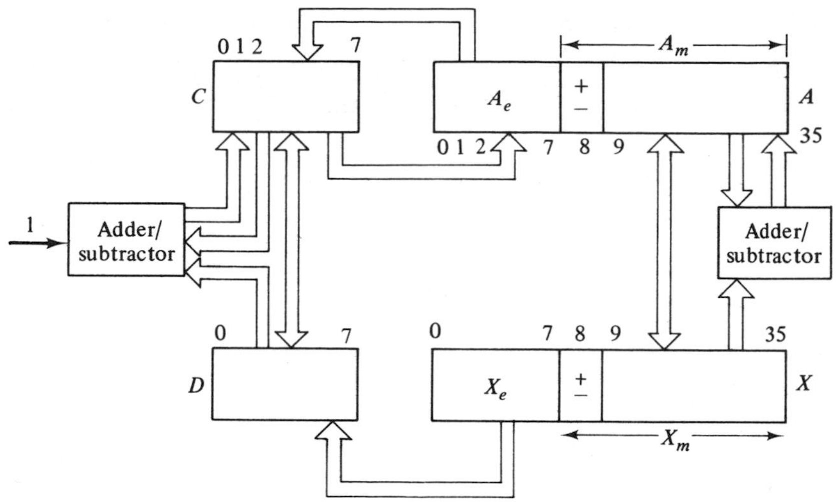

Figure 8.19 Layout for floating point addition and subtraction.

Fig. 8.19. The arrows shown in the diagram indicate the transfer paths needed. What is missing from the diagram is the control signals which initiate these transfers. The transfer path between A_m and X_m is needed in case the operands have to be exchanged, as we shall see later.

The first step after the operation starts is to transfer the exponents into the auxiliary registers and clear their space in the *A* and *X* registers. These steps can be implemented as follows:

START	$1 \longrightarrow G_A$ for add, $1 \longrightarrow G_S$ for subtract, $0 \longrightarrow \sigma$
$(G_A + G_S)\sigma_0$	$A_e \longrightarrow C$, $0 \longrightarrow A_e$, $X_e \longrightarrow D$, $0 \longrightarrow X_e$, $(\sigma + 1) \longrightarrow \sigma$

The second step is the initial alignment of the coefficients. This requires comparison of the two exponents and then appropriate shifting of one of the operands. Usually the *A* register (accumulator) has shifting capabilities, and therefore the alignment can be achieved by shifting the *A* register. Also it is always better to shift to the right rather than to the left so that we do not lose the most significant bit. Therefore, the alignment is to be based on the capability of shifting *A* to the right. In other words, the alignment must start with register *A* containing the coefficient of the number with the smaller exponent. If this is not the case, an interchange of the operands is necessary before the alignment is performed. The possible interchange of operands is described by the following operations:

$(G_A + G_S)\sigma_1$	$S(C, \bar{\bar{D}}) \longrightarrow C, (\sigma + 1) \longrightarrow \sigma$
$(G_A + G_S)\sigma_2$	$C_0 A_m + \bar{C}_0 X_m \longrightarrow A_m,$ $C_0 X_m + \bar{C}_0 A_m \longrightarrow X_m,$ $C_0(\sigma + 4) + \bar{C}_0(\sigma + 1) \longrightarrow \sigma$
$(G_A + G_S)\sigma_3$	$S(C, D) \longrightarrow C, (\sigma + 1) \longrightarrow \sigma$
$(G_A + G_S)\sigma_4$	$C \longrightarrow D, D \longrightarrow C, (\sigma + 1) \longrightarrow \sigma$

So far the exponents have been examined, and an exchange of the operands, if necessary, has been performed. Register A now contains the coefficient of the operand with the smaller exponent (which is stored in C).

The initial alignment (if needed) can be performed as follows:

$(G_A + G_S)\sigma_5$	$S(C, \bar{\bar{D}}) \longrightarrow C, (\sigma + 1) \longrightarrow \sigma$
$(G_A + G_S)\sigma_6$	$C_0 L(A_m) + \bar{C}_0 R(A_m) \longrightarrow R(A_m),$ $C_0(C + 1) + \bar{C}_0 C \longrightarrow C,$ $\bar{C}_0(\sigma + 1) + C_0\sigma \longrightarrow \sigma$

At σ_5, the difference of the two exponents is derived in C. At σ_6 the mantissa of the small number is shifted to the right a number of places equal to the magnitude of the difference in the two exponents. All these shifts and increments take place at the clock pulses within the duration of the control signal σ_6. When the contents of C is 0 (indicated by $C_0 = 0$), the counter is incremented, and consequently the shift and increment process stops.

The next step is to form the sum or the difference of the two aligned coefficients in A_m. At the same time we transfer the exponent of the result into C. This step could be described by

$(G_A + G_S)\sigma_7$	$G_A S(A_m, X_m) + G_S S(A_m, \bar{\bar{X}}_m) \longrightarrow A_m,$ $\overline{A_8 \oplus (G_A X_8 + G_S X'_8)} \longrightarrow \alpha, D \longrightarrow C, (\sigma + 1) \longrightarrow \sigma$
$(G_A + G_S)\sigma_8$	$\alpha(A_8 \oplus (G_A X_8 + G_S X'_8)) \longrightarrow V, (\sigma + 1) \longrightarrow \sigma$

The next step is to check the result. If it is not in normalized form, normalization is necessary.

$$
\begin{array}{l|l}
(G_A + G_S)\sigma_9 & V \cdot L(A) + \bar{V} \cdot R(A) \longrightarrow R(A), \\
 & V \cdot (C + 1) + \bar{V} \cdot C \longrightarrow C, \\
 & V(\sigma + 2) + \bar{V}(\sigma + 1) \longrightarrow \sigma \\
(G_A + G_S)\sigma_{10} & \overline{(A_8 \oplus A_9)}R(A_m) + (A_8 \oplus A_9)L(A_m) \longrightarrow L(A_m), \\
 & \overline{(A_8 \oplus A_9)}(C - 1) + (A_8 \oplus A_9)C \longrightarrow C \\
 & \overline{(A_8 \oplus A_9)}\sigma + (A_8 \oplus A_9)(\sigma + 1) \longrightarrow \sigma
\end{array}
$$

At σ_9, normalization of a result which is too large is performed. This is done in one clock pulse since a maximum of one shift would normalize the result. The transfers associated with σ_{10} normalize a result which is too small. The normalized form is either (0.1xx . . .) for positive numbers or (1.0xxx . . .) for negative numbers. Therefore, the number is in normalized form only when the digit to the left of the binary point (the sign) and the one to the right (MSD) are different. This normalization process might require more than one clock pulse depending on the size of the number to be normalized. In the above transfers, we shall keep shifting the mantissa and decreasing the exponent until $A_8 \neq A_9$. When this condition is satisfied the counter is incremented and the process ends. In some computers the above shift and decrement process is limited to a fixed number of times after which, if the number is still not normalized, the number is obviously too small and is considered to be a 0.

The last step in performing the floating-point addition and subtraction is to attach the exponent to the mantissa of the result and clear the addition and subtraction start-stop flip-flops:

$$
(G_A + G_S)\sigma_{11} \quad | \quad C \longrightarrow A_e,\ 0 \longrightarrow G_A,\ 0 \longrightarrow G_S
$$

The above implementation of floating-point addition and subtraction utilizes a layout of registers and adders which is different from that for fixed-point arithmetic. First, two extra registers, C and D, required for alignment of the coefficients are used for processing the exponents. In other implementations, these registers might not be needed and the alignment could be done using the main registers A and X and the main adder.

If both registers A and X have shift capabilities, then the comparison of the exponents is not necessary. We could always subtract the exponent of X from that of A and then shift the register with the smaller exponent to the right. Consequently, the smaller exponent is incremented with each shift and the common exponent would be the larger one.

Another variation of the above implementation scheme is to perform the addition or subtraction of the two aligned coefficients in an identical arrangement to that of fixed-point addition or subtraction. In this case, the fixed-point addition or subtraction operation would be part of the floating-point addition or subtraction. In other words, one assumes that the fixed-point operation capability exists and then implements the floating-point operation utilizing the former capability. This is an economic way, since usually any computer which implements floating-point arithmetic already has fixed-point arithmetic capabilities.

Let us try to implement a floating-point addition and subtraction scheme which has fixed-point addition and subtraction, described in Section 8.2.1, embedded in it. The processing of the exponents is performed through the use of the auxiliary registers C and D and the auxiliary adder. The transfer equations for this implementation are described below:

$(Q_A + Q_S)P_0$	$A_e \longrightarrow C, X_e \longrightarrow D, (P+1) \longrightarrow P$
$(Q_A + Q_S)P_1$	$S(C, \bar{\bar{D}}) \longrightarrow C, (P+1) \longrightarrow P$
$(Q_A + Q_S)P_2$	$C_0 A_m + \bar{C}_0 X_m \longrightarrow A_m,$ $C_0 X_m + \bar{C}_0 A_m \longrightarrow X_m,$ $C_0(P+4) + \bar{C}_0(P+1) \longrightarrow P$
$(Q_A + Q_S)P_3$	$S(C, D) \longrightarrow C, (P+1) \longrightarrow P$
$(Q_A + Q_S)P_4$	$C \longrightarrow D, D \longrightarrow C, (P+1) \longrightarrow P$
$(Q_A + Q_S)P_5$	$S(C, \bar{\bar{D}}) \longrightarrow C, (P+1) \longrightarrow P$
$(Q_A + Q_S)P_6$	$C_0 L(A) + \bar{C}_0 R(A) \longrightarrow R(A),$ $C_0(C+1) + \bar{C}_0 C \longrightarrow C$ $C_0 P + \bar{C}_0(P+1) \longrightarrow P$
$(Q_A + Q_S)P_7(\tau_0 + \tau_1 + \cdots + \tau_7)$	$R(A) \longrightarrow L(A), R(X) \longrightarrow L(X),$ $0 \longrightarrow A_{35}, 0 \longrightarrow X_{35}, (\tau + 1) \longrightarrow \tau$
$(Q_A + Q_S)P_7\tau_7$	$(P+1) \longrightarrow P, 0 \longrightarrow \tau$
$(Q_A + Q_S)P_8$	$Q_A \longrightarrow G_A, Q_S \longrightarrow G_S, (P+1) \longrightarrow P$
$(Q_A + Q_S)P_9$	$(Q_A\bar{G}_A + Q_S\bar{G}_S)(P+1) + (Q_A G_A + Q_S G_S)P \longrightarrow P$
$(Q_A + Q_S)P_{10}$	$V \cdot L(A) + \bar{V} \cdot R(A) \longrightarrow R(A),$ $V(C+1) + \bar{V} \cdot C \longrightarrow C,$ $V(P+2) + \bar{V}(P+1) \longrightarrow P$
$(Q_A + Q_S)P_{11}$	$\overline{(A_0 \oplus A_1)}R(A) + (A_0 \oplus A_1)L(A) \longrightarrow L(A),$ $\overline{(A_0 \oplus A_1)}(C-1) + (A_0 \oplus A_1)C \longrightarrow C$ $\overline{(A_0 \oplus A_1)}P + (A_0 \oplus A_1), (P+1) \longrightarrow P$
$(Q_A + Q_S)P_{12}(\tau_0 + \tau_1 + \cdots + \tau_7)$	$L(A) \longrightarrow R(A), (\tau + 1) \longrightarrow \tau$
$(Q_A + Q_S)P_{12}\tau_7$	$(P+1) \longrightarrow P, 0 \longrightarrow \tau$
$(Q_A + Q_S)P_{13}$	$C \longrightarrow A_e \cdot 0 \longrightarrow Q_A, 0 \longrightarrow Q_S$

The above transfers differ from the previous scheme at three points. The first point is that before the addition or subtraction of the coefficients, they are shifted to the left eight positions so that the coefficients occupy the same positions in A and X as the fixed-point operands. This is illustrated by the transfers corresponding to the control signal P_7. The second difference, indicated by transfers at the control signals P_8 and P_9, is that the addition or subtraction of the coefficients is performed by setting either the G_A or G_S flip-flop, respectively, as when fixed-point addition or subtraction is done. The third difference is the need for shifting the sum or difference eight positions to the right to evacuate space for the exponent of the result. This step is shown by the control signal P_{12}.

Notice that the left shift at P_7 and the right shift at P_{12} are shown as one-position shifts every clock pulse. If the registers have the appropriate multiple-shift capabilities, then the multiple shift can be performed during one clock pulse.

8.4.2 Floating-Point Multiplication

Multiplication of two floating-point numbers requires the addition of their exponents and the multiplication of their fractional parts. When 2's complement representation is used for negative exponents, the addition becomes the same as that for fixed-point numbers. Similarly, let us assume 2's complement representation for the fractional part so that the multiplication can make use of the fixed-point multiplication discussed previously. The layout of registers and adders is shown in Fig. 8.20. The C and D registers, in which the exponents are processed, and the parallel adder are those defined for the floating-point addition. The counters P and τ are both modulo-8 counters which will be used to control the multiplication operation. They are assumed to be cleared at the beginning of the floating-point multiply instruction. The remainder of the diagram shown at the right-hand side of the figure represents the layout for a fixed-point multiplier shown in Fig. 8.8. We shall assume the same format for the operands as described for floating-point addition and subtraction, with both the exponent and the coefficient of any operand represented in the 2's complement form. The floating-point instruction starts with the multiplicand in register A, while the multiplier is in register X. Before the actual multiplication takes place, the exponents will be transferred to the auxiliary registers C and D, while the multiplier and the multiplicand fractional parts will be positioned at the left 24 cells of X and A, respectively. One key part of the operation is a fixed-point multiply of the fractional parts. Therefore, the positioning of the fractional parts as described above is the same as that at the beginning of the fixed-point multiply described in Section 8.2.2. At this point a fixed-point

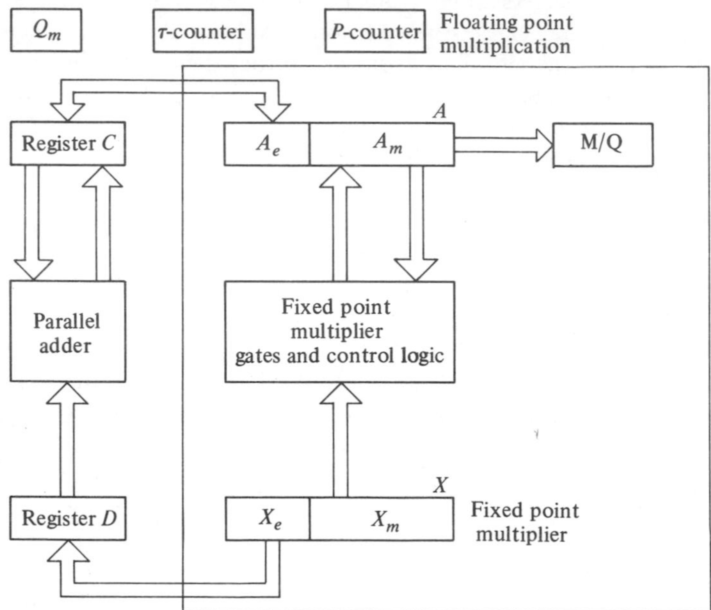

Figure 8.20 Register layout for floating-point multiplication.

multiply start signal is initiated, and all the steps of a fixed-point multiply are performed. At the end of the fixed-point multiply (which is detected by the contents of the fixed-point multiply start-stop flip-flop G_m being 0) the product of the fractional parts, including its sign bit, is stored in the left cells of the combined register (A, Q). Before the sum of the two exponents can be transferred to the register (A, Q), the leftmost eight cells of (A, Q) must be evacuated. This requires shifting the (A, Q) register eight positions to the right and then transferring the exponent. The last step of the multiplication involves a check on the fractional part for a possible result which is not in normalized form. In this case overflow will not occur, since the magnitude of the product of the two normalized fractions lies between $\frac{1}{4}$ and 1. The only possible result which is not normalized is that which lies between $\frac{1}{4}$ and $\frac{1}{2}$. This is indicated, if it occurs, by a 0 to the right of the sign bit. Normalization requires shifting the fractional part one place to the left and decreasing the exponent by 1. The timing signals and the associated sequence of transfers for the floating-point multiplication are described in the following paragraphs.

The operation starts by setting the multiplication flip-flop Q_M and clearing counters P and τ.

START	$1 \longrightarrow Q_M, 0 \longrightarrow P, 0 \longrightarrow \tau$

The next step is to add the two exponents and check for overflow. This can be described by

$Q_M P_0$	|	$A_e \longrightarrow C, X_e \longrightarrow D, (P+1) \longrightarrow P$
$Q_M P_1$	|	$S(C, D) \longrightarrow C, \overline{(C_0 \oplus D_0)}, \longrightarrow \alpha,$ $(P+1) \longrightarrow P$
$Q_M P_2$	|	$\alpha(C_0 \oplus D_0) \longrightarrow V$

An overflow signal will stop the process. If no overflow occurs, the process continues.

The next step is to get ready for the multiplication of the two coefficients. As stated before this multiplication is no different from the multiplication of fixed-point fractional numbers, and consequently it is not going to be repeated here. Instead we shall assume the availability of a fixed-point multiplier. But before we call upon the fixed-point multiplier, the operands must be stored in A and Q in the appropriate positions acceptable to the fixed-point multiplier. This means that both A and X must be shifted eight positions to the left. The shifting operation can be described as

$Q_M P_3 \tau_0$	|	$R(A) \longrightarrow L(A), R(X) \longrightarrow L(X),$ $0 \longrightarrow A_{35}, 0 \longrightarrow X_{35},$ $(\tau + 1) \longrightarrow \tau$
$Q_M P_3(\tau_1 + \tau_2 + \cdots + \tau_7)$	|	The same transfers as τ_0
$Q_m P_3(\tau_7)$	|	$(P+1) \longrightarrow P, 0 \longrightarrow \tau$

The transfers shown at P_3 assume one place shift capabilities. This shift can be done in one clock pulse period if the registers A and X are provided with the appropriate multiple-shift capability.

Once the coefficients are shifted into the left portions of A and X, their multiplication can be performed by sending a signal which sets the start-stop multiplication flip-flop (G_m). The multiplication is completed when a signal is detected which resets the G_m flip-flop to the 0 state. The transfers which describe the above steps can be written as

$Q_M P_3 \tau_7$	|	$1 \longrightarrow G_m, (P+1) \longrightarrow P$
$Q_M P_4$	|	$\bar{G}_m(P+1) + G_m P \longrightarrow P$

After the product of the two coefficients is formed in the (A, Q) register, it must be shifted to the right eight positions to leave space for the exponent of the result to be transferred to the leftmost eight bits. The implementation of this step follows the transfers:

$Q_M P_5(\tau_0 + \tau_1 + \cdots + \tau_7)$	$L(A, Q) \longrightarrow R(A, Q), (\tau + 1) \longrightarrow \tau$
$Q_M P_5 \tau_7$	$(P + 1) \longrightarrow P$

The last step, before the multiplication is completed, is to check the result for the normalized form. The only possible nonnormalized result is between $\frac{1}{4}$ and $\frac{1}{2}$ and requires only one shift. This is achieved by shifting the resulting coefficient of the product one place to the left and decreasing the exponent by 1. The necessary condition to do so is when the bit to the right of the binary point is a 0. Thus, the normalization process can be described by

$Q_M P_6$	$\overline{(A_8 \oplus A_9)}R(A_m) + (A_8 \oplus A_9)L(A_m) \longrightarrow L(A_m)$, $\overline{(A_8 \oplus A_9)}(C - 1) + (A_8 \oplus A_9)C \longrightarrow C$, $(P + 1) \longrightarrow P$
$Q_M P_7$	$C \longrightarrow A_e, 0 \longrightarrow Q_M$

8.4.3 Floating-Point Division

The division of two floating-point numbers can be performed by subtraction of the exponent parts and division of the two coefficients. Assuming the same number representation and format as for the multiplication operation, the implementation of the division operation is compatible with the floating-point subtraction and fixed-point division discussed previously. The basic hardware layout is shown in Fig. 8.21.

The division starts with the dividend and the divisor stored in A and X, respectively. The exponents A_e and X_e are then transferred to C and D, respectively. The parallel adder to the left of the diagram is used to derive the difference of the exponents in C and D. If the coefficients are shifted to the left portions of A and X, then the division of the coefficient can be accomplished through the transfer sequence of the fixed-point division, as described in Section 8.2.3. Remember that the fixed-point division operation derives the quotient in Q and that A contains the remainder. Thus, a transfer of Q into A_m and C into A_e results in the quotient stored in A being a floating-point number.

From another point of view, floating-point division requires actions similar to those required by floating-point multiplication with four differences. First, division requires subtraction of the exponents instead of addi-

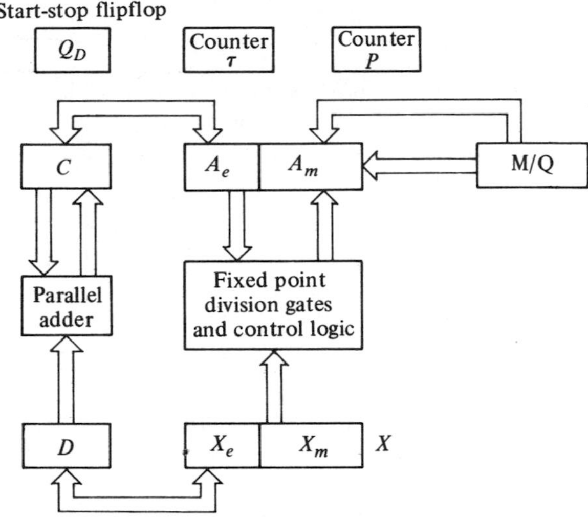

Figure 8.21 Layout for floating-point division.

tion. Second, in division, the coefficients have to be divided instead of multiplied. The third difference is that the fixed-point division leaves the quotient in Q, while the multiplication leaves the product in (A, Q). The fourth difference is that normalization of the quotient is never required if the operands are scaled such that the dividend is smaller than the divisor. All the other actions required by the floating-point multiplication applies to the floating-point division. The transfer equations for floating-point multiplication, with the above four changes, are repeated below to represent the implementation of floating-point division. The operands are assumed to have normalized fractions with the dividend smaller than divisor.

START	$1 \longrightarrow Q_D, 0 \longrightarrow P, 0 \longrightarrow \tau$
$Q_D P_0$	$A_e \longrightarrow C, X_e \longrightarrow D, (P + 1) \longrightarrow P$
$Q_D P_1$	$\mathrm{S}(C, \bar{\bar{D}}) \longrightarrow C, \overline{(C_0 \oplus D'_0)} \longrightarrow \alpha,$ $(P + 1) \longrightarrow P$
$Q_D P_2$	$\alpha(C_0 \oplus D_0) \longrightarrow V$
$Q_D P_3 \tau_0$	$R(A) \longrightarrow L(A), R(X) \longrightarrow L(X),$ $0 \longrightarrow A_{35}, 0 \longrightarrow X_{35}, (\tau + 1) \longrightarrow \tau$
$Q_D P_3(\tau_1 + \tau_2 + \cdots + \tau_7)$	Similar to τ_0
$Q_D P_3 \tau_7$	$(P + 1) \longrightarrow P, 0 \longrightarrow \tau,$ $1 \longrightarrow G_D$
$Q_D P_4$	$\bar{G}_D(P + 1) + G_D P \longrightarrow P, \bar{G}_D Q + G_D A \longrightarrow A$
$Q_D P_5(\tau_0 + \tau_1 + \cdots + \tau_7)$	$L(A, Q) \longrightarrow R(A, Q)$
$Q_D P_5 \tau_7$	$(P + 1) \longrightarrow P$
$Q_D P_6$	$C \longrightarrow A_c, 0 \longrightarrow Q_D$

8.4.4 Floating-Point Arithmetic Unit

A floating-point arithmetic unit refers to a piece of hardware which is capable of performing arithmetic operations of floating-point numbers. We have seen the implementation of the individual floating-point operations such as addition, subtraction, multiplication, and division. If the necessary logic circuits for each of the operations are combined, we obtain a floating-point arithmetic unit. This process has been illustrated before in the presentation of the serial and parallel fixed-point arithmetic unit. Putting the pieces together to form a floating-point arithmetic unit can follow similar thinking, and therefore it is not presented here.

If the reader tries to combine the implementation of the individual operations given in this section, the resulting floating-point arithmetic unit contains a complete fixed-point arithmetic unit embedded in it plus some extra hardware required to process the exponents of the operands. Other designs which do not require a fixed-point arithmetic unit are also possible and might be cheaper to implement.

8.5 LOGIC OPERATIONS

Digital computers perform other operations besides the arithmetic operations previously discussed. In fact, if we look into the instruction repertoire of a computer, we find many instructions which require operations other than arithmetic to be performed. Operations such as testing, comparing, complementation, etc., do not involve any arithmetic and are referred to as logic operations. They constitute an essential part of the instruction repertoire of a computer and serve to select, sort, rearrange, or reformat information within the machine. Some of these logic operations are also used as part of performing arithmetic operations. For instance, a complementation may be used as part of a subtraction operation, or a comparison between two numbers may be used as part of performing a division process. Logic operations can be implemented within the arithmetic unit or as additional control logic. The definition and implementation of some logic operations are presented below.

8.5.1 Testing and Comparison

Testing is usually performed to execute conditional jump instructions. These instructions are executed based on whether certain conditions exist or not. The simplest test involves the content of a particular flip-flop such as the sign or the overflow flip-flop. This can be performed by simply examining the state of the flip-flop under consideration. A more complicated test condition might require that the contents of a register be 0. One way, which employs the arithmetic unit, for such a test is to add -1 to the content of the register and test the sign change. A more expensive implementation is to use a logical-AND gate connected to all the 0 outputs of the register flip-flops.

Some instructions require a comparison between the contents of two different registers. Consider the numbers A and B. Figure 8.22(a) shows a serial comparator for $A > B$. Figure 8.22(b) shows a serial comparator for $A = B$. The initial state of the flip-flop (*FF*) is 0. After all the bits of the numbers (the LSB first) are tested the final state indicates the test result, as shown in the diagram.

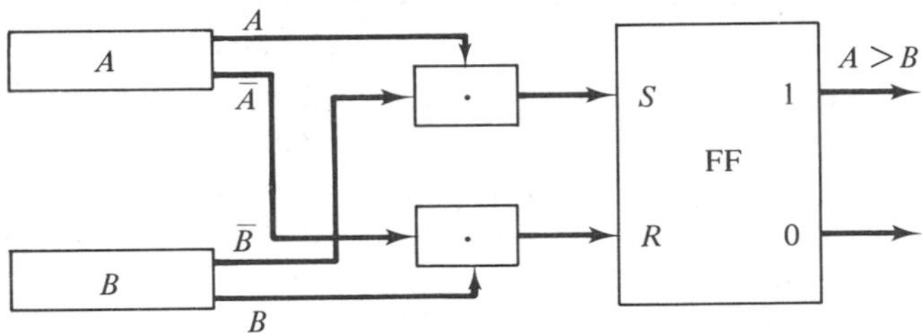

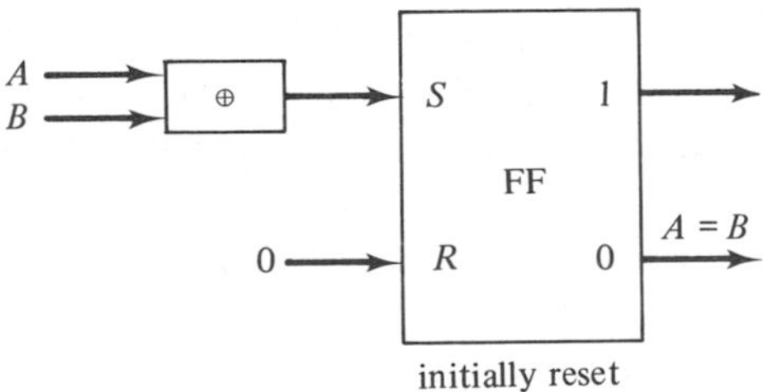

Figure 8.22 (a) Test for $A > B$; (b) test for $A = B$.

In a parallel machine, a test to determine if the contents of two registers are equal can be implemented by using exclusive-OR gates. Figure 8.23 shows such a test for registers A and B. In this case the output of the exclusive-ORs are fed back to A. Register A must, then, be tested to see if $A = 0$ or $A \neq 0$. If $A = 0$, then $A = B$; otherwise, $A \neq B$. This implementation is faster than that using the arithmetic unit, since no carry propagation is involved.

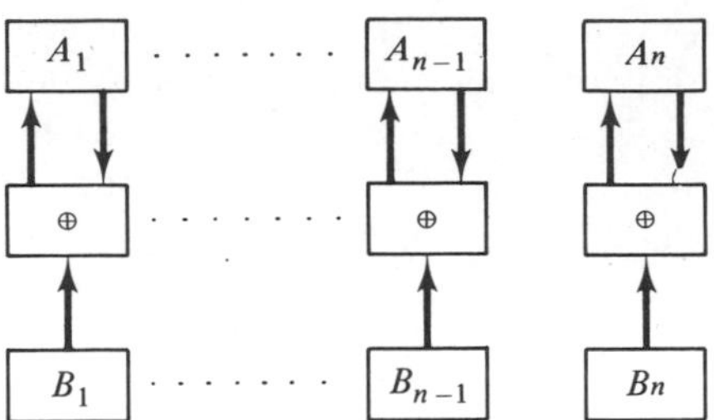

Figure 8.23 Test for $A = B$.

8.5.2 Complementation

Complementation is the most frequently used logic operation. When negative numbers are represented in a complement form, most arithmetic operations involve complementation. The 1's complement of a binary number requires changing every 0 into a 1 and every 1 into a 0. This operation is also referred to as the logical-NOT operation. The complementation of a binary number can be implemented in various ways depending on the type of register where the number is stored. For example, if flip-flops with a complement input are used, then complementation can be achieved easily by applying a signal to all the complement inputs of the flip-flops. In the case where the complement of the contents of a register A is to be transferred into another register B, this might be implemented by simply connecting the 1 output and the 0 output of each flip-flop of register A into the "clear" and "set" inputs of the corresponding flip-flop of register B. Another example of complementing a serial register is to let the complement signal interrupt the regular recirculating loop and let the complement output be recirculated instead. Figure 8.24 shows a complementation scheme for a recirculating register. If the complement signal is held for a complete recirculating cycle,

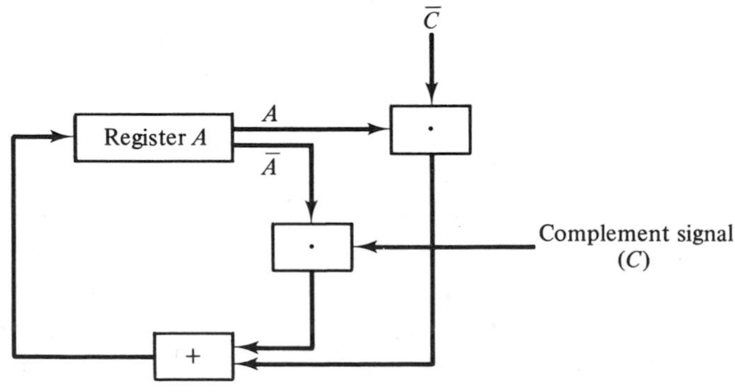

Figure 8.24

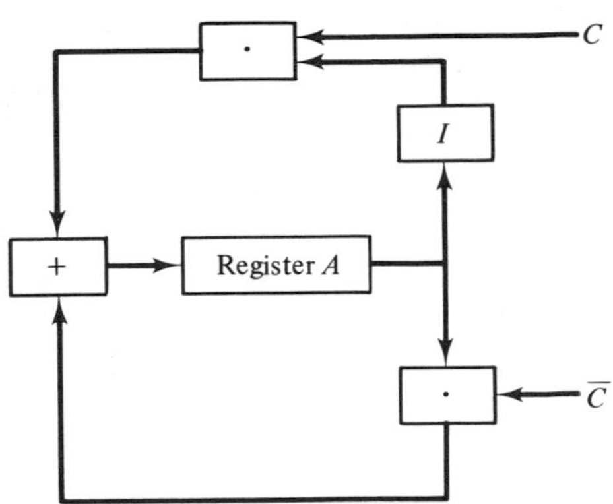

Figure 8.25

then all the bits of the register will be complemented. If the register does not have a complement output, then the arrangement shown in Fig. 8.25 is a possible implementation.

8.5.3 Logic Summation and Logic Product

The logic summation refers to the logic operation OR, while the logic product refers to the logic operation AND. Consider two registers $A = (A_1, A_2, \ldots, A_n)$ and $B = (B_1, B_2, \ldots, B_n)$; the logic summation of $A + B$ is defined as $(A_1 + B_1, A_2 + B_2, \ldots, A_n + B_n)$, where the $+$ operation signifies the logical-OR operation—and similarly for the logic product, except for the OR replaced by AND. The logic summation or product of the contents of two registers can be implemented using OR gates or AND gates, respectively. Figure 8.26 shows one stage of a possible implementation in which the logic summation of the parallel registers A and B is formed in the parallel register C. Figure 8.27 shows an implementation for forming the logic product of A and B in C, where all the registers are serial.

8.5.4 Summary

In this chapter, we have shown the design, including the timing and register transfer equations, for various implementations of the arithmetic functions which can be found in different digital computers. We have not designed a complete arithmetic and logic unit from an integrated point of view. We have combined several separate designs into more complicated functional arithmetic units. In Chapter 12 we shall design the logic for a fixed-point parallel binary arithmetic and logic unit under microprogrammed control.

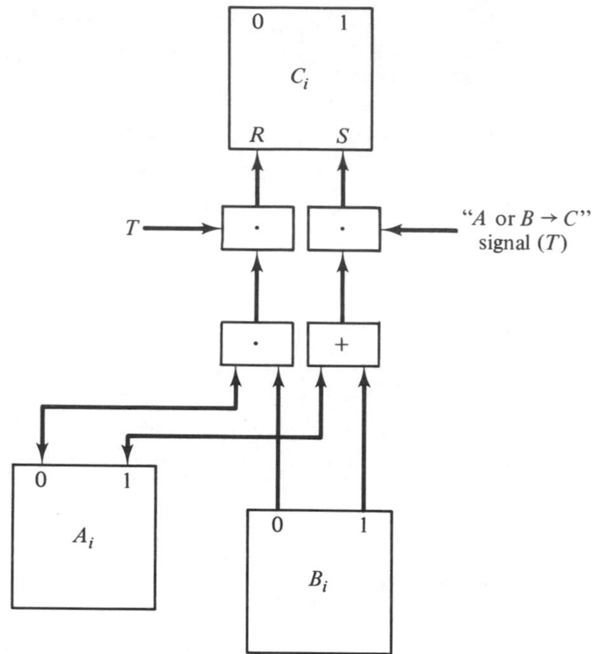

Figure 8.26 Logic summation for parallel registers.

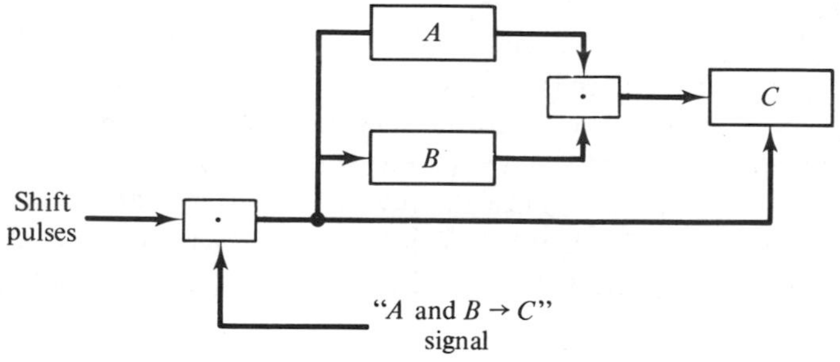

Figure 8.27 Logic product for serial registers.

REFERENCES

BARTEE, T. C., I. L. LEBOW, and I. S. REED, *Theory and Design of Digital Machines.* New York: McGraw-Hill, 1962.

CHU, Y., *Digital Computer Design Fundamentals.* New York: McGraw-Hill, 1962.

FLORES, I., *The Logic of Computer Arithmetic.* Englewood Cliffs, N.J.: Prentice-Hall, 1962.

GSCHWIND, H. W., *Design of Digital Computers.* New York: Springer-Verlag New York, Inc., 1967.

MALEY, G. A., and J. EARLE, *The Logic Design of Transistor Digital Computers.* Englewood Cliffs, N.J.: Prentice-Hall, 1963.

———, and E. J. SKIKO, *Modern Digital Computers.* Englewood Cliffs, N.J.: Prentice-Hall, 1964.

RICHARDS, R. K., *Arithmetic Operation in Digital Computers.* New York: Van Nostrand Reinhold, 1955.

———, *Digital Design.* New York: Wiley, 1971.

STEIN, M. L., and W. D. MUNRO, *Introduction to Machine Arithmetic.* Reading, Mass.: Addison-Wesley, 1971.

WARE, W. H., *Digital Computer Technology and Design*, Vol. II, New York: Wiley, 1963.

WEINBERGER, A., and J. L. SMITH, "The Logical Design of a One-Microsecond Adder Using One-Megacycle Circuitry," *IRE Trans. Electronic Computers*, EC-5, No. 2 (June 1956), pp. 65–73.

PROBLEMS

8.1. Derive transfer equations similar to those in Section 8.1.1 for serial binary addition and subtraction when the numbers are represented in 1's complement.

8.2. Repeat Problem 8.1 for numbers in signed-magnitude form.

8.3. Derive transfer equations, similar to those in Section 8.1.2, for binary serial multiplication when 1's complement is used instead of 2's complement.

8.4. Repeat Problem 8.3 for numbers in signed-magnitude form.

8.5. How many clock pulses does it take to multiply two 36-bit numbers in the serial multiplier discussed in Section 8.1.2? Calculate the number of clock pulses from the time when the multiplier is in MAR and the multiplicand is in the accumulator until the time when the result in its final form is available in the combined (A, Q) register.

8.6. Derive the transfer equation for serial binary division compatible with the equations in Section 8.1.3 but using the restoring division algorithm.

8.7. Given two numbers x and y stored in registers A and B, respectively, design a serial arithmetic unit which computes $|x| - |y|$. Assume that the numbers are in signed-magnitude representation.

8.8. Repeat Problem 8.7 when the numbers are in 2's complement representation.

8.9. Design an adder/subtractor compatible with the one shown in Section 8.2.1 when the numbers are represented in 1's complement form. Show all the transfer equations and a block diagram for the registers' layout.

8.10. Repeat Problem 8.9 for numbers in signed-magnitude representation.

8.11. Design a binary multiplier similar to that shown in Section 8.2.2 when the numbers are represented in 1's complement.

8.12. Show the transfer equations defining the restoring division of binary numbers. Additions and subtractions are performed by a parallel adder, and the numbers are in 2's complement.

8.13. Given two numbers, x and y, stored in registers A and B, respectively, design an arithmetic unit to compute $|x| - |y|$. The numbers are in 1's complement, and the arithmetic unit utilizes a parallel adder.

8.14. Repeat Problem 8.13 when the numbers are in signed-magnitude representations.

8.15. Design a BCD adder which uses the (2, 4, 2, 1) code.

8.16. Repeat Problem 8.15 for the (5, 4, 2, 1) code.

8.17. Let registers A and B contain 12 cells each in natural binary coded decimal (NBCD) representation, with each decimal digit represented by 4 bits. Write a sequence of transfers defining the storage of the sum of A and B in B. Assume serial-by-character, parallel-by-bit operations where one parallel 4-bit adder is employed.

8.18. Repeat Problem 8.17 for a strictly serial operation.

8.19. Consider the layout for a floating-point arithmetic unit to perform addition and subtraction (Fig. 8.19). The operations of addition and subtraction can be performed without the auxiliary registers C and D. Derive transfer equations defining the floating-point addition and subtraction without the use of C and D registers. Show all the other necessary hardware.

8.20. Consider a comparator flip-flop such as that shown in Fig. 8.22. Define the logical relation $A \supset B$ to be true if B contains 1s only in those places where there is also a 1 in A. (There may be 1s in A where there are 0s in B.) What set and reset inputs to the comparator flip-flop would you use for a comparison of

(a) $A \supset B$ (b) $A \geq B$

8.21. Modify Fig. 8.27 to implement the function $C = A + B'$.

9

Memory

In this chapter we shall develop the structure and logic of memory devices and subsystems. In particular, we shall place our main emphasis on the magnetic core memory. We shall also develop some fundamentals of newer electronic memories. Finally we shall develop the mechanical, electrical, and data structure of magnetic drums and disks which are used for on-line secondary storage.

9.1 MEMORY HIERARCHY AND CHARACTERISTICS

9.1.1 Flip-flop Registers

The fastest form of memory in a digital computer is the electric flip-flop registers contained in the control unit, arithmetic unit, and channels. It is through these registers that the computer performs its fundamental operations on data and instructions in order to transform this information into results. The faster these registers are, the faster the machine can perform its set of instructions. These registers are an integral part of the computer design and in fact are the control unit of the machine.

9.1.2 Main Memory

The next level of storage is the main memory. This is the part of storage which is directly addressable. This characteristic permits the control unit to write into or read from each location of the memory. In reality, the main memory is a collection of named registers which has been constructed in an organized fashion in order to reduce its costs. The speed of this memory should be as fast as possible and be consistent with the cost of such a fast memory. Since the faster the technology used for memory is, the more expensive per bit it is, a compromise on the size and speed of addressable memory is made in every computer system. Main memory is usually constructed from magnetic cores or large-scale integrated electronics.

9.1.3 Secondary Storage

Because of the high cost of main memory, it is not economically practical to have all the memory needed by a computing system in that form. As a result, additional memory called secondary storage is usually available. This type of memory is characterized by lower costs per bit of storage and an access time which is much slower than that of main memory. *Access time* is the amount of time it takes between a request to a storage unit to produce a piece of information and when it finally arrives at the output of the unit to be used by the requester. The access time of main memory is usually in the range 0.1–2.0 microseconds, while the access time of secondary storage is usually in the range 5–100 milliseconds. There is this difference in access time because main memory is all electronic and obtains information at electronic speeds, while secondary storage is generally mechanical in nature such as drums or disks and requires the movement of an access mechanism.

9.1.4 External Storage

The outermost part of the memory hierarchy is the punched cards, punched paper tape, and reels of magnetic tape which are used for the basic input to the computer. These devices have a very low cost per bit of storage and a very slow access time since they require a human to fetch the information from a shelf and mount it on an input/output unit. Because of the input-output function of these media, rather than the storage aspect of them, we shall postpone discussion of them until Chapter 10 under the heading "Data, Input/Output, and Channels." Thus, in this chapter we shall deal with the *internal* memory available to the control unit which can be directly controlled by the machine and available to a program. This internal storage is sometimes called *on-line* storage.

9.1.5 Types of Access to Storage Devices

To access information in storage it is first necessary to locate the information. There are two primary methods of locating information on storage devices. One method is called *random* access. A random access storage device is one where it takes the same amount of time to retrieve information immaterial of where it is stored in the device. The flip-flop registers in the control unit and the magnetic core main storage to be studied in this chapter are random access storage devices.

The other access method is called *sequential* access. In a sequential storage device the time to locate a piece of information is not a constant but depends on where the information is stored in the device. Many such devices require a sequential search of information in order to locate the particular piece of information which is of interest and is to be obtained. An example of a sequential storage device is a reel of magnetic tape. To retrieve a piece of information from the middle of the tape, we must sequence through all the tape from the beginning until we find the piece of information we want. Both magnetic disk storage and magnetic drum storage, which we shall study in this chapter, have an aspect of sequential access in their design.

9.1.6 Static and Dynamic Storage

Another method of characterizing storage devices is whether they are *static* storage or *dynamic* storage devices. In a static storage device the information does not change position. Flip-flop registers, magnetic core memories, and electronic MOS memories are examples of static storage. In a dynamic storage device the information is continually changing position. Circulating registers using glass delay lines, rotating magnetic drums, and magnetic disks are examples of dynamic storage devices. In general, a static storage device also has random access characteristics, and dynamic storage devices have sequential access characteristics. The reason for the sequential access characteristics in dynamic storage devices is that usually we have to wait for the information to move under the read heads of the device.

9.1.7 Fixed and Erasable Storage

Another characteristic of storage devices is whether the information recorded therein is permanent, i.e., fixed or erasable. Information stored in flip-flop registers, magnetic core memories, magnetic drums, magnetic disks, and magnetic tapes is erasable. However, information stored on punched cards, punched paper tape, or holographic plates is not erasable. Since the digital computer is designed to modify much of the information which it stores, it usually used erasable-type storage devices for main memory but fixed devices for input and output. Third-generation computers also use special forms of fixed storage called *read-only memory* (ROM), also called *read-only storage* (ROS), to implement the functions of the control unit through a technique called microprogramming.

9.2 SWITCHING PROPERTIES OF MAGNETIC CORES

9.2.1 Physical Construction and Properties

Magnetic core memories are constructed with many small toroidal pieces of ceramic ferrite material called a *core*. There may be more than 32 million cores in a large memory system. Figure 9.1 shows a core with two windings through it. The core has a high permeability and can be easily magnetized by a magnetic flux produced by a current in the write winding. The direction of the current will determine the direction of magnetization, either clockwise or counterclockwise. The core also has a large retentivity and retains its magnetization after the current has been removed. The size of magnetic cores can be as small as 18-thousandths of an inch in outside diameter and as large as many inches for other than memories.

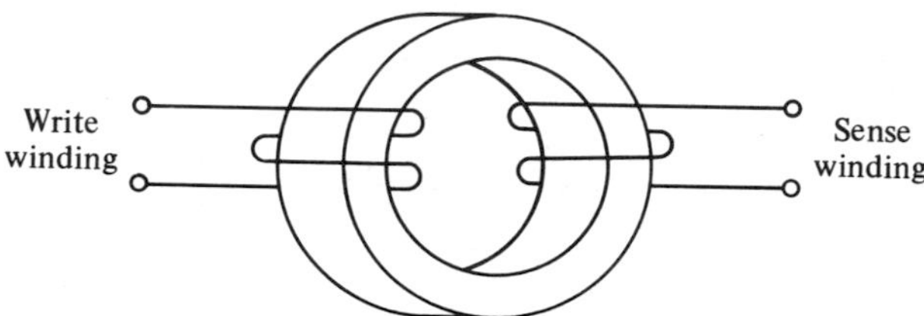

Figure 9.1 A magnetic core with windings.

9.2.2 Hysteresis Loop

The electrical and magnetic properties of magnetic cores are studied by the use of the *hysteresis loop*. This is a graph of the flux density B produced in the core due to the magnetization H produced by the current in the write windings. We can write a relationship between the magnetization H and the current i through the write windings as

$$\oint \mathbf{H} \cdot d\mathbf{l} = Ni \tag{9.1}$$

where $\mathbf{H}$ = magnetic field intensity in the core
$d\mathbf{l}$ = element of the path around the core
N = number of turns of wire in the write winding
i = current in the write winding

For a toroidal coil with an average radius of R we get

$$H = \frac{Ni}{2\pi R} \tag{9.2}$$

The relationship between the magnetic field intensity, H, and the magnetic flux density, B, is

$$\mathbf{B} = \mu\mathbf{H} \tag{9.3}$$

where μ is the permeability of the ferrite material of the core. If the core has a constant cross-sectional area of A square meters, then the total flux around the core is

$$\phi = BA \tag{9.4}$$

If we apply a current through the write winding and plot the values of B and H obtained for a ferrite core, we obtain a graph similar to the one shown in Fig. 9.2. This shows the values of B obtained for each value of H including negative values of current, that is, current going in the opposite direction in the winding.

An inspection of the hysteresis loop reveals the following. When a positive current is applied to the core, the material will saturate at a value of magnetic field intensity shown as $+H_m$, with a magnetic flux density of $+B_m$. Any increase in the positive current applied will not increase the flux density

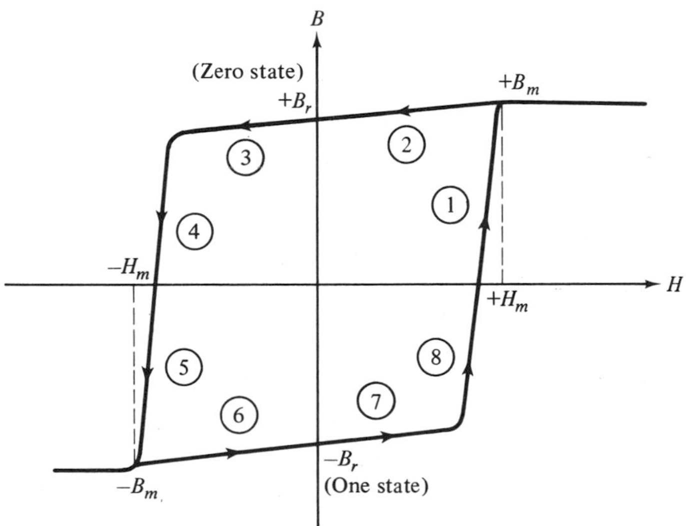

Figure 9.2 Hysteresis loop for a magnetic core.

through the core. If we now decrease the current, we shall decrease the magnetic field intensity until it finally reaches zero. The magnetic flux density will not return down the original path 1 but will go down path 2 until it reaches the point $+B_r$, which is the positive *remanent* point. That is, with no current applied and no field intensity, H, there is still a magnetic field circulating in the core with a flux density equal to $+B_r$. If we now apply a negative current, i.e., a current in the opposite direction, a negative H field will build up and the B field will travel down paths 3, 4, and 5 until the core saturates with a field in the opposite direction with a value of $-B_m$. Once the current is removed and the H field is reduced to zero, the B field will settle at $-B_r$. Thus, with no current applied, a magnetic field with an opposite direction will exist inside the core. If once again we apply a positive-going current, the flux will travel along paths 7, 8, and 1, and once again the material will saturate in the positive sense at $+B_m$. Thus, if we apply a sufficiently large positive current, a magnetic field intensity $+H_m$ of sufficient size to saturate the core will be established with a magnetic flux density of $+B_m$. When we remove the current the magnetic state of the core will settle at $+B_r$. Likewise, if we apply a sufficiently large negative current, one in the opposite direction, an opposite field $-H_m$ will be created and saturate the core with a flux density of $-B_m$. When the current is removed the core will settle at $-B_r$. We have a magnetic device with two stable magnetic states which can be switched to the opposite state with the application of a sufficiently large current of the correct polarity. Let us call the state $-B_r$ the 1 state and the state $+B_r$ the 0 state.

9.2.3 Core-Switching Waveforms

Although we have a method to switch a magnetic core between two stable magnetic states, we must develop a method to sense in which state the core was. To accomplish this we have shown a second winding on the core in Fig. 9.1 called a *sense winding*. We can use this winding to sense changes in flux when moving from one state to the other. A change in flux will induce a voltage on the sense winding which we can detect. It was shown by Faraday that a change in flux will induce a voltage in a winding which links the changing flux, that is,

$$v = N\frac{d\phi}{dt} = NA\frac{dB}{dt} \tag{9.5}$$

where v = induced voltage on the sense winding
N = number of windings linking the flux
$d\phi/dt$ = time rate of change of the flux
A = constant cross-sectional area of the core
dB/dt = time rate of change of the flux density in the core

Assume that the core is in the 1 state at $-B_r$. If we apply a positive current sufficient to create a magnetic field intensity of $+H_m$, then the magnetic flux density will move up to $+B_m$. The net change in flux will be from $-B_r$ to $+B_m$ or approximately $+2B_r$. If, on the other hand, the core was in the 0 state at $+B_r$ and we apply the same positive current, then the flux density will move from $+B_r$ to $+B_m$, which is a negligible change in flux if the *hysteresis loop* is nearly a square. This will induce a negligible voltage on the sense winding. Thus, the material used for the core must have a square *hysteresis loop*. A negligible waveform on the sense winding indicates that the core was in the 0 state, while a large voltage on the sense winding indicates that the core changed its state from the 1 state to the 0 state and therefore was in the 1 state. Figure 9.3 shows the sense-winding voltage outputs for the two conditions as well as the input write-winding current.

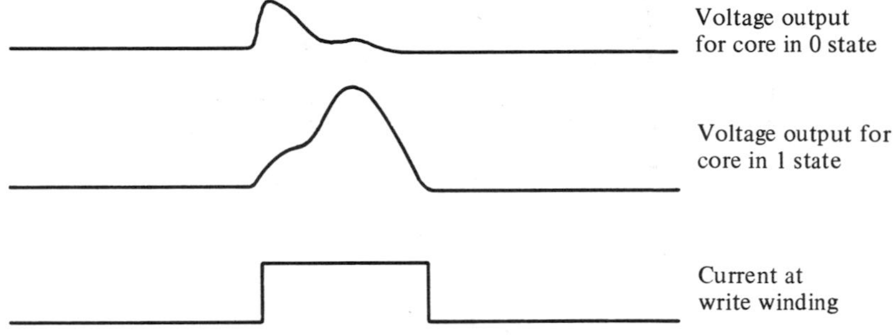

Figure 9.3 Core switching waveforms.

One of the disadvantages which we see with this method of determining the state of a magnetic core is that if the core was in the 1 state, it has changed to the 0 state. Thus, this sense technique is *destructive* since it can change the state of the core. Since we wish to retain the information which is read from the core, we must amplify the output of the sense winding and set a flip-flop; then we can use the state of the flip-flop to reset the core back to its original state.

9.3 COINCIDENT-CURRENT MAGNETIC CORE PLANE MEMORY

9.3.1 Selection Technique

To utilize the magnetic core as a storage device for the main memory of a computer, a method must be developed to read from and write into any given selected core. Such a method is called a *selection technique*. The selection technique to be described is called the *coincident-current* technique. It is the most commonly used method for large main memory systems.

The coincident-current selection technique is based on the physics of Eq. (9.2). The magnetic field intensity, H, is directly proportional to the writing current in the write winding through the core. Thus, to produce the magnetic field intensity $+H_m$ requires a current of $+I_m$. Figure 9.4 shows a modified hysteresis loop for a ferrite core with the abscissa changed from H to I, the write current.

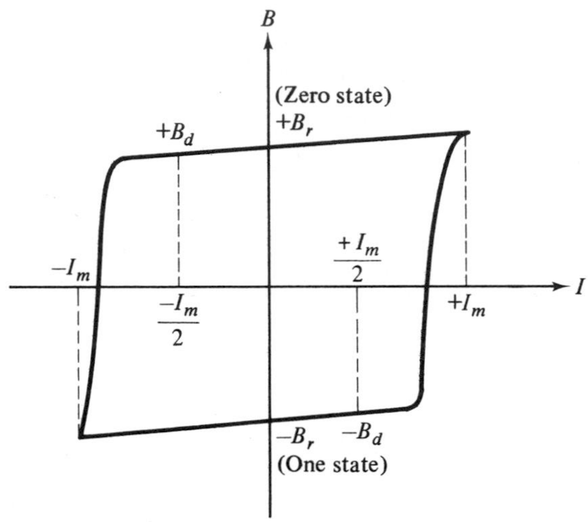

Figure 9.4 Current versus flux density.

Let us assume that the core is in the 0 state at $+B_r$. If a current of $-I_m$ is applied to the write winding, then the core will switch its state to the $-B_r$ point. If, however, the core is in the 0 state at $+B_r$ and a current of only $-I_m/2$ is applied to the write winding, the core will not change state but only move on the hysteresis loop to $+B_d$. When the current is removed the core will move back to $+B_r$ and stay in the 0 state. This performance is due to the square hysteresis loop characteristic. Likewise if the core is in the 1 state at $-B_r$ and a current of $+I_m/2$ is applied to the write winding, the core will not change to the 0 state at $+B_r$ but will only move to $-B_d$ and will return to $-B_r$ when the current is removed. The core will switch if a *full-select current*, I_m, of the correct polarity is passed through it but will not switch state if only a *half-select current*, $I_m/2$, passes through it.

Using the above characteristics we can construct a selection technique on a matrix of cores. Figure 9.5 shows such a matrix or core plane. Through each core of the plane, we thread two wires, the x wire and the y wire. Assume that all the cores are in the 0 state. If our technique is to be useful, we must be able to change the state of any one of the cores without changing the state of any other core. Let us try to change the state of the core located at $x = 2$, $y = 3$.

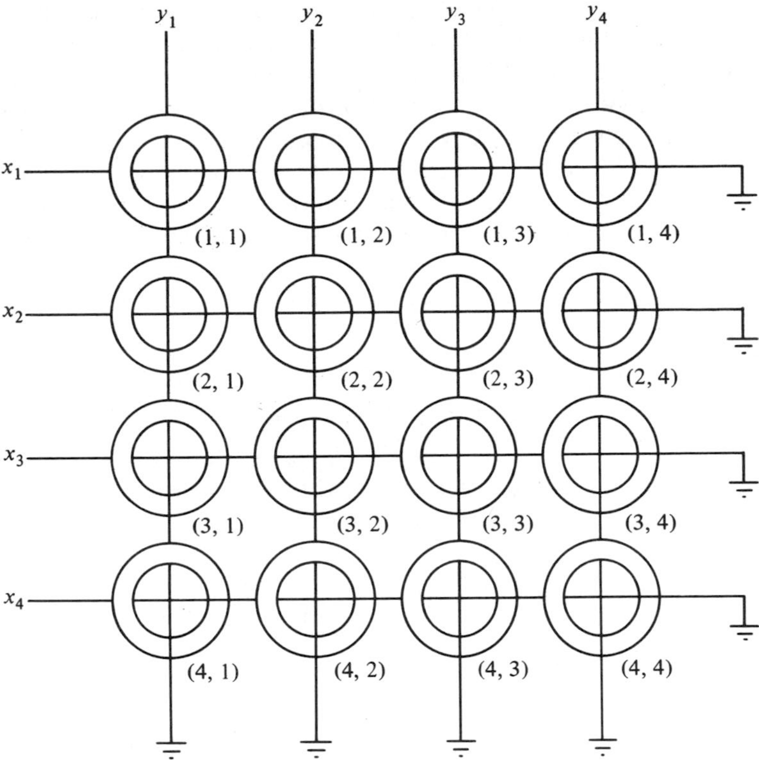

Figure 9.5 A magnetic core matrix.

If we apply a current of $-I_m/2$ on the x_2 line and simultaneously apply a current of $-I_m/2$ on the y_3 line, then the total current passing through the core at location (2, 3) will be $-I_m$ and it will change from the 0 to the 1 state. All the other cores along the x_2 line as well as all the other cores along the y_3 line will have a total current through them of only $-I_m/2$. This amount of current will move their flux density from $+B_r$ to $+B_d$. When the currents are removed, all these cores will return to the $+B_r$ point and only the core at location (2, 3) will have moved around the hysteresis loop and have its flux density reversed so that it now sits at $-B_r$, the 1 state. The core which receives the two half-select currents coincidently will have a full-select current passing through it and therefore change its magnetic state. All the other cores in the plane will have either no current passing through them or only half-select currents passing through them and therefore not change state. The name of this selection technique is therefore the *coincident-current* technique.

The state of any given core in the plane can be sensed by applying coincident half-select currents of positive polarity, $+I_m/2$, to the appropriate x and y lines which pass through it. The coincident full-select current, $+I_m$, will cause it to change its state to the 0 state if it was in the 1 state but remain in the same state if it contained a 0. The fully selected core is the only one in the plane which can possibly change its state and therefore the only core which can possibly develop an appreciable voltage output on a sense winding. Since we know which core we wish to sense, it is only necessary to have a single sense line which threads through all the cores in the planes. A voltage output or lack of such will therefore tell us whether the selected core was in the 0 or 1 state. As before, once we read the state of a core we may change its state in the process; hence, we must save the information in a flip-flop register and restore the core to its proper state by writing back in the state which we destroyed.

9.3.2 Core Plane Wiring and Memory Organization

Figure 9.6 shows a small core plane with three wires threaded through each core. There are 64 cores in the plane with 8 y select lines and 8 x select

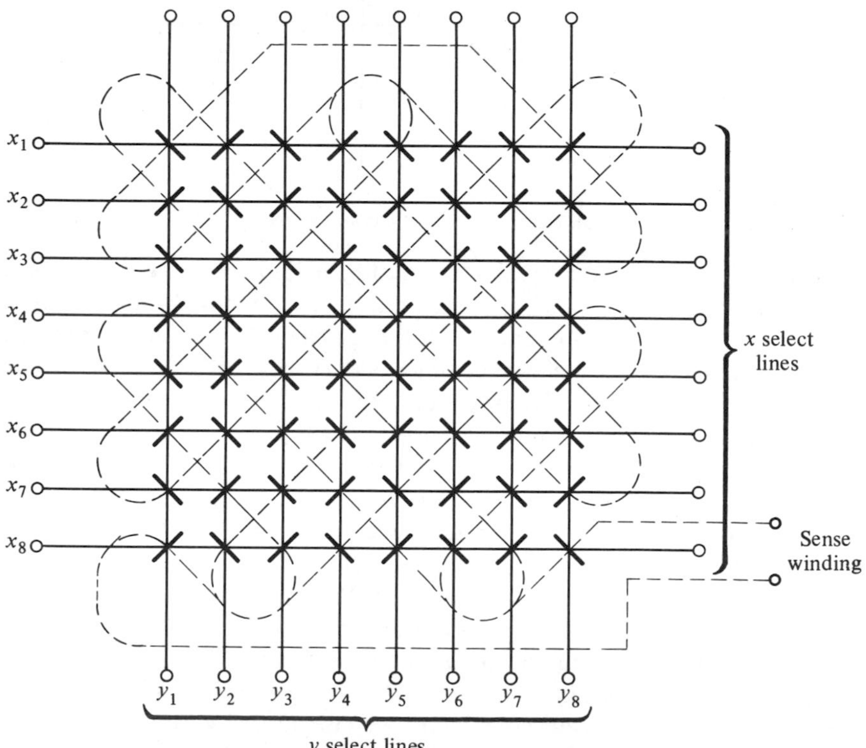

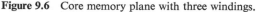

Figure 9.6 Core memory plane with three windings.

lines. The third wire which threads diagonally through all 64 cores is the *sense line* for that plane. Notice that the axis of each core is orthogonal to the axis of each of its nearest neighbors. Also notice the threading of the sense lines; it is diagonal to the select lines. This is done to obtain a minimum coupling from the select lines to the sense line and an optimum cancellation of half-select noise voltages.

Let us now use the core plane to construct a magnetic core memory for a 16-bit word length computer, with 64 words of storage in its main memory. We shall need 16 core planes like the one shown in Fig. 9.6. Each plane contains all the bits for a given bit position for all of the 64 words. Thus, the first plane contains all of the first bit for every word. The planes are stacked into a three-dimensional array. The x lines of each plane are connected in series with the same x line on the two adjacent core planes. Thus, x_1 on the first plane connects in series to the x_1 select line on the second plane and so on all the way on to the x_1 line on the sixteenth plane. The same is true for the x_2 line, etc. Likewise the y lines of each plane are connected in series with the same y line on all of the 16 planes. Each plane maintains its own sense line, and they are not connected together. Each separate sense line for each plane is connected to an individual sense amplifier. Figure 9.7 shows such a core memory array. If we wish to read from the ($x = 3$, $y = 4$) location (address) on one core plane, we shall select that location on each of the 16 core planes. Thus, the current pulse on the x_3 select line will pass through the 8 cores on all 16 planes for a total of 128 cores.

In general there will be as many core planes as bits in the word length for the computer. In our example we assumed a 16-bit word length; hence, there were 16 planes. Had the word length been 32 bits, there would have been 32 core planes. When we write a 1 on the first plane with a coincident write current at a given core location, address (x_i, y_i), we write a 1 at that location on all the core planes. However, this represents a problem since we do not want to write all 1s into a given address of memory. There are times where some of the bits of the word should contain 1s and some should contain 0s. However, as we stand now we can write all 1s into an address or all 0s but not any other data. Therefore, we must develop a new wiring scheme which allows us to write any data at any bit of an address.

The solution to our write problem is to add a fourth wire to each core plane called the *digit winding* or the *inhibit winding*. There is one inhibit winding per core plane, as shown in Fig. 9.7. It threads through each core of a single plane. There are as many inhibit windings as there are planes, and they are *not* connected together. Thus, if we do not wish to write a 1 at a given bit location of an address, we apply a $+I_m/2$ on the inhibit winding of this core plane. The net coincident current at the selected address will be an $-I_m/2$ from the x_i line, plus another $-I_m/2$ from the y_i line plus an $+I_m/2$ applied on the inhibit line. Thus, the total net current of $-I_m/2$ will not be sufficient to

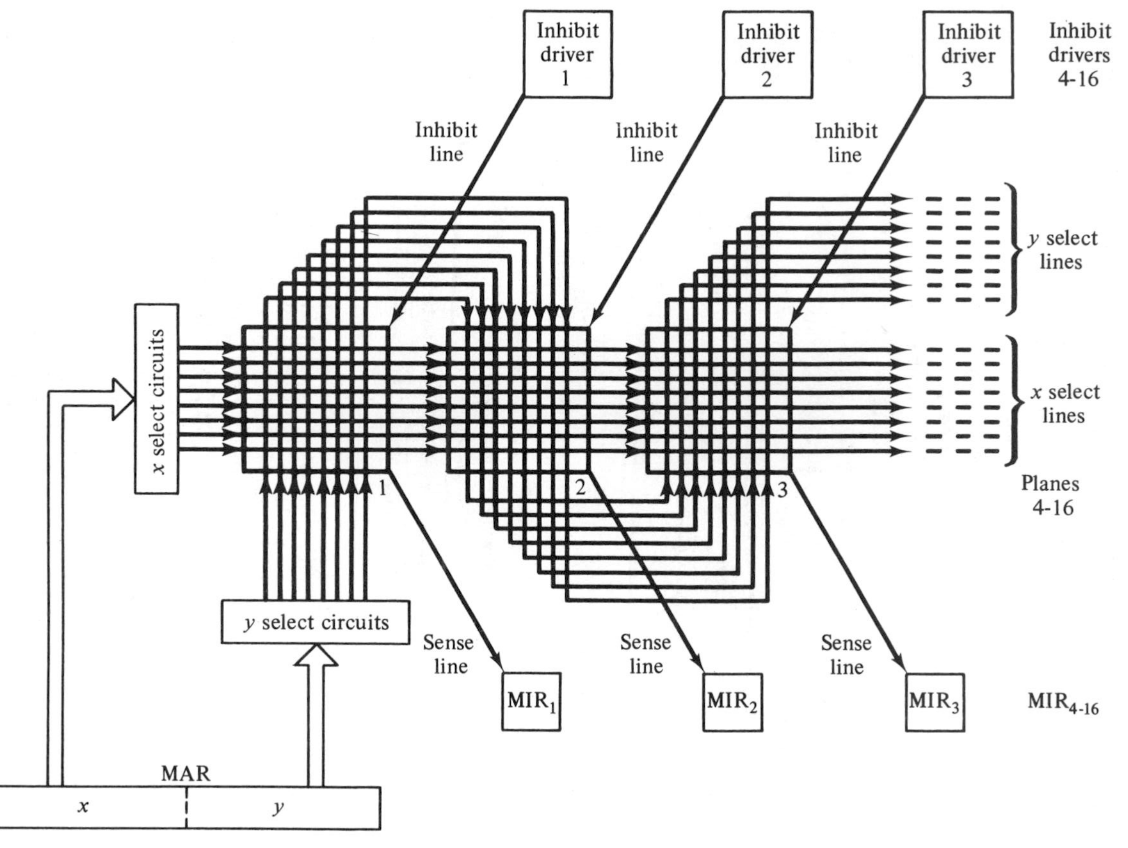

Figure 9.7 Unfolded view of 3D four-wire memory (sense and inhibit lines not shown on planes).

switch the core to the 1 state and it will stay in the 0 state. The above memory organization is called a 3D (three-dimensional) four-wire system.

9.3.3 Read and Write Cycles

The above memory organization usually uses the same cycling of the cores whether data are being read from the memory or written into memory. The total time taken to complete the timing sequence is called the *cycle time* of the memory. It is composed of two parts, the *access time* or *read time* and the *write time* or *regeneration time*. Figure 9.8 shows the sequence of pulses on the four lines through a selected core of the plane for a complete memory cycle. It is assumed that the core was initially in the 1 state. Let us now investigate the write operation and read operation for the memory.

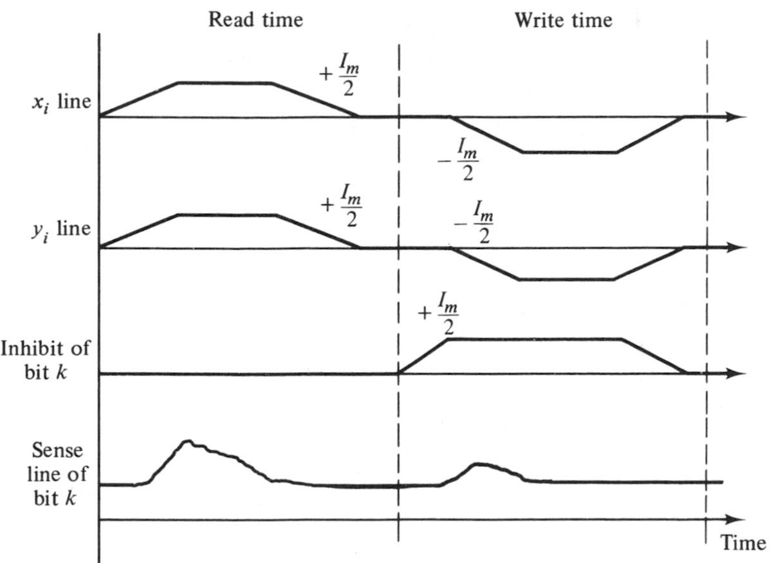

Figure 9.8 Timing sequence of a memory cycle.

9.3.3.1 The Write Operation

Let us assume that we wish to write the binary word 1101100111000011 into location 26 of our words of memory. Further assume that location 26, which corresponds to an octal address of 32, is located at $x = 3$ and $y = 2$ in each of the 16-bit planes of the core. Thus, we select the correct write drivers x_3 and y_2 of the memory. During the read time the coincident select currents are applied and drive each of the cores at location $x = 3$, $y = 2$ of all 16 planes to zero. After the read time is completed, the inhibit drivers on all planes on which we wish to write a 0 are turned on, that is, planes 3, 6, 7, 11, 12, 13, and 14 counting the planes from plane 1, which holds the most significant bit of the binary word. The coincident write currents $-I_m/2$ are applied to the x_3 and y_2 lines during the write time. This will drive the cores on planes 1, 2, 4, 5, 8, 9, 10, 15, and 16 to the 1 state. Of course, those planes on which the inhibit line was driven with $+I_m/2$ will not have an $-I_m$ coincident at any core, and they will remain in the 0 state. At the completion of the write time the word 1101100111000011 will have been written into address 32 (octal) of the memory.

9.3.3.2 The Read Operation

At some time later, say 2 seconds, we wish to read the information from location 26 (decimal) of our memory. Assume that it is the same binary number which was stored during the explanation of the write operation above. During the read time of the access cycle, coincident select currents of $+I_m/2$ are applied to the write lines x_3 and y_2 of the memory. If a large voltage signal is received at the sense amplifier of a given plane during the read time, the select core of that plane contained a 1. If a small voltage signal is received, the selected core contained a 0. Since the cores of planes 1, 2, 4, 5, 8, 9, 10, 15, and 16 contained a 1, we shall obtain a large sense signal from these planes, indicating that the number stored at address 26 (decimal) = 32 (octal) is 1101100111000011. Since during the read time all the cores at the selected address were driven into the 0 state, it is necessary to write the information back into the address during the write time. Thus, the core memory has destructive readout but gives the appearance of nondestructive readout to its external environment. To restore the information, the output of each sense amplifier of each core plane is connected to a flip-flop and sets it to the 1 state with the large sense signal during the read time. The contents of these storage devices is used to control the inhibit line drivers for each plane during the write time. Thus, the inhibit lines connected to those storage devices still in the 0 state will be enabled with $+I_m/2$ currents and will prevent the writing of 1s into those planes. At the completion of the write time the number 1101100111000011 will be restored at address 32 (octal) of memory.

9.4 3D CORE MEMORY ASSOCIATED ELECTRONICS

In the previous sections the 3D/four-wire magnetic core memory selection technique and information storage and retrieval methods were developed. To implement these methods several electronic registers, selection decoders, line drivers, amplifiers, and power supplies will be required as associated electronics. In this section we shall develop the design of this required equipment.

9.4.1 Memory Electronic Registers

The external environment communicates with the magnetic core memory via two electronic registers. One is the memory address register (*MAR*), which contains the address from which data will be written into or read from. The second is the memory information register (*MIR*), which contains the information read from the memory after the read time or the information which is to be written into the memory. To select an address in memory, the address is first loaded into the *MAR*. If we wish to read from that address, the states of the cores will be sensed and written into the *MIR*. Thus, the *MIR* is composed of the electronic flip-flops referred to in Section 9.3.3. The computer can read the information which is in the *MIR*. If we wish to write into the memory, again we first load the address in the *MAR*, and the data to be stored are placed into the *MIR*. After the memory cycle is completed, the information will be stored at the specified address.

9.4.2 Memory Information Register

Figure 9.9 shows a block diagram of the operation of the memory information register (*MIR*) for 1 bit location in the data word or one plane of the memory. Each plane of the memory has the same circuitry. Let us investigate the operation of this circuitry during a read operation and a write operation.

9.4.2.1 The Read Operation

1. The address is loaded into the *MAR*.
2. The *MIR* is reset to zero.
3. The read cycle is commenced in the memory.
4. If the core was in the 1 state, a voltage pulse will appear on the sense line.
5. During the sense line pulse, the sense amplifier is *strobed* with a narrow pulse to minimize the effect of the half-select noise due to all the half-select pulses on those cores.
6. The flip-flop of the *MIR* is set to 1 if the core contained a 1 and a pulse is detected on the sense line. The *MIR* may be read.
7. The cores at the selected address have all been driven to the 0 state.
8. The write cycle is commenced in the memory.

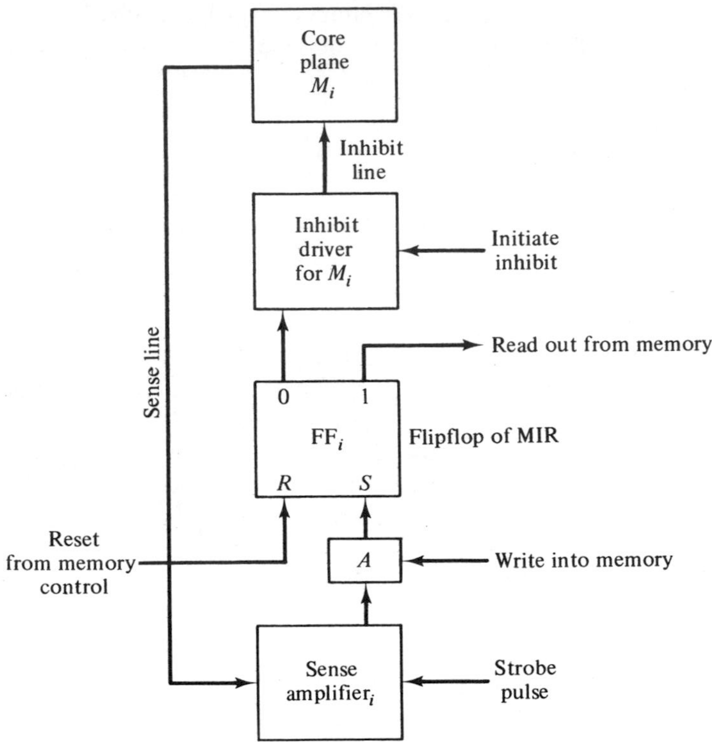

Figure 9.9 Block diagram of the operation of MIR.

9. The flip-flops which remain in the 0 state are used to enable the inhibit drivers to those planes whose core is to remain in the 0 state.

10. At the completion of the write time the information has been restored at the address.

9.4.2.2 The Write Operation

1. The address is loaded into the *MAR*.
2. The information to be stored is loaded into the *MIR*.
3. The read cycle is commenced in the memory.
4. The cores are driven to the 0 state.
5. The sense amplifiers are *not* strobed; thus, the contents of the *MIR* is not changed.
6. The write cycle is commenced in the memory.
7. The planes in which a 0 is to be written have their corresponding flip-flops of the *MIR* in the 0 state, and hence the inhibit lines to those planes are enabled, preventing the writing of a 1.

8. At the completion of the write time the information has been stored in the address.

9.4.3 Memory Address Register, Decoders, and Drivers

The location in the memory from which information is requested or stored is called the *address*. The specific address is transferred from the arithmetic unit, the control unit, or the I/0 channels to the memory unit and held in the memory address register (*MAR*) while the memory completes its current cycle. The block diagram of Fig. 9.10 shows the electronics associated with the *MAR*. These associated electronic blocks are required decoders and line drivers needed to select the specified x_i and y_i select lines of the core planes and to supply the correct polarity select current pulses. The diagram assumes a 16-bit word length and 4096 addresses in the memory arranged so that each core plane is a 64- × 64-bit array. The numbers next to the circles on the lines of the figure present the number of parallel lines needed.

Since we can address 4096 locations in the memory, this will require a

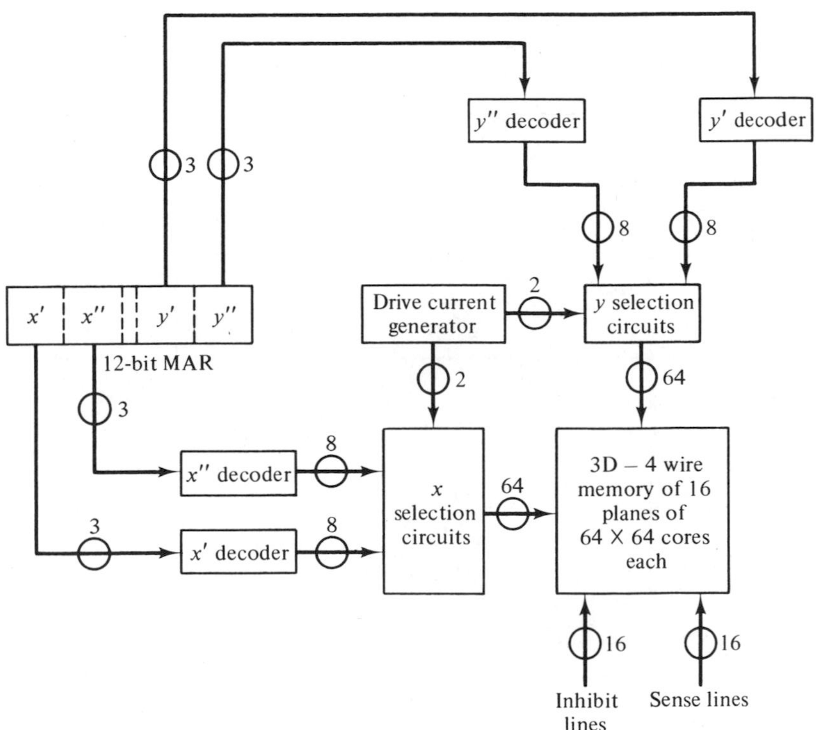

Figure 9.10 Memory address register and associated electronics.

12-bit address. Thus, the *MAR* must be 12 bits. The problem is to determine a good decoding method to obtain one of 4096 cores from the 12-bit address. We should observe that the 3D coincident current core plane represents a level of decoding inherent in the structure of the plane. That is, we need to select only 1 of 64 x lines and 1 of 64 y lines to select 1 of 4096 cores on the plane. Thus, the problem reduces to selecting 1 of 64 lines. We therefore divide the 12 bits of the address into two groups of 6 bits each—one group for the x lines and one group for the y lines. There are several ways to obtain 1 line from 64 from a decoding circuit with 6 bits of input. Chapter 6 included several such methods. If we restrict the fan-in of the gates to 2, we obtain the following numbers for three such methods:

Method	*Logical levels*	*Total number of gates*
Min terms	3	320
Tree	5	124
Dual tree	3	88

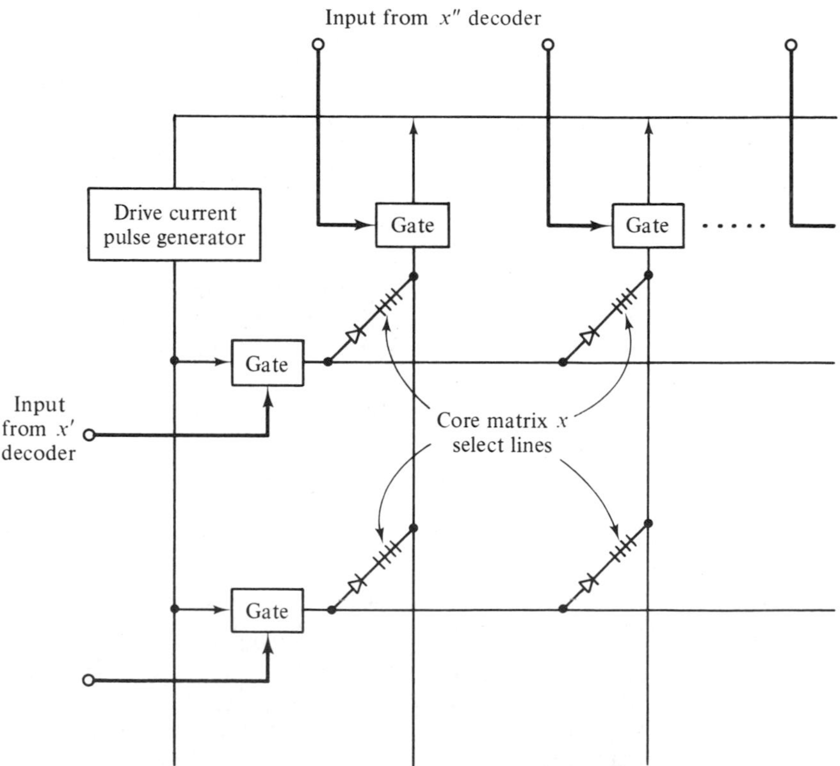

Figure 9.11 8×8 diode matrix selection for X circuit.

Thus, the dual-tree decoder requires the fewest number of two-input gates to select 1 line out of a possible 64 lines. We must use this method on both the x and y lines, and therefore it requires a total of 176 dual input gates for the proper line selections. Figure 9.10 shows the dual-tree decoders needed for the line selections.

The specific y or x selection circuits use a diode matrix scheme. This scheme is shown in Fig. 9.11. One of each set of gates is enabled to permit a passage of a drive pulse through 1 of the 64 possible lines. The diodes are needed to prevent *sneak paths* through lines which were not selected. One of the requirements of any such selection network is that it must provide a path for both positive and negative selection pulses. The above network is for one direction only. Figure 9.12 shows the selection circuit modified for dual polarity pulses.

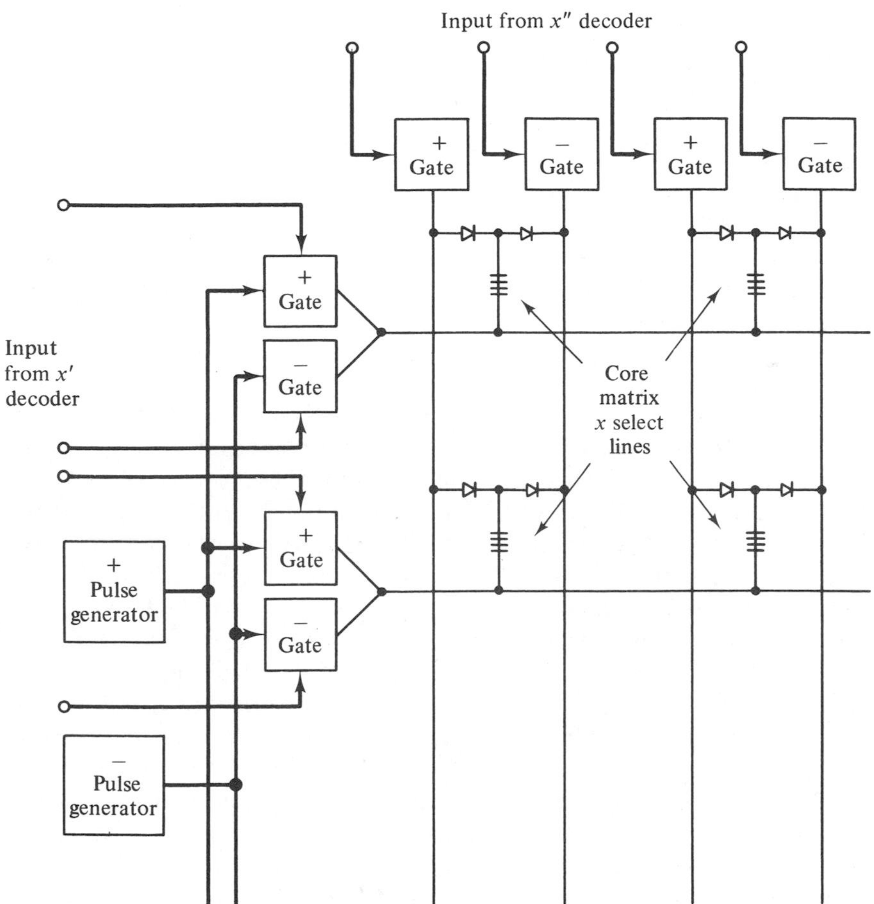

Figure 9.12 Dual polarity selection matrix.

The read and write current pulses are supplied by a high-impedance current generator. The pulses supplied from this generator must be of high accuracy at $I_m/2$. The usual circuit is an emitter-stabilized circuit which clamps the current. Figure 9.13 shows the circuit.

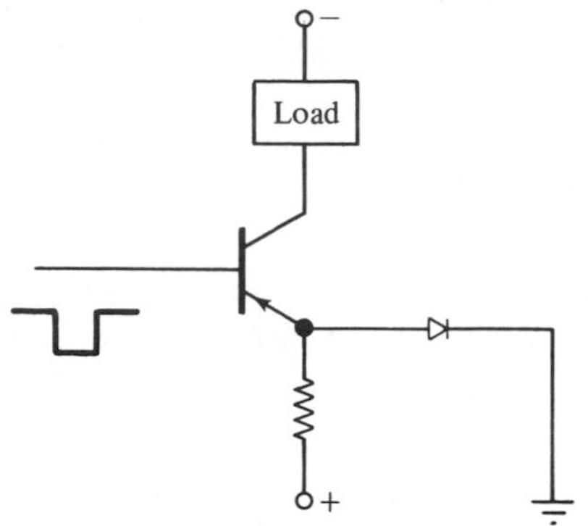

Figure 9.13 Read-write pulse generator.

9.4.4 Sense Amplifier

There is a sense amplifier used for each bit plane of the memory. These circuits are used to amplify the output signal from the selected core and determine if the core stored a 1 or a 0. The amplifier is used to set a flip-flop which is the bit position of the memory information register (*MIR*); see Fig. 9.9.

The basic components of this amplifier are shown in Fig. 9.14. The input transformer is used to match the impedance of the sense line, which behaves like a transmission line with a characteristic impedance of 100 to 300 ohms. It also permits a voltage gain by its turns ratio when terminated with a high impedance obtained by using an emitter follower. The signal is then amplified and transformer-coupled to a rectifier. Rectification is required because the signal may be positive for some cores of a plane and negative from other cores in the same plane. This is so because no attempt is made when fabricating the planes to maintain the cores with the same polarity. The cores are wired as they fall. The rectified pulse is then strobed with the strobe signal as one of the inputs to an AND gate. The timing of this strobe pulse is critical. Figure 9.15 shows the output from a core plane for both a 0 and at 1 output. The output for a 0 stored in the core plane is due to noise which is caused by half-selected cores and other pickup from the drive lines. The strobe pulse is applied to the AND gate at the time when the large output from the 1 state is anticipated and after the noise has decreased. Thus, the gate is enabled at the time of maximum signal to noise. The output of the gate is sampled by a transistor which is adjusted to an optimum level for discriminating between 1s and 0s. The output of this transistor is used to set the flip-flop of the *MIR*.

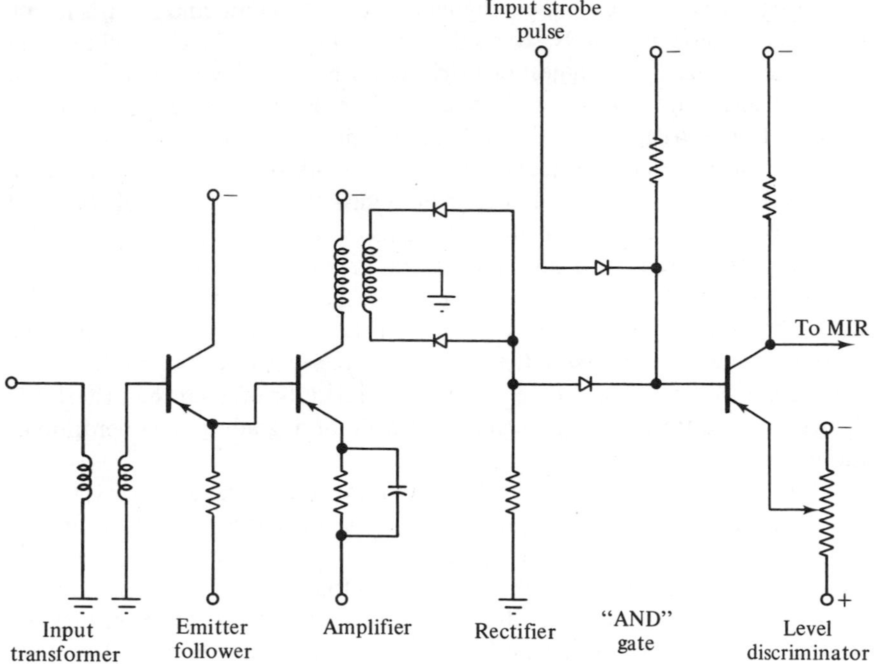

Figure 9.14 Sense amplifier.

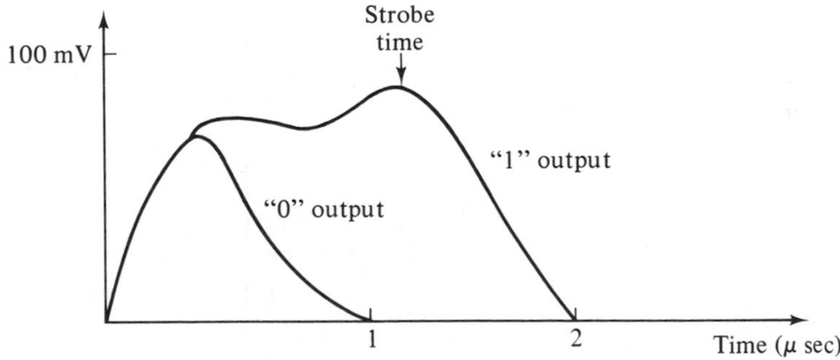

Figure 9.15 Output waveforms of core plane.

9.5 TOTAL MEMORY SYSTEMS

9.5.1 Memory Control Unit

The memory control unit must route requests for addresses to the proper bank of memory in an interleaved memory system. In addition, it must schedule requests to the banks on a priority basis from the various subsystems of the computer which can gain access to the memory. These subsystems are the I/O devices via their channels, the control unit, and the arithmetic unit. The priority is usually set so that the I/O device with the highest data rate of transfer has the highest priority. The other I/O devices have descending priorities based on their data rate. After the I/O devices the priority is given to the arithmetic unit, and the last priority is given to the control unit. A unit has exclusive access only to the bank of the memory in which it wishes to transfer one word of information. After the transfer of the word is completed, the requesting unit relinquishes its control to another unit which is waiting. When the first unit is again ready to transfer information it requests control, and upon the completion of the transfer then in progress, it will gain control if its priority is the highest then waiting. In this manner the units time-share or *multiplex* the memory banks. Because the bandwidth of the memory is so much greater than the bandwidth of even the fastest I/O device, this multiplexed system will work without degrading the operation of the machine.

There are two basic ways of determining the highest priority source which is enabled. One method is to perform a sequential scan starting with the highest priority source and proceeding to the lowest priority until an enabled request line is found. When the line is found, the output of a decoder indicates the line found and switches the bus selection lines to the requesting source. The second method uses the combinational circuit shown in Fig. 9.16.

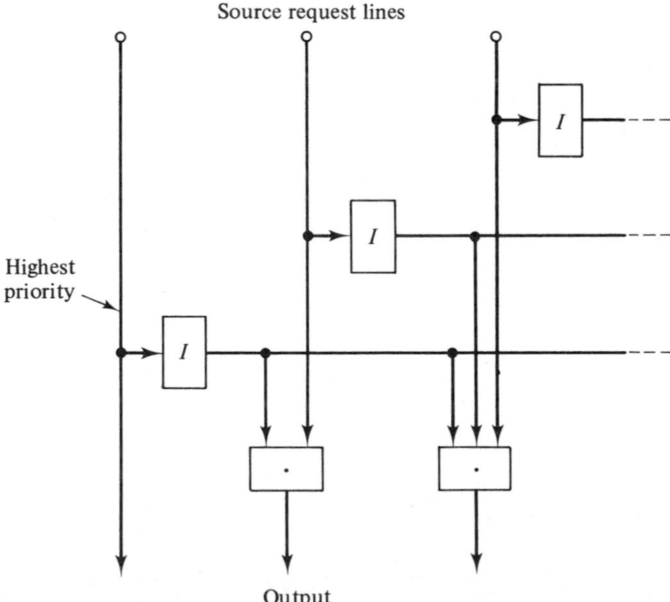

Figure 9.16 Circuit for priority selection.

Both of the methods are for noncritical time requests. That is, no time constraint is placed between the request for a memory word and the granting of the request.

9.5.2 Different Memory Organizations

In addition to the 3D/four-wire memory organization which has been explained in the previous sections of this chapter, there are several other memory organizations which are in use. One of these is a 3D/three-wire magnetic core memory. In this organization the sense winding performs both the functions of sensing and inhibit. If we reexamine the read and write cycles of the 3D core organization, we see that the sense line is used only during the read part, while the inhibit is used only during the write part. Thus, the two functions can be placed on the same line threaded through the core. The reduction in the number of wires threaded through the core reduces the cost of the three-wire memory as compared to the four-wire organization. In addition the three-wire core can have a smaller inside diameter and is therefore smaller. A smaller core can have a faster switching time.

Another organization which has been used in both magnetic core and electronic memory systems is the two-dimensional (2D) or *linear select* organization. In the organization which is shown in Fig. 9.17, only two wires need to be threaded through each core. Thus, smaller and faster cores can be used. The 2D organization does not provide an inherent decoding, as does the 3D organization, and is therefore much more costly to fabricate in larger sizes. However, since it does not require a coincident of current to switch the cores, a large switching current can be used and fast switching can occur.

The 2D core shown in Fig. 9.17 has three words of 6 bits each. To read information from this memory array an address is decoded and a word line

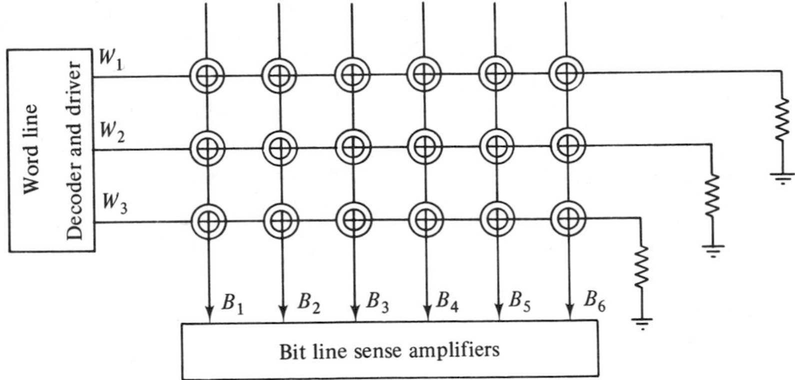

Figure 9.17 2D core organization.

W_i is selected. A positive current pulse is placed onto that line, and any core which is magnetized at $-B_r$ will be switched to $+B_r$ and a voltage pulse will be sensed in each of the sense amplifiers connected to the 6-bit lines B_1 through B_6 where a core has changed its state. Thus, a single current on a word line is sufficient to read the cores. After the state of the cores has been determined all the cores of that word will be in the $+B_r$ state. To restore the cores to their previous state a negative pulse is applied to the selected word line W_i. To prevent those cores which were not in the $-B_r$ state from switching to that state an inhibit positive current pulse of one-half the required switch current is applied to those bit lines. Thus, those cores which are not to be switched back to $-B_r$ will have a net negative current of only one-half the required switching current, while all the other cores on those bit lines will have a positive current of one-half the required switching current and will not be affected. Those cores of the word which have no current in their bit lines will be restored to $-B_r$. Therefore, the 2D memory requires a coincident current system for the writing of information but only a single current on the word line to read from the cores. The 2D organization used with electronic components to produce integrated circuit memories is described in Section 9.6.

9.6 ELECTRONIC MEMORY—LARGE-SCALE INTEGRATION

Electronic memory in some form has always been used in electronic digital computers. In first-generation computers only the registers in the central processing unit were electronic. With second-generation computers the main memory was magnetic core, and the registers were transistor electronics. With the advent of the third-generation computers, the cost of electronics has decreased appreciably, and high-speed scratch-pad memories were introduced to supplement the registers and immediately store used operands. The basis for the main memory remained the magnetic cores, as the cost for electronic storage was still not competitive. Since 1970, however, the developments of technology in integrated electronic circuits have been such that a larger number of storage flip-flops per unit of production have been possible. This technology has been named *large-scale integration* (LSI) and has resulted in cost/performance characteristics of *random access memory* (RAM) that is competitive with magnetic core for main memory.

9.6.1 The Bistable Flip-flop

The basis for an electronic storage cell is the bistable flip-flop, as shown in Fig. 9.18. Two implementations are shown. One uses two transistors and is called a *bipolar storage cell*, and the other uses four metal-oxide semiconductor

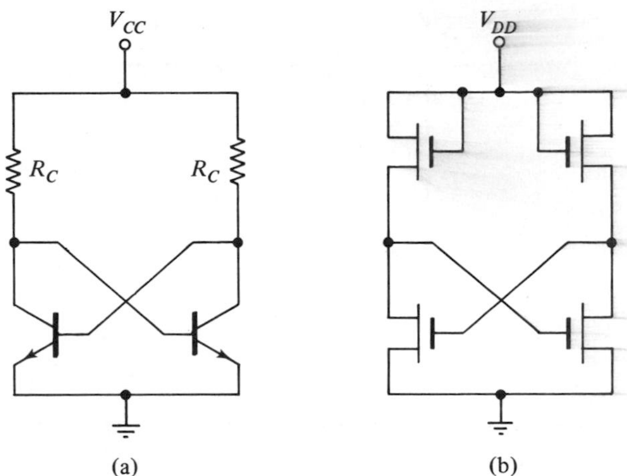

Figure 9.18 Basic flip-flop memory cell: (a) bipolar; (b) MOS.

field effect transistors (MOS). The cell is simple to design, is inherently very high speed, and provides high-yield/low-cost LSI circuits. Either of these two elements can be fabricated in 256 cells on a *chip*. The chips are packaged and used to build large main memory. A chip is a piece of silicon on which the memory cells are fabricated. Each cell occupies about 20-thousandths of a square inch and the entire 256 cells are packaged in a 16-pin dual-in-line package measuring about .870 × .250 in. The cycle time of bipolar cells is about 200 nanoseconds or less, while for MOS devices it is about 400 nanoseconds. These times will decrease in the coming years as this technology continues to develop. In addition the number of storage cells per chip will increase. By 1974 memory chips with 4096 bits of storage were available, and it is predicted that memory chips with more than 64,000 storage cells will be produced.

9.6.2 Selection and Sensing of a Storage Cell

Selection of a single flip-flop from a square array of such storage cells which have been fabricated on a single chip is performed by a linear selection technique. Each cell has four lines connected to it. One line is the dc power line. A second line connects to the emitters and is called the *word line*. Two addition lines are each connected to each side of the flip-flop and are called *bit lines*. Figure 9.19 shows a block diagram of a single memory cell, while Fig. 9.20 shows a diagram of nine such cells from a section of a chip. A particular memory cell may be addressed for either reading or writing by applying the proper signals to both the bit lines to which it is connected and the word line to which it is connected.

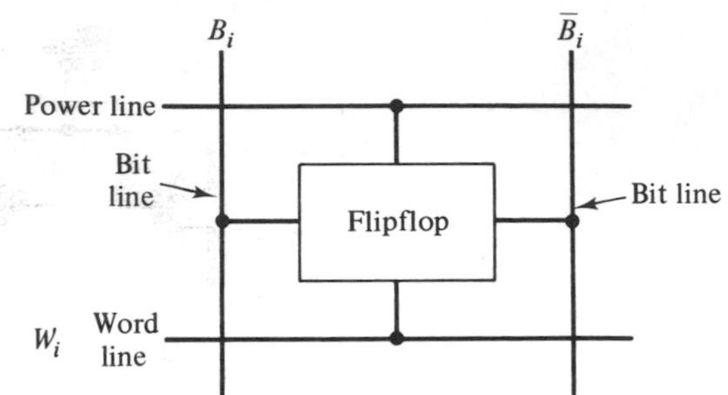

Figure 9.19 Representation of a single memory cell.

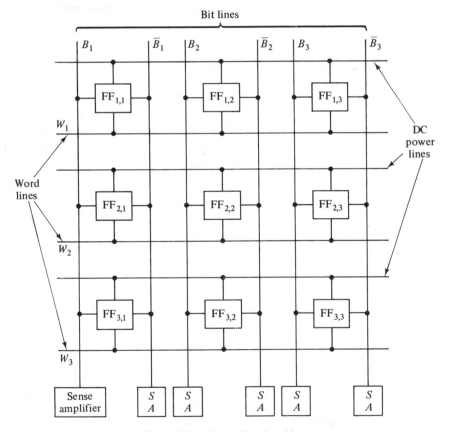

Figure 9.20 Nine cells of a chip.

To *read* from a cell, the voltage on the word line W_i to which it is connected is raised. This causes current to flow through the bit line on the side of the flip-flop which is in the high-voltage state. By sensing the change in current on all the bit lines the state of the flip-flop can be determined. The state does not change during the read cycle. However, since many flip-flops are connected to a given word line W_i, all cells attached to it will be read. This is called a linear select or two-dimensional (2D) memory. If we are interested only in the contents of a single cell, we must use a decoder to select the output of a given sense amplifier to determine the state of only one of the cells on the chip.

To *write* into a given cell on the chip, a coincident select system is used. The word line connected to the addressed cell has its voltage raised; coincidently the bit line to the side of the cell which is to be placed in the high-voltage state is also raised. This causes that side of the cell to raise its voltage if it is not already in that state. When the voltages on the word line and bit line are returned to their quiescent values, the cell will retain the state to which it has been set.

Thus, if we wish to read the state of flip-flop $FF_{3,2}$ in Fig. 9.20, we raise the voltage on word line W_3. This will cause a current to flow from the high-voltage side of flip-flop $FF_{3,2}$, which will flow in the bit line, and we can determine it from the sense amplifiers. It will also cause currents to flow in the bit lines of flip-flops $FF_{3,1}$ and $FF_{3,3}$. We shall only select the sense amplifiers on the B_2 and $\bar{B}_2$ bit lines to read the information from flip-flop $FF_{3,2}$. To write information into flip-flop $FF_{3,2}$, we raise the voltage on word line W_3, and if we wish to write a 1 bit, we raise the voltage on bit line B_2. This causes flip-flop $FF_{3,2}$ to go into the 1 state. None of the other cells is affected during this process.

9.6.3 Organization of an LSI Chip

To make LSI electronic memory compete with magnetic core memory it is necessary to fabricate a large number of memory cells on a single chip. One of the constraints on the chip is the number of connections which can be made to the chip. Because of its small area, only a small number of connections can be made. Another reason to limit the number of connections to the chip is the cost of making them. To obtain the objective of a small number of interconnections to a chip, a level of address decoders is fabricated on the chip along with the memory cells. This is particularly true for chips which use the MOS flip-flop as the storage element.

The MOS device has a limited current output capability. If the load on an MOS storage cell is another MOS device, it will function quite well and have a switching speed of 100 nanoseconds or less. If, however, the MOS cell must drive devices in another package, the extra load results in a significant performance degradation. To overcome this deficiency, a hybrid approach to semiconductor RAM, based on the use of MOS storage cells and bipolar devices for driving, sensing, and decoding logic all on the same substrate, has been developed.

The MOS storage cell arrays are less expensive to produce then bipolar flip-flop storage arrays. The power requirements are less for MOS cells. The bipolar drive and sense circuits connected on the same substrate enable high-speed operations which the MOS cells alone cannot achieve. In addition, the storage chips must interface to the logic circuits of the computer, which are generally bipolar electronics. Thus, using bipolar electronics for input and output circuits on the substrate provides electronic compatibility with the rest of the computer. However, the interconnections between the bipolar and MOS portions of the circuits were expensive to make. To reduce the number of interconnections, part of the decoding of the select circuits was fabricated in MOS technology. Figure 9.21 depicts this type of integrated circuit LSI hybrid device.

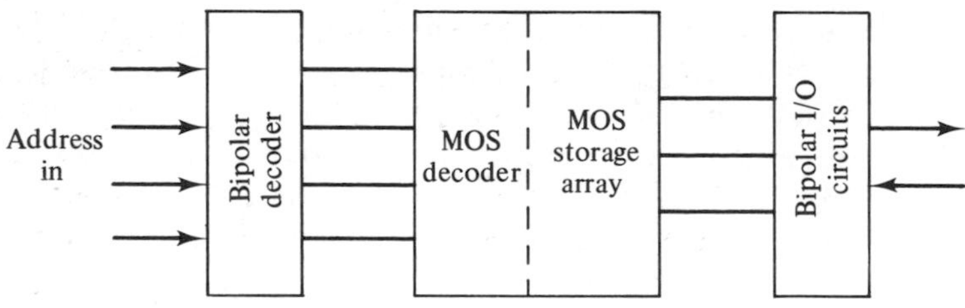

Figure 9.21 Hybrid LSI device.

9.6.4 Charge-Storage MOS Memory

The major constraint on the use of LSI electronic memory in computers is the economic cost of producing the memory chip. To produce a cheaper memory on a price per bit basis requires a higher number of memory cells per chip, since the cost of producing a chip is essentially independent of the density of the cells. The major restrictions on the density of bistable flip-flop chips are as follows: (1) They dissipate too much power and therefore cannot be packed tightly; (2) they occupy too much area, which limits the number which can be placed onto the chip. To overcome these two restrictions an MOS technology based on the storage of charges at a semiconductor junction has been developed. Since to store a charge a flip-flop is not needed, a three transistor per bit MOS charge-storage memory cell has been developed and is shown in Fig. 9.22. This structure has a small geometry, and therefore a large number of storage cells per chip can be fabricated. This technique has the disadvantage that the charge will "bleed off" the junction and must be periodically refreshed.

In Fig. 9.22 the MOS transistor Q_1 stores the charge on its gate. If a negative charge of a sufficient amount is present, then the device is in the 1 state. An insufficient amount of charge means that the device is in the 0 state.

Write
enable
Read
enable
Q_3
Q_1
Q_2
Write
bus
Read
bus

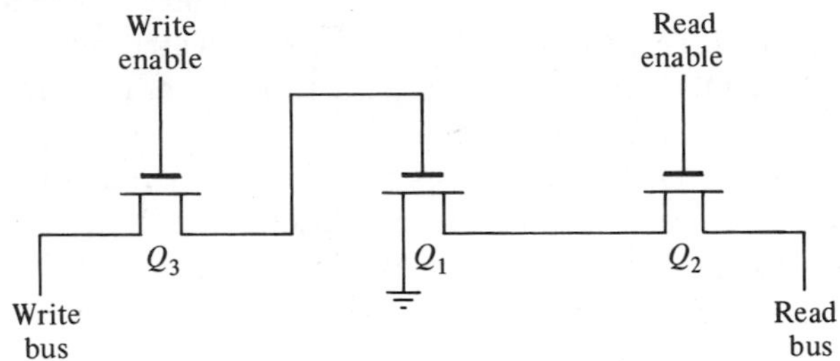

Figure 9.22 MOS charge-storage memory cell.

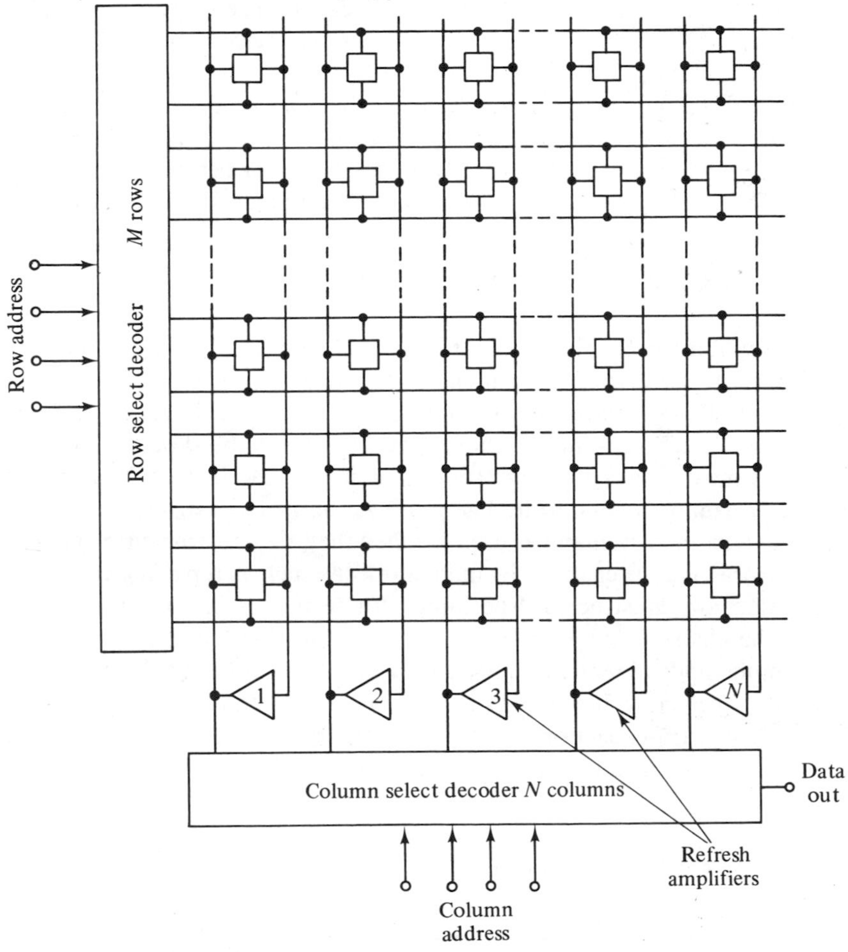

Row address
Row select decoder M rows
1
2
3
N
Column select decoder N columns
Data
out
Refresh
amplifiers
Column
address

Figure 9.23 An M row by N column charge-store memory array.

In addition a means of interrogating the cell for its charge content is provided by Q_2, which connects it to the read bus when activated by the read-enable signal. Transistor Q_3 provides a write path to the charge storage transistor when activated by the write-enable signal.

To supply the periodic refreshing of the charge stored at the junction of transistor Q_1 when in the 1 state, a refresh amplifier is deposited onto the chip. Figure 9.23 shows a typical memory array where each column of M bits has its own refresh amplifier. There are N columns in the array. Thus, there are $M \times N$ storage cells. The array works on a linear select addressing basis. Activating any one of the M read-enable row lines writes the content of the storage cells in that row into their respective refresh amplifiers. Thus, N cells will be affected. Following this by a write-enable signal on that same row will refresh all the memory cells in that row. To refresh the entire array requires a progression of the above steps for all M rows. The refreshing has to be done only for about 2% of the total accesses to the chip.

These dynamic charge-store memories have an access time of less than 200 nanoseconds with less than 100 microwatts of power dissipated per bit in the active mode and less than 10 microwatts of power per bit in the standby mode. They have a density four times that of a bipolar flip-flop memory.

9.7 READ-ONLY MEMORIES (ROM)

Many third-generation computers have their control units designed by a technique called microprogramming. This technique will be described further in Chapter 11, but since the microprograms are stored in special types of memories called read-only memory (ROM), it will be described here. A read-only memory has just that property, namely it can be read from, but the computer cannot write into it. The information in these memories is introduced in a permanent manner, usually when they are constructed. Sometimes the memories can be changed by manual means such as replacing cards from which they are constructed. The important feature is that the information cannot be changed by programming. In this section we shall study three different methods of fabricating read-only memory. Although there are many different types now being constructed, we shall study the diode matrix memory, the card-capacitor memory, and the transformer-coupled read-only memory. They will show the fundamentals of this type of memory organization.

9.7.1 Diode Matrix Read-Only Memory

Recent advances in LSI have permitted the use of a diode matrix as a ROM. The diodes are placed between word lines and orthogonally placed bit lines, as shown in Fig. 9.24. This type of memory organization is known as

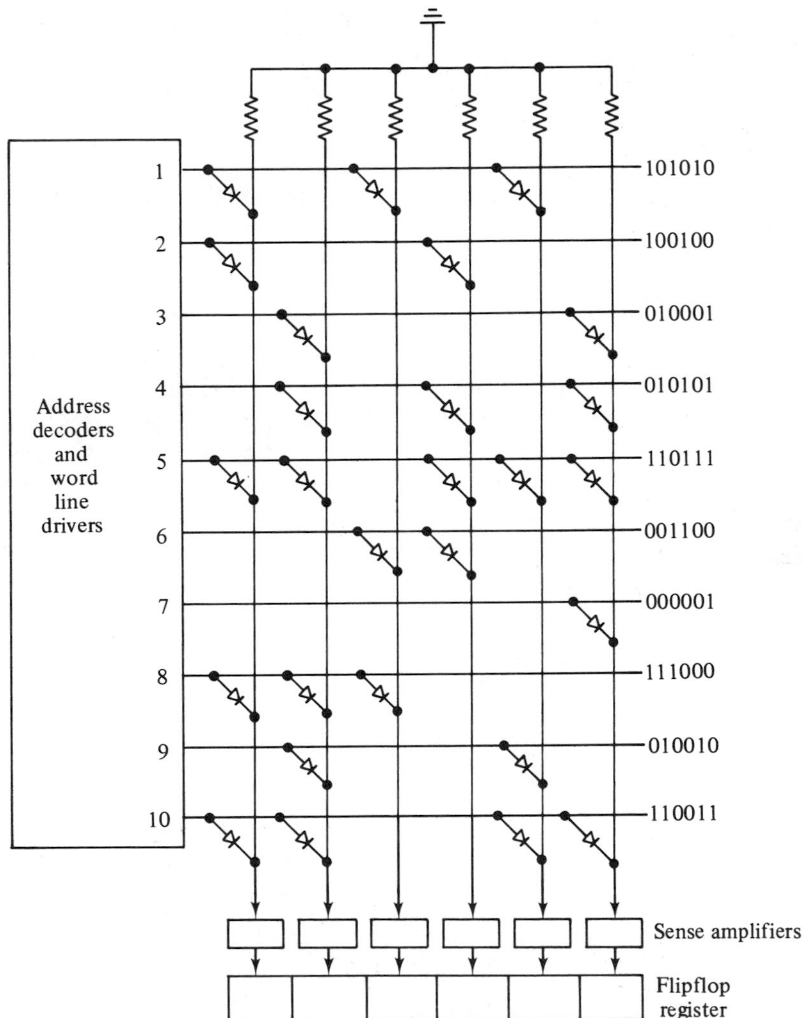

Figure 9.24 Diode matrix ROM.

linear select or two-dimensional (2D). Essentially an entire word of memory is read by placing a read pulse on the desired word line. Since the diodes are reverse-biased in the quiescent state, no current flows into the bit lines. When the read pulse is applied to a word line all the diodes connected to that line become forward-biased and a current flows into all the bit lines where the diodes make a connection. The bit lines are connected to sense amplifiers which detect the current and amplify the signal. The sense amplifiers set the corresponding bits in an output flip-flop register to the 1 state.

One of the problems inherent in the diode matrix ROM is that the input to the sense amplifier is in effect an *n*-input OR gate. This limits the number of word lines which can be used with a given sense amplifier due to the fan-in restriction of the gate.

9.7.2 Card-Capacitor Memory

Card-capacitor memories can provide a fast and economical read-only memory. This type of memory is used in the IBM 360 Model 50 as a micro-programmed control store. The card-capacitor memory can read out a word in as fast as 90 nanoseconds with a cycle time of 200 nanoseconds. The basis for this type of ROM is the coupling of the word lines to the sense lines by a capacitor. Thus, at those bit positions where no coupling is to be made, no capacitor is permitted. The structure of such a ROM is shown in Fig. 9.25. Two cards are arranged face to face, with two sets of lines, one on each, crossing orthogonally. An aluminized ground card covered on both sides by a Mylar dielectric is placed between them to act as a shield. Because of the aluminum ground plane, no coupling can exist between the word lines and sense lines. At those locations where a 1 bit is desired, a hole is prepunched through the metalized ground plane to permit a capacitive coupling.

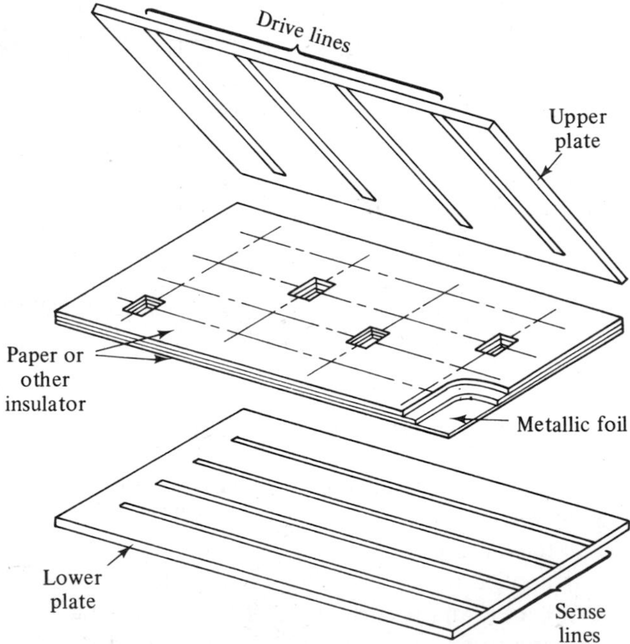

Figure 9.25 Card capacitor ROM. (Samir S. Husson, *Microprogramming: Principles and Practices*, © 1970, p. 156. By permission of Prentice-Hall, Inc.)

The grounded metal plate is the document which carries the information. In most systems this plate can be manually removed from between the other two plates which carry the word card sense lines and replaced by a new plate with different holes. Hence, the ROM is modifiable at manual speeds.

For the card-capacitive ROM to function properly it is essential that the difference between the capacitance at a punched and unpunched intersection be at least an order of magnitude. It is also essential that a good ground reference exist to shield any unwanted stray capacitance between word and sense lines. To study the problem of stray capacitance, let us assume that C_0 represents the value of capacitive coupling between a sense and word line. Let C_s represent the sum of the stray capacitance that exists between sense and ground at an unpunched intersection plus any residual coupling capacitance due to fringe effects. Because of the Mylar dielectric present at an unpunched location as compared to air at a punched location, $C_s \gg C_0$. Assume that there are N sense lines attached to each sense amplifier; i.e., there are N cards in parallel but only one word from one card is to be read. If V_{in} is the magnitude of the input pulse voltage and V_{out} is the magnitude of the voltage at the input to the sense amplifier, then the output will be across a capacitive voltage divider, as shown in Fig. 9.26.

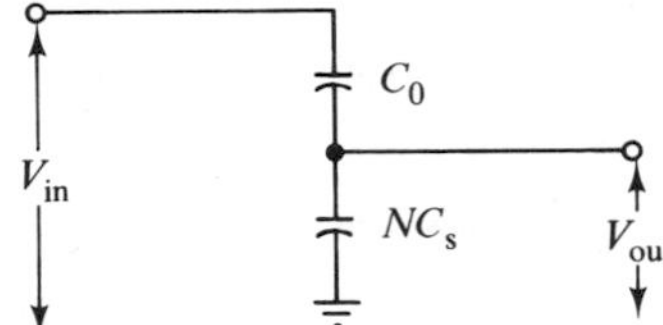

Figure 9.26 Capacitive voltage divider.

We can write a relationship between the input and output voltages as

$$V_{out} = V_{in}\left(\frac{C_0}{C_0 + NC_s}\right)$$

but $C_s \gg C_0$; therefore,

$$V_{out} = V_{in}\frac{C_0}{NC_s}$$

Thus, the input voltage will be attenuated by a factor of C_0/NC_s until it reaches the output. If $C_s/C_0 = 200$ and $N = 30$, then

$$\frac{C_0}{NC_s} = \frac{1}{(30)(200)} = \frac{1}{6000}$$

or the output signal will be 1/6000 that of the input signal.

To overcome the problem of sneak paths due to unwanted capacitive coupling of the simple card-capacitive ROM, the balanced system has been developed. In this system shown in Fig. 9.27, each sense line is divided into a pair of sense lines, one for the 1 bits and the other for the 0 bits. One effect is that each word line now sees a fixed number of capacitors at all times. The pair of sense lines now feeds into a differential amplifier which amplifies only the difference between the signals on these lines.

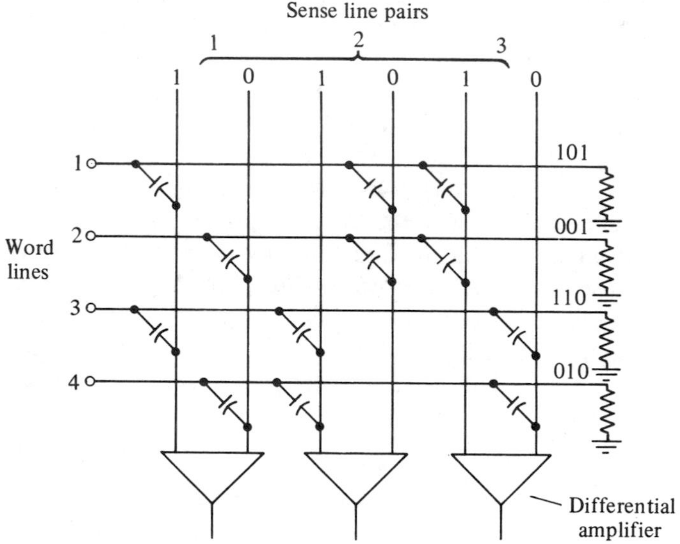

Figure 9.27 Balanced sense line capacitor ROM.

A second improvement is to balance the word lines by dividing them into a pair and using two capacitors per bit. This gives a completely balanced system in which each word driver sees a constant number of capacitors for a load and each sense amplifier sees a constant number of capacitors. This also tends to eliminate the effects due to sneak paths by the use of a differential sense amplifier. Figure 9.28 shows this configuration.

A completely balanced capacitive ROM of the type shown in Fig. 9.28 is used in the IBM System/360 Models 50, 65, and 67 to contain the microprograms of the control unit. The unit has a 200-nanosecond cycle time and a 90-nanosecond access time. Its capacity is 2816 words of 100 bits organized into 16 planes of 176 words per plane.

9.7.3 Transformer-Coupled ROM

Transformer-coupled read-only memories also provide an inexpensive method of storing constant data. It is slower than electronic or capacitive-coupled systems but provides a better coupled signal to the sense lines.

The transformer-coupled ROM has a linear select organization. A magnetic core has a bit line which passes through it. Through each of the

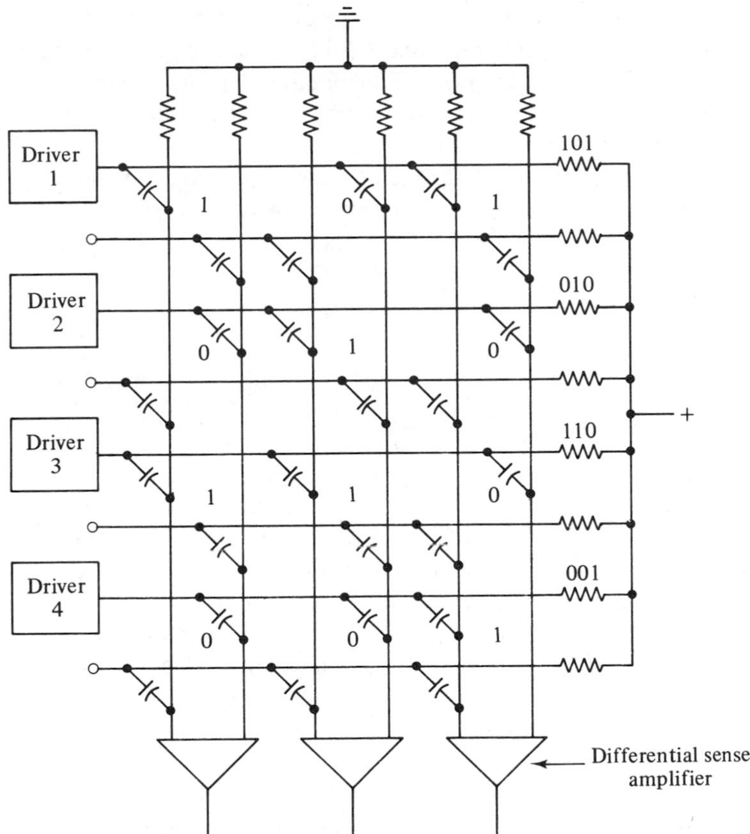

Figure 9.28 Completely balanced capacitor ROM.

cores there are orthogonal word lines. Figure 9.29 shows a three-word by 4-bit ROM. Each of the word lines links through the magnetic core if a 1 is to be produced on the associated bit line. If a 0 is to be produced on the bit line, the word line skips the core. Thus, in Fig. 9.29 word A stores 0110, word B stores 1001, and word C stores 1101. To read from the memory a current is passed through the selected word line. The current will cause a flux to pass

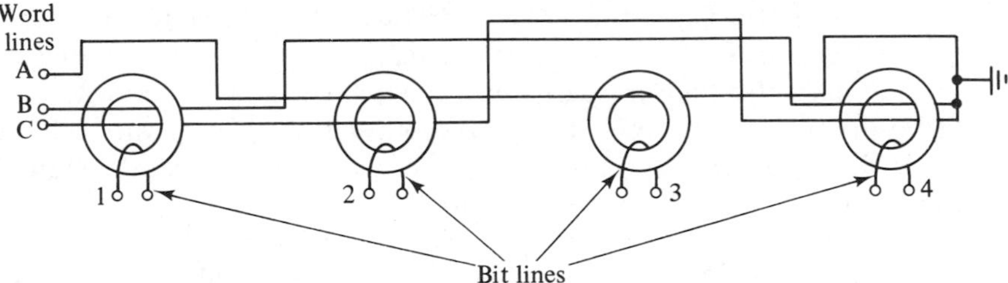

Figure 9.29 Transformer-coupled ROM.

through the core if the word line passes through the core; otherwise no flux will be developed. The changing flux will induce a voltage onto the bit line. This is sensed by a sense amplifier. Since many word lines pass through the same cores, there are not many cores needed. Thus, the transformer-coupled ROM is inexpensive and provides a large output signal. Its disadvantage is its slow speed. The IBM 360 Model 40 uses a microprogrammed control store made from a transformer-coupled ROM. It has 50 magnetic transformers with 256 word lines which can pass through them. The word lines are photoetched on flexible Mylar tapes. The paths are made through the transformers by breaking current paths either through the cores or around the cores.

9.8 MAGNETIC DRUM STORAGE

The memory systems which we have described in the previous sections were based on a bistable cell, i.e., the core and the flip-flop. Access to memories constructed from these devices is random; that is, once an address is given, the location can be read in a fixed amount of time. Its access time is constant for all locations. The limitation of such memories is their price. Usually a large computer cannot afford to have all its data stored in random access memories. Magnetic drums provide a less expensive means to store relatively large amounts of data and still provide a reasonably short access time. These devices have a sequential access characteristic and therefore present an average access time in the 4–100-millisecond range. Most contemporary computers use drums for secondary store and transfer blocks of data to their main memory for execution and data handling.

9.8.1 Mechanical Characteristics of Magnetic Drums

A magnetic drum is basically a rotating cylinder coated with a thin layer of a ferrite magnetic material which has a square type of hysteresis loop similar to that of the magnetic core. Recording heads are mounted on a fixed outer casing along the surface of the drum and are adjusted until they almost touch the rotating surface. The magnetic heads are used to record and read digital information from the surface of the drum by magnetizing or sensing the magnetization of small areas. Figure 9.30 shows two views of a magnetic drum, and Fig. 9.31 depicts the outside of a drum with all the magnetic head connections. Some magnetic drums have several hundred recording heads around the peripheral surface.

The drum rotates at a constant speed and is not stopped or started during the accessing and data transfer process. Each head can access a small

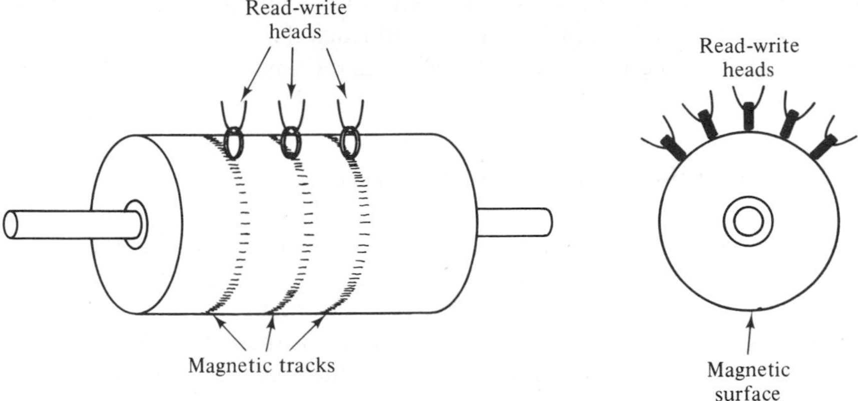

Figure 9.30 Two views of a magnetic drum.

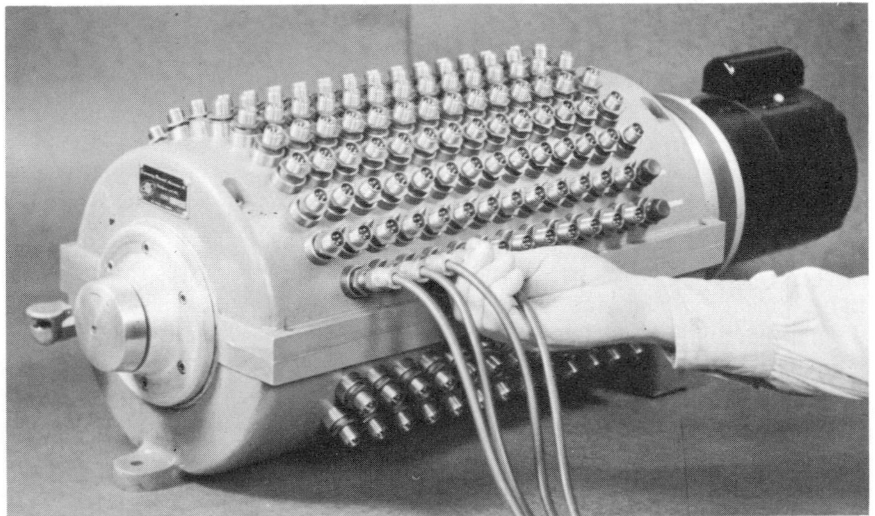

Figure 9.31 A magnetic drum—exterior view. (Courtesy Sperry-Rand Corp., Sperry UNIVAC Division.)

band on the cylinder surface; this band is called a *track*. Data are stored as magnetized spots on the magnetic surface in the tracks. The tracks are therefore subdivided into cells which can store 1 bit of information. All the cells which are under a set of heads at the same time are called a *slot*. Accessing a drum consists of waiting for the addressed slot to rotate under the read-write heads and then reading the information. If data are stored in a random manner on the surface of a drum, then over a long period of time the average access time to a datum will be equal to one-half the rotational period of the drum. The *access time* is composed of two components; one is called the *latency time*, which is the time it takes to reach the first bit of the information, and the second is *transfer time*, which is the amount of time it takes to transfer the requested data.

9.8.2 Electrical Characteristics of Magnetic Drums

Usually an additional track is added to the drum to provide timing signals for the operation of the drum; it is called a *timing track*. The timing track contains a series of permanently recorded timing signals which determine a time unit for the drum system. The drum is therefore self-timed, and any change in rotational speed will not prevent the operation of the drum. The timing track is used to determine the location of each set of slots around the tracks. For example, if the circumference of the drum is 72 inches and timing pulses are recorded at a density of 200 per inch, there will be 14,400 slots per track. If the drum has 50 tracks in addition to the timing track, then the capacity of the drum is 720,000 bits.

Information is written onto a drum by passing current through the write head winding. This causes a magnetic flux to pass through the magnetic material of the head. The head has a high permeability with a small air gap just at the point where it is tangent to the recording surface of the drum; see Fig. 9.32. The gap in the head presents a high reluctance path to flux traveling around the head. Since the magnetic material on the surface of the drum is passing very close to the gap, some of the magnetic flux passes through this high-permeability material. This causes a small area of the drum surface to be magnetized, and since the material has a square hysteresis loop, the magnetic field remains after the area has left the vicinity of the recording head. The heads do not touch the rotating drum in order to prevent wear. The distance between the head and the drum is kept very small, and therefore the drum must be of a constant diameter to maintain the proper spacing between drum and head. If the drum moves farther from the head due to a low spot on the drum, the signal may be too weak to be recorded because of too large a gap. Likewise, if the drum has a high spot, it may touch the head and wear it out.

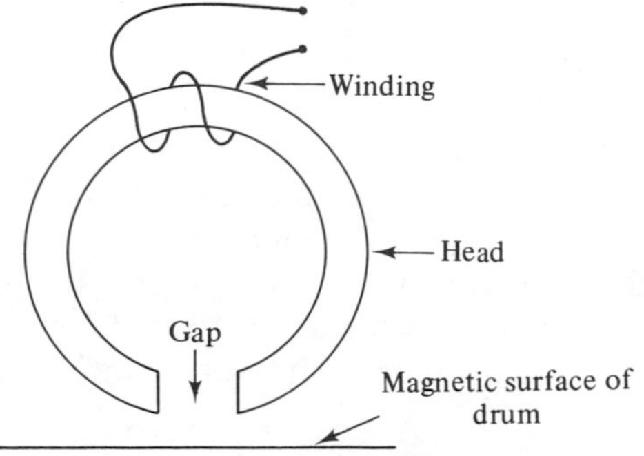

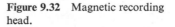

Figure 9.32 Magnetic recording head.

Since the drum rotates on bearings, they must be very good to prevent any wobble in the drum.

Signals recorded on the surface of the drum are read when the slot which has been magnetized passes under the head. Some of the magnetic flux is coupled into the head, and changes in this flux due to the movement of the drum induce a voltage in the head winding. These weak signals are then amplified and used to set flip-flops in buffer registers. Some drums use separate heads for reading and writing, and others use combined read-write heads.

The size and capacity of drums vary over a wide range. Small drums with capacities of less than 25,000 bits have been constructed. Drums this small have from 15 to 25 tracks and 15 to 50 heads. One method of decreasing access time is to have more than one head per track. If there are two heads per track, they are set 180° apart. This decreases the access time by one-half but increases the amount of electronics needed to determine which one of the heads to access. Larger drums can store in excess of 30 million bits and may have up to 1000 tracks. Drums have been designed to rotate from 120 rpm up to 75,000 rpm. The packing density of many drums is 1000 bits per inch, but packing densities in excess of 4000 bits per inch have been achieved by close head spacing and slow rotational speed. Since drums are usually used as extensions for addressable main memory, their data rate is usually high, in the area of 1 to 10 million bits per second. Table 9.1 shows some characteristics of

Table 9.1 *Typical Drum Characteristics*

	IBM 2301	*IBM 2303*	*Byrant PHD*	*IBM 2305/1†*
Average access time (milliseconds)	8.6	8.6	70	2.5
Capacity (megabits)	32	32	340	48
Data rate (megabits per second)	9.6	2.5	1.2	24
Record density (bits per inch)	1,250	1,250	1060	4000
Track density (tracks per inch)	67	67	80	
Track capacity (bits per track)	41,000	39,000		
Number of tracks	800	800		384
Number of tracks accessed in parallel	4	1		2
Rotational speed (rpm)	3,500	3,500		6000
Diameter (inches)	10.7	10.7	20	14
Length (inches)	12	12	43	
Latency (milliseconds)	17.5	17.5		5.0

†Not a drum but a head per track disk file consisting of six disks.

different drums. Information is usually recorded in the NRZI or phase modulation method on newer drums (see Section 9.10).

9.8.3 Parallel Operation of a Drum

When operating a drum in the parallel mode, all the bits of a word are read or written simultaneously. If the computer has a word length of 48 bits, the drum reads 48 data tracks and 1 timing track simultaneously and thus obtains the information in 1 bit time. Since all bits will appear simultaneously, a read amplifier must be used for each bit; thus in the above example 49 read amplifiers must be used. Likewise 48 write amplifiers will be needed for a parallel write operation.

The slots of a drum when used in the parallel mode are located by counting the pulses of the timing track. If each track can hold 2^{14} bits (2^{14} = 16,384), then a 14-bit counter can be stepped by 1 each time a timing pulse appears in the timing track. A new track which has only one permanently recorded pulse in its track is used to set the counter to zero. Using this method, the 3245th slot can be given the address 3245 and be located by counting. When the counter compares with the drum address register the data are transferred between the drum and its buffer register. Figure 9.33 shows a block diagram of the data transfer by the above-described method. The data rate available with this mode of operation is equal to the number of bits in the word length multiplied by the reciprocal of the bit time. This assumes that having waited for the first addressed word to reach the read-write heads we can transfer the next sequential words as they reach the heads in each successive bit time.

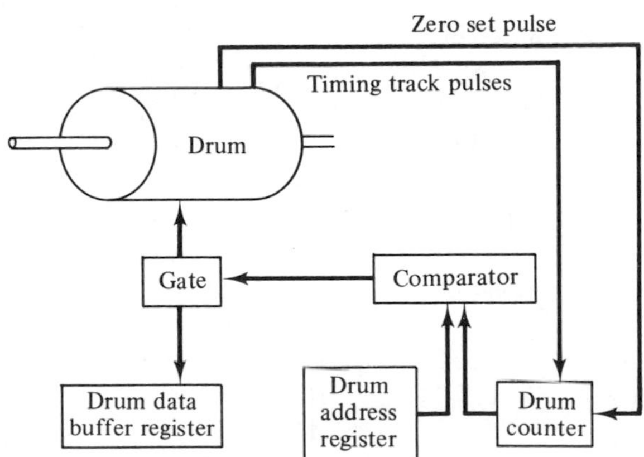

Figure 9.33 Parallel drum system.

9.8.4 Serial Operation of a Drum

When operating a magnetic drum in a serial mode, data will be transferred from or to only one track at a time. Thus, the correct track must be selected from all the possible tracks, and the correct bits within that track must be selected. Each track is assigned a number which is the address of the track. Within each track the bits are grouped into *sectors*. A sector is usually equal to the number of bits in the computer word. Thus, if the computer has a word length of 32 bits and the drum can have 2^{14} bits per track, there will be $2^9 = 512$ sectors around each track of the drum. Each sector will contain one computer word of 32 bits in consecutive sequential slots of the track.

To obtain a word from the drum, or write a word, an address consisting of the track number and the sector number with that track must be given. If we consider a drum with 64 data tracks, each track containing 512 sectors of 32 bits each, then we shall need 6 bits in the address to select 1 of the 64 tracks and 9 additional bits in the address to specify 1 of the 512 sectors. Thus, in the address of an I/O instruction for the computer, the first 6 bits of the address are interpreted as the track number and the remaining 9 bits as the sector number within that track.

The first 6 bits of the drum address register, which contains the track number, are connected to a decoder matrix in order to select the correct read-write head of the addressed track. To locate the addressed sector a counting scheme is used with three timing tracks. One time track contains bit timing pulses—one per slot or bit. The second timing track contains permanently recorded impulses spaced one per sector and indicates the beginning of a sector. The third timing track contains only one pulse and is used to reset the sector counter to zero. The sector is selected by comparing the address of the selected sector with the sector counter. When a comparison is obtained, the proper sector is rotating under the previously selected read-write head of the track and data transfer can begin in a serial-by-bit mode.

A third mode of operation can now be developed for a drum. This is a combined series-parallel mode. In this mode more than one track is transferring data during a given bit time, but a full computer word is not transferred as in the parallel mode. For example, if a computer word is 32 bits and the operands are binary coded decimal numbers, then the 32 bits will represent eight 4-bit BCD characters. These can be transferred one character in parallel from four tracks of the drum and eight characters in serial from the track. Thus, a sector for this drum operation will be 8 bits long, but we shall have to address four read-write heads simultaneously to read the 4 bits in parallel which comprise the BCD character. This mode will provide a higher data rate than the strictly serial mode but a lower data rate than the strictly parallel mode.

9.8.5 Control Unit and Data Organization

Most of the electronics associated with a drum storage unit are contained in a separate unit called a *control unit*. Thus, the electronics to perform the address comparisons, select certain tracks, check parity bits, or check for mechanical speed are included in the control unit. Usually a manufacturer will standardize on the commands to and from the control unit and on the record formats used to record on the drum. The control unit has sufficient electronics to control several drums. This is done to minimize the amount of common electronics, such as power supplies, which would be needed with separate control units. By doing this a lower price per unit can be obtained.

Early drums used fixed record lengths. A *record* is a block of computer words or characters which are related and handled as a unit for purposes of input and output. Since the records were of fixed size, the programmer had to break his records into this fixed size. Modern devices can use variable-length records whose size is indicated at the start of each record. Between each record is a space called the *record gap*. To minimize the amount of space used as record gaps, it is necessary to group records together into a block. At the beginning of each block is a *header* which tells which records are in the block, where they are located, and the length of each record as well as the total block. If the drum is being used to move data into and out of main memory with fixed block size, then the early drum data organization is preferable.

When updating information recorded in a blocked format on a drum, the programmer may search the drum for a *key* contained in the header of a block until a match is found. Then the block is read from the drum. The record within the block is updated in main memory, and the block is reconstituted and written back onto the drum. In addition to the key on which a search is made, the header may contain information used for file protection, i.e., who may access the file and the identification of bad tracks. In addition, special characters are used to identify the beginning of headers, the start and end of records, etc. Usually the control unit can perform the search for the key in the header while being electrically disconnected from the channel. The channel is free to service another I/O request. When the block has been found, the control unit signals the channel and, if it is free, reconnects. The block is then moved by a block transfer into main memory. The part of main memory allocated to receive the information from the I/O devices is usually called the I/O buffer pool.

9.9 MAGNETIC DISK STORAGE

Although magnetic drums provide a relatively large amount of storage at a reasonable price when compared to main core memory, they still do not provide sufficient on-line secondary storage for contemporary computing systems. Another type of electromechanical storage device which permits large amounts of storage at significantly lower costs is the *magnetic disk.* Of course, you have to pay some price for this additional low-cost, on-line storage, and the price you pay is in access time. These devices have longer access times than magnetic drums. Magnetic disks are also sequential access devices and have average access times in the 25–100-millisecond range. Because of the characteristics of their access mechanisms, these devices are called *direct access* storage devices.

9.9.1 Mechanical Characteristics of Magnetic Disks

A magnetic disk memory device is composed of from 1 to 20 disks coated with a ferromagnetic material and stacked with space between the disks on a common spindle. In appearance it looks like a set of phonograph records stacked on an automatic record player, except that there is about $\frac{1}{2}$ inch of space between each record. Information is recorded on each side of the disks by magnetic recording heads which are positioned by a moving arm between the surfaces. The heads move together on one arm as a "comb"-like structure. Figure 9.34 shows this structure. There are six disks shown with five arms. There are recording heads on both sides of each arm, and therefore there are 10 recording surfaces. Information is recorded in concentric tracks around each disk. This is unlike a phonograph record where information is recorded in one long spiral on a surface. Figure 9.35 shows a top view of a disk and depicts concentric tracks. There may be from one hundred to several hundred tracks on each surface of the disks. The disks rotate, and thereby all the bits of a track will pass under the read-write heads. The heads move forward and backward and can reach all the tracks on the disk. Thus, to reach a specific piece of datum on a disk the arms must be positioned to the proper track. The time to perform this is called the *seek* time. Then the disk will rotate until the desired datum reaches the arm. This time is called the *latency* time, just as it

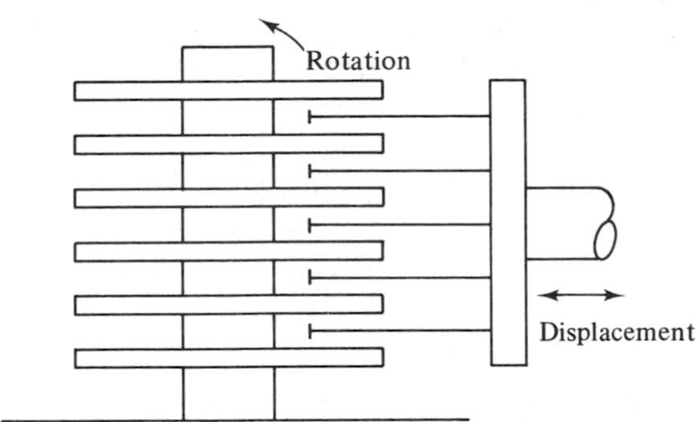

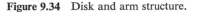

Figure 9.34 Disk and arm structure.

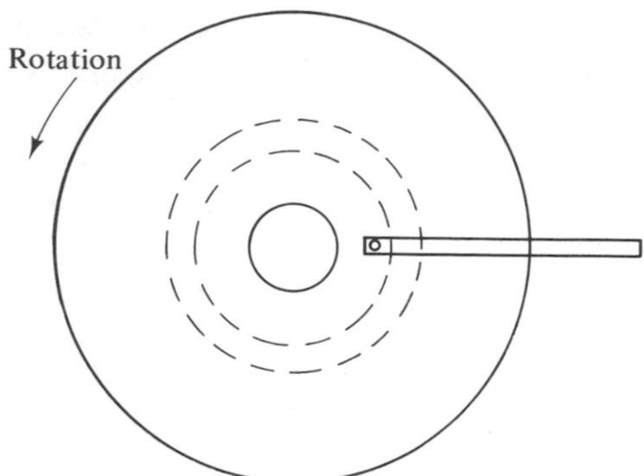

Figure 9.35 Tracks on a disk.

is for drums. Thus, the access time for a block of data on a track of a disk consists of three components: the seek time, the latency time and the time to transfer the block of data to or from memory.

Since all heads move together on the comb-accessing mechanism, they all are aligned on the same track of a different disk. This set of tracks is called a *cylinder*. Thus, once the comb is positioned, to, say, track 127 on disk number 1, it is at this same track for all the other disks. This is therefore cylinder 127. Notice that it is possible to transfer information from each of the tracks of one cylinder without having to reposition the heads. We need only electronically switch from one head to another in order to access a different track in the cylinder.

Some disk units permit the removal of the set of disks called a *disk pack* and its replacement by a different previously recorded disk pack. Thus, disk packs can be handled similarly to reels of magnetic tape. They can be stored off-line in a library and loaded onto the disk units when the owner of the pack wishes to use the computer. Some disk devices have only one access arm, which first must be positioned between the proper two disk surfaces. The access arms of these devices have a vertical motion capability. To move to a track on a different disk the arm must first be retracted, and then moved vertically to the proper addressed disk; the seek operation is then performed for the addressed track and then the latency time until the data arrive at the head. This access time is much longer than the comb structure, *head per disk system*, but it is also less expensive. At the other extreme some disk systems have a *head per track system*. They have no seek time since each track has an associated fixed read-write head. These disk systems have all the accessing characteristics of magnetic drums but usually have much larger bit storage capacities. There are many variations in size, speed, capacity, accessing mechanism, and portability of disk devices.

One of the chief advantages of a disk or a drum is the ability to retrieve any desired record without having to read the intervening records which are recorded in other tracks. Any record can be written without operating on any other records stored in the device. Thus, records in mass secondary storage devices can be updated in place. For this reason these devices are sometimes called random access storage. This is a misnomer which has come about when they are compared to magnetic tape devices. In magnetic tape storage, all intervening data must be scanned before the record of interest is located. Hence, magnetic tape is truly sequential. Magnetic disks are "random-like" during a seek operation since the intervening tracks can be passed but are sequential within a track, and a better name is *direct access* storage.

9.9.2 Electrical Characteristics of Magnetic Disks

Each disk of the disk pack has the same number of tracks. Tracks can be recorded with a spacing of 200 tracks per radial inch on the disk. The track width varies from about 2- to 10-thousandths of an inch. The linear density of recording is in the range of 1000 to 6000 bits per inch. Because of the closeness of the tracks, the positioning of the read-write head during a seek operation is extremely critical. The access arm containing the heads is activated by a moving voice coil which operates similar to a loudspeaker system. This moves the arm to approximately the correct track location. After this a servomechanism system moves the heads back and forth until the track is located. The servomechanism has a set of special signals recorded on one of the disks which is not used for data. The system uses this special "servoing" signal to make the final location of the track and also to maintain the heads on the track while they are reading or writing data.

The read-write heads used on disk systems are also extremely critical in their tolerance. These heads do not touch the disks but ride on a cushion of air just above the surface of each disk. This distance varies among disk systems but ranges from about 50- to 500-millionths of an inch. Because of the riding action, such magnetic recording heads are called *floating* heads. When the disk rotates at high speed, a thin boundary layer of air builds up on the surface of the disk and rotates with it. The floating heads are so shaped that they ride on this air layer, which permits a small distance to exist between the head and the disk. This prevents wear to both of them. The magnetic head is really very small and is mounted in a special assembly called a *pad.* The pad and the recording surface are highly polished to within a few millionths of an inch of surface roughness. The pad is mounted so as to have freedom of translation in a direction perpendicular to the recording surface. The boundary layer of air keeps the head at a relatively constant distance from the surface even though there is pressure applied urging the pad against the surface. It is important that this distance be kept constant since at the extremely high recording densities used by current disks, the readback signal decreases rapidly as the separation between head and surface increases.

Without this simple method of maintaining close and constant spacing, the high recording densities could not be possible and the cost per bit would become very expensive.

The recording heads are made of ferrite as thin as 1-thousandth of an inch with gaps of from .1- to .25-thousandth of an inch. In disk devices using replaceable disk packs, the write head is wider than the read head. This permits a larger tolerance in the position of the head on reading relative to the track which was recorded. Most new disk systems use phase modulation recording. This system of recording will be explained in Section 9.10.

9.9.3 Series and Parallel Operation of a Disk

Usually a disk is operated in a serial-by-bit mode of recording. The heads are positioned to a cylinder, and a track of that cylinder is selected. On that track is a "home" location which defines the beginning of the track. This home location is the same on all tracks, and therefore the tracks are synchronized. After the home location, there is usually a track identification header with a track address and information about what is recorded in that track. There is usually a gap after the header, and then the data records are recorded. If the record is longer than the length of a track, it can spill over onto the next track of the cylinder. Since it is not necessary to move the arm, this is done by the electronics of the control unit which just switches to the new track head.

An alternative method of recording which achieves a higher data rate is to record on several tracks of the same cylinder in parallel. Since the bits are written in parallel, a high data rate is achieved. However, no decrease in seek or latency time occurs. Since these are the predominate times involved in accessing data from the disk, this type of recording is not done often.

A third method of organizing the data on a disk is to divide the tracks into zones. A separate read-write head is used in each zone; however, they are all attached to the same arm. For example, if there are 200 tracks per disk and 50 tracks per zone, then there would be 4 heads on the arm. If the 200 tracks occupied 2 radial inches on the disk, each of the 4 zones would occupy only $\frac{1}{2}$ inch. Thus, the arm would only have to move a maximum of $\frac{1}{2}$ inch to access any track. This method is a compromise between one head per disk and a head per track. It can decrease the seek time significantly.

Some manufacturers record more data in the tracks near the outer edge of the disk than those tracks nearer to the center. Others record a constant amount of information in a track regardless of its location on the disk. Some disks have timing tracks recorded on the disk. If there are several zones on the disk, each zone would have a timing track. The timing tracks are used to synchronize the read-write operation. Some disks also record a set of addresses on a track, which permits addressing of the records within the tracks. These either require a fixed block length for the record size or each record must begin after the start of a new address or sector on the disk. Just as there are many different drum formats, there are many different disk formats. Some typical disk characteristics are given in Table 9.2.

Table 9.2 *Typical Disk Characteristics*

Characteristics	*IBM 2311*	*IBM 2314*	*IBM 3330*
Date of first installation	1962	1967	1971
Capacity (megabits per pack)	58	233	800
Maximum seek time (milliseconds)	135	130	55
Minimum seek time (milliseconds)	25	25	10
Average seek time (milliseconds)	75	60	30
Average latency time (milliseconds)	12.5	12.5	8.4
Data rate (megabits per second)	1.25	2.5	6.5
Recording density (bits per inch)	1,000	2,000	4,000
Track density (tracks per inch)	100	100	200
Track capacity (bits per track)	28,000	59,000	104,000
Rotational speed (rpm)	2,400	2,400	3,600
Diameter of disk (inches)	14	14	14
Number of disks per pack	6	11	11
Tracks per disk	200	200	404
Number of recording surfaces per pack	10	20	19

9.10 RECORDING METHODS ON MAGNETIC SURFACES

The primary method for storing large amounts of information both on-line to the computer and off-line in a library is magnetic recording. On-line methods use moving magnetic drums and disks, while off-line methods use magnetic tape reels. The use of removable disk packs also provides an off-line storage method. Although the characteristics and construction of storage devices differ greatly among the manufaturers, the method of recording the information on a dynamic recording medium is fundamental. The storage medium is usually a relatively thin layer of magnetic oxide which is attached to a magnetically inert support such as an aluminum drum or disk or to a plastic tape. The support is used to achieve the motion required in dynamic storage. The magnetic oxide is uniformly deposited and of uniform density and thickness.

Considerable research has gone into developing the various techniques for recording information onto magnetic surfaces. There are two basic requirements. One is that the writing, storage, and reading of the magnetically recorded signals be as accurate and reliable as possible. The second is that the packing density of recorded bits be as high as possible in order to maintain as cheap a price per bit as possible. These two requirements conflict since as the density increases the reliability of accurately determining the playback signal decreases. Digital information is supplied to the recording head, and the magnetic flux patterns are retained by the material as it moves past the recording head. At some later time, the material is passed under a read head where the change in magnetic flux induces a voltage into the head. These voltage signals are translated by the read circuitry back into digital information. The current passed through the recording head provides a strong magnetic field in the gap of the head. This field penetrates the magnetic oxide and saturates the material with a magnetic flux whose polarity is dependent on the polarity of the current in the head. The method of recording can be divided into two general categories. One is the return to zero (RZ) technique and the other is the nonreturn to zero (NRZ) technique. Both techniques will be described along with the write current waveforms, the magnetic flux waveforms, and the waveforms read by the read head and translated.

Before we discuss the specific recording method we should be sure that we understand the fundamental physics of recording and reading from a moving magnetic material. A current in the record head will induce a magnetic field onto the magnetic surface which is essentially of the same form and polarity as the current. Figure 9.36 shows this. Upon playing the recorded

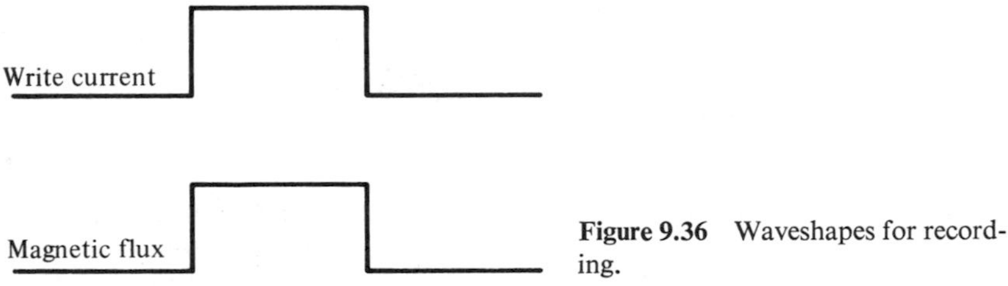

Figure 9.36 Waveshapes for recording.

signal back, the change in flux passing the read head will induce a voltage pulse into the head. A positive-going change in flux will induce a positive pulse, while a negative-going change in flux will induce a negative pulse. Figure 9.37 shows the magnetic flux recorded as shown in Fig. 9.36 and the corresponding voltage waveforms at the playback.

In the discussion of the five different recording techniques that are of practical interest, the information recorded will be 0110010.

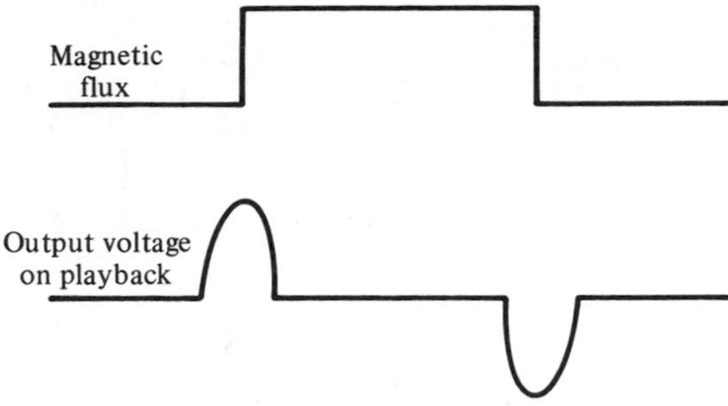

Figure 9.37 Waveshapes on playback.

9.10.1 Return to Zero Technique (RZ)

The return to zero techniques are a misnomer since it is not necessary to return to zero. These techniques, of which we shall discuss two, return to a quiescent state of current through the record head after writing a pulse of current to indicate a 1 bit or a 0 bit or both. The quiescent state does not have to be zero current. The first technique, called the *bipolar* recording technique, does have a zero current quiescent state, but the second technique, called the *recorded 1s*, has a negative current, which indicates a 0 as the quiescent state.

9.10.1.1 Bipolar RZ

Figure 9.38 shows the recording head signal current, the recorded magnetic flux pattern, and the signal developed at the output of the playback head. In addition, a timing signal is also shown. In this system if a 1 is to be recorded, a positive pulse is applied to the recording head, and if a 0 is to be recorded, a negative pulse is applied. In both cases the current returns to zero after the pulse and remains there until the next bit time. The recorded flux has essentially the same waveform as the applied write current. When this flux pattern is passed under a read head, the output voltage shown in part (c) of the figure results. Since a voltage pulse results for each rise and fall of the magnetic flux, there are twice as many pulses in the output voltage waveforms as in the magnetic flux pattern. Thus, the output does not look like the original write current waveform. For a 1, the output signal will be positive during the first part of the bit time and negative during the second half. For a 0, the signal will be negative during the first part of the bit time and positive during the last half.

One method of distinguishing between a 1 or a 0 output is to strobe the output with a positive timing pulse which is derived from the timing track of the drum or other constant timing method. If the output from the read

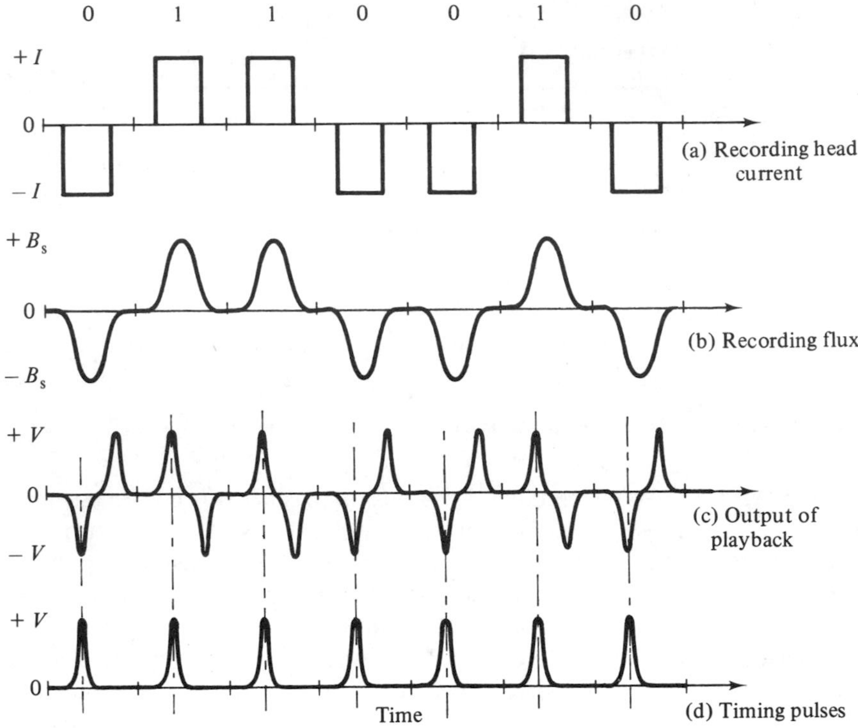

Figure 9.38 Bipolar RZ waveforms.

amplifier, used to amplify the playback head voltage, is sent to one input of an AND gate and the strobe pulse is sent to the second input of the gate, an output from the gate will occur only when a 1 bit was written onto the magnetic surface. For this system to work the timing pulse must be quite narrow and occur precisely at the correct time.

There are two additional disadvantages of the bipolar recording method. One is that the output voltage is only one-half that of the other four methods to be discussed, because the change in flux is from zero to $+B_s$ or to $-B_s$, while in the other methods the flux change is between $+B_s$ and $-B_s$. Since the output voltage is proportional to the time rate of change of flux density induced into the playback head winding, the voltage output will only be half as large as for the other methods. The other disadvantage is that it is extremely difficult to overwrite information which has previously been recorded unless each new bit exactly overlaps the previous bit. This is extremely difficult unless a precisely timed track is available, as on a drum. To use the medium again it is usual to erase all the flux before overwriting with the new information.

9.10.1.2 Recorded 1s RZ

Figure 9.39 shows the input record current waveform, the magnetic flux waveform, the output signal from the playback head, and a synchronizing timing signal for the recorded 1s method of magnetic recording. In this method the current in the recording head is maintained at a negative quiescent value and changed to a positive value when a 1 is to be written. It returns to zero, i.e., the negative current value during the bit time.

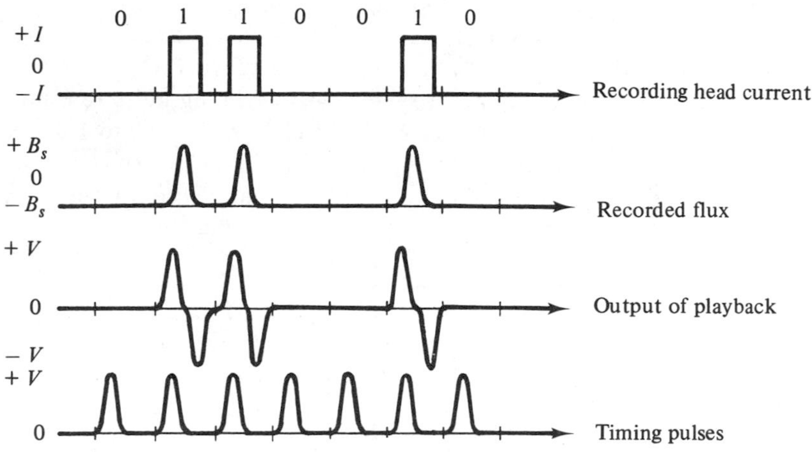

Figure 9.39 Record 1s RZ waveforms.

The output from the playback head will have pulses appearing only when a 1 has been recorded. The output can be amplified and strobed as in the bipolar method. Lack of an output from the AND gate during a bit time is interpreted as a 0. Timing is critical to obtain the proper output at the correct time. Since the method records a negative magnetic flux if a 0 is to be written, it is possible to overwrite the information, as the negative flux will magnetize the surface in the correct direction regardless of its previous state. Also, this method provides twice as much output voltage from the playback head since the magnetic flux changes are twice as large as in the bipolar method.

9.10.2 Nonreturn to Zero Technique (NRZ)

The nonreturn to zero method for recording information on magnetic surfaces only records changes in the signal when the information changes. It makes no arbitrary changes to return to a zero condition. Because there are fewer transitions in the recorded waveform than for RZ, the bandwidth of the NRZ system is less than for the RZ system with the same recorded bit density. In this section we shall study three types of NRZ techniques. Since even in the worst case the NRZ systems have one-half the flux changes of the RZ systems, it is capable of twice the packing density.

9.10.2.1 NRZ Change

This method only records changes in the bit pattern. It remains at $-B_s$ as long as consecutive 0s are recorded and switches to $+B_s$ as long as consecutive 1s are recorded and changes back to $-B_s$ upon encountering the first 0 to be recorded. Figure 9.40 shows the waveforms for this mode of recording. This technique requires a synchronizing timing pulse to establish the bit times. The transitions will establish the changes from 0 to 1 and vice versa. The output can be recovered by setting a flip-flop with the positive output pulse and resetting the flip-flop with the negative output pulse.

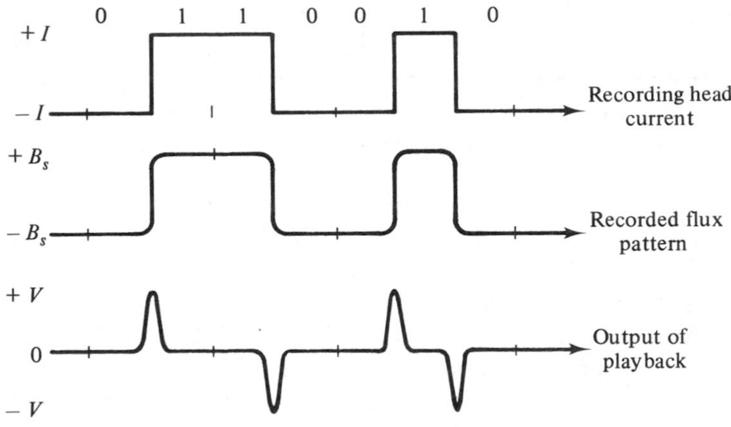

Figure 9.40 NRZ change waveforms.

9.10.2.2 NRZ—Inverted

With this recording method, NRZI, the polarity of the current through the recording head is reversed each time a 1 is recorded and remains constant when a 0 is recorded. If a series of 1s is encountered, the polarity of the recording signal will change for each bit. If a series of 0s is encountered, no change will occur in the recording signal. The specific polarity of the current has no relationship to the value of the bit being recorded; only changes in polarity contain the information. Figure 9.41 shows waveforms for this recording method. Recovery of the output requires only amplification and inversion. This will yield a positive pulse for each 1 recorded. Strobing this

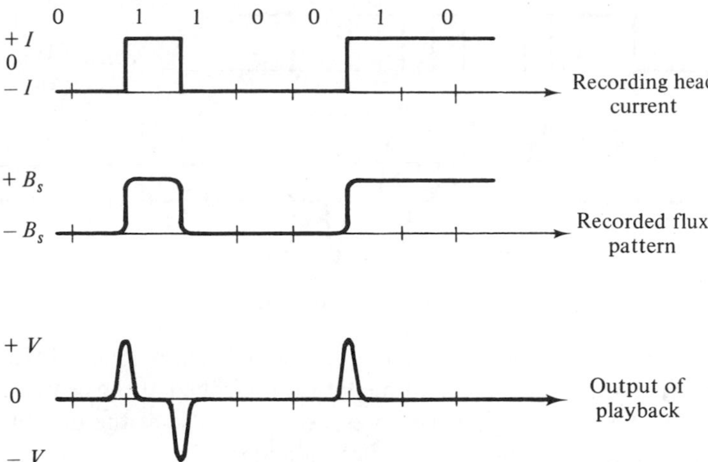

Figure 9.41 NRZI waveforms.

with a positive pulse obtained from the timing track will produce an output for each 1. No output from the AND gate at the time of a strobe pulse is therefore a 0.

An advantage of the NZRI recording technique is a self-clocking capability if odd parity is used and characters are recorded and read in parallel. Consider the case of 8 bits (a byte) recorded in parallel by eight read-write heads and a ninth head used to record odd parity on a parity track. Since there is at least a single 1 bit in each character (odd parity), if the output of all the tracks are put through an OR gate, a 1 bit (change in flux) will occur and a timing signal can be derived for each byte.

9.10.2.3 Phase Recording

Phase recording, also called Ferranti or Manchester recording, is an NRZ-type technique. This type of recording is based on phase changes in the signal to record a 1 or a 0. A 1 is represented by a positive signal followed by a negative signal, while a 0 is represented by a negative signal followed by a positive signal (opposite phase). The change in signal occurs in the center of a bit time. Figure 9.42 shows the waveforms for this type of recording. The lowest frequency of flux changes occurs for an alternating bit pattern of 1 and 0, while the highest frequency occurs for an unchanging bit pattern, i.e., all 0 or all 1s. The second is twice the frequency of the first, and therefore the bandwidth of this system is easily determined. Also, there is no dc component in this signal. Thus, ac amplifiers and high-frequency heads can be used in this system with a well-defined bandwidth. This limits the noise component and permits high-speed recording. The other NRZ techniques have low-frequency

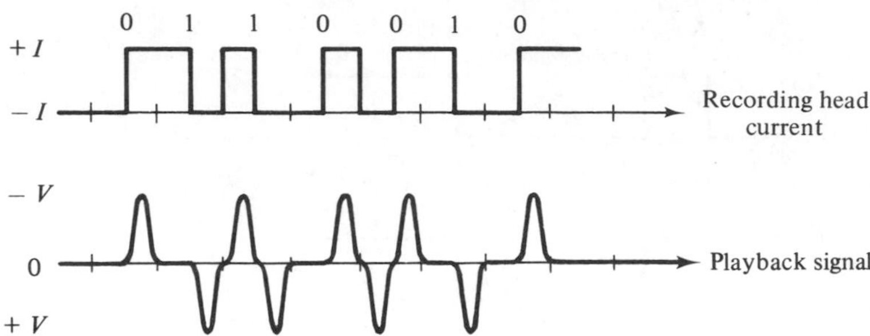

Figure 9.42 Phase recording waveforms.

components and require wide-bandwidth design of their associated electronic components. To recover the signal the output is strobed at the center of each bit time with a positive signal. The presence of a pulse at the output of the strobe gate indicates a 0 bit; the absence indicates a 1.

REFERENCES

BARTEE, T. C., *Digital Computer Fundamentals*, 3rd ed. New York: McGraw-Hill, 1972.

BONN, T. H., "Mass Storage: A Broad Review," *Proc. IEEE*, 54, No. 12 (Dec. 1966), pp. 1861–1869.

BREMER, J. W., "A Survey of Mainframe Semiconductor Memories," *Computer Design*, 9, No. 5 (May 1970), pp. 63–73.

BRYANT, R. W., et al., "A High-Performance LSI Memory System," *Computer Design*, 9, No. 7 (July 1970), pp. 71–77.

Burroughs Corporation, *Digital Computer Principles*, 2nd ed. New York: McGraw-Hill, 1969.

DE ATLEY, E., "The Big Memory Battle: Semis Take on Cores," *Electronic Design*, 18, No. 15 (July 19, 1970), pp. 70–77, 113–114.

HILL, F. J., and G. R. PETERSON, *Digital Systems: Hardware Organization and Design*. New York: Wiley, 1973.

LONDON, K. R., *Techniques for Direct Access*. Philadelphia: Auerback Publishers Inc., 1973.

RILEY, W. B., ed., *Electronic Computer Memory Technology*. New York: McGraw-Hill,

WALTHER, T. R., "Dynamic N-MOS RAM with Simplified Refresh," *Computer Design*, 12, No. 2 (Feb. 1973), pp. 53–58.

WARE, W. H., *Digital Computer Technology and Design*, Vol. II. New York: Wiley, 1963.

PROBLEMS

9.1. To drive a 3D memory, we suggested splitting the required switching current into $I_x = -I_m/2$ and $I_y = -I_m/2$. Is it possible to use other combinations of I_x and I_y? If yes, then show some of these combinations and discuss the advantages or disadvantages of your suggestion when compared to the above choice.

9.2. For a 4096-word, 8-bit, coincident-current memory, how many flip-flops are needed for
(a) The memory address register?
(b) The memory buffer register?

9.3. Consider a 3D core memory unit containing 4096 words, 32 bits each. Each word is directly addressable. Compute the following:
(a) The number of cores.
(b) The number of sense amplifiers.
(c) The number of selection drivers.

9.4. Repeat Problem 9.3 for a 2D organization.

9.5. Compare the size and cost of the decoders required for the 3D memory and the 2D memory which contain 4096 words, 36 bits per word.

9.6. Compare a 3D/four-wire and a 3D/three-wire core plane with respect to
(a) Cost. (b) Speed.

9.7. Compare the following memory arrays with respect to speed and cost per bit:
(a) 2D core memory. (b) 3D core memory.
(c) Bipolar flip-flop memory. (d) MOS memory.
(e) Charge storage MOS memory.

9.8. Draw a diode read-only memory as in Fig. 9.24 which stores the 10 decimal characters in excess-3 code.

9.9. Draw a transformer-coupled read-only memory as in Fig. 9.29 which stores 0101 in word A, 1100 in word B, 0011 in word C, and 1010 in word D.

9.10. Draw a capacitor read-only memory as in Fig. 9.28 which stores 0011 in word 1, 1011 in word 2, 0101 in word 3, and 1010 in word 4.

9.11. (a) What drum diameter is required for a 4096-word parallel memory if the recording density is set at 500 bits per inch? (b) What is the average access time of this memory if the drum rotates at 2000 rpm? (c) What is the clock rate?

9.12. A magnetic drum has 16 tracks of 2048 bits per track. If the drum rotates at 12,000 rpm, compute the following:
(a) The average access time.
(b) The data transfer rate (bits per second) if the drum is organized bit-serial.
(c) The data transfer rate if the drum is organized bit-parallel.

9.13. For the disks described in Table 9.2, find how many binary digits can be recorded on one surface of a given disk.

9.14. Draw the flux waveform and the playback signal for return to zero recording of the following data:
(a) 1101101110110 (b) 1010110011000

9.15. Repeat Problem 9.14 for the following recording techniques:
(a) Nonreturn to zero change.
(b) Nonreturn to zero inverted.
(c) Phase recording.

10

Data, Input/Output, and Channels

To make use of the processing subsystems of the computer, i.e., the control unit, memory unit, and arithmetic unit, there must be an interface between the processing system and the human user. The user supplies a program which tells the computer what it is to do. The user must also supply input data on which the program operates. The computer must be able to communicate the results to the user. In addition there must be a two-way conversational capability between the computer and the operator. All this communication takes place through the computer's input/output subsystem.

The I/O subsystem consists of a software system, the input/output control supervisor (IOCS), which controls the entry and placement of user programs and data as well as the output of results; the channels and control units, which provide a path to and from main memory to the I/O devices; and the I/O devices, which provide a translation capability between human notation and the binary system of the computer.

Since the processing capability of the computer is based on its ability to perform its simple operations at very high speed, the problem in the design of the I/O subsystem is to ensure rapid access to the required data and the ability to communicate with the user or operator when information must be transacted between the man-machine interface. The computer is too expensive and performs too fast to rely upon the human to supply data whenever needed to perform the algorithm. Thus, the data must be prepared beforehand in machine-compatible form and be available to the computer when called. One basic medium used for input of programs and data in machine-compatible form is the punched card. This medium is prepared off-line from the computer on a *key-punch* machine. The deck of cards is placed into a *card reader* and the computer can use this as input. It can read the cards at a rate of 2000 per minute. This provides a much faster rate of input to the machine than trying to type directly into the computer. However, for a large amount of data such as the weekly payroll for a large corporation, this method of input may still be too slow. A faster input medium must be found. The *magnetic tape* is such a medium but requires a second level of off-line preparation. Thus, the punched cards must first be prepared; then these are read by a small off-line unit onto the magnetic tape. The magnetic tape is then used as the direct medium for the input to the computer. An alternative scheme which has become available in recent years is to key the information directly onto a magnetic tape or a magnetic disk pack in an off-line operation. This eliminates the intermediate punched cards.

Input data sometimes are obtained from *source documents* which can be utilized by humans but not by the computer. These documents can be programs, reports, sales slips, or accounting records. An off-line process is performed to place the required information into a machine-readable form and this becomes the input to the computer via an I/O unit. Some types of source documents are developed which are already machine-readable. For example, certain forms can be read by *optical character readers* which translate alphanumeric typed characters on sales slips or bills directly into machine codes. Another type of character reader is the *magnetic ink character reader* which is used to read the funny-looking characters on bank checks. Both these systems are an attempt to relieve the requirement of preparing an intermediate form compatible with the computer's I/O capability.

Even with the capability of reading input data from high-speed tapes at 320,000 characters per second, this may be too slow a speed for large amounts of input activity if the computer would have to stop processing each time it wished to perform an I/O operation. To overcome this possible deficiency, second-generation computers developed an *overlap* capability by which the I/O activity is performed by a *channel* with a minimum amount of disruption to the remainder of the computer. This simultaneous I/O operation with CPU compute operations has lead to the *multiprogrammed* method of operation for third-generation computing systems. Under this method of operation a minimum amount of CPU time is spent on I/O and an attempt is made to maximize the number of programs executed per unit of time (i.e., throughput). This results in more diverse and faster units needed for both I/O and secondary storage and the removal of human intervention from on-line control of the system. The programming system controls the computer operation and informs the operator of procedures to follow. The scheduler attempts to optimize the use of resources and schedules I/O to increase system throughput.

10.1 I/O SUBSYSTEM AND I/O INSTRUCTIONS

10.1.1 I/O Subsystem

A large part of any large general-purpose digital computer system is the I/O subsystem. This subsystem is used to move programs, data, and results into and out of the main memory under control of the CPU. In addition a large amount of data and programs is retained by the system and stored on the drums, disks, and tapes which comprise the on-line secondary storage and off-line library. Much of the work of the operating system is directly related to the use, control, cataloging, and storage within the I/O subsystem. We see that the I/O subsystem is a misnomer since with large modern computers much of its activity is related to the secondary storage subsystem.

The hardware of the I/O subsystem is divided into three levels. At the bottom are the I/O and storage devices such as punched-card readers, line printers, disk storage packs, and drives and magnetic tape transports. Above them sits the control units. Each such device must have a control unit to interface with the computer. The control units interface to the channels, which relieves the CPU from having to monitor I/O activity. The channels interface to the memory control unit, which provides for an information path in and out of main memory. Figure 10.1 shows a typical block diagram of this hierarchy.

Some features of this structure should be observed. One is that a single control unit may control several devices of the same type, i.e., the magnetic tape units. That more than one control unit may connect to the same channel is also shown. In fact, the IBM System 360 permits 8 control units to connect to a channel and 8 devices to connect to a control unit. Thus, 64 devices can be connected to a single channel. If the computer had 6 channels, as many as 384 I/O and storage devices could be connected to the machine. This is generally not done. Why? The answer is contained in the method by which this subsystem operates.

The various channels can all communicate with the memory control unit (MCU) simultaneously. If two channels should request the same memory bank simultaneously, the MCU will break the tie on a preassigned priority basis. At the next level down, only one control unit can communicate with the channel to which it is connected at any time. Thus, if we have 2 control units each with 8 magnetic tape units connected to them, only one of the 16 magnetic tape units can be reading or writing at any one time. This is the reason channels are not usually fully loaded with devices. Too

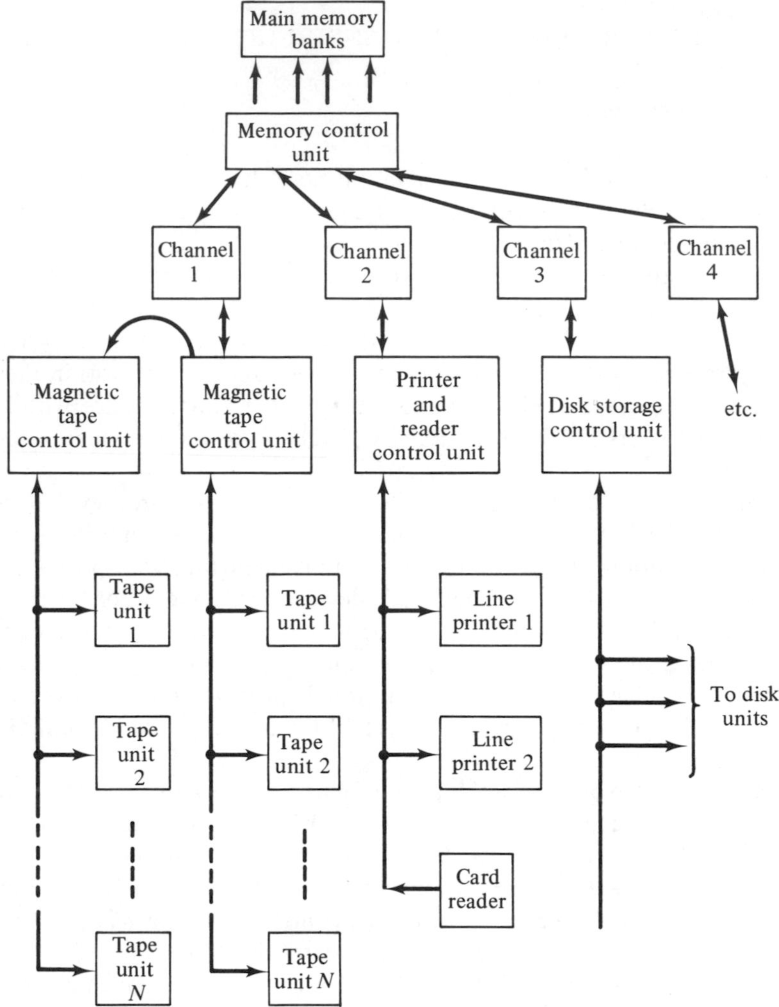

Figure 10.1 I/O subsystem for a large digital computer.

much activity on a single channel will overload the capability of the channel and slow the computer down.

Although the I/O subsystem for large computers may be complex, this is not true for all computers. Small computers may have very simple channels, i.e., the CPU itself, and they control the devices directly. These simple channels do not permit an I/O-compute overlap. The CPU must stop computing and must monitor the I/O process. Although not well suited for optimizing the utilization of the computer system, the simple channel can be quite inexpensive and be well suited for control of a dedicated application. Simple channels are discussed, in detail, in Section 10.3. More complex channels are discussed in Section 10.4 and in Volume II of this text.

The control units for each type of device are different and conform to the needed requirements of the device which they are controlling. However, they all interface to the channels. This interface is generally the same for all the channels of the machine and is called the standard I/O interface. The control unit provides a control and data path between an I/O device and the standard I/O interface. Part of the standard I/O interface may be the status of the device; the control unit supplies this information to the channel.

Manufacturers have two different methods for connecting control units to the standard interface of the channel. One method supplies all communication between all the control units connected to a given channel over a single common bus. Any signal supplied by the channel is available to all the control units connected to it. However, at any given instant only one of the control units is logically connected to the channel. The selection of a particular control unit for communication with a channel is controlled by a signal from the channel that passes serially through all the control units connected on the bus and permits each selected control unit to respond to different signals from the channel. A selected control unit remains physically connected to the channel until the information which is to be transferred has been completed or until a signal from the channel requests a disconnect. This type of channel structure is called a *daisy chain* and is shown in Fig. 10.2. The other method of connection provides a separate set of lines from

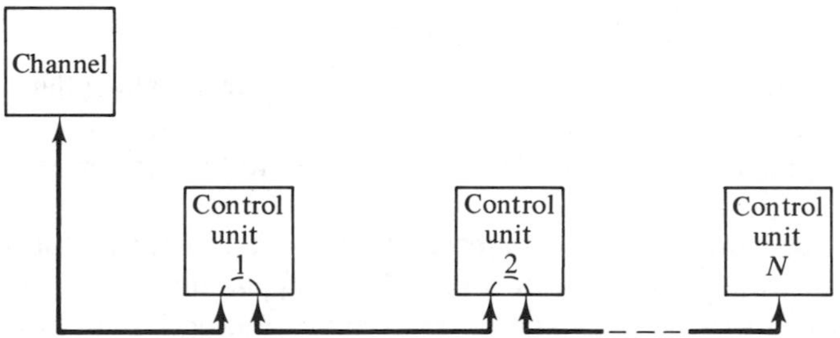

Figure 10.2 A daisy chain connection.

the channel to each of the control units connected to it. Each control unit sees only those commands which are sent to it by the channel. This structure, called a *radial* interconnection, is shown in Fig. 10.3. One of the advantages of the radial structure is that for a sufficiently intelligent channel it is possible to have more than one control unit physically connected to the channel at any instant of time and transferring data. The daisy chain, because of its

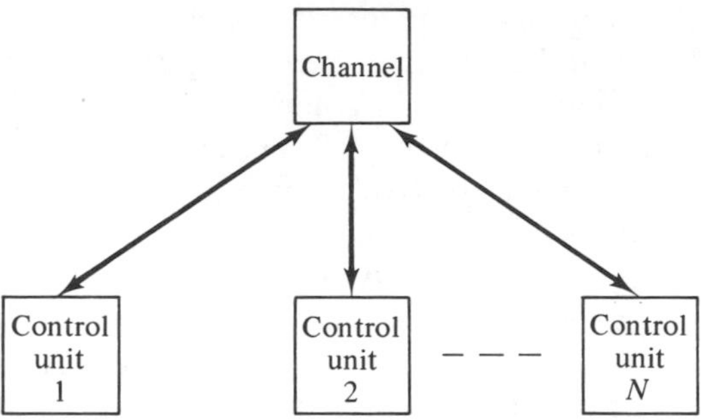

Figure 10.3 A radial connection.

single-bus structure, can transfer data between only one control unit at any instant.

The control unit may be used to control several devices of the same type. These devices are usually connected to the control unit in a daisy chain structure. Thus, since there is only one bus, only one device can be transferring data to the control unit at any given instant.

10.1.2 I/O Instructions

To use the I/O subsystem, the computer has instructions which deal with both the channels and the devices. The fundamental part of the I/O instructions must perform the following functions:

1. Select a particular device.
2. Specify the first address in memory to or from which data are to be transferred.
3. Specify the number of words which are to be transferred.

The manner in which these fundamental parts of I/O are performed varies among the present types of computers. They depend on the CPU design, the complexity of the channels, and the software operating system supplied with the computer. For example, instead of specifying the number of words which are to be transferred, some computers specify the address of the last word of the block of information to be transferred.

In addition to specifying the above three pieces of information, the I/O instruction usually specifies an operation or function to be performed, such as read, punch, print, rewind, or request status. This information is device-dependent and is passed along by the channel to the particular control unit. For a simple channel where the CPU actually performs the transfer of data between the I/O device and memory, under program control, only the selection of a device and the function to be performed need to be specified as an I/O instruction. The other two pieces of information are supplied by the programmer as part of the routine which provides for the transfer of data. Thus, a minicomputer with a simple register used for I/O, as explained in Section 10.3, can have I/O instructions of the following form:

EXTERNAL FUNCTION X The contents of address X contains the code for one of the I/O devices which is connected to the I/O register and also a code for the operation to be performed.

READ X Transfer the contents of the I/O register to memory location X.

WRITE X Transfer the contents of memory location X to the I/O register.

CONNECT X This is another form of the external function instruction. If a machine has several channels, this instruction may also specify which channel is used as well as the device to be connected and the function.

DISCONNECT X This disconnects the device from the computer and terminates the I/O operation.

COPY STATUS X Transfer the contents of the I/O control register, *IOCR*, to memory location X. This contains the status of the I/O device.

STATUS REQUEST X This requests the status of a device to be placed into the *IOCR*. The COPY STATUS X instruction then transfers the status to a location in memory where it can be examined by a program.

READ X, Y Transfer X number of words from a device which has been connected to the CPU by a previous CONNECT X instruction into consecutive memory locations starting with location Y. This instruction is usually used with a computer which has a direct memory access type of channel; see Section 10.4.

In large computers which have several channels, as explained in Section 10.1.1, the I/O instructions come in two parts. The first part is interpreted by the central processing unit as either an input or output instruction and has an address which contains the second part of the I/O instruction. These second parts are commands to the channel which specify all of the four pieces of information needed for an I/O operation. With the I/O instructions separated into two parts it is possible to overlap computing and data transfers. The first instruction is decoded by the CPU, which passes the address to the specified channel. The channel then takes over the I/O operations, while the CPU continues to compute. At the location specified, the channel will obtain the device to be connected, the operation to be performed, and the number of words to be transferred as well as the first address where the data are to be placed in memory. This information, called channel commands, may be contained in one or more words of memory. The channel has access to memory via the memory control unit and can fetch and store commands and data as needed to perform the I/O operations.

10.1.3 Memory and Buffers

The architecture of contemporary digital computers requires that both the instructions, in binary form, and the data on which they operate be simultaneously in the high-speed main memory. If the data are not in main memory, they first must be brought there under program control from a secondary storage device or an I/O device. After that the program can operate on the data to transform them into useful results. To make efficient use of the high speed of the machine, it is necessary that most, if not all, of the instructions of a program be in main memory when the program is executed. Likewise, most of the data should be in main memory; however, this is not usually possible for a large business-type program since it may have a vast amount of data to be used in its computations. If the program is also very large, it may not fit into the finite size main memory. The solution to this problem of large program or data is to segment the information into logical parts. Program segments are called *overlays*, while data segments are called *blocks*.

When a program gets too large to fit into its part of main memory, called a *region*, it is subdivided by the programmer into several parts. Each part represents a logical part of the program with a minimum of interaction between the parts. These segments are then stored in the secondary storage of the computer and under program control are called into the region of main memory to be executed. As each segment enters it overlays on top of the previous segment of the program which was there—and hence the name. Since it overlays in the same part of main memory, it destroys the instructions which were in the first segment.

The data are also brought into memory via the channels in segments. These segments, called *blocks*, are placed into areas of main memory which are designated by the IOCS as *buffers*. The operation occurs in the following manner in a computer which is multiprogramming two independent programs.

1. Program A requests that a block of data be placed in a buffer in memory.

2. A channel under control of the IOCS starts to move these data to a buffer which has been assigned for use to program A.

3. Program A is waiting for data and is no longer computing. The CPU is allocated to program B and is executing the instructions of program B concurrently with performing step 2 above.

4. Program B has used all the data in its buffer and requests that the next block of data be brought to its buffer.

5. A channel under control of the IOCS starts to move this data to a buffer which has been assigned for use to program B.

6. If Program A's buffer has been filled, the CPU is allocated to program A and it now processes the data it requested concurrently with a channel filling program B's buffer. If the buffer of program A has not yet been filled via the channel, the CPU remains idle until the buffer is filled.

Using this method, two programs can utilize the resources of the computer in an overlapping manner. Of course, if the buffers are not filled before a request is made by the other program for data, the CPU will be idle. To minimize the idle time of the CPU, many computers operating in the multiprogrammed mode have three or more user programs sharing their main memory and overlapping the computer's resources.

An alternative scheme which uses only a single program with two buffers allocated to its data is called *double buffering*. The two buffers are filled with consecutive blocks of input from the input unit. As the user program needs data, it obtains them via the IOCS, from the first buffer. When that buffer is empty, i.e., all the data have been utilized by the program, the user program requests that another block of data be brought into main memory. The IOCS places this third block of data into the empty first buffer. In the meantime, the user program continues to execute on the data which are in the second buffer. If the buffers can be filled before the user program has finished utilizing the data, the CPU will not have to be idle but simply switches between the data in the two buffers in memory.

10.2 ADDRESSABILITY, DATA FORMATS, AND DATA MOVEMENT

The hardware instructions provided with a computer to perform input/output are relatively simple when compared to the variations in I/O devices and types of data formats which are involved in a large computer system. The difference between hardware provided and complex programming uses is provided by the operating system and in particular the input/output control supervisor. It is this piece of software which provides an efficient and flexible I/O system for the user.

The user program in many large computer systems may not directly control the movement of I/O devices and data. The I/O instructions are considered privileged and may be executed only by the operating system in the master mode. When the user wishes to move data, he supplies this information to IOCS subroutines which provide for

1. Control of the I/O units.
2. Checking the path via channel, control unit, and device.
3. Supplying adequate main memory buffers.
4. Checking the legality of the operation to be performed on the particular device.
5. Checking for priority and errors.
6. Correction steps due to error or unusual conditions.

Another reason that the I/O operations are performed by the operating system and not the individual user is due to the multiprogrammed operation of large computers. With several users simultaneously in the computer, it is certainly possible for them to interfere with each other on I/O operations. Thus, two could easily try for the same device, etc. The IOCS can schedule the I/O operations and prevent such interference.

10.2.1 Fields, Records, and Blocks

The smallest piece of digital data is the *bit*. Although this is a very useful size from the viewpoint of logic and electronic design, it is too small a unit of information from the viewpoint of data handling. Bits are encoded into characters as shown in Chapter 4. The size of alphanumeric information comes in two popular sizes. One size is the 6-bit *character*. The 6 bits allow for 64 different characters to be represented in a binary code. The other size is the 8-bit *byte*. This allows for 256 characters to be encoded. In this text we shall call both the 6-bit and 8-bit encoded information a character.

The characters are grouped together to form a *field*, which represents a name or a number of a FORTRAN variable, etc. Thus, in FORTRAN the FORMAT statement can define the fields on a card. Groups of fields are combined to form a *record*. A single punched card is called a unit record. At this point we arrive at a conflict in the use of the word *record*. Historically, a record was used to mean a recorded block of information on magnetic tape. We shall use the word record to mean a logical group of information typically handled as a unit by the user. We shall call the physically recorded block of information a *block*. Thus, a block may be a set of records. Sometimes, to differentiate between the double use of the word record, the logical group of fields is called a *logical record* and the recorded information is called a *physical record*.

A record might, for example, contain all the employment information about one single employee. The unit of information which contains all the employee records is called a *file*. The file is generally given a name, such as the payroll file or inventory file. The file may comprise part of a storage device or all of it or overflow onto many storage devices.

The physical storage device such as a reel of magnetic tape or a magnetic disk pack is called a *volume*. Thus, a file is a logical name which may occupy part of or all of a volume or many volumes. Another name for a file is a *data set*. Thus, a group of logical records is a data set.

We see that there are two ways of discussing the organization of data within the secondary storage subsystem of a computer. One way is the logical structure of the data, i.e., records, files, data sets, while the other is the physical structure of where and how the data are recorded, i.e., blocks and volumes.

10.2.2 Blocking and Buffers

Data in the main memory are addressed a word at a time and manipulated in this fashion. Some computers which are character- (byte-) addressable can manipulate the data in main memory a character at a time. Some instructions even permit the execution of instructions on a bit of data within main memory. However, when data are moved into or from main memory, it is generally done in blocks, where there are many words to a block. This is done because of the slow access capabilities of the mechanical devices which constitute the secondary storage, i.e., disk, drums, and magnetic tape, or I/O devices, i.e., card readers, printers, and paper tape.

The large amount of time to access a piece of data on a disk requires that, for efficient operation of the system, more than a single word of datum be transferred. Thus, the programmer moves a record or several records which have been formed into a block from the disk to a buffer in memory. He works on the data from each record and may update these records and restore the new information back to the same location on the disk. Thus, the I/O subsystem requires the blocking and unblocking of data before they can be efficiently stored.

For large multiprogrammed computers, there is a need for many buffers in main memory to handle all the I/O devices. To conserve costly main memory space it is not possible to allocate a fixed part of memory to each possible device. Therefore, the buffers must be allocated dynamically when needed. The IOCS has a large amount of main memory for its use which is called a *buffer pool* area. From this area it allocates a buffer to an I/O request as it is needed and returns the buffer to the pool when the program is completed. An even more dynamic scheme is to return the buffers to the pool after each use and to create buffers from the pool as they are needed.

The IOCS has a set of subroutines which the programmer can invoke with the use of subroutine calls. He can tell the IOCS that he wishes to use a particular file by issuing an OPEN statement. This will allocate buffers for use by that file. Using a subroutine call, such as GET, he can fill buffers with data from I/O devices and have additional buffers allocated to his job. With the use of a subroutine call, such as PUT, he can have the contents of a filled buffer written back onto the I/O device and the buffer released back to the buffer pool by the IOCS. When he has finished using that file he issues a CLOSE subroutine call, which releases all buffers from use with that file.

10.2.3 Catalogs and Directories

The computer programmer creates a file or data set from a group of logical records. He uses this file at some subsequent time to read certain records it contains. He may change some of the records and rewrite the file. The new file can be saved permanently in secondary storage, or it may be saved only until he makes another run, or it may be destroyed when no longer needed. The programmer is not usually interested in exactly where his file is being stored within secondary storage. He is not interested in the physical location. He only wishes to obtain the file by name when he wants it. This is similar to using a book in a library. The user is not interested in where the book is stored on the shelves but only that it can be brought to him when he requests it by name and author. The method used to perform such a retrieval in a library is the card catalog. The reader looks for the book in the card index, and this gives him a call number which defines the location of the book on the shelves.

A large digital computer system with a vast secondary storage can provide the same service with a catalog of files. The computer's IOCS maintains the status of the catalog. Each time a user creates a new named file, the IOCS inserts it into the catalog and records the physical location on which the file was written. Each time the user tells the system to destroy a file, its name is removed from the catalog. When a user returns and requests access to a file, the IOCS obtains its physical location from the catalog. If it is on an attached storage unit such as a magnetic disk unit, it would be loaded into the assigned buffers in blocked format. If the catalog indicated that the file is on a disk pack which is not on-line at the time, the operator would be instructed, via a message on the console typewriter, to load the needed disk pack onto a disk unit of the machine.

Each volume of the secondary storage has as its first record a label. This uniquely defines the volume by name. If the operator has had to mount the volume to bring the required file on-line, the operating system checks the volume label to be sure that the correct one was mounted. Another record near the beginning of the volume records all the files or data sets which are recorded on that volume and where they are on the volume. Thus, once the correct volume has been mounted, the IOCS can determine where on the volume the required file is located. At the beginning of each file is its unique identifying label name. The IOCS checks this name to be sure that the correct file is being accessed, and then it loads the correct buffers with the data blocks.

10.2.4 Data Organization and Access Methods

The organization of the data records on a volume is dependent on the type of data and how the data are to be used as well as the characteristics of the volume. There are four basic ways the data can be organized: (1) sequential, (2) indexed sequential, (3) direct, and (4) partitioned.

1. *Sequential.* This is one of the simplest ways to store records. It is the only way permitted on sequential-type devices such as magnetic tape or punched cards. The records are placed in physical rather than logical sequence. Thus, given a record, the location of the next record is determined by its physical position in the data set.
2. *Indexed Sequential.* When the records have control information, such as a key, on which the records are to be retrieved this organization can be used. The records are arranged in collating sequence according to the key. A separate index maintained by the system gives the location of principal records. The access to a record can be either in order of the keys (that is, sequentially as stored) or directly to a particular record via a search on the index of keys.
3. *Direct.* The actual physical location of the record is used for accessing, and the records are organized in any manner the user wishes. This is a very fast method of accessing the data, but the user must maintain the physical address of the record.
4. *Partitioned.* The data set is divided into a number of sequentially organized subsets. This is a combination of sequential and indexed sequential. Each subset is given a name, and it is placed in an index at the beginning of the entire data set. Thus, each subset can be referenced as an indexed sequential type of organization. Within the subset the data are organized sequentially. This organization is useful for keeping libraries of programs. Each program can be indexed by name, but the program is fetched sequentially.

Since the data can be organized on a standard basis, it is possible to allow the system to organize the data on secondary storage devices and to access these data through subroutines which provide for the automatic buffer control in main memory. The combination of accessing and data organization is called an *access method.*

10.3 SIMPLE CHANNELS

Let us first consider a simple communication path between the I/O control units and the computer main memory. It is shown in Fig. 10.4 and consists of only two registers: the *IOIR*, input/output information register, and the *IOCR*, input/output control register. The *IOIR* is used to move

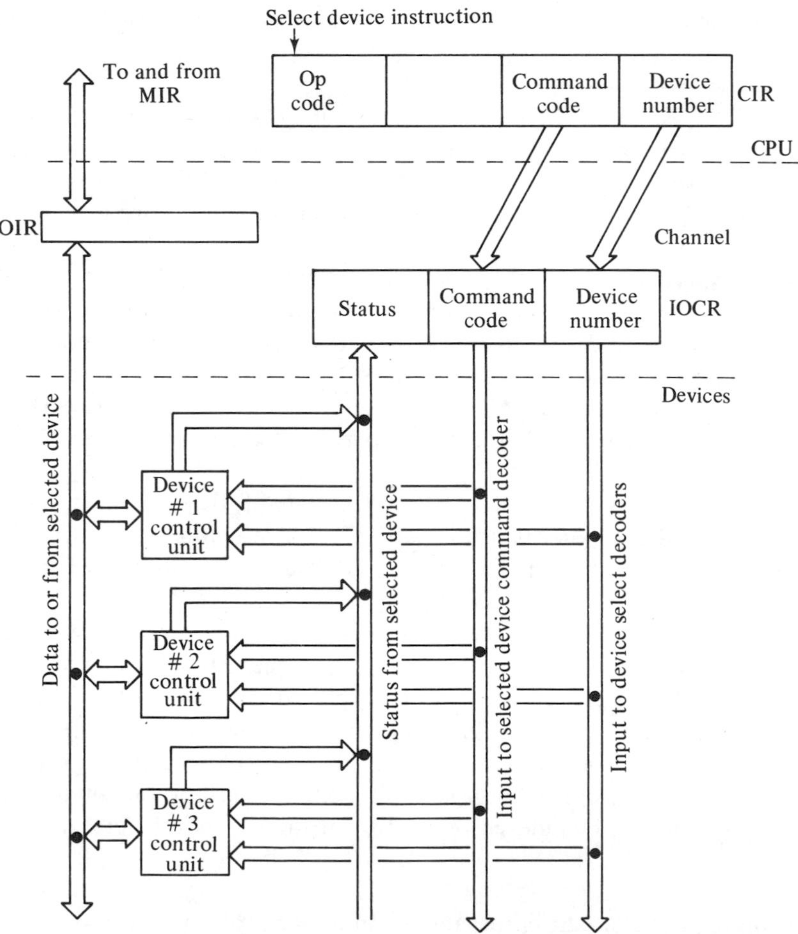

Figure 10.4 I/O bus structure.

information between the I/O units and main memory. One word at a time is transferred over this path. During an input operation (read) the selected device's control unit places one word into the *IOIR* when it is available. The execution by the CPU of an external read instruction transfers this information from the *IOIR* into the *MIR* for storage in main memory. During an output operation (write), the execution by the CPU of an external write instruction moves one word from main memory into the *IOIR*. The word is transmitted to the device control unit whenever the control unit is ready to accept the information The *IOIR* operates as a one register buffer unit.

The CPU instruction which selects an I/O device has three parts, as shown in Fig. 10.4: the op-code, i.e., select device; the device code number; and the command code which specifies what the device is to perform. The last two fields are transferred from the *CIR* into the *IOCR*. The *IOCR* is used to control the exchange of information over the *IOIR* between main memory and the external device control unit. The bits of the *IOCR* are organized into fields with certain interpreted meanings. The fields may be

1. Code to select a particular I/O device.
2. Code for the particular I/O operation.
3. Code for a mode of data transfer.
4. Code for device status.

Both the CPU and the I/O device control unit can place information into some fields of the *IOCR*. The CPU places the information into the first three fields when it executes the select device instruction. The I/O control unit places information into the device status field as it performs the I/O operation.

When the CPU starts an I/O operation by executing the select device instruction it enters the I/O state. The CPU control unit will generate the proper timing signals to perform I/O operations. These signals are used to strobe data lines and gate command lines.

The *IOCR* establishes a "hand-shaking" procedure between the computer and the external I/O device control unit. This hand shaking provides for proper selection and timing of the transfer of information. The selection is performed by addressing the selected I/O device via the device selection lines. These lines are connected to all the I/O device control units. Each I/O device control unit has a device select decoder which interprets the specific codes on the select lines. The device select decoder which compares with the select code presents a select control signal which connects the selected device to the command and data lines.

The command lines contain the code of the operation to be performed by the selected device. The command lines are also connected to all the I/O device control units. However, the select control signal of the selected device will couple only that device to the command bus. Hence, the command will be received only by the selected device.

The information transfer is performed between the *IOIR* and the I/O device via the data lines. Again all the devices are connected to the data bus, but only the device which has been selected will have its gates connected to the I/O data bus open for the transfer of data.

Part of the status field of the *IOCR* is a device-available status bit. This single-bit field is used to establish a timing control between the CPU and the I/O device. The execution of a write instruction is controlled by the status of this field. If the I/O device is ready to receive data, it sets this bit and the write instruction is permitted to execute. If the device is busy, it resets the bit and prevents the transfer of data from main memory to the *IOIR*. Likewise on a read instruction the I/O device control unit sets the bit after it fills the *IOIR*. This permits the execution of the read instruction, which also resets the bit to 0. This 1-bit field establishes the hand shaking. All the I/O device control units are connected to the device-available status bit, but only the selected device can change the status of the field since it is the only device whose gates are opened to this one-line bus.

10.3.1 The I/O Device Control Unit

There are four I/O buses to which all the I/O device control units are connected:

1. The I/O data bus: Data moves between the device and the *IOIR* over this bus.
2. The I/O device selector bus: The selected device code number is presented on this bus.
3. The I/O command code bus: The operation which the selected device is to perform is presented on this bus.
4. The device available status bus: The selected device presents its timing signals to the CPU over this single-line bus.

The control unit of each device listens on the device selector bus to see if its code is present. If it is, it opens the gates of its lines to the remaining three buses comprising the I/O bus system. Figure 10.5 shows a block diagram of the general I/O device control unit.

10.3.1.1 Device Selector Decoder

Each I/O device connected to the I/O bus subsystem has a device selector decoder. The device selector bus is connected to all the device selectors.

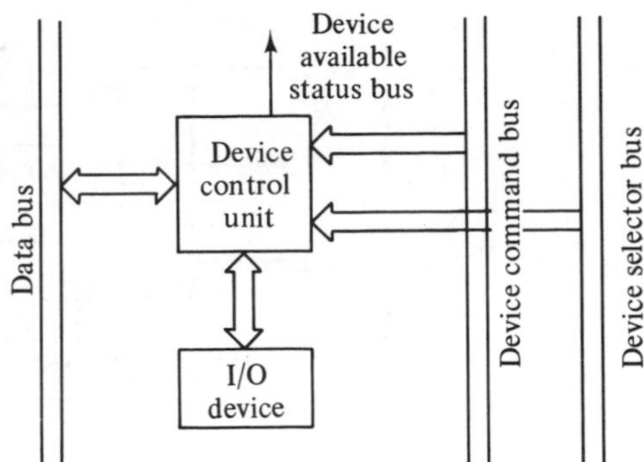

Figure 10.5 Connection to I/O device.

Each device selector decoder is assigned a unique select code and is enabled only when its assigned code is present on the selector bus. The selector with the proper code will open the gates on the command code bus to that I/O device control unit.

The device selector decoder need be nothing more than an AND gate which produces a 1 output when the proper code is supplied on its input. If we assume a six-line device selector bus and wish to detect the code for device number 13 (001101), then the output of the selector should be a 1 if the above bit pattern is detected on the six lines. Assume that SB0 represents the least significant bit line and that SB5 represents the most significant bit line; then

$$\text{DETECT} = \overline{\text{SB5}} \cdot \overline{\text{SB4}} \cdot \text{SB3} \cdot \text{SB2} \cdot \overline{\text{SB1}} \cdot \text{SB0}$$

The following AND gate using inhibit inputs will perform this logical function:

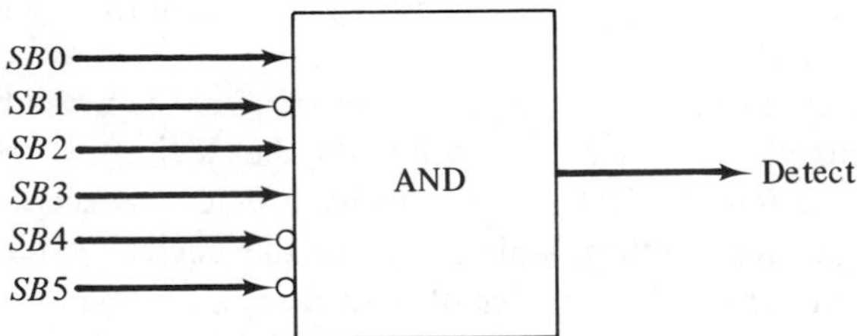

10.3.1.2 Command Decoder

A block diagram of a command decoder is shown in Fig. 10.6. The output of the device selector decoder enables the AND gates of the command decoder as shown, where a 4-bit command code bus is assumed. This permits a maximum of 16 possible commands to each of the I/O devices. It is up to the designer of the I/O device control unit to define the code for each command he desires. Some devices need a very few commands to perform ade-

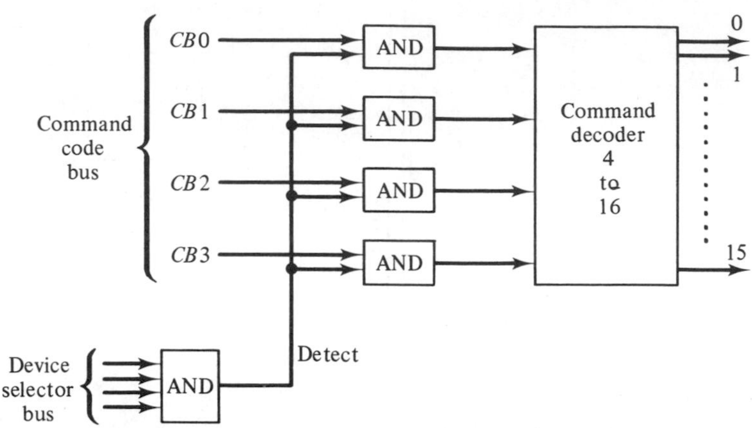

Figure 10.6 Command decoder for I/O device.

quately. Others, such as magnetic tape transports, require many different commands in order to function properly. The programmer needs to know the command code and its definition in order to provide the proper code in the device selection instruction.

10.3.1.3 Device-Available Flip-flop

Each device control unit has a 1-bit register flip-flop which is used to determine the status of the device during the transfer of data. The output of this flip-flop gated with the output of the device selector is connected to the device-available status bus which controls the device-available bit in the status field of the *IOCR*. Figure 10.7 shows the logic of this connection in the device control unit.

If the selected device is busy, it will reset its device-available flip-flop in its device controller. This in turn will reset the device-available bit in the status field of the *IOCR*. If the selected device is ready to accept data from the *IOIR*, it will set its flip-flop, which will set the device-available bit in the *IOCR*; this will permit the transfer of data from main memory to the *IOIR* and the subsequent transfer from the *IOIR* to the device. On an input operation (read), the device controller sets its device-available flip-flop when it has completed the transfer of data to the *IOIR*. This will set the device-available status bit in the *IOCR* and will permit a store instruction to execute, thus storing the data in main memory. After the information in the *IOIR* has been transferred to the *MIR* of main memory, the *IOIR* is cleared and the status-available bit of the *IOCR* is cleared to 0. A signal is also sent to the I/O device controller to clear its device-available flip-flop.

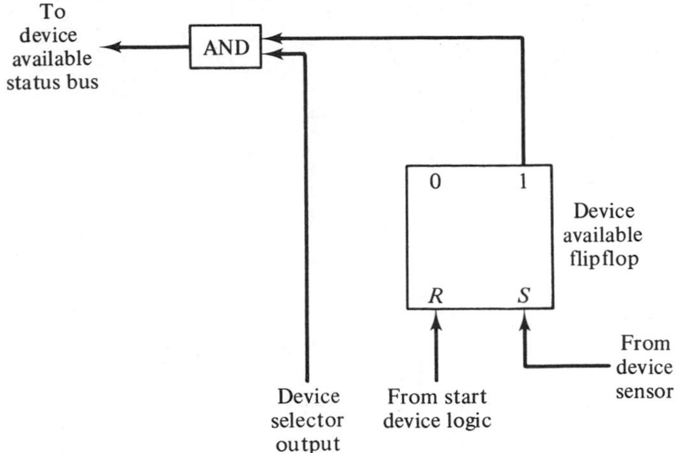

Figure 10.7 Device available status.

10.3.1.4 Data Transfer Control

The transfer of information is done between memory and the device control unit via the *IOIR* as a one-word buffer. The movement of data is controlled by

1. The command code of the select device instruction; i.e., read a word into main memory or write a word from main memory.
2. The availability of the device as specified by the device-available flip-flop in the device controller and the device-available status bit in the *IOCR*.
3. Timing signals used to open gates.

Figure 10.8 shows a block diagram of the data path and control signals needed to perform the write (output) command. Figure 10.9 shows a block diagram of the data path and control signals needed for the read (input) command.

In some computers there is no *IOIR* as presented in this section, but the data are transferred between main memory and the I/O device via the accumulator. Thus, these machines, usually minicomputers, use the accumulator to serve as the one-word I/O buffer register.

The *IOCR* can also be omitted, and the device selection code and command code can be sent directly to the device selector and device command decoder. The device-available status bus sends the signal to a flip-flop in the control logic of the CPU directly.

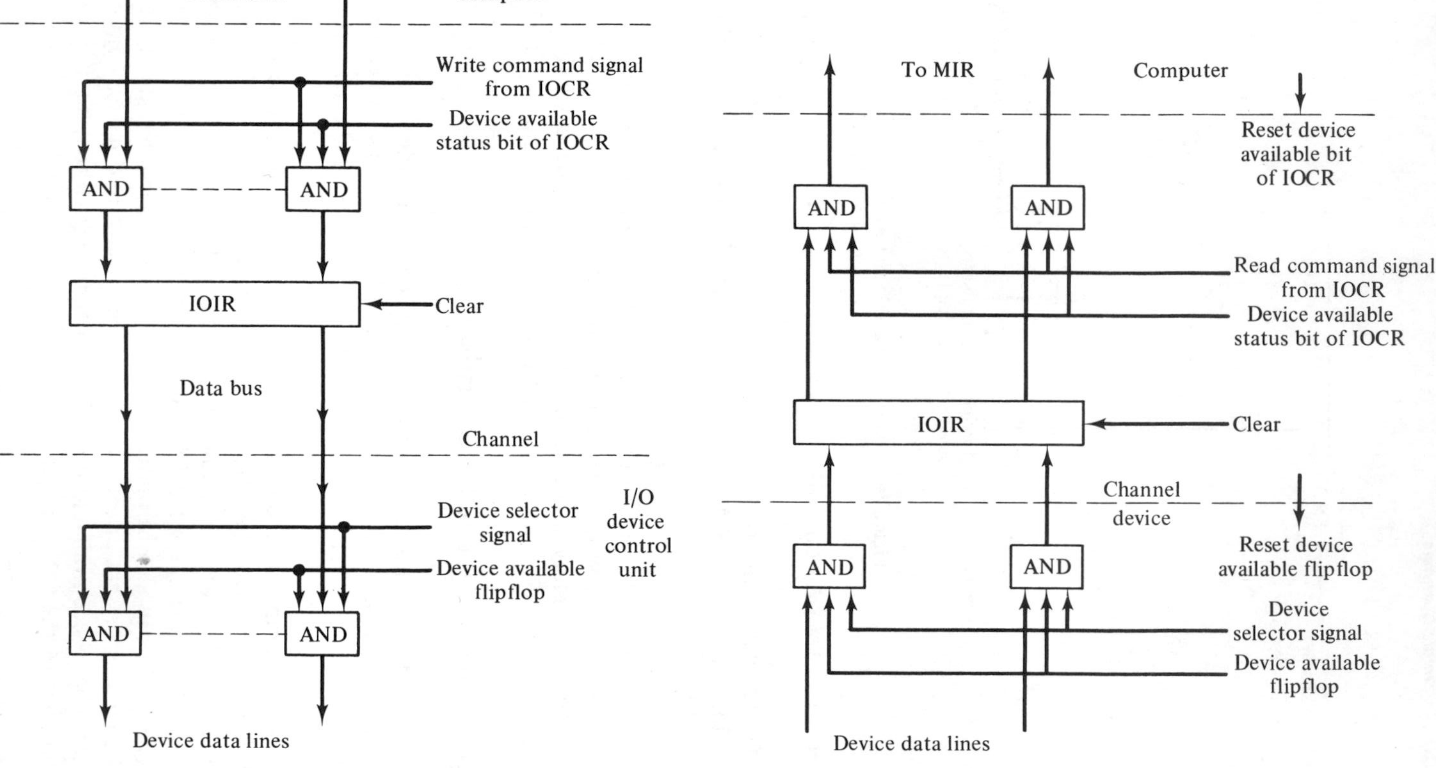

Figure 10.8 Output (write) data path.

Figure 10.9 Input (read) data path.

10.3.1.5 Status Codes

Part of the *IOCR* is called the status field. However, we have used only a single bit of this field, i.e., the device-available status bit. In many computers that may be the entire status field. However, many computers permit for more information to be passed back to the *IOCR* from the device controller about the status of the I/O operation. The codes are a function of the particular type of I/O device which is connected to the channel. Some of the types of status information are

1. Device ready or busy.
2. Device disconnected.
3. Device power not on.
4. Parity error detected during transmission.
5. Operation completed.
6. Tape reel not mounted.

These status codes are sent via the status bus to the status field of the *IOCR*. The CPU can check the status field and transfer the status field to the accumulator where it can be interpreted by software subroutines which handle I/O abnormalities.

10.3.2 Delay Execution Programmed Transfer

The interaction between the CPU executing an I/O program and the simple channel connected to a selected device has a critical timing constraint. The transfer of information between the *IOIR* and the memory is under the control of the program. The movement of information between the *IOIR* and the I/O device is under the control of the device. There must be a method of synchronizing these movements so that the computer will supply a word only when the device can accept the word. If this is not the case, the memory could supply data faster than the device could accept them. Thus, the device must indicate to the channel when it is ready to receive or transmit information. This, in turn, must be used to initialize a request to the program for the movement of information to or from memory. There are several different programmed techniques used to perform this synchronization with a simple channel.

One of these techniques is called *delay execution.* A sequence of I/O instructions required for the particular device is embedded in the program. The execution of each of these instructions is delayed until the I/O device's control unit signals the CPU to execute an instruction. This is usually accomplished by checking the status of the device. If the device is ready, the transfer takes place by execution of the I/O transfer instruction. If the device is not ready, the program remains in a loop which keeps checking the device status until the device is ready. This process takes place every time an information word is to be transferred. Using this method the CPU is constrained to perform I/O operations moving blocks of information in and out of memory, and it runs at the speed of the device which is transferring the information. Thus, the computer is wasting time by being slowed down to the speed of the I/O device. It should be observed that with this simple channel it is the program which has the control information of where to store or fetch data in memory. If the program executes I/O instructions one after another and therefore synchronizes its I/O instructions to the speed of the device, it wastes time. The program could be written so as to perform other useful instructions during the waiting time and then jump back to perform the instructions associated with the I/O when the device commences to transfer data. This is, however, tricky business and requires considerable care on the part of the programmer to ensure that all intermediate instructions do not take excess time and therefore lose data when the device is ready by not getting to the I/O instructions in time.

Some of the disadvantages of the delay execution scheme for I/O on a simple channel are

1. Slow down of the CPU to synchronize with the I/O device.
2. Inability of having more than one I/O device in operation at any one time. Even if more than one *IOIR* is present, it is extremely difficult to synchronize the CPU to two different asynchronous I/O devices.
3. Inability to recover from an unexpected I/O fault condition. If the programmer wrote the program to transfer 100 words of data from the device to memory but there are only 98 words in the block of data, then the computer will "hang up" waiting for data which will never arrive.

Figure 10.10 shows a program which utilizes delay execution to process input data from an I/O device. The code is in-line with the program. The

Address	*Contents*		*Comments*
			Program
	STA	SAVE	Save accumulator [*ACC*] ⟶ SAVE.
	LXI	99	Load index register immediate 99 ⟶ XR.
	CND	21	Connect I/O device #21.
TEST	SKP		If device #21 is ready, skip next instruction.
	UNB	−1	Unconditional branch back one instruction.
	READ		Move one word of data ⟶ ACC.
	STA, X, DATA		[*ACC*] ⟶ DATA + [*XR*]
	DECX		Decrement index register by 1, [*XR*] − 1 ⟶ *XR*.
	BXP	TEST	If [*XR*] ≥ 0; next instruction is at TEST.
	DCD		Disconnect I/O device #21.
			Remainder of program

Figure 10.10 Program using delay execution.

device has a code number of 21. The SKP instruction is a skip the next instruction if the device signals that it is ready to transfer data; otherwise, perform the next sequential instruction. That instruction is an unconditional branch back to the SKP instruction. This is a loop which keeps the CPU waiting until the I/O device is ready to send data. It causes the CPU to synchronize with the speed of the I/O device. The program is designed to place 100 words of data into memory. The first word of data is placed at location DATA +99 and the last word of data is placed at location DATA. It does this by using an index register which is initially loaded with 99. The program decrements the index register by 1 each time it executes the loop and jumps back to address TEST if the index register is not negative.

10.3.3 Status Technique

This method of interaction between the device and the computer is accompanied by the I/O device's control unit setting the status field in the *IOCR*. The state of this field can be tested by programmed test instructions. The execution of I/O instructions is attempted only after a test indicates a successful completion is possible. A program is written so as to test the status of the device when I/O is anticipated during the execution of the main program. If the status condition is not present in the *IOCR*, the test instruction has no effect, and the main program continues executing as if the test had not been made. If the test condition is present, the test instruction forces a jump to a subroutine which handles the I/O with appropriate instructions. After the completion of the I/O, the subroutine returns to the main program just after the test instruction to continue execution.

An advantage of the status technique as compared to the delay execution technique is the ability to execute I/O instructions only when the I/O device has made a request for service. Therefore, no unnecessary waiting time is encountered. A disadvantage is the loss in computer time because the status must be checked periodically by the main program. Another disadvantage is that one cannot predict when the status will change and therefore exactly when the program will branch to the I/O subroutine.

A very important advantage of the status technique is its ability to recover from unexpected errors. This can be obtained by having several different status code conditions. If the I/O subroutine checks for a *ready* condition before each transfer of data but instead finds an *operation completed* condition, then it can determine whether the correct number of words have been transferred. Again assume that we wish to read 100 words and that before we read a word from the device we check its status to see if it is ready. After transferring 98 words the device changes its status to operation complete. Since the I/O program knows that we wish to read 100 words and since it keeps track of the number of words transferred, it knows that there is an error and can enter an error correction routine. Thus, the computer will not hang up and require operator intervention but has a method to attempt to correct the error.

10.3.4 Program Interrupt Technique

The program interrupt technique uses a hardware system which causes the CPU to stop executing the current computer program and forces it to jump to an interrupt subroutine and commence to execute that routine. The interrupt subroutine is programmed to handle the I/O device which created the interrupt. The interrupt is caused by a signal from the I/O device control unit, and the jump to the subroutine is made automatically by the CPU and not by any programmed jump instruction. Once the I/O subroutine has completed its work, control is sent back to the original program at the point at which it was interrupted.

The advantages of the interrupt technique are

1. The ability to recover from unanticipated error conditions. This can be done by having the I/O device control unit provide different interrupts to a subroutine which can interpret and possibly correct the error conditions.
2. The avoidance of having to go and check the I/O device to determine if it is ready to supply or receive data, as in the status technique. The interrupt signal from the device tells the CPU when the device is ready.

A disadvantage of the interrupt technique is the loss of control of when the I/O unit will cause an interrupt. In certain applications it may be necessary not to permit the I/O interrupt to transfer control. For example, in a real-time control application, there may be critically timed parts of the program which if interrupted would have disastrous results on the application. Another time when an I/O interrupt would not be wanted is when the computer is already responding to an interrupt from another I/O device. To prevent these interruptions at inconvenient times most computers which use the interrupt technique, have instructions which can disable the interrupt hardware and thus prevent a new interrupt while the present interrupt is being serviced by the CPU. Once the servicing is complete, the CPU enables the interrupt system; the pending interrupt becomes active, and the CPU services this new interrupt.

Other interrupt system architecture provides for a selective interrupt disable by the use of a mask register. Each possible interrupt is associated with a bit position of the mask register. If the associated bit is set to a 1, the interrupt is enabled; if it is a 0, the interrupt is disabled. The bits may be set or changed by a programmed instruction. Thus, which interrupts may provide interruptions can be changed over the course of a program under the program's control.

The hardware for a simple program interrupt feature is shown in Fig. 10.11. Each device is connected to its device interrupt flip-flop in its control-

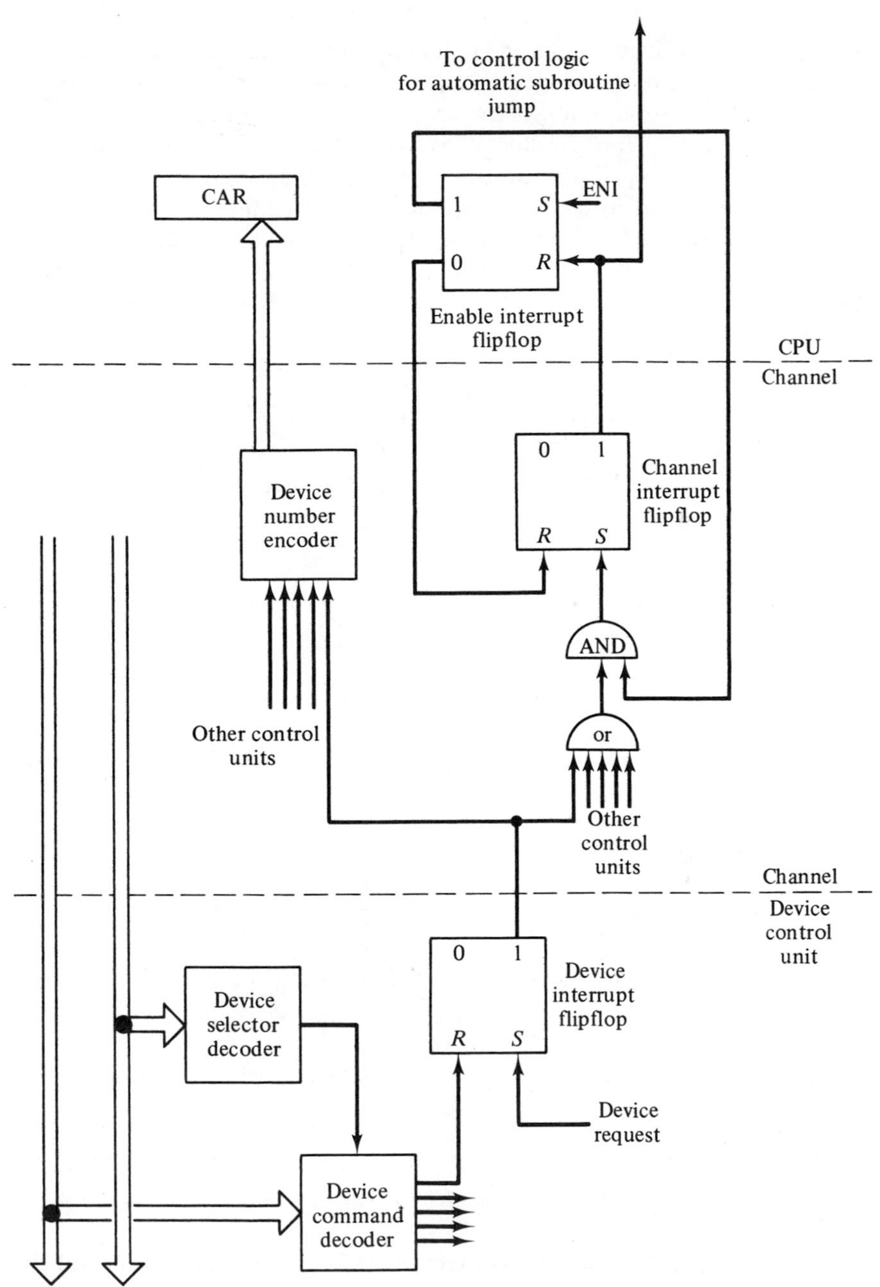

Figure 10.11 Simple interrupt structure.

ler. This is connected to the interrupt flip-flop in the channel. All the I/O devices connected to that channel share the channel interrupt flip-flop. When a device requires service, it sets its device interrupt flip-flop, which, in turn, sets the channel interrupt flip-flop. This signals the CPU to enter the interrupt state. The CPU stops executing the current program and transfers control to an I/O subroutine to serve the specific device. The interrupt line from the device to the channel activates the device number encoder, which produces the binary number code of the particular device which caused the interrupt. The operation of this simple interrupt structure is shown in the example of Fig. 10.12.

Memory address	Contents		Comments
O		O	Keeps return address
15	UNB	SERVE	Unconditional branch to I/O service routine for device # 15
⋮			
SERVE	STA	SAVE	[ACC] → SAVE
	LDA	O	[O] → ACC
	STA	RETURN	[ACC] → RETURN
	ENI		Enable interrupt system
	⋮		Remainder of service routine
	LDA	SAVE	[SAVE] → ACC
	UNB,	I, RETURN	
SAVE	O		Save accumulator location
RETURN	O		Return address location

Figure 10.12 Program for I/O interrupt technique.

Assume that the computer is executing a background program and that the interrupt flip-flop is set while the CPU is executing the instruction located at address 723. The CPU will complete this instruction and then enter the interrupt state. At that time the contents of the *CAR* have been updated and contain the address of the next sequential instruction, i.e., 724. The interrupt logic causes the contents of the *CAR* to be stored at a specific address. In the example, it is location 0. The interrupt hardware then places the output of the device number encoder into the *CAR*. In the example the interrupt is caused by device number 15; the output of the device encoder would be 01111, which would be placed into the *CAR*. The interrupt state causes an unconditional branch instruction to be executed. Thus, the instruction contained at location 01111 will be executed. This is another unconditional branch instruction, which jumps to the beginning of the subroutine which services device number 15. It is located at address SERVE (symbolic). When the channel interrupt flip-flop is set, it disables any other device from interrupting. Therefore, an instruction must be executed to enable the interrupt system.

All service routines perform the following actions:

1. Save the accumulator which contains information used by the interrupted program. This will be restored at the end of the device service routine just prior to returning to the interrupted program.
2. Move the return address, which has been placed by the interrupt hardware at address 0, to a location within the I/O device's service routine. This will permit another device to interrupt this routine without losing the trail back.
3. Enable the interrupt system so that another device can interrupt the presently executing routine.
4. Perform the required I/O instructions and routine needed to service this interrupt.
5. Restore the accumulator.
6. Perform an unconditional branch using indirect addressing with the indirect address being the address where the return address was stored.

10.3.5 Combination of Simple Channel Techniques

Most computers which have simple channels, where the CPU is involved in the transfer of data between memory and an I/O device, utilize a combination of the three types of I/O control techniques which were just discussed. For example, a computer may employ the delay execution technique for the transfer of information across the *IOIR*, while it may utilize the program interrupt technique to indicate that a new status code has been placed into the status field of the *IOCR*. Many computer architectures permit the programmer to decide which of the three techniques he wishes to use for a particular I/O situation or application. The programmer may or may not test for status conditions. Likewise, he could shut off the interrupt system and not permit any interrupts.

Some computers have several different types of channels which have different I/O devices connected to them. A paper-tape reader may be connected to a channel which requires a delay execution technique in order to transfer information, while a magnetic tape may be connected to another channel which utilizes an interrupt mechanism for the control of the transfer of data. Thus, the I/O technique needed may depend on the channel to which the device has been connected.

10.4 DIRECT MEMORY ACCESS CHANNELS

Most computers of the third generation and most minicomputers use a more sophisticated I/O path than the previously described simple channels. These multiple-register channels permit the overlap of computation and independent I/O transfer. They are called *direct memory access* (DMA) channels because they have a direct data path to main memory. Typically these channels contain the following registers:

1. An input/output information register (*IOIR*), which acts as a buffer between the device control unit and main memory. All information is moved via this path.
2. An input/output control register (*IOCR*), which contains the I/O device selection code, the code of the operation to be performed by the device, and a status field which holds device and or channel status codes.
3. An input/output address register (*IOAR*), which contains the address in main memory where information is to be transferred. It is initially loaded by the computer with the location of the first word to be transferred and is automatically incremented after each transfer of a word. Thus, consecutive words in memory are transferred over this type of channel.
4. A word counter (*WC*), which counts the number of words transmitted. Its contents are automatically decremented after each word is transferred. When the contents have reached zero the number of words to be transferred has been completed and the DMA channel ceases the transfer of words and places an operation complete status in the *IOCR* and may cause an interrupt to the CPU. The number initially loaded into the word counter by the program is the number of words which will be transferred from this device. An alternative to the word counter is to place the address of the last word which is to be transferred into a *last word register*. With this method no words are counted, but the contents of the *IOAR* are compared with the contents of the last word register after each word is transferred over the channel. When they compare, the last word in the block of data is being transferred and the channel ceases I/O operations to this device just as if the word counter had reached zero.
5. A byte counter (*BC*), which is used to communicate with a control unit on a byte basis instead of a word basis. The data path between main

memory and the DMA channel has a 1-word width. Most devices have a data path of 1-byte width. The channel disassembles a word into bytes and transfers a byte at a time to the device's control unit. To accomplish this byte movement, the *IOIR* usually has shift register capability. The byte counter is used to count the number of bytes transferred from a given word in the *IOIR* and to control the shifting of the *IOIR*. For example, if the word length is 32 bits, composed of 4 8-bit bytes, the BC would count modulo-4 to control the shifting of the *IOIR*. Figure 10.13 shows a block diagram of the registers in a DMA channel.

The method by which a program initiates an I/O operation using a DMA channel varies among computers. Some computers have a special instruction which initiates an I/O operation on a channel. It requires that three additional words be transferred to the channel in order to fill the *IOCR*, *IOAR*, and *WC*. Thus, this instruction requires four memory accesses to complete its execution. After the information has reached the DMA channel, it commences the I/O operation, and the CPU can continue processing a program. Another method requires only a single memory word as the start I/O instruction to the channel. Part of this instruction is an address where the channel will find the first of the information words needed to fill the three registers. The next two sequential addresses contain the other two words. Thus, once the CPU executes the start I/O instruction, it is free to continue processing, and the channel will fetch the needed information which permits it to perform the I/O requirement.

Another variation is for the CPU to execute a start I/O instruction to a channel which then accesses a fixed location in memory. The contents of this address specify the starting address of the information to fill the *IOCR*, *IOAR*, and *WC*. When the channel finishes the current I/O operation, it again accesses the fixed location in memory to see if another address has been placed there. If it has, the channel starts a new I/O operation. Thus, the channel can *link* or *chain* a sequence of I/O operations together without CPU program assistance. This requires a more complicated channel but permits a continuous overlap between channel operations and independent programs executing in the CPU.

Since both the CPU and the channel require access to main memory, there must be a method for each to gain access in case of a tie. Usually the priority is set so that the channel will obtain the memory cycle before the CPU. Thus, the channel "steals a cycle" from the CPU, executing its program. This happens relatively infrequently when considered from the CPU's viewpoint. If the CPU executes an instruction on the average of 1 per microsecond while the I/O device delivers a word on the average of 1 per 10 milliseconds, then the channel will steal one cycle in every 10,000 memory cycles.

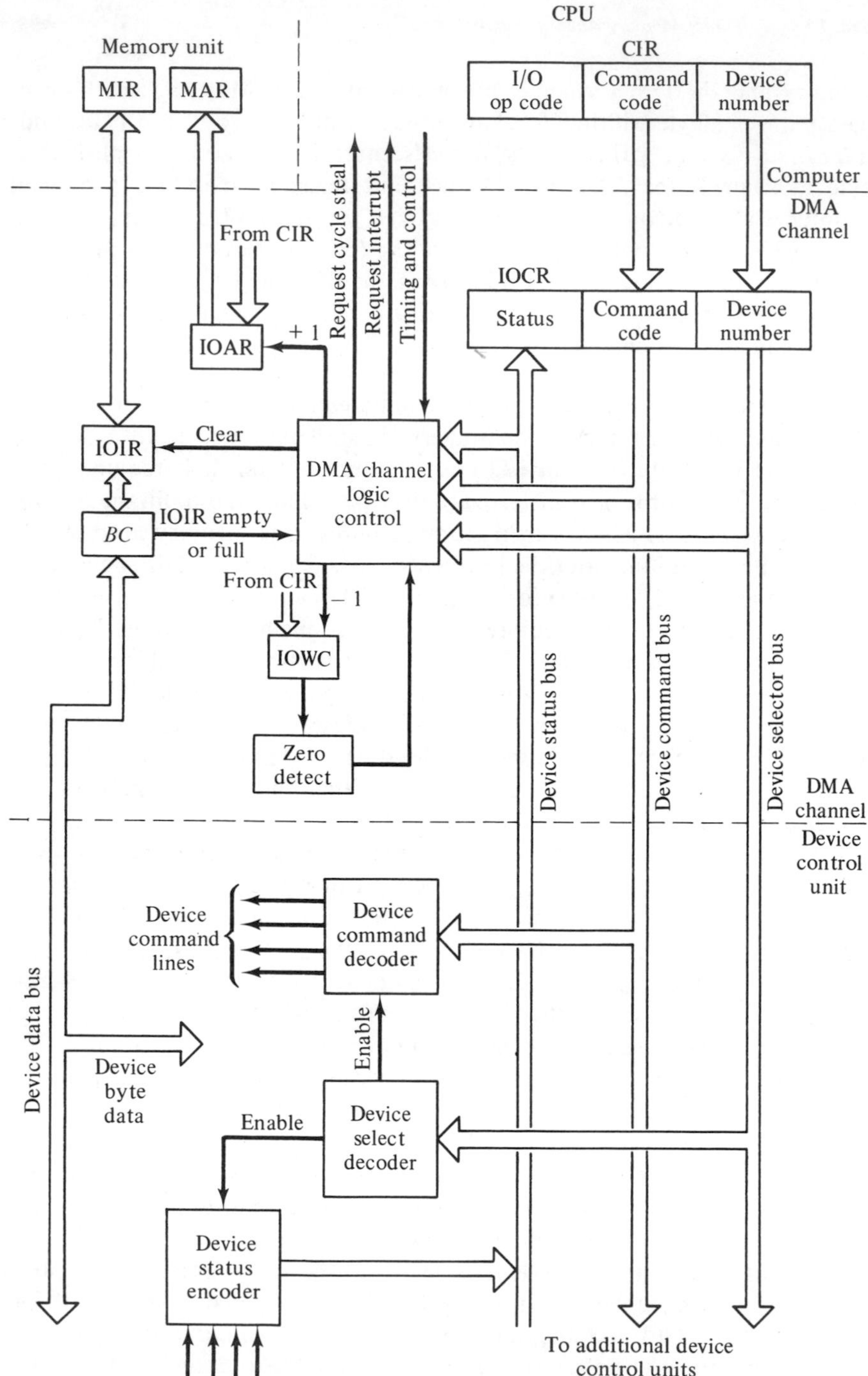

Figure 10.13 DMA channel interfaces.

Even though, under normal conditions, the CPU and channel operate under independent control, the program can monitor the channel and the control unit of the selected device. This is done by status checks to the status field of the *IOCR* or by receiving program interrupts. The device controller can place status information into the channel. This information can be obtained by an interpretive program by the use of a *copy status* instruction. The use of this instruction moves the status to an arithmetic register where it can be interpreted by test instructions. Channel status codes may be provided for such conditions as channel busy, channel ready, illegal operation, operation completed, parity error, and device inoperative.

In addition to the use of channel and device status, the DMA channel may produce program interrupts. Errors or conditions in the device can be conveyed to the channel, which can signal by an interrupt for a particular program to analyze or correct the situation. The channel could interrupt if it receives a new start I/O instruction before it has completed the previous operations. The external device could signal the channel if the controller detects a parity error; the channel could then interrupt the program to force an entry into a subroutine to request a reread of the data.

10.5 I/O DEVICE CHARACTERISTICS

The communication with the memory and central processing unit of the computer is done via the I/O devices. These devices provide a method of translating alphanumeric information which the human uses for information into binary and binary codes which the computer uses for information. This section is concerned with the mechanical and electrical characteristics of these I/O devices. These devices include punched-paper-tape readers and punched-card equipment, keyboard equipment, line printers, cathode ray tube (CRT) visual readout, and magnetic tape units. The last device can also be considered as an extension of memory or a secondary storage device; however, we shall treat it both ways but include it in this section on I/O.

10.5.1 Punched-Paper-Tape Equipment

Punched tape is one of the oldest media for storing programs and data to be read into a digital computer. The first-generation digital computers all used this medium for their basic I/O. This early use of punched tape on computers was due to its development in the telegraph industry where the devices for punching and reading the tapes had advanced to a high degree of reliability. Today, paper tape is used as the input medium for small- and medium-sized computers. It is inexpensive and easily stored; however, it is not easily corrected if punched incorrectly.

10.5.1.1 Punched Tape

Punched tape comes in various widths from ½ inch to several inches and can be a medium-weight paper, an oiled paper, or a plastic. In general, the information is punched a character at a time across the width of the tape, which creates holes in the tape. A character can be represented by either five, six, seven, or eight holes depending on the code used to represent the characters. Figure 10.14 shows the various tape configurations in use, and Fig. 10.15 shows a segment of a punched paper tape. Each of the tapes has an additional smaller hole termed the *sprocket hole*. This hole is used to align the tape and feed it into a tape reader device.

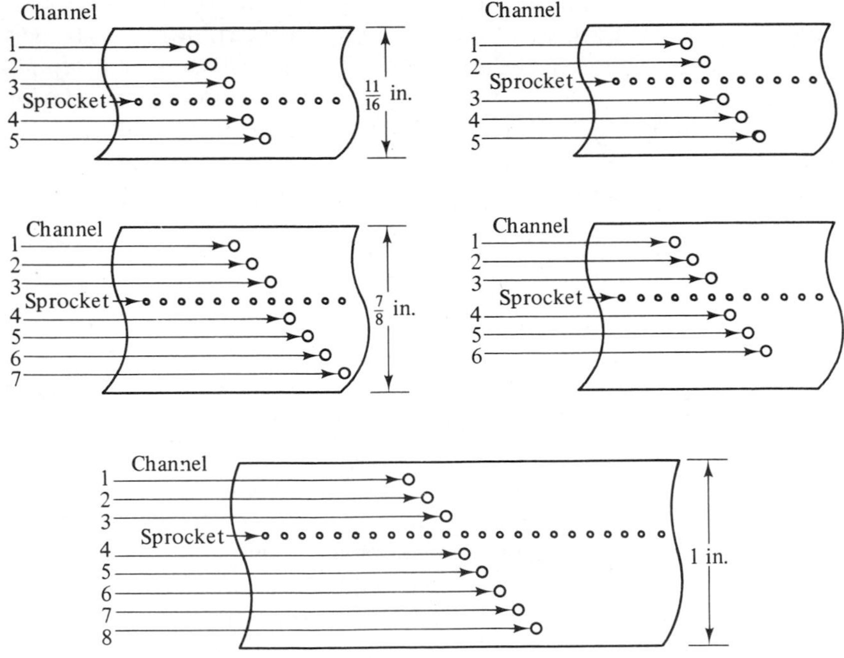

Figure 10.14 Punched paper-tape formats.

10.5.1.2 Punched-Tape Reader

The paper-tape reader is an electromechanical device capable of reading punched paper tape at speeds up to 2000 characters per second. Reading is performed by a light source and photocells. There is one photocell for each level used in the paper tape. The hole in the tape represents a binary 1; the absence represents a binary 0. To read data from a paper tape it is mounted on a paper-tape reader which moves the tape past the photoelectric read station; the sprocket hole is read to generate a timing pulse.

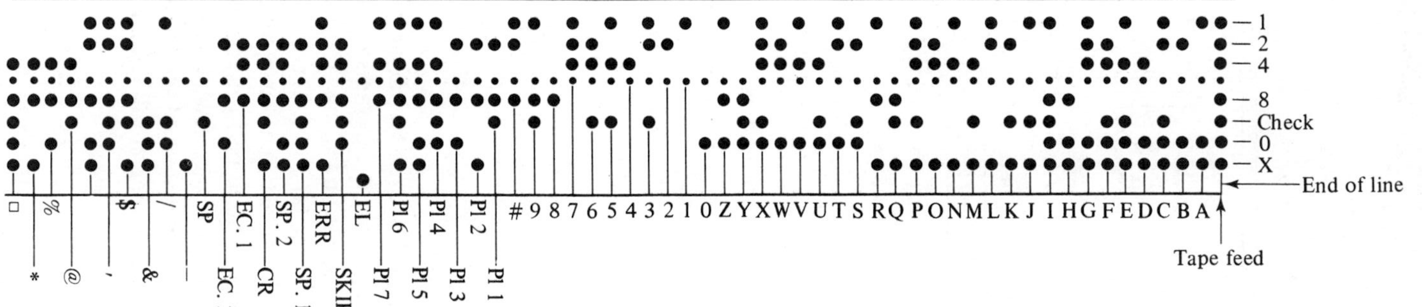

Figure 10.15 Segment of punched tape.

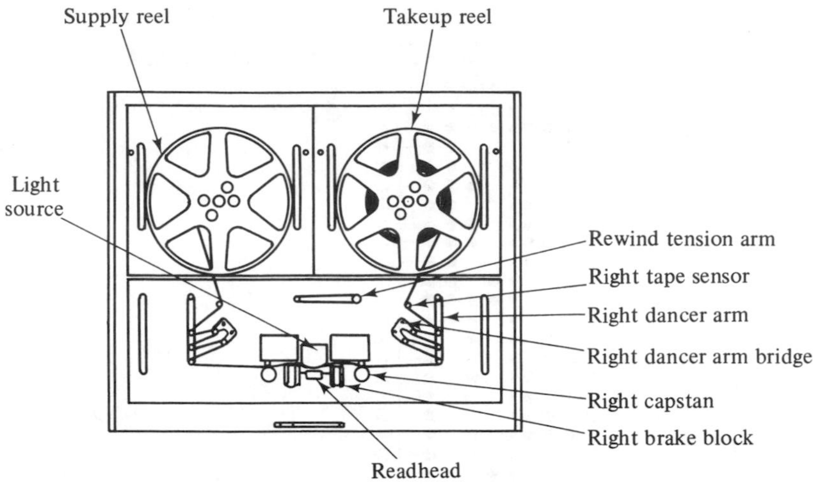

Figure 10.16 Paper-tape reader.

Figure 10.16 shows a simplified front panel of a paper-tape reader. The reader is symmetrical about the read head. One side consists of a supply reel from which the paper tape is threaded through the dancer arms and tape sensor, over a capstan and through the read station. It then passes over the capstan of the takeup side, through the dancer arms and tape sensor, and onto the takeup reel. During the read operation, the tape is moved by the forward capstan past the read station. This causes a tension in the supply dancer arms, which causes the servo motor which drives the supply reel to gain speed and supply tape. The excess supply of tape on the takeup side causes the dancer arms to move, and this causes the takeup reel to begin to spool tape. Thus, the servo system attached to each reel attempts to equalize the supply and takeup of the tape.

The tape can be stopped on a single character by the high-speed magnetic brake which pinches the tape; the capstan stops and the servo motors, and dancer arms prevent the tape from tearing. The tape sensors are used to detect the beginning and end of the tape so that the unit may stop before the tape unthreads. Tape may be moved in reverse by activating the reverse capstan and causing the supply servo to perform the takeup function and the takeup to act as supply. Usually a faster rewind can be performed by unthreading the tape from the dancer arms and read station and allowing the supply reel to act as takeup reel running at full takeup speed.

10.5.1.3 Tape Punch

Paper tape may be punched at speeds in excess of 120 characters per second under computer control. The tape can be punched in either the code used internally by the computer, or by the use of a hardware translator it may convert the incoming code to another one. A character is punched across the width of the tape with all channels punched simultaneously as well as the sprocket hole.

The information is fed from the channel a character at a time to the punch buffer until the punch mechanism is set for punching. The data control a set of code magnets which operate the punch pins. One character is punched at each cycle of the punch with the selected pins being driven through the tape, which creates the holes. The sprocket holes are punched with each cycle. The pins are withdrawn, and when they are clear of the tape, the tape is advanced for the next character to be punched. The channel receives a signal from the punch and supplies the next character to the punch buffer.

10.5.1.4 Keyboard, Reader, and Punch

The use of the typewriter or keyboard is to produce punched tape off-line. This must be done in conjunction with a punch. In this type of configuration the depression of a key on the typewriter sends an electric signal to the punch which activates the proper punch pins. Thus, the typewriter produces the hard copy of alphabetic characters, while the paper-tape punch produces a copy which can be used as input to the computer.

At other times it is desirable for an operator to communicate directly with the computer operating system. This can be done through the keyboard, which is connected directly to an I/O register. Likewise, the computer can communicate to the operator through the keyboard by activating the keys. This is also done through the I/O register to which the keyboard is attached.

A third method of operating a keyboard is in association with an off-line paper-tape reader. This connection provides an off-line method for printing the output provided by punched tape. Thus, the computer can use the higher-speed punched paper tape as output and use an off-line slower typewriter to prepare printed pages, i.e., hard copy. The paper-tape reader supplies electrical signals to the typewriter, which activates the correct key depending on the punched character which the reader has read from the tape prepared by the computer on the punch. Figure 10.17 shows a typewriter with reader and punch.

10.5.2 Punched-Card Equipment

The use of the punched card as an input and output medium is widespread and is used especially in business data processing. The punched card is the most widely used medium for input, especially in batch-oriented systems. After the programmer writes his program, it is usually punched into cards, read into the computer, and executed on test data which are supplied

Figure 10.17 Keyboard, reader, and punch. (Flexowriter automatic writing machine, courtesy Singer Business Machines.)

along with the program. Once *debugged*, the program may be stored on a disk of the secondary storage if it is to be used again. The input data are usually supplied on punched cards; the program is retrieved from secondary storage and operates on the data read in from the card reader. The output is usually done on a line printer, but some cards may be punched as output if they are to be used as subsequent input to another program executed at a later time.

Punched-card equipment consists of a card reader and a card punch, which are attached to the computer via a channel and are under the control of the *IOCS*. In addition a manual punch called a key punch is used to prepare punch cards off-line.

10.5.2.1 Punch Cards

The IBM punched card contains 80 vertical columns and 12 horizontal rows. The rows are subdivided into two groups: The unlabeled zone at the top constitutes rows 12 and 11, while the numbers are shown as 0 through 9. Each column of the card is used to represent one alphanumeric character by the unique placement of holes punched into that column. The row labeled 0 is considered both a zone and a numeric. It is a numeric when the only hole in that column occurs in row 0. Several different codes are used to represent alphanumeric characters. Table 10.1 shows the locations of the holes to represent the 48 characters in the Hollerith code. Table 4.11 shows the location of the holes for the EBCDIC code.

Table 10.1 Card Punches for Hollerith Code

	No zone	*12*	*11*	*0*
Zone only	Blank	+	—	
0	0			
1	1	A	J	/
2	2	B	K	S
3	3	C	L	T
4	4	D	M	U
5	5	E	N	V
6	6	F	O	W
7	7	G	P	X
8	8	H	Q	Y
9	9	I	R	Z
8-3	=	.	$	,
8-4	'	)	*	(

10.5.2.2 Card Reader

The card reader is connected to the computer and may be capable of reading cards at a rate of 2000 per minute. Cards are placed in a hopper and are fed one at a time past a read station. Reading is performed a column at a time. A light source is detected by photocells whenever a hole in the card is passed between the source and the cells. After being read the cards are sent to a stacker area. Each column read on the card must be decoded and the character code changed to the internal bit representation used to store information in the computer before it is sent to the memory via the channels. The path of the card in the reader is from the hopper through the read station and into the stacker. The major operations performed within the reader are feeding a card, reading it, and then stacking it.

There are two methods used to feed a card. One method uses a knife edge which moves back and forth and feeds one card at a time toward the read station. To ensure that only one card moves toward the read station, a slit is placed in the output of the hopper, which is adjusted to allow only a single card to pass. Figure 10.18 shows this method and also the second method which uses a revolving belt which is forced against the card and pulls it into the read station. Once the card enters the read station the belt is pulled back until the next card is to be read. The reader monitors the card to ensure against jamming. Both of the methods are started by a *read a card* instruction from the processor.

A card is read as it passes through the read station. The card is read a

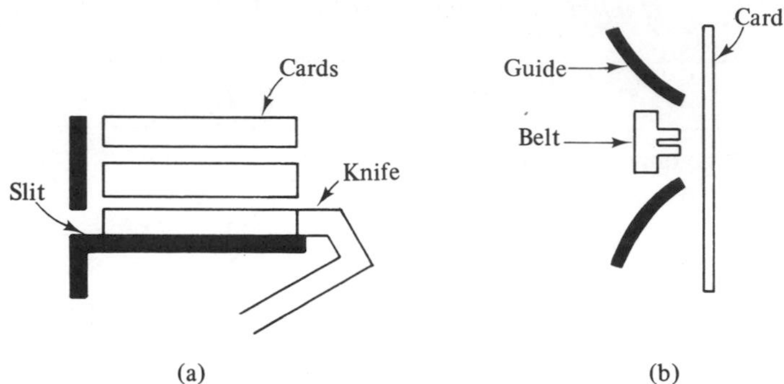

Figure 10.18 Card feeding methods.

column at a time. The information is checked for a legal character, decoded and encoded into machine code form, and sent to the computer via the channel. A set of 12 photocells and light sources scans each column. A hole in the card will activate the photocell by the light beam. The absence of a hole will block the beam and indicate a 0. Timing is supplied as the card moves past the read station.

After the card has been read it is transported to the stacker. There the cards are collected and held in the same order in which they were read.

10.5.2.3 Card Punch

The card punch is an electromechanical device which can punch 80-column cards at a rate up to 500 cards per minute. The card punch is connected to the computer via a channel and operates under processor control. The cards are placed into a feed hopper from where they are fed a card at a time to the punch station. The cards are punched a row at a time starting with row 12. This requires 80 parallel punch knives controlled by the buffer register contained in the card punch. Thus, the characters to be punched on one card must be transferred to the punch prior to any punching. After each row is punched into the card, it is advanced, and the next row is punched. Once the punching is completed, the card moves to a read station where it is read and compared against the original data for accuracy. After verification, the card moves to the stacker area.

10.5.2.4 Key Punch

To enter programs or data into a computer requires the off-line preparation of the information into computer-compatible mode. In most batch-oriented systems this is done by punching holes in cards on a key-punch device. This device has a hopper where unpunched cards are stacked and which are fed a card at a time by the key-punch operator to the punch station. By depressing the keys on the keyboard, the operator punches the code into the cards one character at a time, starting with column 1. Usually the character is also printed at the top of the column while the information is punched. This is to allow for easier reading of the information and detection of errors in punching. The card moves toward the left one character after each column is punched. Upon the completion of key punching the data into the card, the card can be released from the punch station by depressing a *feed* key. This also feeds a new card into the punch station. Thus, punching is similar to typing with one line per card.

10.5.3 Typewriter and Terminals

In recent years, systems have been developed which permit the user to *interact* with the computer. He inputs his programs and data directly into the computer via a *terminal.* The terminal may take several forms. It may be a typewriter which can be used by the programmer for input and by the computer for output or it may be a keyboard and cathode ray tube display. The latter device will be discussed later.

The remote typewriter terminal is often connected to the computer over telephone lines. These lines are designed to carry audio frequency and are not well suited to transmit digital signals. To convert the digital signals to a form usable on the telephone lines the terminal is connected to a *modem* (modulator-demodulator). At the computer is another modem which converts the signal back to digital form. A simple typewriter terminal can be connected via an *accoustic data coupler* modem which converts the binary signals into a series of audio tones for transmission over the telephone lines. The system is similar to the touch-tone telephone system which uses accoustical tones for conveying dial information. Many terminals require only that you dial the computer center on a conventional telephone and place the handpiece into a cradle in the accoustic data coupler, which then forms a modem.

10.5.4 Line Printer

The line printer produces a printed line of alphanumeric characters which is similar to the output achievable with a typewriter. Thus, the printer converts the internal code of the computer to printed characters which the human can comprehend. Modern high-speed line printers are capable of printing over 2000 lines per minute with a line being as large as 200 characters. There are two general classes of printers. One is the mechanical printer which utilizes the contact of print type and paper with ink and a hammer similar to a typewriter, while the other is the electrostatic printer which electronically forms and records characters.

10.5.4.1 Mechanical Printers

Mechanical line printers come in two types. One is a drum printer, and the other is a chain printer. The drum printer consists of a metal drum with a character set cast on its surface. If the printer can print 132 characters to a line, the drum would have 132 of the same characters in each row. Figure 10.19 shows this type of a drum printer. The drum rotates, and therefore

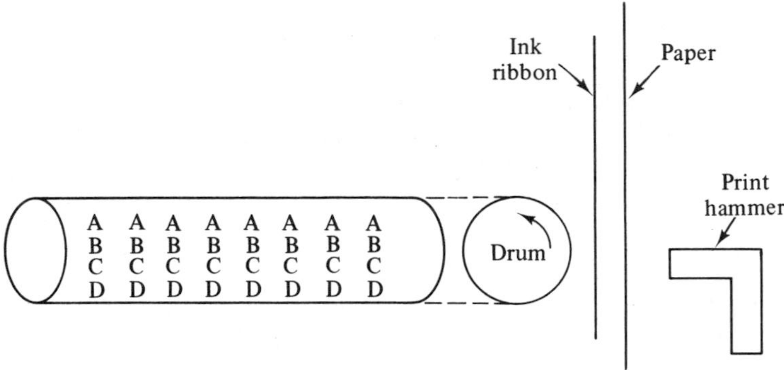

Figure 10.19 Drum printer.

each character of the print set will pass the print position once each rotation. Thus, any character of the set can be printed at any location on a line. A block diagram of a drum line printer is shown in Fig. 10.20. There is a separate hammer located at each position of the line. As the characters come past the hammers they are compared with the characters held in a buffer in the printer. This buffer holds a full line of characters. When a character is in front of the hammers, all the locations which compare with the character are printed by the individual hammers. The hammers force the paper and ink ribbon against the character on the drum much as a typewriter key does. A code wheel is rotated with the print drum and contains the binary code for each character in the character set. As the code wheel spins past a set of photocells, the characters are detected and the electrical signals are sent to the 132 comparison circuits. All the locations in the print line which compare are printed. Once all the characters are printed a new line of characters is inserted into the printer buffer. Thus, the line printer does not really print a complete line at one time but builds a line up as the print drum rotates. In the worst case, it requires one complete rotation of the drum to complete the printing of a line. Figure 10.21 shows how a line is developed. Assume that the buffer has been filled and that the drum is aligned at letter M.

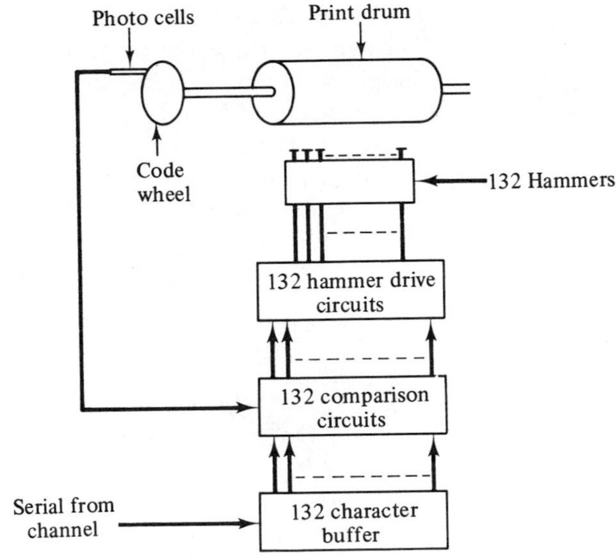

Figure 10.20 Block diagram of a line printer.

Aligned character							
N		N		N	N		
O		N		N	N		O
P		N		P N	N		O
R		N		PR N	N		OR R
S		N	S	PR N	N	S	OR R
T		N	S	PR NT	N	T S	OR R
A	A	N	S	PR NT	N	T S	OR R
D	A	N	S	PR NT D	N	T S	ORD R
E	A	NE	S	PR NTED	N	T S	ORDER
H	A	NE	S	PR NTED	N	TH S	ORDER
I	A	INE	S	PRINTED	IN	THIS	ORDER
L	A	LINE	IS	PRINTED	IN	THIS	ORDER

Figure 10.21 Development of line of print.

10.5.4.2 Electrostatic Printers

The electrostatic printer eliminates the mechanical motion associated with the printing operation and thereby achieves a high printing speed. The only mechanical motion is the feeding of the paper. The characters are formed by electronic means and formed onto the moving paper. A developing system is required to fix the characters onto the paper.

One type of electrostatic printer is manufactured by Stromberg Datagraphics, Inc. and uses a special electronic cathode ray tube called a CHARACTRON®. This tube shapes the electron beam of a cathode ray tube into characters which are displayed on the face of the tube. The printer uses a special electrostatic copy paper which is negatively charged. A character is formed on the paper by discharging an area of the paper in the form of the character when it is passed in from the CHARACTRON tube. The paper is passed through a developing section to fix the character to the paper.

The CHARACTRON tube uses a beam-forming technique to generate characters. An electron beam is passed through a metallic stencil matrix which is in the neck of the tube; see Fig. 10.22. The matrix is a thin metal

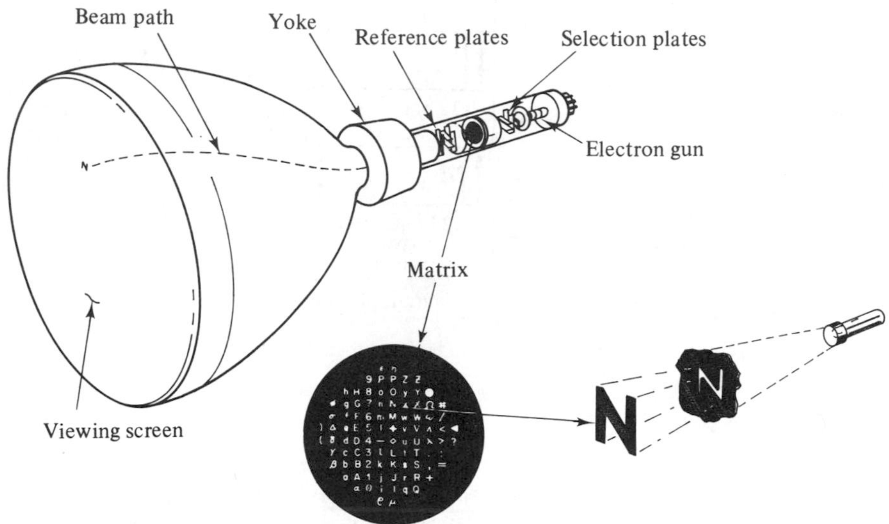

Figure 10.22 A CHARACTRON® tube. (Courtesy of Stromberg Datagraphics, Inc.)

disk which contains the alphanumeric character set. It can contain up to 200 characters. The selection plates deflect the beam through one of the characters where the beam is formed into the shape of the character. The potential on the selection plates is controlled by the code contained in the buffer register of the printer. Each character references a different location in the matrix. After passing through the character matrix, the shaped beam is directed by the convergence coil to the reference plates, which align the beam back to the axis of the tube. The beam is focused and deflected onto the face of the tube at the location at which that character is to be displayed and therefore printed. This tube can form and display up to 10,000 characters per second.

The special paper is polarized as it is fed to the tube. It is positively charged on its uncoated side and negatively charged on the side which faces the tube. A lens system projects the displayed characters onto the sensitized paper, which reduces the negative charge where the image forms, and an electrostatic image in the shape of the characters is created. The paper moves to the developing area where the charge attracts particles of thermoplastic material into the shape of the characters. The paper then passes through a heat fuser area which causes the thermoplastic to bond to the paper, forming a permanent image.

10.5.5 Cathode Ray Tube Displays

Information display devices which permit communication between the computer and the operator have increased the speed of man-machine interaction. Each participant sends and receives information in his own language. The information is translated between the computer and operator by encoding and decoding hardware. Thus, the operator sees alphanumeric characters and graphs on a visual display which is similar to a television tube, while the computer sends and receives binary coded information. Since a full screen of information is available for the operator, he can quickly obtain large amounts of information needed to monitor the computer's operation.

The cathode ray tubes (CRT) used in computer display devices are similar to those used in oscilloscopes and television sets. The information displayed on the face of the tube is created by electronically positioning an electron beam and electronically turning the beam on and off. These displays are sometimes called *scope* displays, a shortening of the word oscilloscope. A keyboard is used by the operator to "type" characters into the computer and simultaneously display the characters on the CRT screen. The combination of CRT and keyboard is often called a display terminal. Display terminals are used by computer operators to monitor and observe the status of large computer systems. Computer users use terminals in an interactive mode in a time-shared, on-line method of operating the system. The terminal is often remotely located from the computer, and the interconnection is made over telephone lines.

The information display system is composed of a CRT and its associated control, which positions the beam; a symbol generator, which electronically forms the characters displayed; a small memory, which stores the digital code of the characters; interface and control electronics, which provides control between the terminal and the computer; and the keyboard. In addition, some terminals have a light pen, which permits the editing of data on the CRT as well as the ability to draw diagrams on the screen. Figure 10.23 shows a block diagram of such a terminal.

The CRT produces an image by directing an electron beam against a phosphor-coated screen (the face of the tube) which emits light when struck by the electrons. The CRT consists of a filament structure, which emits

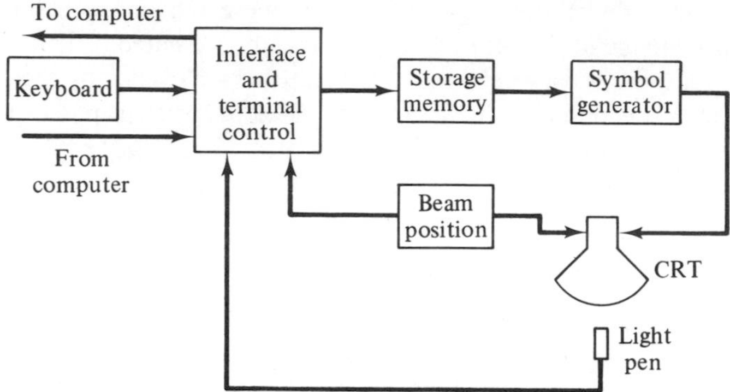

Figure 10.23 Block diagram of a display terminal.

electrons; a focus control, which forms the electrons into a beam with a narrow cross-sectional area; an intensity control, which limits the number of electrons which reach the screen and thus controls the brightness of the spot; a deflection system, which positions the beam to a place on the screen; and the phosphor-coated face of the tube. The intensity control can turn the beam completely off or allow no electrons to reach the screen. This is called *blanking* of the scope. The deflection system can position the beam along both the horizontal and vertical axes with independent circuits used for each direction.

When the electron beam strikes the phosphor coating on the face of the tube, light is given off at that spot. The type of phosphor coating determines the persistence or the period of time for which the spot will emit light after being struck by the electrons. Typically this period is in the microsecond to millisecond range. To provide a constant image and eliminate flicker in the image, the electron beam must rewrite the picture many times a second. This is called the *refresh rate* and is 30 times per second for commercial television.

Positioning of the beam by the deflection system is usually done by either the random positioning method or the raster scan method. In the random positioning method, the beam is deflected from one point on the screen to any other point on the screen selected at random. If the blanking is on during the movement of the beam, then only the two end spots will be seen. If the blanking is off, then a straight line will be seen on the screen. Figure 10.24 shows these two conditions. In the raster scan method, the electron beam is positioned in the upper left corner of the tube face, and the beam moves horizontally across the screen with the blanking off. Thus, a straight horizontal line is produced. The blanking is turned on at this point, and the beam is retraced back to the left side of the tube and one line lower. The blanking is turned off, and a second line is traced onto the screen. Thus,

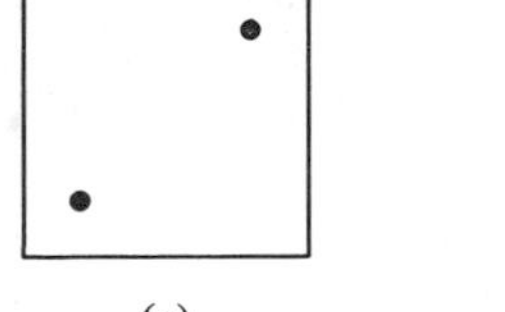
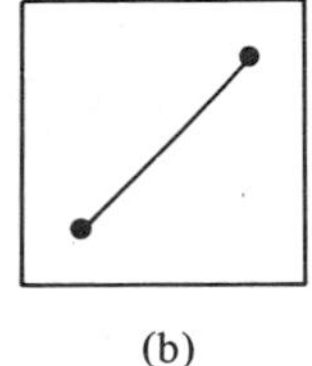

Figure 10.24 Random positioning method: (a) blanking on; (b) blanking off.

the beam scans the face of the tube similarly to reading a page in a book. This method is used in home television sets with 525 lines per complete picture scan and 30 pictures per second. Figure 10.25 depicts this method, with the dotted lines showing the retrace which would not be seen since the blanking is on during those periods.

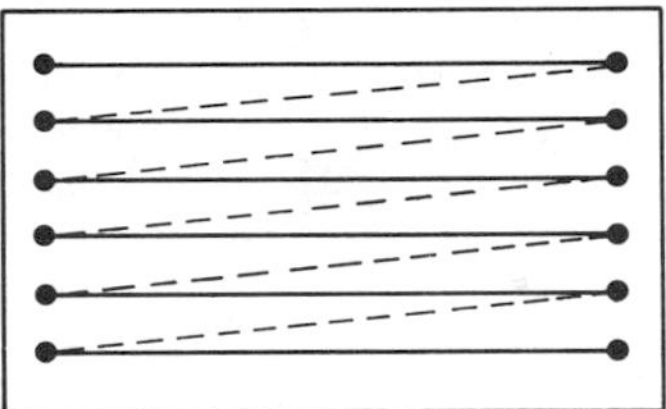

Figure 10.25 Raster scan method.

There are two major methods of forming one alphanumeric character on the face of the tube; one is termed *beam forming* and the other is termed *signal generation*. Beam forming uses the CHARACTRON tube and was explained in Section 10.5.4.2. It can also be used in graphic terminals.

The signal generator method uses a combination of blanking and control voltages applied to the deflection system to form characters. The deflection voltages are used for positioning, and the blanking voltages form the image. Signal generators are either raster-type or function-type generators.

In the raster-type signal generator the tube is divided into a series of rectangles, each just large enough to display a single character. Thus, for a sufficiently large screen there may be 80 characters per line and 80 lines permitted on the screen. This would yield 6400 possible character positions. Within each small rectangle a raster scan is developed. The character is developed by blanking the beam on or off by either the dot or line method of character formation.

The dot method of character formation uses a 5×7 matrix of dots within each of the small rectangles. Each character is formed by lighting a prearranged set of the 35 dots to form a character. Figure 10.26 shows this method and the line method. In the line method the blanking voltage is used to produce a character as the raster scan is produced in each of the small rectangles. During the raster scan, the beam is off and is turned on only when the character is to be produced. The pattern of blanking is controlled from

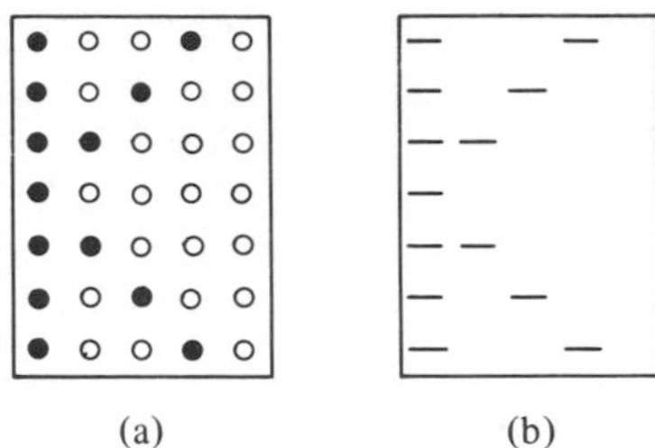

Figure 10.26 Character formation by (a) dot method; (b) line method.

the local memory of the graphic terminal. It is also necessary to refresh the entire display many times each second in order to produce a flicker-free display.

The function-type generators do not use a raster scan within each character rectangle but use the deflection system to trace the outline of the character directly. Characters are formed by three methods: (1) programmed dot, (2) Fourier method, and (3) programmed strokes. The programmed dot method is similar to the raster dot character generation method except that the character outline is prdouced on the deflection system, and the blanking produces the dots. The Fourier method produces characters utilizing Lissajous patterns by applying combinations of different frequencies to both the horizontal and vertical deflection plates. The stroke method uses a programmed series of straight-line strokes to trace out each character. This is similar to a human doing block printing.

Every operation of the graphical display is under the control of the computer. Instructions are required to initiate each function or operation. The data to be displayed are contained in the computer memory. There are several methods used to encode data for use by the display. Let us consider the simple system where every point to de displayed is stored in the computer's memory. The computer sends the (x, y) coordinates in binary to the display where they are converted to analog voltages by digital-to-analog converters. The following example considers such a simple system.

Example 10.1

Let us consider a display which can have 80 characters per line and 80 lines per picture on the screen. There are a total of 6400 characters per picture. Assume that each character is formed by a 5×7 dot matrix.

$$\text{Total number of dots} = 80\frac{\text{ch.}}{\text{line}} \times 80\frac{\text{lines}}{\text{picture}} \times 35\frac{\text{dots}}{\text{ch.}}$$

$$= 224{,}000\frac{\text{dots}}{\text{picture}}$$

y Coordinates:

$$80\frac{\text{lines}}{\text{picture}} \times \left(7\frac{\text{dots}}{\text{line}} + 2\frac{\text{dots}}{\text{interch. gap}}\right) = 720\ y \text{ coordinates require 10 bits to encode}$$

x Coordinates:

$$80\frac{\text{ch.}}{\text{line}} \times \left(5\frac{\text{dots}}{\text{ch.}} + 2\frac{\text{dots}}{\text{interch. gap}}\right) = 560\ x \text{ coordinates require 10 bits to encode}$$

This assumes two dots between each character as an intercharacter gap (space).

$$\text{Total coordinates per picture} = 720 \times 560 = 403{,}200$$

Assume that 80% of the picture is blank. Most of the 35 dots for each character are not used, and many characters are blank characters.

$$403{,}200\frac{\text{dots}}{\text{picture}} \times 20\% = 80{,}640\frac{\text{dots}}{\text{picture}}$$

Assume a refresh rate of 30 pictures per second.

$$80{,}640\frac{\text{dots}}{\text{picture}} \times 30\frac{\text{pictures}}{\text{second}} = 2{,}419{,}200\frac{\text{dots}}{\text{second}}$$

This requires an average bandwidth of more than 48 megahertz between memory and display and is not an economically viable system. It requires about 2.5 million accesses per second to memory (20 bits per access). If the display is remotely located, the cost of the transmission facilities would be quite expensive.

Example 10.1 shows that refreshing the display from the computer memory is very expensive in both bandwidth and memory accesses. To overcome this, most displays have a local storage buffer memory used for refreshing. The communication between computer and display is reduced to only changes and updates.

Instead of transmitting the location of the dots to be displayed, greater communication efficiency can be obtained if the binary coded characters are transmitted. The symbols in the form of ANSCII code are stored in the buffer memory. The memory is organized so that each location contains the data to be written with a particular position on the CRT. When the information is fetched from the buffer memory its location provides the necessary information needed by the deflection system to locate the character on the CRT face, and by decoding the character the proper character generator can be used to write that character in script form. The computer main memory must now contain only the binary character codes for the information being displayed on the CRT display.

Example 10.2

From Example 10.1, consider a display with a local buffer memory.
Main memory storage requirements are 6400 characters.
Data rate: Assume that the entire picture is changed every 2 seconds.

$$6400\frac{\text{ch.}}{\text{picture}} \times 8\frac{\text{bits}}{\text{ch.}} \times \frac{1 \text{ picture}}{2 \text{ seconds}} = 25{,}600 \text{ bits/second}$$

Thus, the use of a local buffer memory for refresh and the transmission of characters instead of dot position drastically reduce the amount of information which must be transmitted between the computer and the display.

10.5.6 Magnetic Tape Unit

Magnetic tape is used as both an I/O medium and as a storage medium. However, magnetic tape is an external storage medium rather than an on-line secondary storage such as drums and disks. Because of the long access time and the need to manually load a reel of tape, it is being considered in this chapter. The cost of magnetic tape per bit stored is the cheapest storage media, and the data rate is also the highest I/O device. Hence, magnetic tape has excellent characteristics for several uses. There are four basic parts to a digital magnetic tape system: (1) the magnetic tape reel, (2) the tape transport mechanism, (3) the reading and writing electronic system, and (4) the switching and control unit.

10.5.6.1 Magnetic Tape

Magnetic tape is generally a flexible plastic tape with a thin coating of some magnetic material on one side. The tape can have a width of from $\frac{1}{4}$ inch to several inches. Most tape used is $\frac{1}{2}$ inch in width. The tape is placed on a reel and usually comes in $10\frac{1}{2}$-inch-diameter, 2400-foot lengths. The tape reels are mounted on the tape transport to be used. Because of the ease of handling and the large storage capability per reel, i.e., about 36 million characters, large volumes of business records can be maintained in a magnetic tape library.

Magnetic tape is recorded with one character written across the width of the tape and generally comes as either 7 channels, 6 data bits, and a parity bit, or 9 channels, 8 data bits, and parity. Densities are 200, 556, 800, or 1600 characters per inch of tape. The recording method is usually NRZI for the first three densities and phase modulation for the higher 1600 bits per inch. Figure 10.27 shows magnetic tape format for both the 7 channels and the 9 channels tapes.

Information is organized on a magnetic tape in the form of files. These files are composed of records. The records may be blocked so that one physical record on the tape is composed of several logical records. This is done to increase the capacity of storage on the tape. Between each physical record is a *record gap*. This gap is the space needed to start and stop the tape between records. Thus, if the records are small, most of the tape would be wasted gap. Thus, small records are generally blocked into larger records. Records can be of variable size, with the record gap and a longitudinal parity character written after each record. The longitudinal parity is an even parity

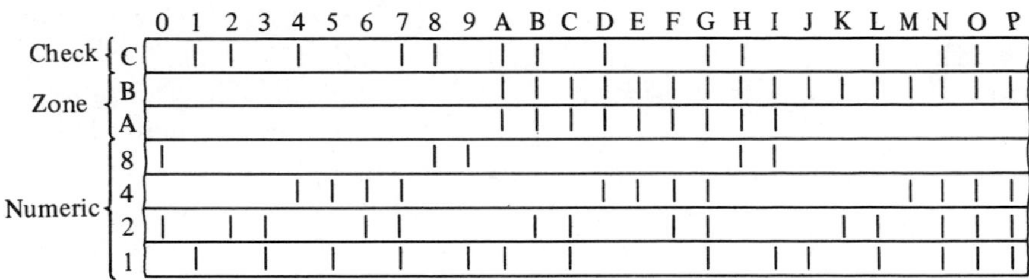

Seven track format (even parity)

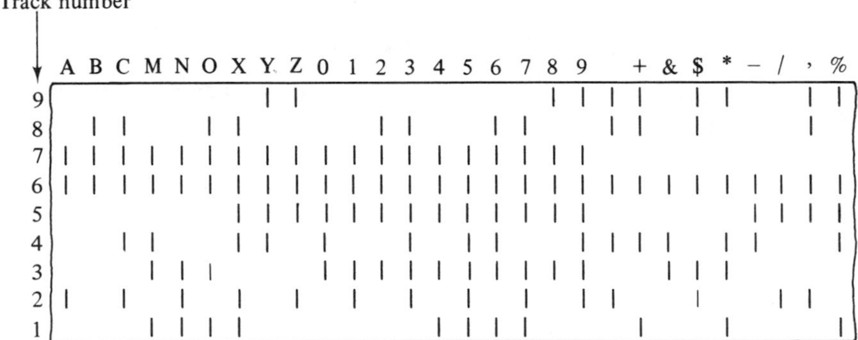

Nine track format (odd parity)

Figure 10.27 Magnetic tape formats.

kept for each channel of the record. It is used by the tape system for error checking. The interrecord gap is a function of the start-stop speed of the tape transport and is usually about $\frac{3}{4}$ inch. Between each file on a reel of tape is an end-of-file gap, which is much larger than the end-of-record gap. The end of a file is usually delineated by a special end-of-file character as well as the long gap.

10.5.6.2 Magnetic Tape Transports

The tape transports used in digital magnetic tape recording have two unique characteristics when compared to analog tape transports: 1) They start and stop very quickly, and (2) they have very high tape speeds. The faster the transport can start and stop the tape, the faster the tape can be brought up to speed, and the smaller the interrecord gap needs to be. A high tape speed provides a high data rate and therefore gets the needed data into memory with less delay.

Since the tape reels and drive have a high inertia, they cannot be started and stopped sufficiently fast. To provide the acceleration and deceleration, a method which isolates the tape reels, used to take up tape, and the tape drive has been developed. The tape movement is controlled by moving the tape against a capstan. The rotation of the capstan moves the tape forward or backward. The tape touches the capstan only on its plastic side. This prevents magnetic tape wear. The tape is looped from a reel into a vacuum column, past the capstan, into the second vacuum column, and onto the takeup reel. Figure 10.28 shows such a tape transport. By having several feet of tape in each of the vacuum columns the capstan can move tape for a short time without having to move the reels. This decouples the movement of the tape from the high moment of inertia of the reels. The reels have separate servo drive circuits and high-speed brakes. This permits the reels to eventually supply sufficient tape once the capstan starts to move tape. It also stops the release of tape from the reels. The speed of the reel motors and its direction are controlled by sensing lights and photocells located in each vacuum column. Thus, the reel systems attempt to maintain a sufficient tape loop in the columns with the capstan actually moving the tape past the read-write heads. The magnetic oxide side of the tape touches only the head.

Start and stop time for the movement of the tape in tape transports is on the order of 1 millisecond. This is the time to bring a standing tape up to

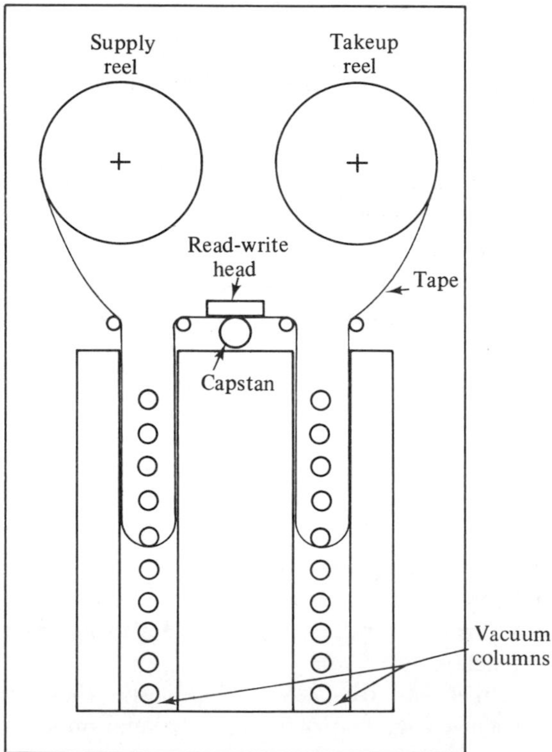

Figure 10.28 Vacuum column tape transport.

speed or the time to bring a moving tape to a complete standstill. The speeds at which the tape moves past the head vary with the tape transport. Some standard speeds are 37.5, 75, 112.5, 100, 150, and 200 inches per second. The tape can be rewound onto the supply reel at higher speeds than it can be read. The read-write head is moved away from the tape, and the servo system runs at full speed. Tape can be rewound at speeds up to 500 inches per second.

10.5.6.3 Read-Write Electronic System

This part of the magnetic tape system consists of the magnetic read-write heads and the electronic amplifiers needed to obtain a signal. In addition certain electronic signal processing must be performed to change the signals obtained from the tape into signals compatible with the channels and CPU.

Most tape systems use a two-gap read-write head. This dual type of recording head is useful since during writing the second head can be used to read the information just written by the first head and compared against this information in a buffer register. Thus, any error which occurs while writing will be detected. Current tape systems can read and write tape in either direction of tape movement. Figure 10.29 shows a dual gap read-write head. The heads are very narrow and are assembled into a complete magnetic head structure with one recording head per track or channel across the recording tape.

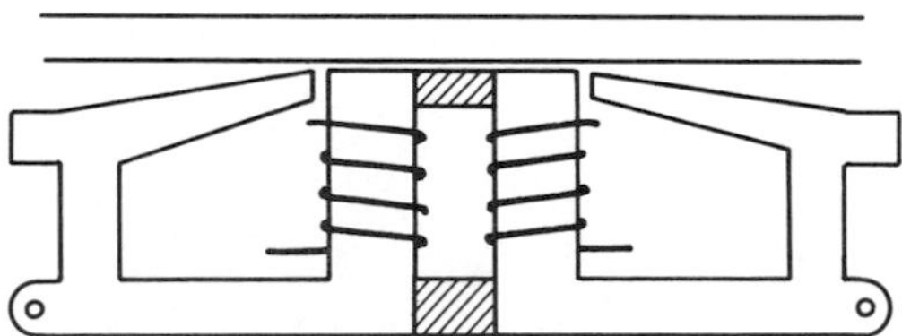

Figure 10.29 A dual gap read/write head.

The signal from the read head is amplified and processed into a form compatible with the rest of the computer logic. Let us assume that we read a tape which is recorded in NRZI mode (see Section 9.10.2). The signal obtained from a single head (track) may appear as shown in Fig. 10.30, where the upper signal was the recorded waveform and the lower signal is the signal obtained from the playback. Let us assume that the computer uses a positive signal to represent 1 and a negative signal to represent 0; we must convert the pulses of the read head to level signals. Let us assume that the tape is recorded with odd parity; then there is at least 1 bit recorded for each character on the tape. If we amplify the signals from the recording heads, rectify them, and put them through an OR gate, we can obtain a pulse out

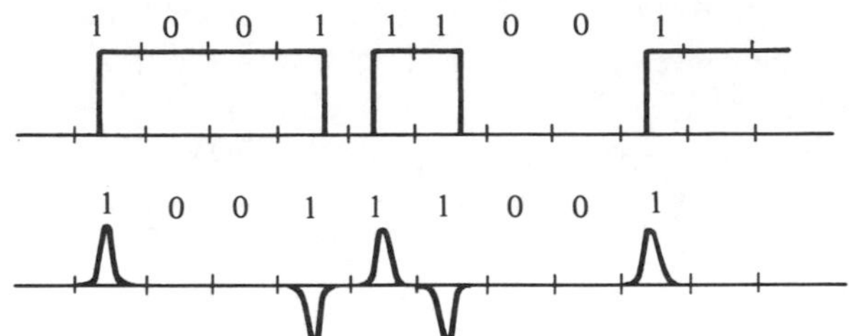

Figure 10.30 Signal from a NRZI tape.

of the gate for each character on the tape. This is a clock pulse, and the NRZI technique is self-clocking. The circuit in Fig. 10.31 shows how we can recover a computer-capatible waveform.

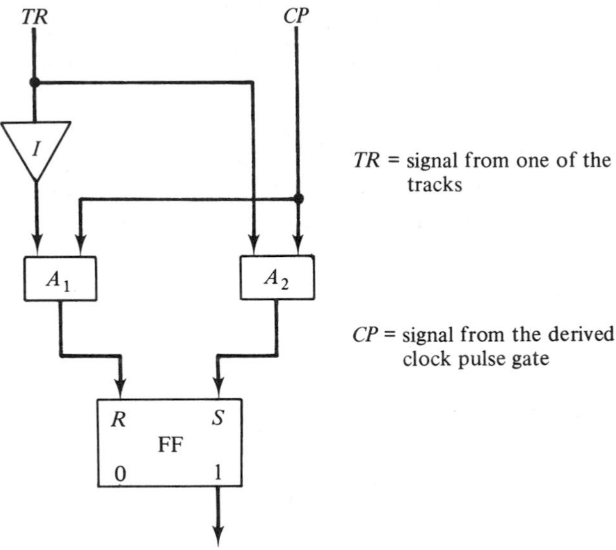

Figure 10.31 Circuit used to obtain computer compatible waveforms from NRZI waveform.

10.5.6.4 Switching and Control Unit

The magnetic tape transports are connected to the channels of the computer through control units. These units have much of the electronics needed to control the movement of the tape. The control unit detects the interrecord gap and the longitudinal parity character which is recorded after each record; it also detects the longer end-of-file gap and the end-of-file (EOF) character. Instructions such as advance or backspace N number of records are interpreted by the control unit, and commands are sent to the tape transport. Likewise movement of the tape forward or backward to the end of a file is performed in the same manner. The control unit checks the lateral parity of each character during a read operation; it also checks the longitudinal parity for each record and performs the comparison. If an error is detected, it informs the channel.

Near the beginning of a reel of magnetic tape is a strip of aluminum-reflective material placed on the tape. Another strip is placed near the end of the reel. Between these two aluminum strips is the usable tape. These markers, if detected by the transport, will stop the movement of the tape and thus prevent complete tape unwinding.

A typical tape system will have the following operations which it can perform under control of the computer via a channel:

1. Read forward: Read a record in the forward direction.
2. Read backward: Read a record in the backward direction.
3. Write forward: Write a record in the forward direction.
4. Advance N records: Move the tape forward N records; no data are transferred.
5. Backspace N records: Move the tape backward N records; no data are transferred.
6. Advance to EOF: Move the tape to the end of the current file; no data are transferred.
7. Backspace to EOF: Move the tape back to the end of the previous file; no data are transferred.
8. Rewind to the beginning of the usable tape: Tape is rewound to the load point (beginning of the tape).

Only the first three of these operations transfer data to or from the memory of the computer. The other five operations position the tape to a desired point. If the I/O command system records special recognition information at the beginning of each record before the data are recorded, then it is possible to call for a record by name. This information is usually called a *key*. With this type of operation the tape unit can search the tape for a given key and stop at the beginning of the requested record. This is also under control of the control unit.

Controls on the front panel of each tape transport permit manual control of the units. These include the operations of rewind, load, and unload. The latter operations can also be performed by the computer.

10.5.7 Key-Tape and Key-Disk Off-Line Data Preparation

One of the most expensive parts of operating a large computer installation is in the preparation of data to be used by the computer. Most data preparation requires the off-line punching of cards on a key-punch machine.

This preparation can be 25–50% of the total cost of operation. Once the cards are prepared and edited for mistakes, they must be read into the computer through a card reader. This is a slow input device when compared to magnetic tape. In addition, the storage of the cards is quite expensive.

An alternative to the use of cards as an input medium is to use magnetic tape. However, until recently, this still required the preparation of punched cards which were read off-line into a reader which was connected to a magnetic tape unit and could transfer the information to the tape. This permitted a higher data transfer rate from the tape into the computer and a lower storage cost since the cards could be abandoned and the tape stored.

The need for the punched cards as an intermediate storage medium is redundant, and it is now possible to key data directly onto magnetic tape. There are several companies which offer key-tape data entry units. These units have a keyboard; when a key is depressed by the operator, a character appears on a display and is also entered onto the magnetic tape. The magnetic tape unit is a special type called an incremental magnetic tape which permits one character at a time to be written or read instead of a full record. If a small memory is included in the system, a more advanced system which permits a full line of data to be keyed, stored, and corrected can be used. When the operator finishes editing the data, he pushes a record button which records the full line of characters onto the magnetic tape. Still more advanced data entry systems connect a number of keyboards to a small special-purpose computer equipped with a magnetic tape transport. This system permits extensive editing by the computer. It can compare inputs, check formats, sort and collate data, and perform many operations on the input data. Thus, the tape which it prepares for use by the large general-purpose machine has been "preprepared."

In addition to the use of magnetic tape as the input medium for the computer, one can use a magnetic disk pack. The magnetic disk permits even more editing capability since it is easier to "randomly" read and write on a disk. Records can be updated or changed in place on a disk. Once the data have been recorded onto the disk, they can be removed from the key-disk data entry system and used as input on the large computer.

10.5.8 Optical Character Reader and Magnetic Ink Character Reader

An ideal input device would be one which could read alphanumeric printed or hand-written characters. Such a device would therefore input information which is also compatible with humans and no intermediate input medium would be needed. Although no inexpensive device has been found which can recognize handwriting, a large amount of research is presently endeavoring to develop the techniques and equipment. At the present time optical character readers (OCR) which recognize certain characters (fonts) are available, as is the magnetic ink character reader (MICR), which recognizes specified characters.

The MICR uses an ink which has magnetic properties and a set of characters which have a special form, as shown in Fig. 10.32. This system was originated by the American Banking Association for use by banks. You have probably seen these characters on your bank checking account. The character set uses a 7×10 matrix with the magnetic ink forming a character. The MICR detects, by magnetic induction, which of the ink patterns has passed under the reader's head.

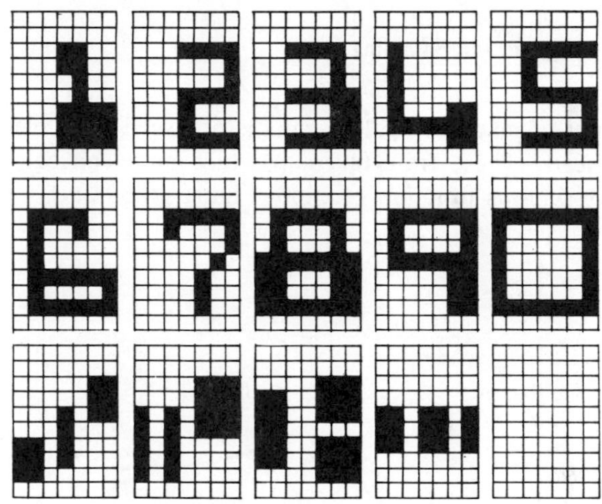

Figure 10.32 Magnetic ink character set.

The OCR usually recognizes one or a few type fonts which are printed on paper with regular ink. The printed characters are examined by passing the paper under a light and lens system which detects inked from noninked areas. A system of logic then determines which of the characters caused the particular patterns. The limitation of this system is the requirement that the documents being scanned must have been typed with the correct type font.

10.6 ANALOG-TO-DIGITAL AND DIGITAL-TO-ANALOG CONVERSION

Digital computers are often used in real-time control systems (see Section 1.3.3). In this type of application the basic data into the computer may not be in digital form but may be a continuous time or analog signal. Likewise these signals may be derived from the motion of mechanical parts of the system which the computer is controlling. An example is the positioning of the rods in a nuclear reactor which controls the amount of fission and thus the temperature of the reactor. The computer monitors both the position of the rods and the temperature of the reactor. Before the computer can use these analog data they must be converted into digital form. The device which performs this is called an *analog-to-digital converter* (ADC). If the computer must control a mechanical device, it will need to convert its digital signals into analog signals. This is performed by a *digital-to-analog converter* (DAC). An example of the use of the DAC is in the positioning circuits of the CRT display under digital computer control.

10.6.1 Analog Signals and Sampling

The output of many control systems is obtained from electrical devices called *transducers*. These transducers produce an electrical analog signal which is a continuous function of time. Figure 10.33 shows such a signal,

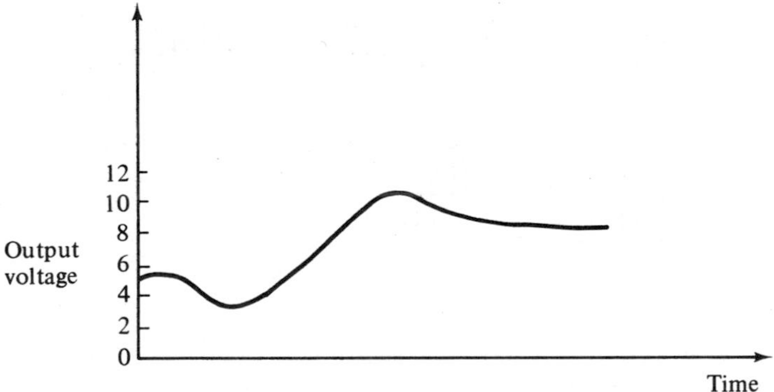

Figure 10.33 An analog signal.

which may vary between +3 and +12 volts. If we want to represent such a signal with a digital number, we shall need at least 4 bits which can represent numbers from 0 through 15. However, at some time the above analog signal is 6.342 volts. This will require many more than 4 bits to represent. Thus, we can never continually represent this analog signal even if we build a 6-bit analog-to-digital converter. The ADC can only approximate the analog signal. First, since it requires some time to make the conversion, the ADC only approximates the analog signal at distinct times or it *samples* the signal. Second, since the analog value can only be approximated, the ADC provides a *quantizing error* due to the finite bit length. The precision of the ADC is equal to the number of bits in the digital approximation, while the amount of error between the analog signal and the digital approximation is a measure of the accuracy of the ADC. Figure 10.34 shows a continuous analog signal and the sampled quantized digital approximation to the signal.

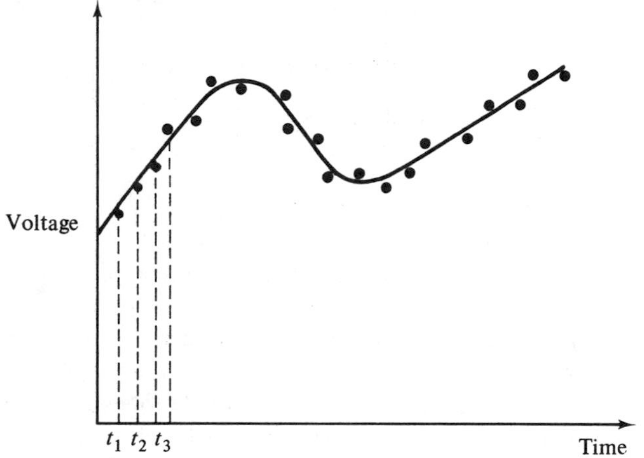

Figure 10.34 Sampled quantized signal.

10.6.2 Digital-to-Analog Converter (DAC)

The basis for converting a digital signal into an analog signal is a resistance ladder network. One such type of network is shown in Fig. 10.35 with four binary inputs labeled b_1, b_2, b_3, and b_4. Each input, b_i, is either at a positive potential $+V$ or at ground 0. The output voltage E_0 will be a dc voltage ranging between 0 and $+V$ and will be proportional to the value of the binary number represented by the inputs, where b_1 represents the most significant bit. Because of the binary relationships among the resistors of the

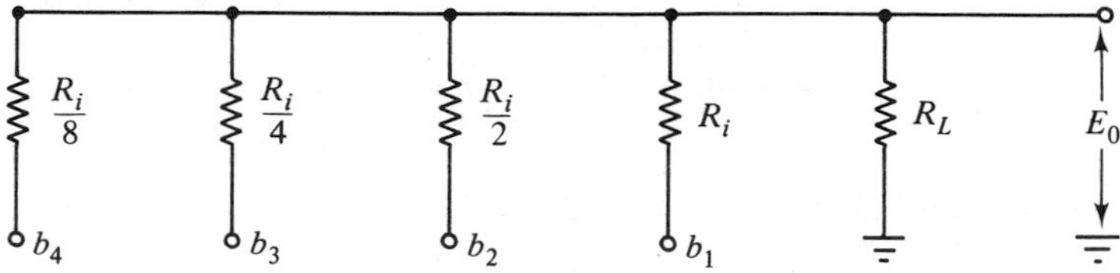

$R_L >> R_i$ $b_1, b_2, b_3, b_4 = +V$ or 0

Figure 10.35 Digital-to-analog resistance ladder.

ladder, each bit contributes its binary weight to the output voltage, E_0. The above ladder network has a shortcoming in that for a 12-bit DAC the spread of resistance values will become very large and the precision resistors will become expensive. The ladder network shown in Fig. 10.36 overcomes this difficulty.

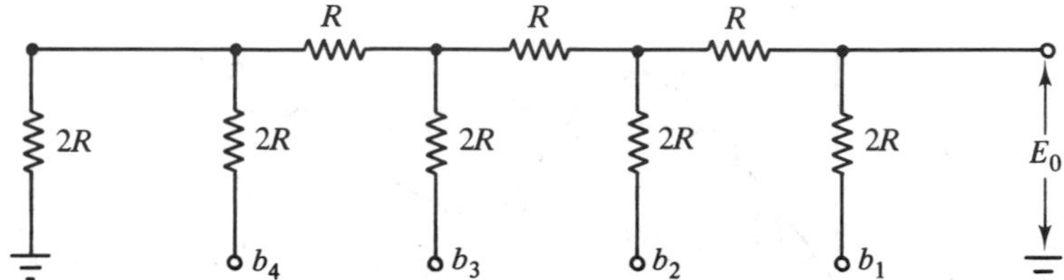

Figure 10.36 Another DAC ladder netword.

Usually the outputs of the flip-flops, which are the binary inputs to the DAC, are not precise enough to be used directly as the inputs. They control the inputs from a precise level amplifier which supplies a voltage obtained from a reference supply. Figure 10.37 shows a block diagram of the more precise DAC.

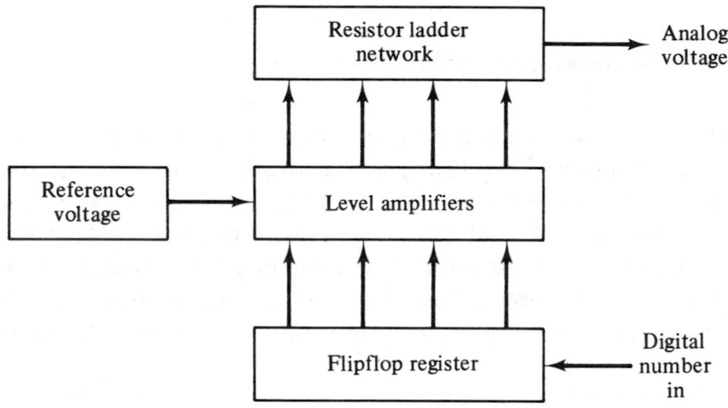

Figure 10.37 Block diagram of DAC.

10.6.3 Analog-to-Digital Converter (ADC)

There are two different types of analog-to-digital converters. One changes a mechanical motion into a digital electric signal, while the other changes an analog electrical signal derived from a transducer into a digital electrical signal.

10.6.3.1 Mechanical Shaft Encoder ADC

This type of analog-to-digital converter can convert a mechanical shaft position directly into a digital signal which represents the shaft angle of rotation. It is possible to first convert the shaft position to an analog voltage by the use of a transducer, such as a potentiometer, and then convert the voltage to a digital voltage. This method is not so accurate as the use of the direct mechanical shaft encoder. The shaft encoders usually use Gray code to encode the digital information. This is done to eliminate possible large errors due to possible ambiguity in the shaft angle. To understand why this type of code is used, first consider an NBCD shaft encoder as shown in Fig. 10.38. This encoder is connected to the shaft and is made of four concentric

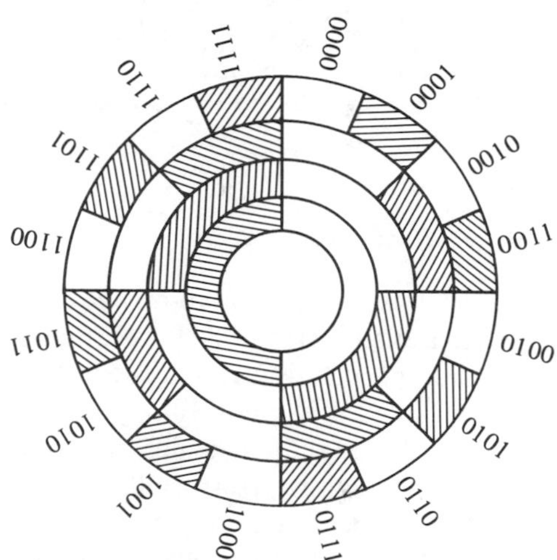

Figure 10.38 NBCD shaft encoder.

bands on which a set of four brushes slide. Each band is constructed of alternate sections of insulators (white area) and conductors (dark areas) which are coded to yield the NBCD code. A voltage is connected to each conductor, and as a brush contacts the conductor a binary 1 will occur at its output. When the brush is over the insulator a binary 0 will occur on the brush output. The four brushes can therefore present all possible binary combinations from 0000 to 1111 as the encoder rotates on the shaft. Only 16 different shaft angular positions are shown, but many more can be encoded if additional bands are added to the disk.

The problem with using an NBCD coded disk can be seen by considering what may happen if the shaft stops rotating such that the brushes are half on those sections which code 0111 and those which designate 1000. This ambiguity can occur because of the finite width of the brushes. Thus, they may land between well-defined binary coded sections. This ambiguity occurs whenever several bits are changing in the code. However, for the NBCD code this occurs quite often. For example, at the position indicated above, the output could be 1111 or 0000 or 0011 or almost any possiblity. To overcome this possible ambiguity, consider the disk which is coded in the Gray code and shown in Fig. 10.39.

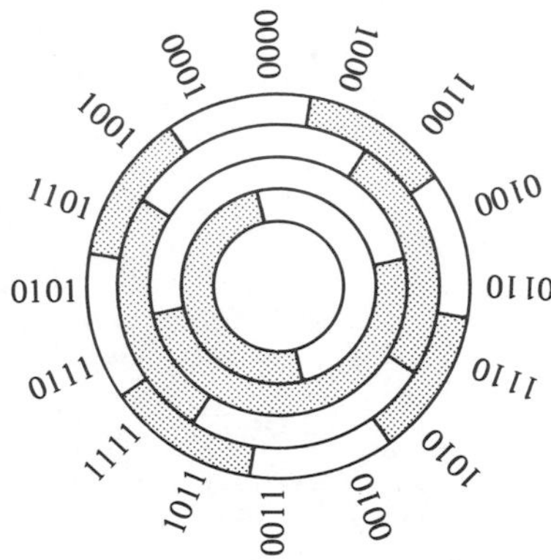

Figure 10.39 A gray code encoder.

10.6.3.2 Electronic ADC

The electronic analog-to-digital converter converts an electronic analog signal to a digital signal. Actually, the ADC converts a fixed voltage into a digital number. However, if this can be done at sufficiently high speed, the ADC can track the analog voltage by sampling it and therefore can produce a series of digital outputs which approximate the time-varying analog signal to within the quantizing error. Thus, the ADC must quantize time by sampling the voltage at discrete times. The number of times per second the ADC can convert is called the *conversion rate.*

There are many types of electronic ADCs which operate under different principles. One method converts a voltage to a frequency difference through the use of a voltage-controlled oscillator (VFO). The shift in frequency of this oscillator is proportional to the voltage at its input. The increment in frequency is measured on a frequency counter and displayed. The counter starts from 0, which represents 0 voltage on the input to the VFO. The larger the voltage on its input, the larger the frequency shift and the higher the count on the display. Of course, the display is calibrated in terms of volts and not frequency shift.

Another type of ADC uses a ramp comparison method. This method starts a voltage which varies linearly with time and also a frequency counter which counts a known precise frequency derived from a crystal. The linear ramp voltage is fed as one input into a voltage comparator. The other input to the comparator is the unknown voltage which is to be converted into a digital signal. When the two voltages compare, a signal is sent to the counter, which stops counting. The output of the counter which is on the display

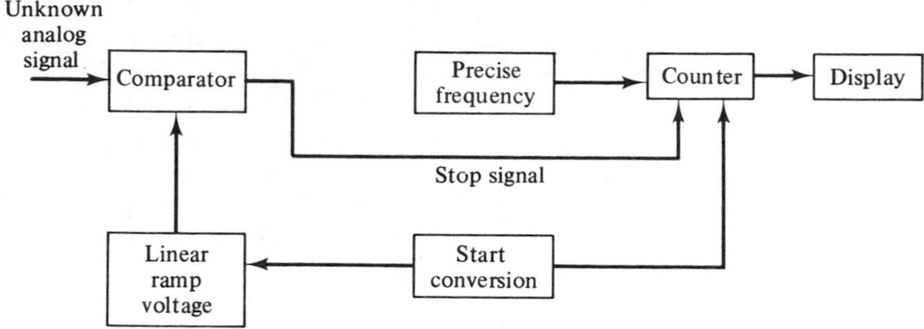

Figure 10.40 Ramp-comparison ADC.

shows the value of the analog voltage. Figure 10.40 is a block diagram of such a converter.

Instead of using a display for the output of the ADC in the above two examples, it is possible to use only the flip-flops which drive the display and thus produce a digital signal as output.

A class of ADCs are built around the use of a DAC and a comparator in a closed-loop feedback system. Their fundamental components and structure are shown in Fig. 10.41. The basis of operation is to compare the voltage output from the resistor ladder network with the input analog signal. If they do not compare, modify the contents of the flip-flop register and thereby change the output from the DAC. The control circuits keep this up until the analog comparator determines that the analog signal and the signal from the DAC are equal to within the quantizing error. The output of the flip-flop register is the digital number. There are several types of such converters. One uses an up-counter for the flip-flop registers. This method starts a conversion with 0 in the counter and increments up binarily until the DAC output equals the analog output. It is called a *counter converter*. Conversion

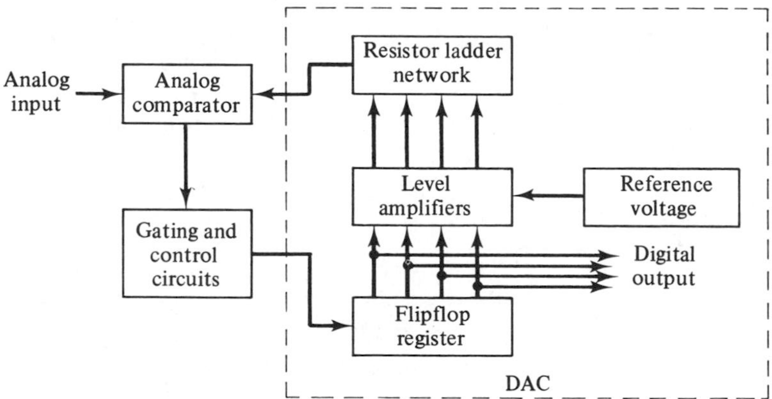

Figure 10.41 Feedback ADC using a DAC.

time for this type of converter has a maximum of 2^N counts for an N-bit number and an average time of $2^N/2$. For N large, this converter may take too much time and have a slow conversion rate.

Another type of feedback converter uses an up-down counter for the flip-flop register of Fig. 10.41. Once the proper binary number has been determined, the converter can continuously follow the analog voltage. For this reason it is called the *continuous converter*. This provides a very fast conversion rate. If, however, the analog voltage changes faster than the counter can change, the readout will be delayed. This method is therefore sensitive to the slope of the analog voltage.

A third method which requires one step per bit of the binary number to convert an analog voltage into a binary number is called the *successive approximation converter*. It is a little more complicated than the other two types of feedback ADCs previously discussed, and a block diagram is shown in Fig. 10.42. It operates by repeatedly dividing the voltage range into halves and determining in which half the analog voltage lies. The system first tries the binary equivalent of 50% of full-scale voltage. Next it tries either quarter scale or three-quarters scale depending on whether the first approximation was too large or too small, respectively. After N approximations an N-bit number has been determined. The successive approximation converter requires a control register to gate the generated signals to the first bit of the output flip-flop register, then the second bit, and so on.

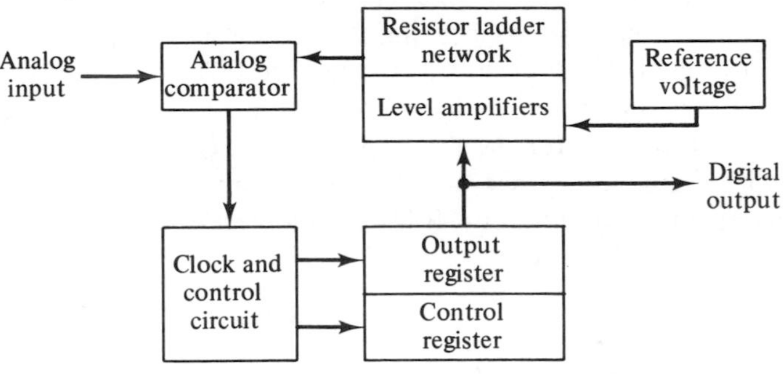

Figure 10.42 Block diagram of a successive approximation ADC.

REFERENCES

BARNES, F. N., "Phase Encoding vs. NRZI for Magnetic Tape Recording," *Computer Design*, 12, No. 2 (Feb. 1973), pp. 80–82.

BARTEE, T. C., *Digital Computer Fundamentals*, 3rd ed. New York: McGraw-Hill, 1972.

BRICK, D. B., and E. N. CHASE, "Interactive CRT Terminals," *Modern Data* (May–June 1970).

BYCER, B. B., *Digital Magnetic Tape Recording: Principles and Computer Applications.* New York: Hayden, 1965.

DAVIS, S., "Printer Selection Factors," *Computer Design*, 11, No. 12 (Dec. 1972), pp. 45–57.

GSCHWIND, H. W., *Design of Digital Computers.* New York: Springer-Verlag New York, Inc., 1967.

HOESCHELE, D. F., Jr., *Analog to Digital/Digital to Analog Conversion Techniques.* New York: Wiley, 1968.

LONDON, K. R., *Techniques for Direct Access.* Philadelphia: Auerback Publishers, Inc., 1973.

SOUCEK, B., *Minicomputers in Data Processing and Simulation.* New York: Wiley, 1972.

WARE, W. H., *Digital Computer Technology and Design*, Vol. II. New York: Wiley, 1963.

PROBLEMS

10.1. Make a list of all the different types of I/O devices you can find by looking through the literature. Determine the data rate for each device.

10.2. Draw a configuration diagram for the main computer center at your school or at your place of employment.

10.3. We wish to copy blocks of data from a magnetic tape T on channel 1 onto a magnetic disk D on channel 2 via main memory. The IOCS can supply buffers for this operation B_i, $i = 1, 2, \ldots$. The I/O instructions for the machine are (1) READ TAPE INTO B_i, (2) WRITE DISK FROM B_i, and, (3) DEVICE X WAIT, where each READ or WRITE transfers an entire block of data to or from a buffer. Each instruction must wait until the previous instruction completes all transfers on the channel before another operation can begin on the channel. The blocks of data are each of different size. The WAIT instruction holds a transfer from beginning until the previous transfer has completed.

(a) Write a program which uses only two buffers to transmit five blocks of data.

(b) Write a program which uses three buffers for the transfers.

10.4. A single-address computer can execute instructions at an average rate of 2.5×10^5 per second. It has a channel with a data rate of 1.0×10^5 bytes per second. A memory word is 4 bytes long, and the memory can complete a word access in 4 microseconds. The memory is interleaved two ways. Assume that each instruction requires two accesses to memory: one to fetch the instruction, and one to obtain the operand. A magnetic tape connected to the channel records 1600 bytes per inch and moves at a speed of 200 inches per second. If the memory is performing at 100% utilization, what is the utilization of the CPU when the magnetic tape is transferring data to a memory buffer? Assume double buffering and a cycle steal channel.

10.5. A single-address computer can execute instructions at an average rate of 2.0×10^5 per second. It has a memory with a cycle time of 2 microseconds. Each instruction requires two accesses to memory. The same magnetic tape as in Problem 10.4 is connected to the machine. If the utilization of the CPU is 100%, what is the utilization of the memory?

10.6. A single-address CPU can execute $.24 \times 10^6$ instructions per second. The operations of the I/O channel and CPU are fully overlapped. A double-buffer I/O system is used. If the block size of the buffer is 1600 words and the programming environment causes 3.0 executions for each input word in the buffer, how many blocks must be transferred each second to keep the CPU fully utilized?

10.7. If the I/O device of Problem 10.6 is a magnetic tape with a read-write capability of 1600 bytes per inch and 200 inches per second, can it keep the CPU in the problem fully utilized? Assume 4 bytes per word and neglect inter-record gap.

10.8. Consider the CPU and I/O device of Problems 10.6 and 10.7. Assume that each instruction requires two accesses to memory: one for fetching, and one for the operand. What is the speed of memory needed to support such an activity?

10.9. For a simple channel, develop the logic needed to tell the CPU when the I/O device has reached its speed and is ready to receive data.

10.10. Using the delay execution technique, what information is received by either the computer or the I/O device if the computer program responds too late to an external request from the device (a) during input and (b) during output?

10.11. A CPU can execute instructions at the average rate of one per microsecond. The I/O structure has a simple channel with an *IOIR* used as a buffer. The data from the I/O device arrive 1 byte at a time with a data rate of 250,000 bytes per second. If there are 4 bytes per memory word, how many instructions can be executed between successive times the CPU must store the contents of the *IOIR*?

10.12. For a computer which uses the interrupt technique for I/O transfer, how does one handle two asynchronous I/O devices which are simultaneously transferring data over the channel?

10.13. What instructions must be added to the computer's repertoire if a selective interrupt disable using a mask register is added to the simple interrupt technique?

10.14. If the interrupt system is disabled, how will interrupts which occur from another device be retained so that they can be answered once the current interrupt service is completed?

10.15. Design a priority interrupt DMA channel with a selective disable mask register.

10.16. Using the code of Fig. 10.15, show how a paper tape will appear if you punch your full name.

10.17. Why does each character of a paper-tape character set have at least one punch?

10.18. What are some of the problems which the designer must solve in order to read a punched card all at once? Is this a practical way?

10.19. Design the logic which will take a seven-level paper-tape character into a register one character at a time but will store four characters in the register before sending the buffered word to memory in parallel.

10.20. Using A/D and D/A components, design a vector generator for use with a CRT display to draw lines on the display.

10.21. How many characters can be stored on a 2400-foot roll of magnetic tape if the density is 6400 bits per inch, the blocking factor is 4000 characters per block, and the interrecord gap is $\frac{1}{2}$ inch?

10.22. High-speed magnetic tape units use phase-encoded data. Design the logic needed to change phase-encoded information read from the tape to a computer-compatible waveform; i.e., positive level equals logical 1, and ground equals logical 0.

10.23. Is there an advantage of having a fixed block size for all data recorded on the magnetic tapes of a computer center? Justify your answer. What should be the size of the block?

10.24. Design the electronics and logic needed so that a magnetic tape unit can read forward and backward. Assume phase-encoded information.

10.25. What are the advantages, if any, of using magnetic ink characters for reading data when compared to optical scan techniques?

10.26. If odd parity is used on the ninth level of a magnetic tape, it is possible to detect a single error in that character but not correct the error. Consider grouping the characters into a fixed-block length of say seven characters. At the end of this block, the control unit adds a block of correction data which shows the number of 1s in each level of the seven-character block. Using these data a single error in the block can be detected and corrected. What should the correction block look like? What is the minimum size of this block?

10.27. Why can't the Gray code be used when designing the resistance ladder network used in a DAC?

10.28. What is the quantization error in a 10-bit ADC?

10.29. What is the needed precision of the reference voltage for a 10-bit ADC?

10.30. Compare the minimum, maximum, and average conversion rates for
(a) The counter converter.
(b) The continuous converter.
(c) The successive approximation converter.

10.31. Derive the relationship between the conversion rate of a continuous A/D converter and the slope of the analog signal being converted.

11

The Control Unit

In the previous chapters we have considered the function of the arithmetic unit, the memory unit, and the I/0 unit. In this chapter the control unit is presented. From our previous discussion, it is clear that the control unit has two major purposes. The first purpose is to control the sequencing of information-processing tasks performed by the machine. The second purpose is that of guiding and supervising each unit in the machine to make sure that each unit carries out every operation assigned to it at the proper time. In a stored program computer, the function of a control unit is to fetch the instructions of a program from memory (according to the sequence desired by the specific program to be executed), interpret every instruction, and sequence the necessary signals to other units of the machine which perform the processing of the instructions.

Any data needed to execute a given instruction may be stored in memory or in some registers of the machine or may be in an input device. When data are needed to execute an instruction, the control unit must know where the data are stored and what operation is to be performed on them. The location of the data is sometimes specified explicitly by the instruction, while at other times it is determined by the addressing method designed in the machine and indicated by some tags in the instruction word.

The details of the control unit circuitry vary widely from one machine to another. This is partially due to the existence of different machine architectures and the variations in the set of instructions to be executed. Since the control unit's major function is to control the execution of programs, its design depends largely on the instruction set that makes up the programs processed by the machine. Therefore, an essential part of designing a control unit is a knowledge of the instruction set and how each instruction is coded in an instruction word.

The discussions in this chapter deal with the basic principles of the design of control units, in a general sense. The application of these general concepts is given in Chapter 12, where a specific machine is designed.

11.1 DISTRIBUTED AND CENTRALIZED CONTROL

Although the function of the control of any computer is controlling the internal operations of the machine, the location of the control circuits within the machine varies from one machine to another. Most early machines were built with distributed control units. That is, the control circuits for one unit of the machine (say, the arithmetic unit) are located in one part of the machine, while the control circuits for another unit (say, memory) are in another. The distribution might even go further into the control circuit associated with any specific machine instruction. For example, the control circuit for the arithmetic operation ADD is located in one part, while the control circuit for a MULTIPLY instruction is in another. The arithmetic units discussed in Chapter 8, with all the associated control signals, can be viewed as an example of a distributed control, with the assumption that the control signals are locally generated within the arithmetic unit. In this case, the control unit only initializes the operation, and the local control takes over until the operation is performed. Then the control is transferred back to the control unit.

Later machines were built with centralized control. All the control signals needed for the various operations of the machine are generated through central control circuits. Thus, the designer has to decide on the timing and control signals needed to control and synchronize all the operations in the various units. Therefore, the circuits which generate these signals are integrated together and located in the same place (control unit). This technique minimizes the necessary hardware for the control and is most suitable for small machines.

For large machines, the design of a centralized control becomes a difficult task. In this case, it is more convenient and desirable to break down the control into small elements, each associated with a specific operation or unit of the computer. The control circuits would be located at various parts of the machine. Since large machines tend to overlap the performance of various operations, the decentralized control approach is most appropriate. Most of the computers today have multiprogramming capabilities. This is the case where two or more programs are residing in core simultaneously. The programs may be run in parallel with the execution of one program being overlapped with the I/O activities of the other programs. Thus, the major part of the control circuitry is controlling the execution of one program, while little attention is given the the other programs which are in the I/O phase. The control of the I/O is supplied mainly by the channels. In this case, a high degree of parallel operation can be achieved if one of the programs requires considerable execution while the others require considerable I/O. A typical example of this is to run a business-oriented program and a scientific program together. The separation of the channel control from the main control is just one example of the use of distributed control in recent computers. Further distribution of the control is more desirable in computers with parallel processing capabilities.

In this text, we are primarily concerned with the design of small computers, and therefore our discussion deals with centralized control. Only the I/O channel and I/O devices are considered to have their own local control.

11.2 WORD FORMAT AND INSTRUCTION REPERTOIRE

11.2.1 Instruction and Data-Word Format

In Chapter 7 we defined a computer *word* as a group of binary bits which are used as the basic unit of information in a machine. A word could represent a piece of data, referred to as a data word, or an instruction; then we speak of an instruction word. Again the meaning of an instruction or a data word depends on several factors. For example, the data word 101101101011 could represent the decimal number 2923 in a binary machine, the octal number 5553 in a binary coded-octal machine, and B6B in a hexadecimal machine. As another example, the data word 01101001 represents the decimal number 105 in a pure binary machine, while it represents the decimal number 69 in an NBCD machine. In the above examples we assumed the word represents a data word; otherwise, if it is an instruction word, it must be interpreted differently.

An instruction word is usually divided into two major fields; one represents a code for the operation (op-code) to be performed, while the other field contains information needed by the control unit to execute the specific operation. Usually, the second field is divided into several sections. The number of divisions depends on the type of computer. For example, the computer might be single, double, or triple address; consequently, the second field is divided into one, two, or three parts, respectively. Sometimes part of the instruction word is used in connection with the type of addressing method used in the machine, as in indexing, base addressing, etc. These fields are referred to as tag fields. Several instruction formats have been discussed in Chapter 7, and they are not repeated here; instead a general form of the instruction format is shown in Fig. 11.1 for single-address instructions.

Op-code	Tags	Address

Figure 11.1 General form of instruction format.

Let us consider a single-address machine. Any program to be run by the machine has to be stored first in the memory. Both instructions and data of the program are stored in memory with each instruction or data word occupying a specific location. Usually the instructions are stored in consecutive locations in memory. Once the first instruction is brought out of memory and executed, the control unit continues to bring other instructions out from the consecutive locations until a fault or branch instruction is encountered.

The word length of the machine determines the size of the addressable memory. For example, in a single-address machine with direct addressing only, the size of addressable memory is determined explicitly by the address field of the instruction word. For an address field of 12 bits, the memory size is $2^{12} = 4096$ words. The size of addressable memory can be increased, for the same address field size, through the use of a base register, or index register. For instance, if a 16-bit base register is used with base addressing, then we can access an instruction in any location of a memory with size $2^{16} = 65{,}536$ words. A similar statement applies to an index register and location of data. Whenever an instruction is to be fetched from memory, the actual location of the instruction is determined by adding the contents of the program counter and the contents of the base register. The base register could be initially loaded by any number between 0 and 65,536. Similar arrangements apply to the use of an index register to locate the data in memory. If the instruction format refers to addressing by indexing, then the actual address of the operands is obtained by adding the contents of the direct address field to the contents of the index register. The result of this addition is placed in the memory address register as the effective address of the information to be referenced by the instruction. This method of increasing the memory size, without having to increase the word length, requires additional instructions to service the base register (or index register). Also some additional hardware to implement the derivation of the effective address is necessary.

The fact that the memory is shared between the instructions and data belonging to the program to be run offers a great flexibility. For instance, if the program is a business-oriented one, then the instructions might occupy a relatively small amount of memory, leaving a larger space for the data. For a scientific application, the allocation of memory might be just the opposite, with the instructions occupying more storage than the data. Most of the present computers have a high-speed main memory and a relatively lower-speed auxiliary or secondary storage. In such a case, usually the instructions and part of the data are stored in main memory, while the remaining part of the data is stored in the secondary storage. After the data in main memory have been processed, more data are brought into main memory (called swapped into main memory), and this process continues until all the data are processed.

The operation code field of the instruction specifies the operation to be performed. The number of bits in the op-code field determines the maximum number of different instructions which can be performed by the machine. The set of instructions performed by a machine is referred to as the instruction repertoire of this machine. For instance, if there are 6 bits in the op-code field, then $2^6 = 64$ different code words can be encoded and the machine can implement up to 64 different instructions. The size of the op-code field also depends on the machine's internal number system. If the machine uses a BCD code, then four binary positions in the op-code field are needed to encode 10 different operations and each operation is identified by a decimal number represented in binary coded decimal form. If the machine uses alphanumeric code internally, with 6 binary bits per alphanumeric character, then the op-code field size is a multiple of 6 bits. For instance, the op-code might consist of two alphanumeric characters, in which case it is expressed by 12 bits in the machine.

If the machine uses the pure binary number system internally, it becomes tedious to refer to the op-code by binary numbers. It is more convenient to refer to the op-code and addresses in octal or hexadecimal. This is done for the reason of compactness, especially when it is easy to convert an octal or hexadecimal number into binary and vice versa. These numbers have been used in Chapter 7 to refer to the op-code and the memory locations in the programming examples.

11.2.2 Instructions Classification

The design of a control unit depends on the type and format of the instructions to be executed by the machine. Most of our discussion about control units applies directly to a control unit which is designed to interpret and control the execution of single-address *memory reference instructions*. These instructions involve operations that require information which is stored (or is to be stored) in one memory location. If the single-address restriction is removed, we still talk about memory reference instructions which involve one or more memory locations. Other classes of instructions which are used in different machines are the *register reference instructions* and the *input/output instructions*. Instructions in these two classes are being widely used in recent machines, and therefore they are described below.

11.2.2.1 Register Reference Instructions

Computers sometimes contain special registers, in the arithmetic and control unit, which are used to store information during program execution. It is desirable to process the information contained in these registers. The instructions which are used to carry out the processing of information stored in the machine's registers are referred to as register reference instructions. The operations to be performed by these instructions are

1. Load the contents of a register with a specific value.
2. Perform a logical or arithmetic operation on the contents of one or two registers, respectively.
3. Shift the contents of one register.
4. Information transfer from one register to another.
5. Modify the contents of a register by adding to it or subtracting from it a constant number.
6. Test the contents of a register and take decisions based on the result of the test. For instance, the decision might be that the sequence of instructions to be executed is to be altered if the test condition is satisfied.

The format of register reference instructions can take different forms depending on the task to be performed by a specific instruction. In some cases the instructions and the operations to be performed are mapped one to one; that is, for every operation to be performed there corresponds one instruction. In other cases a number of operations can be performed by a single instruction. Figure 11.2 shows two general types of register reference instructions.

Figure 11.2(a) shows the instruction format when the information to be processed is stored in registers. For instance, the operation might be to compare the contents of two registers to find out if they are equal. The decision field might refer to the location of the next instruction if the condition is satisfied; otherwise the instruction sequence is not changed. Another example would be to perform arithmetic or logical operation on the contents of registers i and j and store the result in a third register indicated by the decision field.

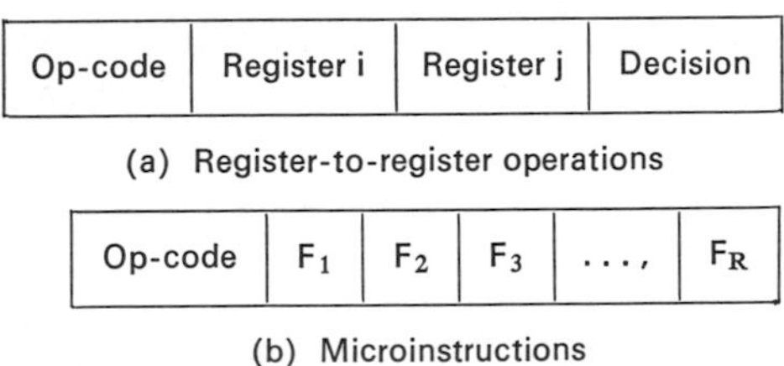

Figure 11.2 Two general types of register reference instructions.

Figure 11.2(b) shows a register reference instruction format with many different fields. This type of instruction format is used when several register operations are performed by one instruction. For example, let us assume that we wish to implement a large number of simple operations. If each operation is to be performed by its own instruction, the number of different instructions in the machine might be too large, and elaborate encoding and decoding becomes necessary. Instead, one can group a selection of simple operations into a class and refer to them by a single op-code. Then divide the rest of the instruction word into different fields as shown in Fig. 11.2(b). For instance, let us assume that fields $F_1, F_2, \ldots, F_R$ are 1 bit each and correspond to a particular simple operation. If the operations dealing with, say, the accumulator are combined into one instruction, then $F_1, F_2, F_3, \ldots, F_R$ might correspond to "clear the accumulator," "complement the accumulator," "increment the accumulator by 1," etc. A 1 in any particular field would indicate that the associated operation is to be performed; otherwise the content of a field is 0. As an example, the 2's complement of the contents of the accumulator could be performed by placing a 1 in the complement (1's complement) field and a 1 in the increment field. When the execution of the various simple operations has to be sequential, as in the 2's complement case, then the fields could be assigned such that when the corresponding operations are executed in order they yield the required sequencing. Instructions of this type are called microprogrammed instructions, and they are often used in microprogrammed control units (see Section 11.6).

11.2.2.2 Input/Output Instructions

The I/O instructions are used to transfer information between the computer (main frame) and the I/O (peripheral) devices. The general format of an I/O instruction is shown in Fig. 11.3. Since there are more than one I/O device associated with a computer, each device can be assigned a number. The device number field in an I/O instruction specifies the number of the particular device to be used. The contents of the device control field vary widely from one device to another. In general, it supplies control information which is necessary to execute the information transfer to or from the device. For some devices the contents of the device control field are very simple (e.g., the beginning and end of a block of information), and for other devices the information in the control field is used by the device controller to generate a complex program in order to control information transfer.

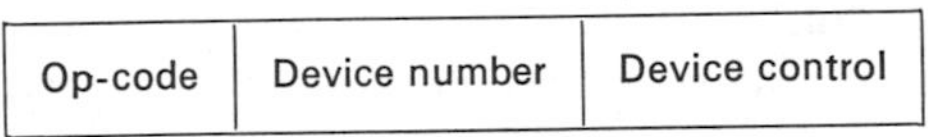

Figure 11.3 I/O Instruction format.

11.2.3 Instructions Repertoire

The instruction repertoire is the set of instructions the machine is capable of performing. The number of different instructions implemented in a general-purpose computer varies from one machine to another. Given a desired capability of a machine, there is a minimum set of instructions which, if implemented, a programmer can use to solve problems within the machine's capabilities. On the other hand, using a minimum set of instruction makes the programming task very difficult. It is always desired to have enough instructions in the instruction repertoire to make programming less tedious. The use of more instructions than the minimum also provides higher operational speed.

The set of instructions implemented by the machine is determined not only by the machine's desired capabilities but also by the instruction format and the machine's architecture. In Chapter 7, we have seen different machine architectures and instruction formats with matching sets of instructions. The instruction set for a single-address machine is different from that for a two-, three-, or four-address machine. The capabilities added to the machine by using special control registers such as an index or base register require extra instructions to service those registers. The instruction set for a machine also varies with the addressing methods used.

Once the instruction set of a machine is determined, then the machine must be built with the capability of performing each instruction in the set. To solve a given problem on the machine, the programmer must devise an algorithm and be able to break his program into a sequence of instructions which are within the machine's instruction set. If the machine implements only a minimum set or *essential set* of instructions, then the programmer has to derive his algorithm for solving the problem such that it could be coded into a sequence of the essential instruction set. For instance, while ADD and SHIFT instructions are essential, MULTIPLY can be considered a nonessential instruction, since it could be performed by a series of ADD/SHIFT instructions. Similarly, floating-point arithmetic operations can be performed by using a number of fixed-point operations. Thus, floating-point arithmetic operations are not absolutely necessary.

To design a machine which would be easy to use, the system designer must provide a relatively redundant set of instructions. In other words, the instruction set must be larger than the essential set. The hardware implementation of an instruction execution means that the control unit generates the internal sequencing for all the logic and transfer operations corresponding to the instruction. If all the possible instructions which a computer could execute were implemented in hardware, then the design of computers would always result in large machines. In that case not only is the hardware design of the machine complicated, but also the cost is very high.

The problem at hand is how to design a machine with a given capability and two opposing constraints. First, it must provide sufficient instructions for convenience in its programming, and, second, the hardware cost must be kept down. The solution to this problem is the concept of *software implementation* of some nonessential instructions. This means that the implementation of a nonessential, more complex instruction is obtained by programming from a set of simple instructions. For example, a small computer may provide hardware sequences for the most elementary operations of addition, subtraction, shift, etc. The more complex operations, including multiplication and division, have to be implemented by software (programming). Another computer might still implement floating-point arithmetic operations by software, while fixed-point multiplication and division are performed through hardware implementation. When we speak of software implementation, we mean that some relatively complex instructions are provided in the instruction repertoire of the machine although they are not implemented in the hardware. Therefore, the *user* is free to include these instructions in his program, and the system programmer must provide a sequence of hardware-implemented instructions for each instruction which is not implemented directly by the hardware.

The concept of software implementation of complex instructions does satisfy the design constraints of programming convenience and low cost. The price paid is speed. A software-implemented instruction usually takes a longer time to execute than if it is implemented directly by hardware (internal sequencing). Therefore, if one assumes that the user's convenience is essential, the trade-off between hardware and software implementation of the complex operations must be based on a compromise between cost and speed. Usually the software takes over where the hardware leaves off. Figure 11.4 shows a map of operations of increasing complexity and the hardware-software trade-off. The cutoff point between hardware and software is determined by the hardware-implemented instructions in the machine's instruction repertoire. A few examples of cutoff points are shown in the map of Fig. 11.4.

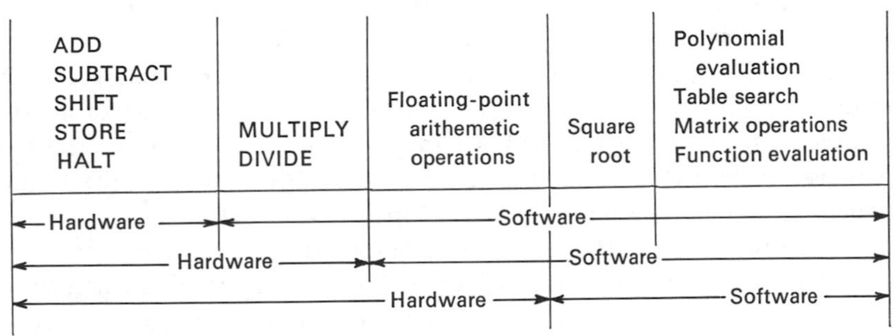

Figure 11.4 Trade-off between hardware and software.

The recent advance in integrated circuit technology has led to a significant drop in the hardware cost and size. Consequently the trade-off between hardware and software implementation is moving in favor of more and more hardware implementation.

The concept of microprogramming is a very powerful tool in the sequencing of complex operations. In microprogramming, a set of essential microoperations are implemented by the machine. These microoperations are chosen such that for every instruction in the machine's instruction repertoire a microprogram is generated by the control unit to execute the instruction. Although this is similar in principle to the software implementation of a complex instruction, the microprograms are already wired in and ready for execution. The microprogramming concept adds a new dimension to computer architecture and is presented at the end of this chapter.

11.3 INSTRUCTION SEQUENCING

The function of the control unit of a stored program computer can be classified into tasks. The first is to control the execution of each of the basic operations required by a specific instruction. In performing this task the control unit generates a sequence of control signals corresponding to each instruction. These control signals govern the operations performed by the functional subunits throughout the computer. The second task is to control the order in which the instructions are executed, as predetermined by the program (or intermediate results). This second task is the major objective of a stored program computer.

Consider a given program stored in the machine's memory, with its instructions in consecutive locations X through X + N. Let us further assume that the machine's instructions are single-address instructions. To execute the program, the control unit reads out the first instruction from location X, and generates the control signal sequence necessary for its execution. Then the control unit reads the next instruction from location X + 1 and issues the necessary control signals for its execution in a similar manner. This process continues until a halt or branch (or jump) instruction is encountered. In the case of a jump instruction, the control unit realizes that the order of instruction execution has to be changed. It simply reads the location of the next instruction to be executed (as indicated by the address field of the branch instruction) and starts a new sequence of instructions.

Therefore, the two tasks of the control unit are to fetch the instructions from memory according to the order set by the program, and it must generate the required sequence of control signals to execute a given instruction. These two tasks are identified as the *fetch phase* and the *execute phase*. Thus, the sequence of operation for most instructions, in a single-address computer, is an alternation of a fetch phase (also referred to as a fetch cycle) followed by an execute phase (or execute cycle).

In most machines, much of the machine's instruction set involves operations that require information which is stored in one or more memory locations. These types of instructions are referred to as memory reference instructions. Other types of instructions which do not involve memory reference (such as register reference instructions and I/O control instructions) are common in large machines. The fetch phase of any instruction involves a memory read operation, while the execution phase of a memory reference instruction requires at least one read or one write operation. The time required to read a word from memory or write a word into memory is referred to as the memory cycle time. The memory cycle time is fixed, and only one command to read or write is satisfied during a cycle time.

Before we describe the fetch phase and the execute phase, let us review the operations to be performed during the read and write cycles. If we are reading from memory, the address of the word to be read is transferred to the *MAR* and the memory is instructed to read. The memory selection circuit selects the location, and the information is delivered into the *MIR*. While the information is being processed by the control unit (during the remaining part of the memory cycle), the memory rewrites the information back from where it was read. If the cycle is a write cycle, the address at which the information is to be written as well as the information itself are transferred into the *MAR* and *MIR*, respectively, and the memory is given a write command. If the memory is required to erase before writing new information, it erases first and then writes the new information in the second half of its cycle. We shall assume that the read and write commands to the memory are implemented by setting either a *read* flip-flop *R* or a *write* flip-flop *W*. The *R* flip-flop or the *W* flip-flop is reset by the memory as soon as the command is satisfied. When both the *R* and *W* flip-flops are reset, it indicates that the memory is available.

11.3.1 The Fetch Phase

In Chapter 7, we introduced two control registers as the minimum set of registers for a control unit. These are the *CAR* (current address register) and the *CIR* (current instruction register). The *CAR* holds the address of the instruction currently being processed. Assume that the instructions of a program are stored in consecutive locations in memory. If the *CAR* has a count-up capability, then the contents of the *CAR*, after it is incremented by 1, would point to the address of the next instruction to be processed. The *CIR* holds the instruction word itself, which is now ready for interpretation and execution.

When the control unit enters the fetch phase of operation, a memory read cycle is required to read out the instruction. The address of the instruction to be fetched is contained in the *CAR*. The fetch phase of the control unit can be described by the following set of transfers:

$$T_0 | \quad CAR \longrightarrow MAR,\ 1 \longrightarrow R$$
$$T_1 | \quad M\langle MAR\rangle \longrightarrow MIR$$

$$T_2| \quad MIR \longrightarrow CIR\text{; decode; restore memory}$$

$$T_3| \quad CAR + 1 \longrightarrow CAR;$$

First a memory read cycle is initiated, and the address of the instruction to be fetched is sent into the memory address register (*MAR*). The word stored in the location indicated by *MAR* is delivered into the memory information register, *MIR*. The contents of the *MIR* are then transferred into the current instruction register of the control unit, and the control unit then increments the *CAR* to point to the next instruction location. The above actions represent the fetch phase for most of the instructions of small machines. The exceptions are the branch instructions. After the instruction is transferred into the *CIR*, the op-code field is decoded, and consequently the control unit knows what type of operation is to be executed. If the decoder output refers to a branch-type instruction, the *CAR* is not incremented by 1, but instead the address field of the *CIR* is transferred into the *CAR* register.

Once the instruction is transferred into the *CIR* and the *CAR* is already modified to contain the address of the next instruction, the fetch phase is completed, and the control unit enters the execute phase of operation. At this point, the control unit knows what to do, and it enters into the execute phase to do the required operations.

11.3.2 Indirect Addressing and the Defer Phase

The instruction sequencing described above assumes direct addressing, where the address field of an instruction refers to the effective address of the operand. When indirect addressing is used, a one-cell field (tag) in the instruction format is set to a 1 to so indicate. Whenever indirect addressing is detected, the execution phase is delayed, and the computer enters into the defer phase where the effective address is fetched from memory. Therefore, assuming one level of indirect addressing, the defer phase is a memory read cycle to fetch the effective address of the operand. The defer phase can be described by the following set of transfers:

$$T_0| \quad 1 \longrightarrow R$$

$$T_1| \quad Ad(CIR) \longrightarrow MAR$$

$$T_2| \quad M\langle MAR\rangle \longrightarrow MIR$$

$$T_3| \quad MIR \longrightarrow Ad(CIR)$$

By the end of the defer phase, the effective address of the operand is available in the address portion of the *CIR*, and the computer enters into the execute phase.

11.3.3 The Execute Phase

At the start of the execute phase, the output of the decoder determines the operation to be performed. If the op-code calls for a memory reference instruction, then the address field of the *CIR* (*Ad*(*CIR*)) contains the address of the operand. In general, each given op-code requires a different sequence of control signals to be generated by the control unit. Examples of the control sequences needed to execute various operations are discussed below.

11.3.3.1 Memory Reference Instructions

For this type of instruction whose execution requires a memory read or write operation, the execute phase sequence must control the required read or write memory cycle. As an example, the transfers for the execution of an ADD instruction are shown below:

$$
\begin{array}{l|l}
T_0 & 1 \longrightarrow R;\ Ad(CIR) \longrightarrow MAR \\
T_1 & M(MAR) \longrightarrow MIR \\
T_2 & MIR \longrightarrow X;\ \text{restore memory} \\
T_3 & X + A \longrightarrow A
\end{array}
$$

For more complex instructions such as multiply, the clock phase T_3 might not be enough to perform the operation. The execution phase will continue through multiple memory cycles.

11.3.3.2 Branch Instructions

Every computer has a number of conditional and unconditional branch instructions. A branch instruction might require (if the condition is satisfied) a change in the sequence of instruction execution. Therefore, it is necessary to transfer the contents of the address field into the *CAR*. As an example the transfers for BAN (branch if Acc. negative) can be written as

$$\bar{A}_0(CAR + 1) + A_0(Ad(CIR)) \longrightarrow CAR$$

Since there is no operand to be fetched from memory to execute a branch instruction, the execution can be overlapped with the fetch phase. The combined fetch/execute for the BAN instruction is shown below:

$$
\begin{array}{l|l}
T_0 & CAR \longrightarrow MAR;\ 1 \longrightarrow R \\
T_1 & M(MAR) \longrightarrow MIR \\
T_2 & MIR \longrightarrow CIR;\ \text{restore memory; decode} \\
T_3 & \bar{A}_0(CAR + 1) + A_0(ad(CIR)) \longrightarrow CAR;
\end{array}
$$

11.3.3.3 Register Reference Instructions

Similar to the branch-type instructions, the register reference instructions do not require any memory accesses to obtain the operands. Once the instruction is fetched, the operands are available in the computer's registers, and the operation can be performed immediately without accessing the memory. If the instruction is simple, it can be fetched and executed in one memory cycle. As an example of overlapping the fetch and execution of a register reference instruction, the transfers for increment the *ACC.* are shown below:

$$\begin{array}{ll} T_0 | & CAR \longrightarrow MAR;\ 1 \longrightarrow R \\ T_1 | & M(MAR) \longrightarrow MIR \\ T_2 | & MIR \longrightarrow CIR;\ \text{restore memory; decode} \\ T_3 | & (A + 1) \longrightarrow A \end{array}$$

Although all the register reference instructions do not require fetching of any operand from memory, it is not always possible to perform the operation in a single clock phase. For example, a shift instruction might require a longer time to execute, and therefore its execution might last for multiple memory cycles depending on the number of shifts to be performed. In such a case, the execute phase takes several memory cycles.

Figures 11.5 and 11.6 show the control registers of a simple machine with the transfers associated with the fetch and execute phases, respectively, of a memory reference instruction.

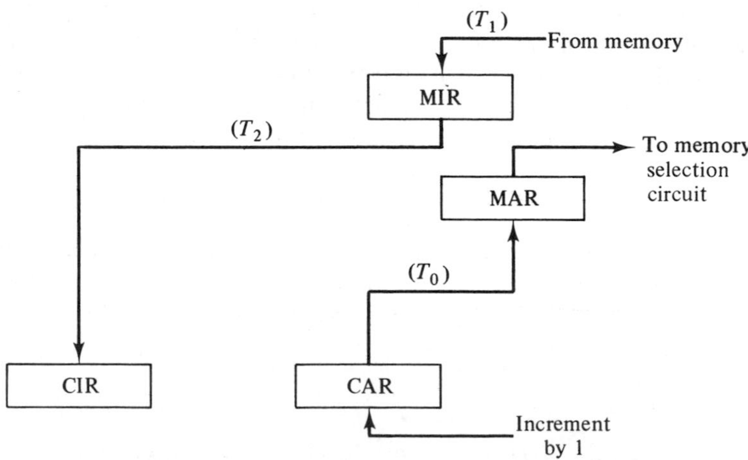

Figure 11.5 Control registers in the fetch phase.

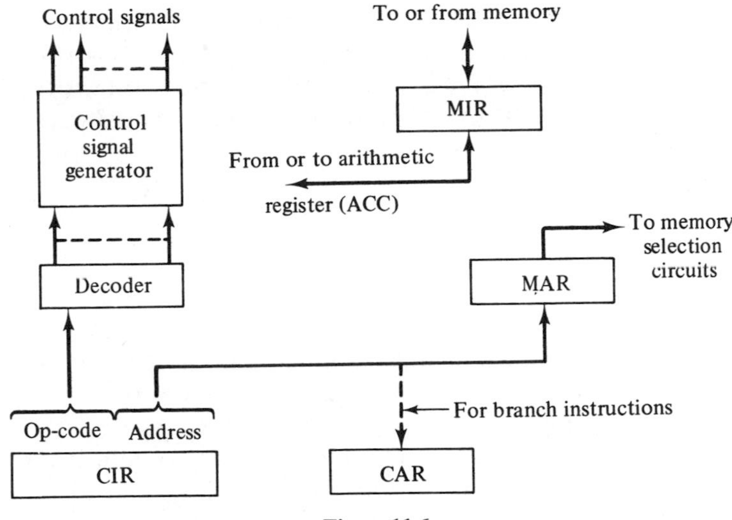

Figure 11.6

T_1, T_2, and T_3 shown in Fig. 11.5 correspond to the time signals which control the necessary transfers of the fetch phase. Figure 11.6 shows the registers in the execute phase of a memory reference instruction. Part of the execution phase is associated with either a read or a write memory cycle depending on the specific instruction. During this period, the address of the operand is transferred from the address part of *CIR* into the *MAR* to access the memory. While the operand is brought out of memory, the op-code is already decoded, and the control unit is ready to perform the specific operation indicated by the instruction. The transfer path from the *CIR* into the *CAR* is used only if a branch-type instruction is being executed. In such a case, instead of having the *CAR* pointing toward the next instruction as usual, it points toward a new sequence of the program which begins with the address indicated by the address part of the branch instruction.

11.3.3.4 I/O Interrupt Handling

Many computers have an interrupt interface with some I/O devices. The interrupt originates from the I/O device which is requesting service. For small computers, the interrupt signals are usually interfaced directly to the control unit, while large computers handle the interrupt interface through their DMA channels. In either case, an interrupt flip-flop in the control unit is set by the interrupt. After the interrupt is serviced, the control unit resets the interrupt flip-flop. Since an interrupt over the I/O interface could be received at any time, the central unit must check periodically to see if such an interrupt has occurred. Usually the control unit checks for the occurrence of an interrupt every time it finishes the execute phase of the currently processed instruction. If the interrupt flip-flop is set, the computer temporarily halts its normal course and handles the interrupt; otherwise the next instruction in the program is fetched. The I/O interrupt structure and the software support to handle it was shown in Section 10.3.4.

11.3.4 Control Flow

In a previous section, we stated that in order to execute a program, the control unit goes from a fetch phase to an execute phase, back to the fetch phase, and so on until a halt instruction is encountered. For machines with indirect addressing and/or interrupt capabilities, the control flow must be modified accordingly. Therefore, before the control unit enters into the fetch phase, it must check for an interrupt. Similarly, before it enters the execute phase of a given instruction, it must check for indirect addressing. If the indirect addressing flag is set, then the control unit goes through the defer phase first and then to the execute phase. Figure 11.7 shows a flow chart for the overall control flow.

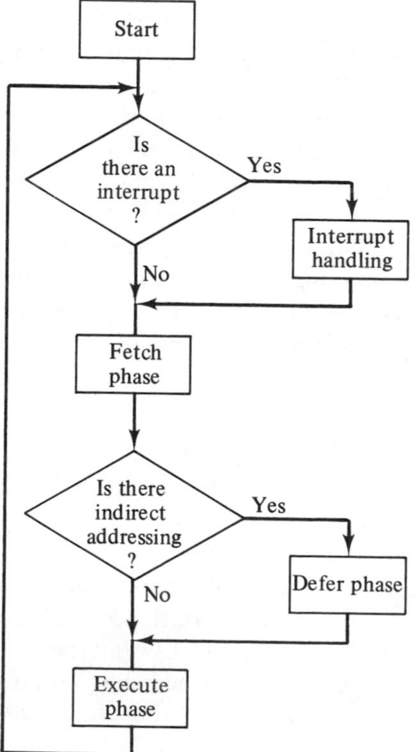

Figure 11.7 Overall control flow.

11.4 INDEX REGISTERS, BASE REGISTERS, AND ADDRESS ADDER

The use of index registers facilitates the writing of iterative programs as described in Chapter 7. It greatly reduces the number of instructions required in an iterative program. This is achieved by the automatic modification of the referenced addresses without altering the instructions themselves.

When index registers are used, an index field must be provided in the instruction word. The computer hardware might provide one or more index registers. The index field in the instruction word indicates if the index register is to be used with the given instruction and, if so, which index register is referenced. Therefore, an index selection circuit and an adder must also be provided. The contents of the index register to be selected must be loaded, before it is used, with a desired index value. During the fetch phase, the contents of the selected index register is added to the contents of the direct address field to obtain an effective address. It is the effective address which is transferred to the *MAR*. The address addition is performed by the address adder.

If there is only one index register, then a 1-bit index field is sufficient. A 0 in the index field indicates that the index register is not used, while a 1 indicates the use of the index register. If three index registers are available, then 2 bits are necessary for the index field. The four different code combinations in the index field are assigned to the four possibilities: no index, index register 1, index register 2, or index register 3. The selection of the index registers is achieved by an index selection matrix. Figure 11.8 shows the arrangement of a control unit with indexing capabilities.

To achieve the objective of using index registers, some additional instructions to service the index registers are needed. These instructions provide the capabilities to load the index register by some number, increment or decrement the index, and test the contents of the index register. The exact number and specific functions of these index instructions vary from one machine to another, but the above three functions are basic to any machine with indexing capabilities.

Some additional registers in the control unit are the base registers. The concept of base addressing has been presented in Section 7.2.2.5. When base addressing is used, the address adder is utilized to produce the effective address. A machine might have more than one base register, for instance, one base register for instructions and another for data. Also, if multiprogramming is used, each program might be assigned a base register. When base addressing is used, the instruction word might provide a field to specify the base register used. If one base register is used with all instructions, then the base register might be implied and no base register field is necessary. In this case, the

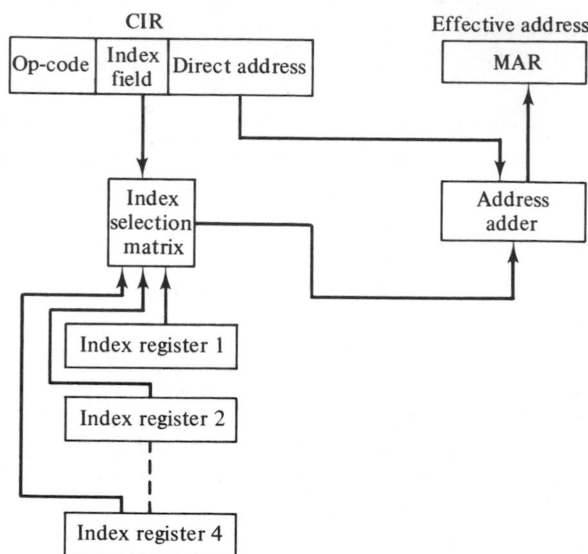

Figure 11.8 Control unit with indexing capabilities.

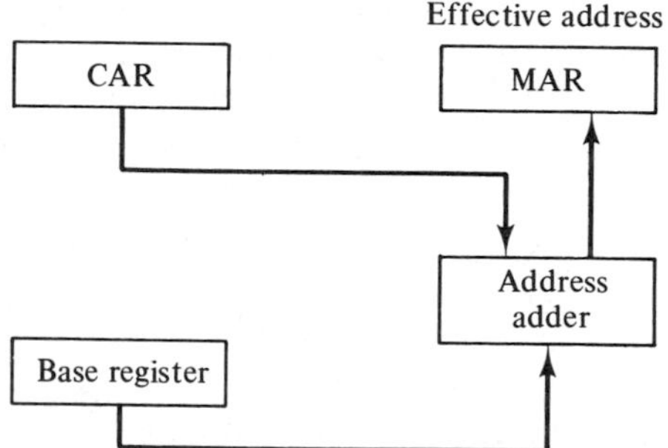

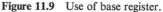

Figure 11.9 Use of base register.

physical address of any instruction is obtained by adding the contents of the base register to the contents of the *CAR*. Figure 11.9 shows such a design architecture.

If both indexing and base addressing are used simultaneously, then obtaining the effective address of an operand could require the addition of three quantities: the direct address, the index register, and the base register. This could be achieved by two addition cycles of the address adder. An alternative scheme is to employ two address adders, as shown in Fig. 11.10. The addition of either the index register or the base register or both is indicated by the corresponding fields in the instruction format. When more than one index or one base register are available, selection matrices are used to decode the corresponding fields and select the specified registers.

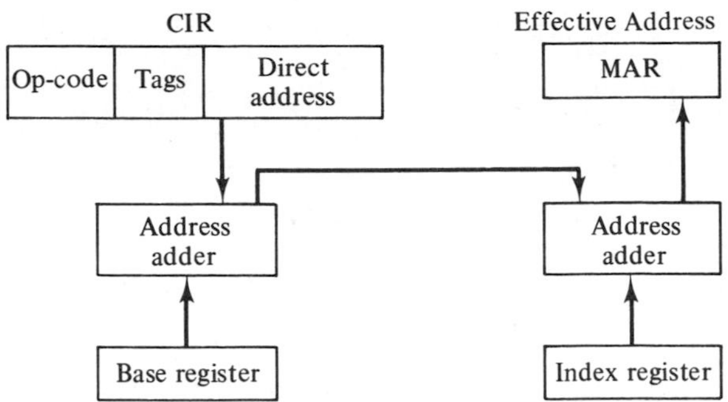

Figure 11.10 Two address adders to obtain the effective address.

11.5 SEQUENCING OF OPERATIONS

We have seen how the control unit controls the automatic sequencing of instruction execution. For each instruction the control unit goes through a fetch phase followed by an execute phase. Each instruction requires a specific sequence of control signals for its execution. Therefore, the control unit must generate a sequence of control signals to continuously fetch and execute the instructions of a given program in the machine's memory.

11.5.1 The Sequencer

The control circuits which generate a sequence of control signals for the execution of each instruction are sometimes referred to as the *sequencer*. Although the function of a sequencer is well defined, the details of its design vary from one machine to another. In principle, a sequencer generates a specific output sequence corresponding to any given instruction op-code as input. In general, the sequencer output might also depend on the status of various units of the computer. This second type of input (status) is needed because the execution of some conditional instructions might depend on the status of one or more functional subunits of the machine. For instance, a branch on an accumulator negative instruction may require for its execution a test of the sign bit of the accumulator (A). If A is negative, a branch into a specified location for a new sequence of instructions is performed. In case A is nonnegative, the control unit fetches the next sequential instruction. A block diagram illustrating the function of a sequencer is shown in Fig. 11.11. The op-code of an instruction is decoded, and the corresponding input to the decoder is energized. If the instruction is a conditional instruction, the corresponding status input is also energized. Once the sequencer receives a

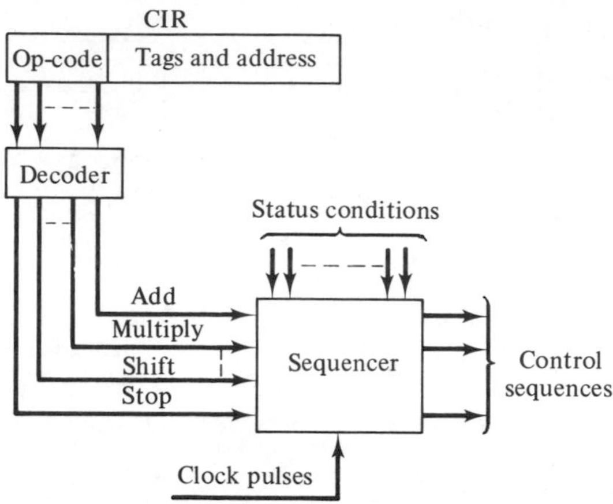

Figure 11.11 Functional diagram of a sequencer.

specific input, it generates the necessary control sequence, which is sent to the functional units of the machine in order to execute the instruction. This control sequence is then followed by the control sequence to fetch the next instruction. The number of control lines and their timing are by no means fixed for all machines but depend on the organization and details of the design of the computer's functional units.

11.5.2 Control Sequences for Simple Operations

To illustrate the basic concepts in the implementation of a sequencer, we shall first consider the sequences of control signals needed for the execution of a few simple instructions individually. A simple sequencer would be the result of combining the individual sequences together.

Each instruction cycle is composed of a fetch phase and an execute phase. For a given machine, the control sequence for the fetch phase is common for all instructions and does not depend on the type of instruction. A typical control sequence for the fetch phase was previously discussed. Figure 11.12

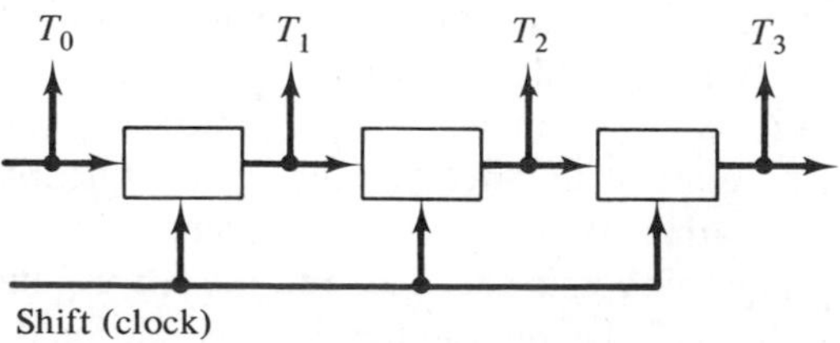

Figure 11.12 A typical fetch sequence.

shows the timing signals of such a sequence generated by the use of a ring counter.

As an example, let us consider the control sequence for the execution of a single instruction such as ADD. Figure 11.13 shows a ring counter which generates the necessary sequence. The last control signal in the sequence represents the add command to a parallel binary arithmetic unit. In this case, the time between successive clock pulses should be long enough for the parallel addition of the contents of the two registers A and X to be completed. The control signals before the add command transfer the operands into the A and X registers. Here we are assuming a single-address add instruction, which means add the contents of A to the contents of the memory location in the address field and put the results into A.

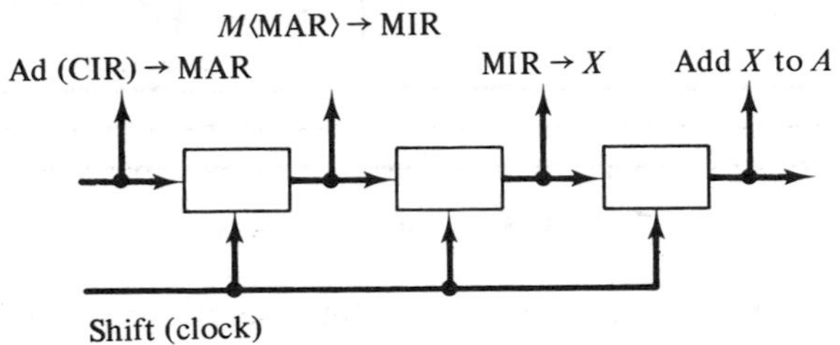

Figure 11.13 Control sequence to execute an ADD instruction.

The control sequence for a BRANCH instruction is another sample of a control sequence which is different from that of a normal instruction. Let us consider a conditional branch instruction such as branch on accumulator negative. This instruction is to be interpreted as follows: Check the sign bit (A_0) of the accumulator, and if it is negative, take the next instruction from the location specified by the address field of the instruction; otherwise the next sequential instruction is to be executed. Figure 11.14 shows a possible implementation for a branch on negative control sequence, which includes both fetch and execute phases.

The concept of generating control sequences can be expanded to provide for different sequences for the fetching and execution of all instructions. Figure 11.15 shows the concept of a simple sequencer.

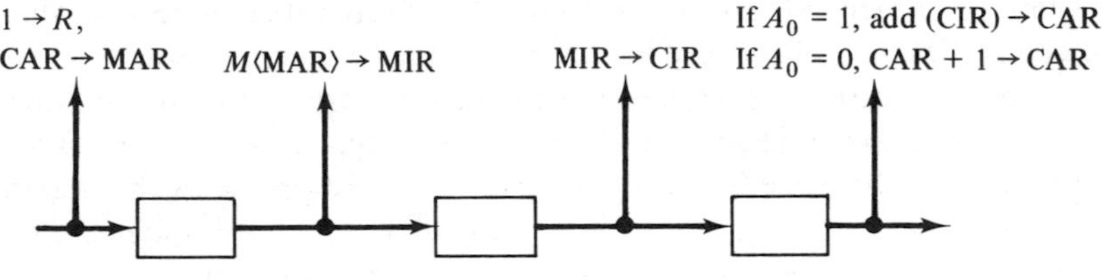

Figure 11.14 Control sequence for BRN.

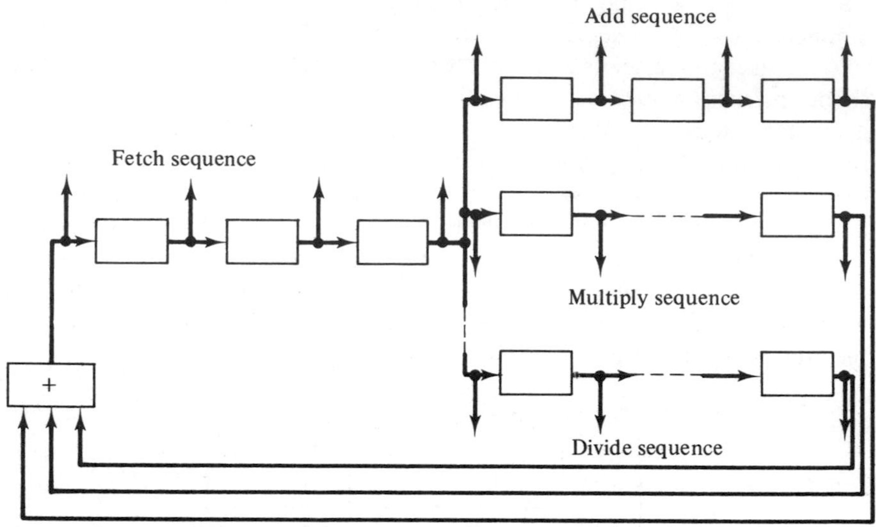

Figure 11.15 A simple sequencer.

The first part of Fig. 11.15 shows the fetch sequence which is common to all instructions and implemented by one ring counter. The second part contains a specific sequence corresponding to every instruction. It contains as many ring counters as there are different instructions in the machine's instruction set. The cycle length of each counter depends on the number of individual timing signals needed to execute the corresponding instruction.

11.5.3 Variations in the Sequencer Design

The simple sequencer described above applies to synchronous machines as long as the flip-flops of the ring counters are synchronized with the machine's clock (clocked flip-flops). If the flip-flops of the counters are replaced by delay elements with variable delays, then the resulting sequencer is an asynchronous one.

The synchronous sequencer in Fig. 11.15 requires a large number of flip-flops to implement. This is due to the use of separate counters to generate the control sequence for every instruction in the instructions set. A relatively slower but less expensive implementation can be achieved by using one fixed binary counter and a decoder at its output. This type of sequencing is shown in Fig. 11.16. The cycle-length of the counter corresponds to the maximum length of the control sequences. The disadvantage of this implementation is the slow speed, since every instruction requires the full count of the counter, regardless of its actual timing requirement. The long instructions might make

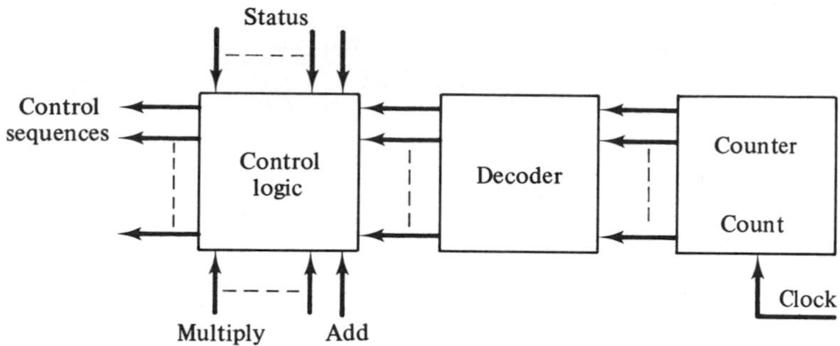

Figure 11.16 A sequencer using a fixed counter.

use of all the counts, while the short ones utilize a few counts only and the machine idles until the end of the counter cycle.

A compromise implementation is possible by using a selectable-modulo counter, where the length of the counter cycle is determined by the instruction to be executed. That is, the counter is reset as soon as the execution of an instruction is completed. Another compromise implementation is to use a fixed counter with a cycle length determined by the fetch phase. The cycle length is equal to the number of sequential operations required by the fetch phase. The execution phase is then controlled by the same counter output. For those instructions which require a longer control sequence, multiple cycles of the counter are utilized. For some register reference instructions and branch instructions, the execution might require only one transfer, and therefore their execution could be overlapped with the fetch phase. This implementation is very common in synchronous machines and is used in the computer designed in Chapter 12.

11.6 BUSING AND TRANSFER PATHS

In Section 6.2.1 we considered only the gating between a single pair of registers. The control unit is responsible for generating the control signals which provide for the gating of any one of several registers into any one of several other registers. One way to provide for the various transfer paths between the registers is a direct extension of the method of Fig. 6.7. For example, suppose that we have the two registers, *MIR* and *MAR*, and that each one requires a transfer path to either one of the two registers *CIR* and *CAR*. The direct extension of the method in Fig. 6.7 is shown in Fig. 11.17. Each transfer path requires a set of AND (or logically equivalent) gates to combine the register outputs with the control signal. An OR gate is required at the input of every cell of the receiving register. The number of inputs of

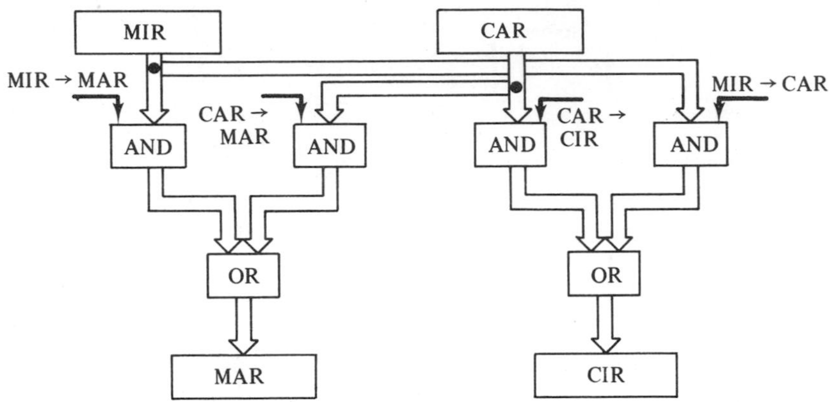

Figure 11.17 Transfer paths between registers.

each OR gate is equal to the number of various paths into the register it is feeding.

When the number of transfer paths increases, this method of direct transfer becomes very expensive. A hardware saving can be achieved by using an interconnection bus. Figure 11.18 shows the bus method applied to the four registers of the above example.

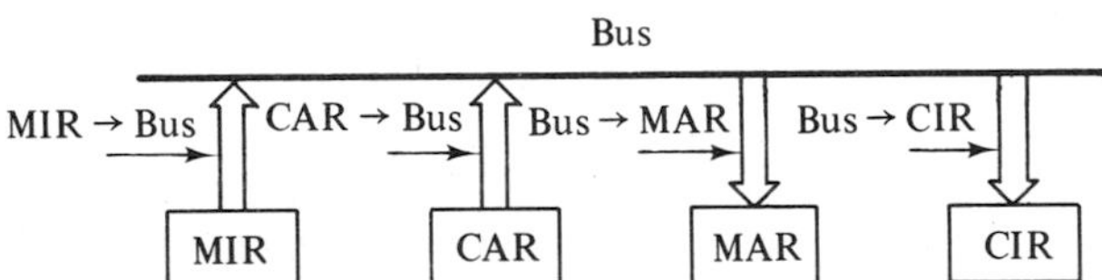

Figure 11.18 Interconnection bus.

In most cases, the gating of register outputs onto the bus is limited to one register at a time. It is acceptable to gate the data on a bus into any number of registers simultaneously. Unless transfers are allowed to overlap, the use of a bus will increase the time required for a register-to-register transfer from one clock pulse to two.

The single-bus interconnection can be extended to allow transfers between any number of registers, as shown in Fig. 11.19. The restriction here is that

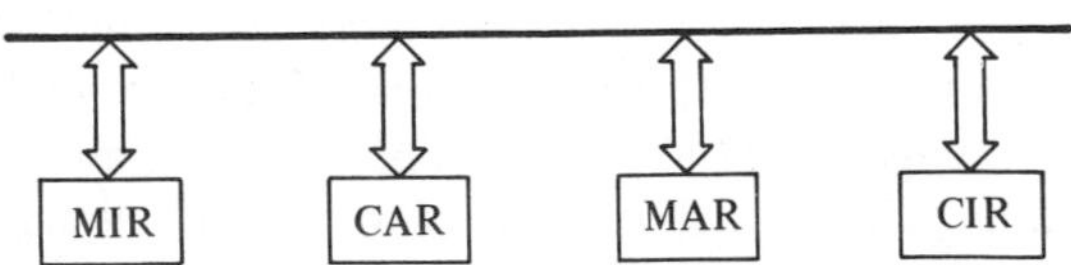

Figure 11.19 Generalized single bus.

only one transfer path is allowed at one time; e.g., $MIR \longrightarrow BUS \longrightarrow MAR$ will transfer the contents of the *MIR* into the *MAR* on two successive clock pulses.

In most machines a combination of methods for transfer paths is utilized. For example, if a single-bus connection is used, the connection to the arithmetic and logic unit (*ALU*) must have at least one direct path since two inputs must be applied at the same time. This principle applies to any functional subunit where more than one input is required simultaneously. When a *DMA* channel is used for I/O transfer, a direct path between the I/O bus and *MIR* is desirable. This path allows the overlapping of instruction execution and I/O transfer by cycle stealing. Otherwise, if the only path to the *MIR* is through the bus of the control unit, cycle stealing would not be possible.

In general, busing is used to minimize the complexity of the routing circuitry associated with the registers. Hardware complexity leads to higher cost and to susceptibility to failures. Such complexity can be reduced by the introduction of time-shared buses. The reduction in cost and complexity is exchanged for time. A bus system is usually slower than that with direct register-to-register transmission. In most cases the low speed of transmission would not affect the overall performance of the system. This assertion is true since the memory cycle time greatly exceeds the clock time of arithmetic and control circuitry. In other words, the memory access is usually the limiting factor in processing, and the use of busing may not degrade the speed of a computer.

In large computers, it is often necessary to house the various units of the system in separate cabinets. These units are connected through cables. These cables are viewed as communication lines and must be designed to maintain the high-speed and noise-free transmission desired. Therefore, special amplifiers are required, and careful consideration must be given to the delays introduced by transmission. All these factors add more cost and make it necessary to time-share the use of these cables in a bus structure rather than as direct register-to-register links.

11.7 CONTROL PANEL SWITCHES AND OPERATING MODES

The control panel of any computer contains several switches, buttons, and display lights. The number and function of these elements on the panel vary widely among various computers. The major function of the control panel elements is to allow manual control of the computer. Although any human intervention slows down the computer operation, some manual control is desired for some debugging and maintenance purposes. In this section some common panel switches and some desirable operating modes are discussed.

11.7.1 Simple Control Panel

The most common elements of a control panel are

1. Power switch: When placed into the ON position, it puts the computer in the wait state until the start button is depressed.
2. Start button: When depressed, it starts the computer operation, assuming that the program is already stored in memory.
3. Stop button: Whenever it is depressed, it halts the computer operation.
4. Register displays: These are light indicators to display the contents of some registers in the machine.
5. Clear buttons: When depressed, they clear the registers on display.

11.7.2 Operating Modes

Stored program digital computers operate mostly in an automatic mode. Programs and data are stored in memory before execution starts. Once a program starts execution, it will run until completion or a natural breakpoint. The completion is detected by a halt instruction. The breakpoints could be a malfunction of equipment, or in the case of complex systems a program could halt if some computer resources, such as memory space, or the CPU are not available.

Operational modes, other than the automatic mode, are desirable for debugging and checking purposes. In general, there are three common modes of operation:

1. Automatic mode.
2. Single-instruction mode.
3. Single-step mode.

To provide for the computer to operate in any one of the above three modes, a three-position mode selector switch is available on the control panel. Any specific mode of operation is activated by setting the mode selector switch to the desired mode position and then depressing the start button.

The automatic mode executes instructions at a high rate, one instruction after another. Execution stops when a HALT instruction is executed or the stop button at the control panel is depressed.

The single-instruction mode allows the computer to execute a single instruction every time the start button is depressed. When the computer is in this mode, one instruction is fetched from memory and executed. The machine then temporarily halts unless the start button is depressed again. Therefore, the single-instruction mode is activated by putting the mode selector switch to single-instruction position and the start button is depressed. For every subsequent instruction in the program to be executed, the start button on the panel has to be depressed.

The single-step mode sequences through every clock phase of a control cycle one at a time. The control cycle could be a fetch cycle, execute cycle, or defer cycle. The single-step mode is activated only when the mode selector switch is at the single-step position. The computer goes through one step and temporarily halts until a button on the control panel, say STEP, is depressed. Upon repeated depression of the STEP button, the computer sequences, one step at a time, from one instruction execution to the next.

11.8 MICROPROGRAMMED CONTROL UNIT

11.8.1 Definition of Microprogramming

It is generally agreed that the term *microprogramming* in its presently accepted sense was introduced by M. V. Wilkes in 1951. Microprogramming was proposed as an orderly method of designing the control unit of a conventional digital computer, as opposed to the ad hoc procedures used in designing control units. The term microprogramming was, then, based on the analogy between the sequence of transfers (between registers and between registers and functional units, such as the adder) required to execute a machine instruction and the sequence of individual instructions in a conventional user program. Each step was called a microinstruction, and the complete set of steps required to process a machine instruction was called the microprogram. For example, the instruction "Add X to A," in a conventional user program, might comprise the following set of microinstructions:

$$CAR \longrightarrow MAR, \langle M \rangle \longrightarrow MIR, MIR \longrightarrow IR,$$
$$CAR + 1 \longrightarrow CAR, Ad(CIR) \longrightarrow MAR, \langle M \rangle \longrightarrow X, S(A,X) \longrightarrow A$$

Except for a slight divergence in meaning by Vanderpoel, who used the term to describe a system which permitted the user a larger repertoire of microinstructions by allowing individual bits in an instruction to control certain processor gates directly, the term has the same basic meaning as in the Wilkes' scheme.

To incorporate the basic idea of a stored program implementation of the control unit and the recent exploitation of dynamically microprogrammable systems, S. S. Husson gave the following definition, which attempts a balance between the systems and the hardware approach:

> Microprogramming is a technique for designing and implementing the control function of a data processing system as a sequence of control signals, to interpret fixed or dynamically changeable data processing functions. These control signals, organized on a word basis and stored in a fixed or dynamically changeable control memory, represent the states of the signals which control the flow of information between the executing functions and the orderly transition between these signal states.

Defining the term strictly from the hardware point of view, microprogramming implements a machine instruction as a subroutine of primitive operations called microinstructions.

11.8.2 Basic Layout of a Microprogrammed Control Unit

The basic layout of a microprogrammed control unit is shown in Fig. 11.20. The basic components of the layout are the control memory, the control address register, and the microinstruction register. The control memory stores a microprogram corresponding to each op-code of a conven-

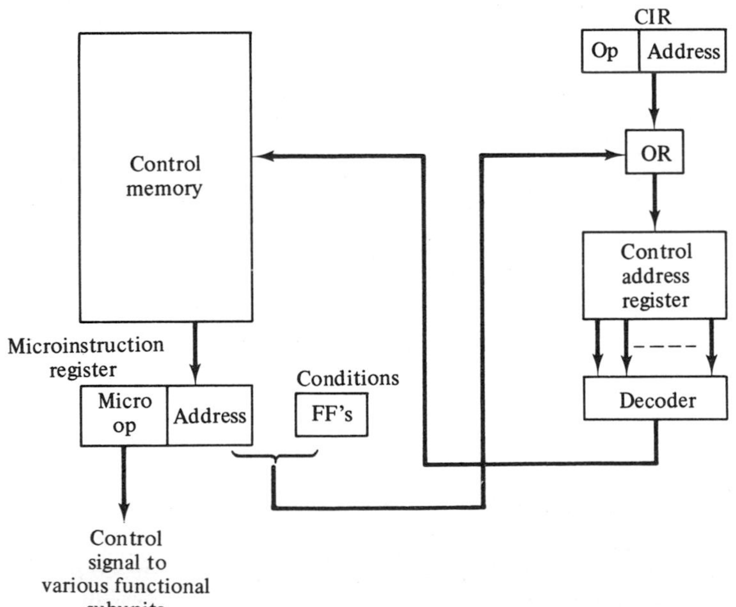

Figure 11.20 Basic layout of a microprogrammed control unit.

tional instruction. Once the op-code is decoded, the corresponding microinstructions are retrieved from the control memory. The control address register and the decoder serve as the address register and selection mechanism for the control memory, respectively. The contents of the control address register determine the location of the microinstruction to be retrieved from the control memory. The microinstruction register holds the retrieved microinstruction, which consists of the micro op-code and the address of the next microinstruction in the control memory. After the current microinstruction is executed, the address of the next microinstruction is entered into the control memory to retrieve the next microinstruction. This process continues until all the microinstruction sequence corresponding to the conventional instruction is executed. Then the op-code of the next conventional instruction is transferred into the control address register. Conditional jumps are implemented by letting the states of some conditional flip-flops modify the address of the next microinstruction to be retrieved. In general the contents of the control address register, and consequently the address of the microinstruction to be retrieved, are determined by the address field of the microinstruction (sometimes modified by conditional flip-flops) or the op-code of the conventional instruction.

11.8.3 The Wilkes' Scheme

The earliest implementation of a microprogrammed control unit is known as the Wilkes' scheme. Figure 11.21 shows the layout of this scheme.

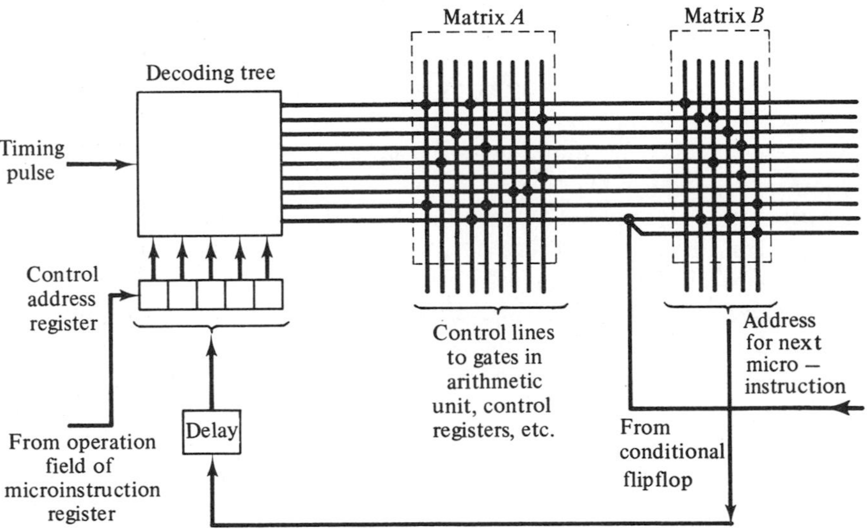

Figure 11.21 Wilkes' implementation of microprogram control.

The microprogram is stored in a nondestructive read-only memory which consists of two matrices, A and B. These were ferrite core matrices with linear selection and may be regarded as corresponding to two fields in the microinstruction, one for control and one for sequencing. The access to the memory is a decoding tree with an associated address register of width n bits as input. A well-defined timing pulse enters the decoder, which then selects one of 2^n outputs, which also serves as a horizontal matrix line in matrix A. The activated horizontal line represents a microinstruction. Connection points between the horizontal line and vertical lines in matrix A determine which control signals will be energized and thus which microoperations will be effected. The activated horizontal line also passes through matrix B, which provides sequencing by setting the appropriate flip-flops for the control address register, R, via delay circuits. The contents of R are then used to select the next microinstruction. When the new contents of R are decoded, new control pulses enter the system from matrix A, and a new address enters register R from matrix B, and so forth.

Decision and branching in the system is provided by branched wires from matrix A to matrix B and conditional flip-flops. Selection of one of alternate paths into matrix B depends on the state of one or more of the flip-flops. For example, if a flip-flop C represents the sign of the accumulator, then making the choice of the next microinstruction might depend on whether the accumulator contents are positive or negative. For each conditional branch a conditional flip-flop can be incorporated into the system.

There are implicit relationships in the above scheme. First, the width of the control address register, R, determines the number of microinstructions which can be stored by the matrix. The number of horizontal lines exiting from A is 2^n, where n is the width of R in bits. The second relationship is that the number of vertical lines in B are equal to the word width of the microinstruction.

In summary, the implementation scheme just described replaces the conventional control logic by a read-only memory (in this case the wired-in matrices) whose outputs become the control signals. The address to the read-only memory (*ROM*) is generated from function codes, sequence count, conditional flip-flop settings, data path status information, and the previous microinstruction. A specific disadvantage of the wired-in matrices' implementation is the difficulty of altering their contents.

Initially microprogramming schemes were limited to the ROM for control storage. But with the advent of fast writable memories, the microprogramming concept came to include user-alterable and dynamically microprogrammable programs. This new concept has been reflected by the S. S. Husson definition of microprogramming stated earlier.

11.8.4 Variations in the Implementation

Wilkes' scheme is the simplest implementation of the basic layout of a microprogrammed control unit. A few possible variations in the operational details are presented in the subsequent discussion. All the variations to be discussed assume the Wilkes' scheme as the basic implementation. The modification can be generally classified into two major categories:

1. Microinstruction address selection.
2. Instruction set strategies (control word organization).

Examples of modifications in category 1 are described in Section 11.8.4.1, while those in category 2 are described in Section 11.8.4.2.

11.8.4.1 Microinstruction Address Selection

A possible modification is the addition of a register to hold the address of the next microinstruction. It is loaded from the address matrix. The content of this register would be sent to the control address register for retrieving the next microinstruction. This modification is illustrated in the basic layout of Fig. 11.20, where the address field of the microinstruction register represents the register described above.

Another variation which concerns the organization of the control memory is to have the control memory organized as two storage matrices each with its own decoding tree, thus providing faster operation at lower cost, since the size of the decoder is reduced. This setup could provide some overlap for sequential microinstruction. The basic layout for a microprogrammed control unit which uses two separate storage matrices is shown in Fig. 11.22.

A third possible variation concerns the manner in which the control address is updated. In the basic scheme it is assumed that the microinstruction which is being executed specifies the address of the next microinstruction. This is not necessary if the control address register is modified to function as a counter when the next microinstruction address follows the current address and as a register when a branch is required. This arrangement is similar to that used in the *CAR* of a conventional control unit.

11.8.4.2 Control Word Organization

Modifications in this category may be described first in terms of whether there is direct or encoded control. In the basic scheme there is a single control line exiting from the control matrix for each control gate associated with a

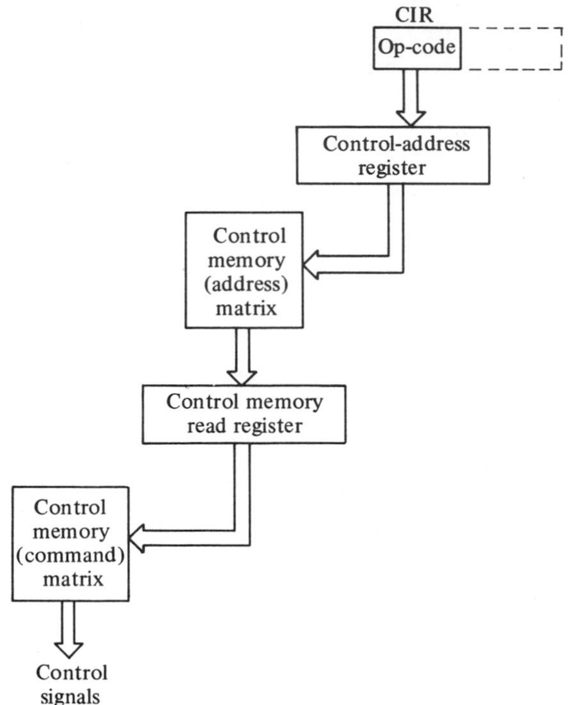

Figure 11.22 Microprogrammed control unit with two storage matrices.

functional unit. This *direct control* is feasible when the number of control gates is small and there is only one data path. In this case the cost/speed ratio is favorable. For example, consider a functional subunit to be an adder with n stages. If direct control is applied, then n bits of output from the control matrix are required to control the input to the n stages of the adder. If instead of using n bits in the control matrix for this adder we use m bits (where $n = 2^m$), then the m bits can be decoded to activate the desired control lines. One decoder would be used for each functional subunit; this is referred to as *encoded control.* In this case, each functional subunit is controlled from a specific field of the control word after it has been decoded. It would be expected that as the size and complexity of the system increased, more gates could be grouped in the same field and encoding would have a greater cost advantage. Also, since each functional unit is controlled by a group of bits in one field of the control word, it would be much easier to follow the function of the various control signals. A disadvantage of encoding is the tendency for the number of microinstructions to be very large.

The encoded microinstructions, as described above, have many fields, each having direct control over a functional subunit of the computer such as the adder, shifter, and memory select. Other fields are used for computing the address of the control storage access. This type of encoded microinstruction is called *minimally encoded.* Another format for the microinstructions is referred to as *highly encoded* microprogramming. In the highly encoded format, the microinstruction contains an operation code field F and several operand fields $O_1, O_2, \ldots, O_n$. The F field indicates the microoperation to be performed, while the operand fields control hardware activities such as adder inputs and destinations. Figure 11.23 shows the format of a minimally encoded and a highly encoded microinstruction. Highly encoded microinstructions are short compared to the minimally encoded ones. Typical lengths are 360 bits for minimally encoded and 32 bits for highly encoded. Minimal encoding allows for more parallel performance than high encoding. Storage space can be saved by high encoding, but there is a time and cost loss in translating to the bit level from the ROM through an encoder.

Another variation which concerns the control words in the command matrix is whether control signals are energized concurrently or sequentially. When the execution of a microinstruction always results in a simultaneous issue of control signals in a single time period, then we speak of *vertical programming.* Control may be direct or encoded, and in either case a series of

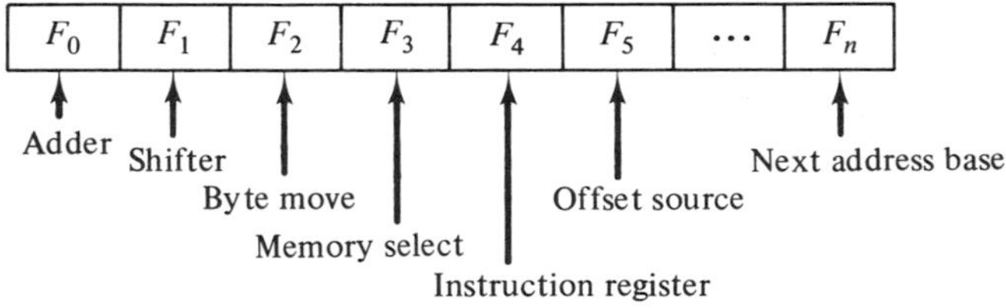

(a) Minimally encoded microinstructions

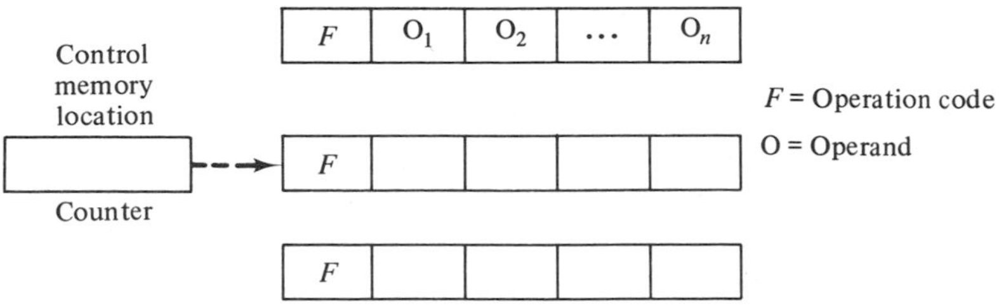

(b) Highly encoded microinstructions

Figure 11.23 Microinstruction formats.

microinstructions is required to execute even the simplest conventional instruction. This is due to the fact that a different microinstruction is required for every microcommand in the conventional instruction. If all microcommands corresponding to a conventional instruction (such as add and multiply) are obtained at once and then executed sequentially, we speak of *horizontal microprogramming*. In comparing the horizontal and vertical schemes, we find that the shorter "word" length associated with the vertical scheme is offset by the greater numbers of accesses to the control memory. The trade-off between memory accesses and the use of storage is often balanced by using a combination of schemes.

11.8.4.3 Use of Read-Write Memory

Technological advances in memories are the reasons for solid improvements in microprogramming techniques. In this section we shall look at control memories briefly, as a background for the application of microprogramming (see Section 10.6.5).

Primarily, the read-only memory (ROM) has been used for microprogram storage because of several advantages over read-write memories. For instance, ROM has fast operating speed (because a write cycle is not needed), high reliability, inalterability of information by error, high information density, and low cost for a given speed. But ROM has a disadvantage in that it is relatively complicated to alter its contents. In addition, there are practical limitations to its size. The most desirable control storage is a read-write one which has a cycle time of the same order of magnitude as the CPU speed with a reasonably low cost.

Large-scale integration (LSI) promises to overcome the prohibitive cost of fast (at least twice as fast and preferably 5 to 10 times faster than main memory) read-write control storage. LSI control memories have a nondestructive readout and are designed so that the read cycle is as short as possible, even at the price of a relatively long write cycle. These characteristics are suitable for control memories since read accesses are much more frequent than write accesses. With a read-write control memory it will be economical to transfer microprograms from main storage to control storage when needed for software support applications, such as compilation, emulation, simulation, and operating systems. These various applications of microprogramming are possible mainly due to the use of read-write storage as control memory. Some of these applications are discussed below.

11.8.5 Applications of Microprogramming

In addition to the systematization of control, microprogramming yields improvement in performance: first, by a high degree of parallelism in data paths so that many microsteps can be performed in one machine cycle (the

90-bit microinstruction of the IBM 360/50 is an example of this); and again, by the high degree of decision logic which can be performed in a single microinstruction. This performance improvement can be seen in table search and sorting routines when implemented in microcode. Also, the instruction fetch time is reduced if a desired subroutine is microprogrammed and reached by a macro. Only the single fetch for the macro is needed. There are many applications of microprogramming which exploit these performance improvements. Some of the major areas of microprogramming applications are computer-series compatibility, emulation, microdiagnostics, software support, special-purpose devices, and dynamic microprogramming.

11.8.5.1 Computer-Series Compatibilities

A major application of microprogramming is providing compatibility of instruction sets between smaller and larger machines of a series. The RCA Spectra 70 and the IBM System/360 are well-known examples of a line of computers with a range of speed, capacity, and cost but with compatible instruction sets achieved by microprogramming.

11.8.5.2 Emulation

Emulation is defined as the combined software/hardware interpretation of the machine instruction of one machine by another. In emulation, the target machine's architecture, hardware and software, is mapped onto the host machine. If the word length of the host machine is a multiple of the target machine while the number of arithmetic registers in the host machine is the same or greater, emulation by microprogramming is facilitated. The key operation is choosing a small set of machine instructions to augment the machine instructions of the host and speed up software simulation. Microprogrammed emulation provides a less costly means of making system changeovers and is 5 to 10 times faster than its nonmicroprogrammed counterpart.

A frontier in emulation might be a universal host machine to emulate a very large variety of second- and third-generation machines.

11.8.5.3 Microdiagnostics

Microprogramming diagnostic routines have allowed refinements and increased the speed of detecting and localizing faults, including error detection and correction of microstorage itself.

Previously, maintenance packages were software routines written for each specific system. Disadvantages of the software approach were the large amount of storage needed (as much as 4K bytes each) and the necessity of choice of routine by the maintenance engineer. Software diagnostics did not achieve the desired high percentage of error detection and error diagnostic capabilities. Another previous approach used in some computer systems was the use of hardware fault-locating tests, which proved to be expensive to implement and difficult to use.

Microdiagnostics have proved superior to software diagnostics for a number of reasons:

1. They are faster by as much as 10 times. Part of the favorable speed ratio is due to the avoidance of the redundancy in testing data paths.
2. They are less costly to produce.
3. They use much less of the system resources, such as channels, I/O, and storage.
4. They can isolate the error to a smaller number of replaceable units.
5. They reduce the amount of training and service equipment needed to maintain the system.

A general procedure in microdiagnostics is to send a test pattern through a given data path and analyze the output from the data path to select a finer set of test patterns to further isolate the error. The microprogram is not necessarily kept in residence, for a large block of control store addresses would be required.

11.8.5.4 Software Support

The excess microprogramming capacity not needed for the basic instruction set can also be used for microprograms designed to support software. In large systems special-purpose processors are usually used to implement the system software operations. If these traditionally software-implemented operations are microprogrammed, more CPU space and time are released, and the software control of the computing system is simplified. This might result in increasing system throughput and easing the programmer's task.

Some software areas amenable to microprogramming are compiling and interpreting of high-level programming languages, floating-point arithmetic, stack manipulation, searching, and evaluating polish strings.

Areas being explored in software support to enhance system performance are IOS macros, control programming, and interrupt handling.

11.8.5.5 Special-Purpose Devices

Special-purpose devices refer to fast processors capable of handling specific processes efficiently. They are used in many applications such as data communication, data acquisition, and device controllers. Application of microprogramming in the design of such systems is a growing field. Special interest in microprogramming in designing efficient data-processing systems arises in military aerospace applications and data management. Optical data processing and radar data processing apply microprogramming for greater efficiency. Process control on a real-time basis is a key application.

From another point of view it is possible to design a general-purpose microprogrammed processor with sufficient flexibility to be adapted to specific processes. Such a general-purpose processor has been built and demonstrated its capability in real-time digital signal processing.

11.8.5.6 Dynamic Microprogramming

A dynamic microprogramming system allows routines to be easily microprogrammed. The computer can be restructured to represent any instruction vocabulary by writing and loading a microprogram. With LSI read-write memories, dynamic microprocessors are practical.

The use of read-write memories as control memories has raised the question of letting the user access microstorage. Now that definite protection policies have become known as a result of experience in time-sharing systems, the possibility of user-alterable microstorage seeems to be viewed with less trepidation.

11.9 AN EXAMPLE OF A MICROPROGRAMMED CONTROL UNIT

In this section the design of a microprogrammed control unit is presented as part of the design of a small computer. This section is not intended to show the complete design of a practical computer. It is only an application to some of the design principles presented in this chapter, and in particular the microprogrammed control. A more detailed design of a medium-sized computer is described in Chapter 12.

First, a small stored program computer is described. The description includes the layout, the architecture, and the transfer equations for the implementation of the given instructions. The design is based on the use of sequential logic control. The microprogramming concepts described in Section 11.8 are then applied to this small computer to illustrate the design of a microprogrammed control unit.

11.9.1 Computer Layout and Registers

Figure 11.24 shows the layout of a small computer with sequential logic control. The memory unit contains 32,000 words, 24 bits each. The memory information register (*MIR*) and the accumulater are each 24-bit registers. The memory address register (*MAR*) and the current address register (*CAR*)

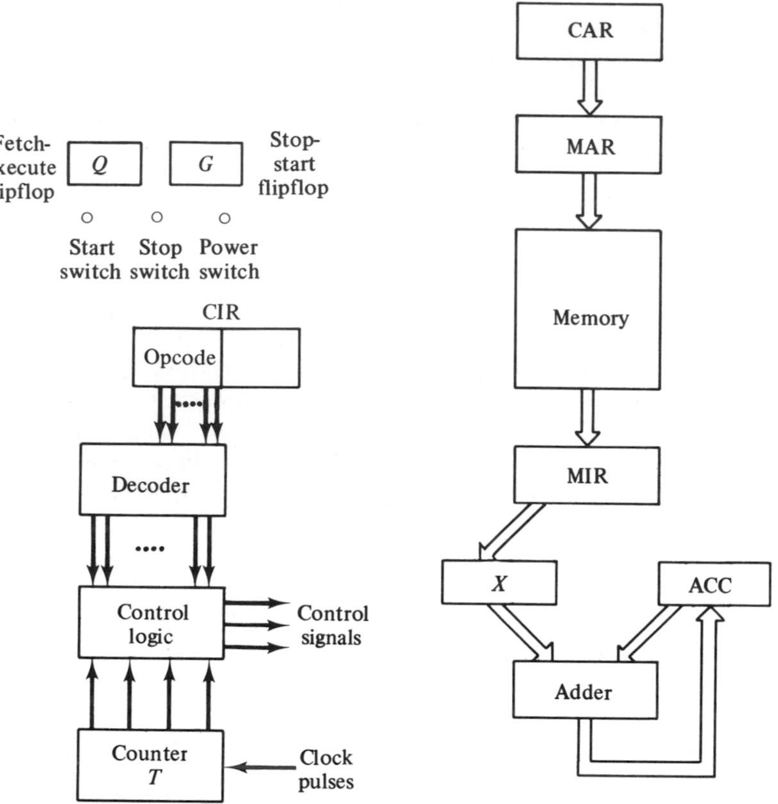

Figure 11.24 Layout of a small computer with sequential logic control.

are 15-bit registers. Therefore, there could be 2^{15} or 32,768 locations of main memory which are directly addressable. The current instruction register (*CIR*) consists of 24 bits with 6 bits for the op-code. This allows for 2^6 or 64 different instructions.

The power switch and the start switch will manually start the computer when sequentially switched to the ON position. The power ON position resets the start-stop flip-flop *G*, which is then set by placing the start switch to the ON position. When *G* is set, the computer goes to the *run state*, and program sequencing starts. The stop switch, when turned on, resets the *G* flip-flop, and the computer goes to a wait state. When the computer is in the wait state, it continuously checks the state of the flip-flop *G*. Whenever the *G* flip-flop is set, the computer goes to the run state.

11.9.2 Instruction Format and Repertoire

The instruction format is shown in Fig. 11.25. Bits 0 through 5 constitute a 6-bit op-code field; bits 6, 7, and 8 could be used as flags for various addressing methods. In our design only direct addressing is implemented, and therefore these bits are not used. For direct addressing, bits 6 through 8 contain 0s and bits 9 through 23 constitute a 15-bit direct address field (D.A.).

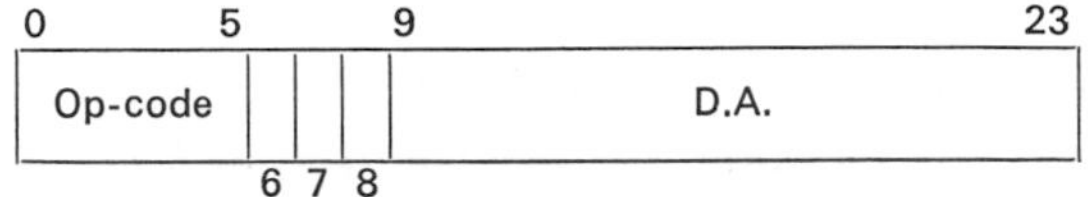

Figure 11.25 Instruction format.

The op-code field allows up to 64 different instructions. For simplicity, only a small number of instructions are considered. Table 11.1 shows the set of instructions which are implemented. The table includes the instruction op-code in octal form, the instruction mnemonic, the name of the instruction, and a short definition of the instruction.

11.9.3 Timing and Typical Instruction Cycles

Fetching and execution of instructions are controlled by the output of the counter T as indicated in the layout of Fig. 11.24. The counter T is driven by 1-microsecond clock pulses and has a cycle length of 4. The memory cycle time is 4 microseconds, and therefore each memory cycle consists of four

Table 11.1 *Instruction Repertoire*

Op-code	*Mnemonic*	*Name*	*Definition*
01	ADD	Addition	$[ACC] + [\text{M(D.A.)}] \longrightarrow ACC$
02	CLA	Clear	$0 \longrightarrow ACC$
03	STO	Store	$[ACC] \longrightarrow \text{M(D.A.)}$
04	UNB	Unconditional branch	$[\text{D.A.}] \longrightarrow CAR$
05	BAN	Branch when negative	if $ACC_0 = 1$, $[\text{D.A.}] \longrightarrow CAR$
06	COM	Complement	$\overline{ACC} \longrightarrow ACC$
07	SHR	Shift right	$\text{L}(ACC) \longrightarrow \text{R}(ACC)$
10	CSL	Circular left shift	$\rho^{-1}(ACC)$
11	STP	Stop	$0 \longrightarrow G$

clock pulses. This timing indicates that each memory cycle coincides with the counter cycle, which in turn is considered a control cycle. The instruction fetch consists of one control cycle, and each instruction execution takes one control cycle. This timing does not constitute an optimal design. In this section simplicity is more important than considering an optimal design. If more speed is someone's concern, the execution of many of the instructions in Table 11.1 can be overlapped with the fetch cycle. Overlapping is not considered here in order to simplify the design of the microprogrammed control later.

Before describing the transfers for the fetch cycle and the various execution cycles, a description of the memory read and the memory write cycles are given below.

Memory Read Cycle

$T_0|$ $1 \longrightarrow R$, $CAR \longrightarrow MAR$ or $ad(CIR) \longrightarrow MAR$
$T_1|$ $M(MAR) \longrightarrow MIR$
$T_2|$ ↓ Restore memory
$T_3|$

Memory Write Cycle

$T_0|$ $1 \longrightarrow W$, $ad(CIR) \longrightarrow MAR$
$T_1|$ ↓ Erase memory
$T_2|$
$T_3|$ $MIR \longrightarrow M(MAR)$

Fetch Cycle

$\bar{Q}\,T_0|$ $1 \longrightarrow R$, $CAR \longrightarrow MAR$
$\bar{Q}\,T_1|$ $M(MAR) \longrightarrow MIR$
$\bar{Q}\,T_2|$ $MIR \longrightarrow CIR$, Decode
$\bar{Q}\,T_3|$ $CAR + 1 \longrightarrow CAR$

With each instruction being executed in one memory cycle, the execution cycles are described below:

ADD Cycle

$Q\,T_0|$ $1 \longrightarrow R$, $ad(CIR) \longrightarrow MAR$
$Q\,T_1|$ $M(MAR) \longrightarrow MIR$

$Q\,T_2|$ $MIR \longrightarrow X$

$Q\,T_3|$ $X + ACC \longrightarrow ACC, 0 \longrightarrow Q$

STO Cycle

$Q\,T_0|$ $1 \longrightarrow W, ad(CIR) \longrightarrow MAR$

$Q\,T_1|$ $ACC \longrightarrow MIR$

$Q\,T_2|$

$Q\,T_3|$ $MIR \longrightarrow M(MAR), 0 \longrightarrow Q$

The execution of each of the remaining instructions requires simple but specific transfers which are performed at one clock period. Therefore, only this time clock is shown. No transfers are required at other timing pulses, and therefore these timing pulses are omitted below:

CLA

$Q\,T_3|$ $0 \longrightarrow ACC, 0 \longrightarrow Q$

UNB

$Q\,T_3|$ $ad(CIR) \longrightarrow CAR, 0 \longrightarrow Q$

BAN

$Q\,T_3|$ $ACC_0(ad(CIR)) + \overline{ACC_0} \cdot CAR \longrightarrow CAR, 0 \longrightarrow Q$

COM

$Q\,T_3|$ $\overline{ACC} \longrightarrow ACC, 0 \longrightarrow Q$

SHR

$Q\,T_3|$ $\mathrm{L}(ACC) \longrightarrow \mathrm{R}(ACC), 0 \longrightarrow Q$

CLS

$Q\,T_3|$ $\rho^{-1}(ACC), 0 \longrightarrow Q$

STP

$Q\,T_2|$ $0 \longrightarrow G, 0 \longrightarrow Q, 0 \longrightarrow CAR, 0 \longrightarrow MAR$

The specific execution cycle which the computer goes to is determined by the output of the decoder which decodes and interprets the op-code field of the instruction.

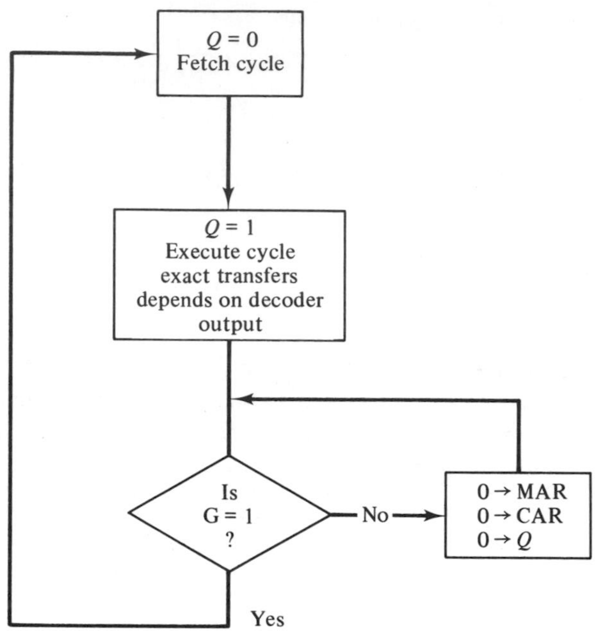

Figure 11.26 Instruction sequencing.

Figure 11.26 shows the fetch and execute sequencing of instructions within a program execution. It is assumed that before a program starts to be executed it is loaded into memory starting at location 0. The first instruction is fetched as soon as the *G* flip-flop is set. If the *G* flip-flop is reset, either by the execution of a STP instruction or by the manual stop, the computer resets the *CAR* and *MAR* and enters a wait state. In the wait state the computer continuously checks the *G* flip-flop. Whenever *G* is set, a new sequence starting at location 0 begins.

11.9.4 Microprogrammed Control

As described in Section 11.8, the concept of a microprogram control unit is a more formalized method of generating the control logic for a digital computer. This method of microprogram control is applied to the control of the small computer described in the previous sections. The layout of the microprogrammed computer is shown in Fig. 11.27. When compared to Fig. 11.24, the layout in Fig. 11.27 contains additional hardware which is necessary for microprogram control. This additional hardware consists of a control memory, *CM*, with its address register *CMAR* and its buffer register *CMBR*. The *CMAR* holds the address in the control memory, which is

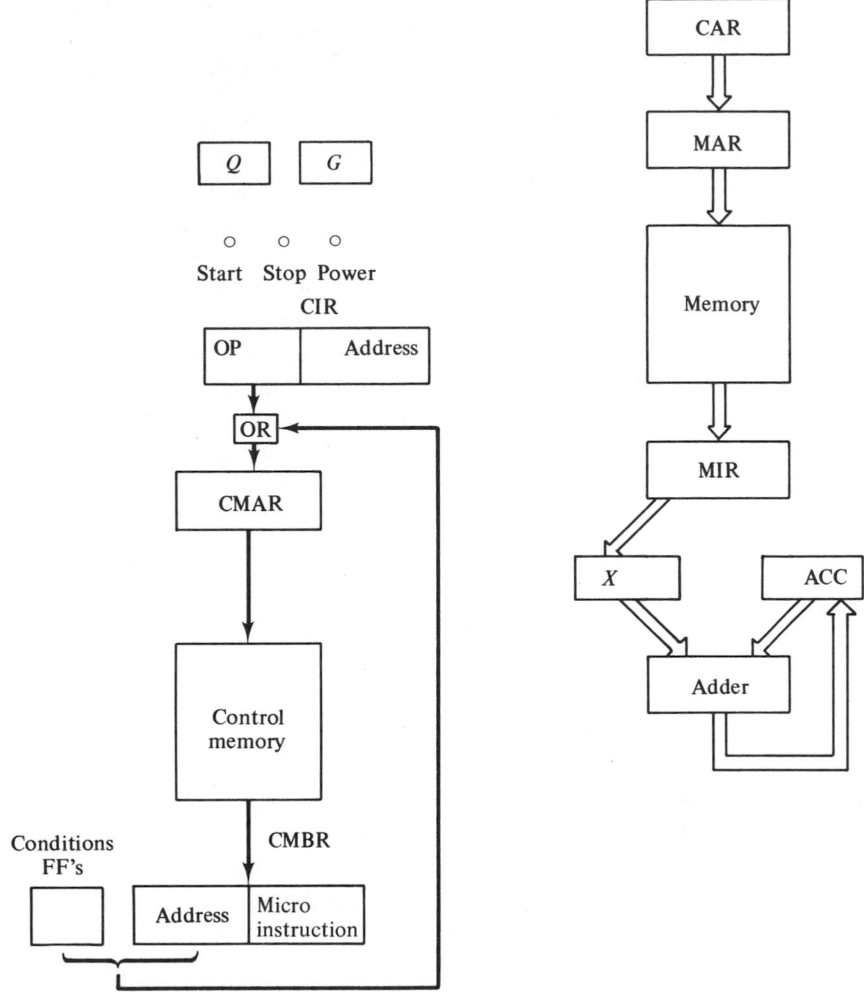

Figure 11.27 Layout for a microprogrammed computer.

being fetched at one time. The *CMBR* holds the information which is retrieved out of the control memory. This information is referred to as the microinstructions. The principal operation of a microprogram control unit is explained in Section 11.8.

Referring to Fig. 11.27, the operation can be described as follows: The instruction to be executed is fetched from main memory, and the op-code part is placed in the *CIR* register. The op-code of the instruction specifies the operation to be performed, and thus it is used to generate a unique sequence of microinstructions for that operation. This is achieved by mapping the op-code into the address of the control memory where the first microoperation corresponding to the macrooperation (instruction) is stored. Thus, the op-code, or a function of it, is entered into the *CMAR*, and a control memory read cycle is initiated, resulting in the transfer of a word (microinstruction) from the control memory into the control memory buffer register, *CMBR*. In this example, the control memory word is divided into two fields. One field contains the bits which are used as the control signals in the execution of the instruction, that is, the microinstruction. The other field specifies the address in the control memory of the next microinstruction to be fetched.

Figure 11.28 shows the control memory word format. The size of the next address field is a function of the control memory size. The microinstruction field depends on many factors, such as the number of instructions in the computer, and to a larger extent on the architecture design of the microcode, as explained in Section 11.8.4. In this example a very simple, although not necessarily optimal, architecture in which almost every bit in the microinstruction field corresponds to one microoperation (register transfer) is used. A few examples where more than one microoperation is controlled by the same bit in the microinstruction field will be illustrated as examples of reducing the control memory word length.

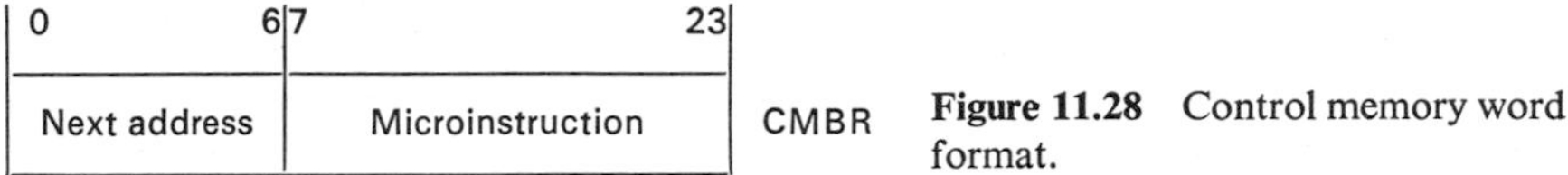

Figure 11.28 Control memory word format.

In a microprogram control computer every instruction fetch or instruction execution requires at least one control memory read cycle. The control memory read cycle must be initiated within the computer cycle at the proper clock period in order to retrieve the required microinstruction before the clock time for its performance. The timing and the transfers for the control memory read cycle are shown below.

Control Memory Cycle

T_0 |

T_1 |

T_2 | $op(MIR) \longrightarrow CMAR$ or $ad(CMBR) \longrightarrow CMAR$

T_3 | $CM(CMAR) \longrightarrow CMBR$

In the above transfers, the control memory cycle is assumed to coincide with the main memory cycle. The *CMAR* is loaded at clock period T_2, and the readout is performed at T_3. The fetch cycle and a typical execution (ADD) cycle for the microprogram control computer are shown:

Fetch Cycle

$T_0|$ $CAR \longrightarrow MAR, 1 \longrightarrow R$

$T_1|$ $M(MAR) \longrightarrow MIR$

$T_2|$ $MIR \longrightarrow CIR, op(MIR) \longrightarrow CMAR$

$T_3|$ $CAR + 1 \longrightarrow CAR, CM(CMAR) \longrightarrow CMBR$

Add Cycle

$T_0|$ $1 \longrightarrow R, ad(CIR) \longrightarrow MAR$

$T_1|$ $M(MAR) \longrightarrow MIR$

$T_2|$ $MIR \longrightarrow X, ad(CMBR) \longrightarrow CMAR$

$T_3|$ $X + A \longrightarrow A, CM(CMAR) \longrightarrow CMBR$

The transfers for other execution cycles can be written similarly by adding the control memory transfers to the transfers of each specific instruction. The timing of the control memory cycle reads a control word at clock pulse T_3 of the fetch cycle so that the execution indicated in the retrieved microinstruction starts at T_0 of the execution cycle. Similarly, the control word corresponding to the next fetch operation is read out at T_3 of the current execution cycle.

The transfers to be performed during a fetch cycle or an execution cycle are controlled by the proper timing pulse and also by the proper control bit in the microinstruction. Table 11.2 shows each transfer with the associated timing signal and control bit of the control memory buffer register. Table 11.2 shows 24 different microoperations, not including operations initiated by the manual control switches. If the control memory word contains a bit for every microoperation, a total of 24 bits are required for the microinstruction field. In our design we choose the control memory word length to be 24 bits. Seven bits (0–6) are used to store the address of the next microinstruction and only 17 bits (7–23) are used for the microinstruction field. This is achieved by using a single bit to control more than one microoperation. This is possible whenever the combined microoperations belong to the same cycle but occur at different clock times as shown for $CMBR_9$, $CMBR_{13}$, and $CMBR_{14}$. Once such a combination is made, it will prevent the execution of any of the component microoperations by itself. Another instance when combined control of microoperations is possible is indicated by $CMBR_7$. In this case the microoperations occur at the same time clock. This will also restrict the use of any of these microoperations at a different clock time. Other methods for further reduction of the control word length which uses decoding are implemented in the design of the larger computer discussed in Chapter 12.

Table 11.2 Transfer Operation for the Small Computer

Control bit (CMBR bit)	Clock period	Transfer operation	Cycle in which required	Decimal numbering
$CMBR_8$	T_0	$1 \longrightarrow R$	Read	1
9	T_0	$CAR \longrightarrow MAR$	FETCH	2
10	T_0	$1 \longrightarrow W$	STO	3
8	T_1	$M(MAR) \longrightarrow MIR$	READ	4
11	T_2	$MIR \longrightarrow CIR$	FETCH	5
11	T_3	$CAR + 1 \longrightarrow CAR$, if $G = 1$	FETCH	6
12	T_0	$ad(CIR) \longrightarrow MAR$	READ	7
13	T_2	$MIR \longrightarrow X$	ADD	8
13	T_3	$X + ACC \longrightarrow ACC$	ADD	9
14	T_2	$ACC \longrightarrow MIR$	STO	10
14	T_3	$MIR \longrightarrow M(MAR)$	STO	11
15	T_3	$0 \longrightarrow ACC$	CLA	12
16	T_3	$ad(CIR) \longrightarrow CAR$	UNB	13
17	T_3	If ACC_0, then $ad(CIR) \longrightarrow CAR$	BAN	14
18	T_3	$\overline{ACC} \longrightarrow ACC$	COM	15
19	T_3	$L(ACC) \longrightarrow R(ACC)$	SHR	16
20	T_2	$op(MIR) \longrightarrow CMAR$	FETCH	17
21	T_2	$ad(CMBR) \longrightarrow CMAR$	All executions	18
7	T_3	$CM(CMAR) \longrightarrow CMBR$, if $G = 1$	Every cycle	19
22	T_3	$\rho^{-1}(ACC)$	CSL	20
23	T_2	$0 \longrightarrow G$	STP	21
21	T_3	if $G = 0$, $0 \longrightarrow CAR$	All executions	22
21	T_3	if $G = 0$, $0 \longrightarrow MAR$	All executions	23
21	T_3	if $G = 0$, $0 \longrightarrow CMAR$	All executions	24

The transfers generated by the manual control switches are as follows:

POWER ON| $1 \longrightarrow G$, $0 \longrightarrow CMBR$, $0 \longrightarrow CMAR$,
$0 \longrightarrow MAR$, $0 \longrightarrow CAR$, $0 \longrightarrow MIR$, $0 \longrightarrow CIR$

START ON| $1 \longrightarrow G$, $1 \longrightarrow CMBR_7$, $0 \longrightarrow CMBR$ except bit 7

The control signals to implement the above transfers are generated directly by sequential logic, and not through the control memory.

The microprogram which is stored in the control memory consists of 10 control words. One control word is for the fetch sequence, and the remaining nine store the sequences for the nine instructions in the computer. The microprogram of the computer is shown in Table 11.3.

Table 11.3 *Control Memory Contents (Microprogram)*

Location (*CMAR*)†	*Op-code*	*Contents of the control word by bit* 0 → 6†	7	8	9	10	11	12	13	14	15	16	17	18	19	20	21	22	23
01	ADD	12	1	1	0	0	0	1	1	0	0	0	0	0	0	0	1	0	0
02	CLA	12	1	0	0	0	0	0	0	0	1	0	0	0	0	0	1	0	0
03	STO	12	1	0	0	1	0	0	0	1	0	0	0	0	0	0	1	0	0
04	UNB	12	1	0	0	0	0	0	0	0	0	1	0	0	0	0	1	0	0
05	BAN	12	1	0	0	0	0	0	0	0	0	0	1	0	0	0	1	0	0
06	COM	12	1	0	0	0	0	0	0	0	0	0	0	1	0	0	1	0	0
07	SHR	12	1	0	0	0	0	0	0	0	0	0	0	0	1	0	1	0	0
10	CSL	12	1	0	0	0	0	0	0	0	0	0	0	0	0	0	1	1	0
11	STP	12	1	0	0	0	0	0	0	0	0	0	0	0	0	0	1	0	1
12	FETCH	00	1	1	1	0	1	0	0	0	0	0	0	0	0	1	0	0	0

†Numbers in these columns are in octal.

REFERENCES

"Annotated Bibliography for Microprogramming," *SIGMICRO Newsletter*, 3, Issue 2 (July 1972), Association for Computing Machinery.

CHU, Y., *Computer Organization and Microprogramming*. Englewood Cliffs, N.J.: Prentice-Hall, 1972.

DAVIES, P. M., "Readings in Microprogramming," *IBM System J.*, 11, No. 1 (1972), pp. 16–40.

FLYNN, M. J., "Microprogramming: Future Prospects and Trends," *1971 IEEE Int. Conv. Dig.* (1971), pp. 318–319.

GSCHWIND, H. W., *Design of Digital Computers*. New York: Springer-Verlag New York, Inc., 1967.

HILL, F. J., and G. R. PETERSON, *Digital Systems: Hardware Organization and Design*. New York: Wiley, 1973.

HUSSON, S. S., *Microprogramming Principles and Practice*. Englewood Cliffs, N.J.: Prentice-Hall, 1970.

"Special Issue on Microprogramming," *IEEE Trans. Computers*, C-20, No. 7 (1971).

TUCKER, S. G., "Microprogramming Control for System/360," *IBM System J.*, 6, No. 4 (1967), pp. 222–241.

WILKES, M. V., "The Growth of Interest in Microprogramming," *Computing Surveys*, 1, No. 3 (1969), pp. 139–145, Association for Computing Machinery.

PROBLEMS

11.1. Assume that a computer has several memory banks, each bank consisting of 4096 words with a 24-bit word length. Design an effective instruction format for this computer which provides for indexing, relative and indirect addressing.

11.2. Assume a computer with a word length of 12 bits and a memory capacity of 4096 locations. Select an addressing technique or a combination of addressing techniques which would allow you to address any location in memory. Demonstrate the above capability by showing how you specify a jump instruction to any location in memory using a 12-bit instruction word.

11.3. A single-address computer has a word length of 24 bits. The main memory consists of 60,000 words. The computer has 64 different instructions and two index registers. Assuming that the computer uses a single word instruction, draw a diagram for the instruction format, allocating the various necessary fields.

11.4. For a single-address computer, determine the relationship between the size of the current address register (*CAR*) and the size of the address field for

each of the following addressing methods:
(a) Direct addressing.
(b) Indirect addressing.
(c) Base addressing (in this case, determine the size of the base register).
Base your comparison on the assumption of having the maximum possible memory size.

11.5. Repeat Problem 11.4 for the following addressing methods:
(a) Indexing with the size of the index register equal to the size of *CAR*.
(b) Self-relative addressing.
(c) A combination of direct, indirect, and indexing.

11.6. Select a small computer you are familiar with and separate its instruction set into the three categories given in Section 11.2.2.

11.7. A SKIP instruction examines the contents of some registers and then transfers to the next instruction in sequence or skips that instruction and proceeds to the next one. Using the transfers for the fetch phase given in Section 11.3.1, write a set of transfers for the execution of the instruction "SKN: Skip on Accumulator Negative."

11.8. Consider a computer with the following interrupt-handling method: "When the central unit receives an interrupt it stores the contents of the *CAR* in location 0000 of the memory and then fetches the next instruction from location 0001." Write a set of transfers for the execution of this interrupt.

11.9. Show the control register's layout similar to that shown in Fig. 11.6 for the interrupt handling described in Problem 11.8.

11.10. Write and explain the timed register transfer equations for a single-address computer with indexing and indirect addressing to complete the FETCH phase. Use a 24-bit word length with a 15-bit direct address field. The memory has 2^{15} words.

11.11. A synchronous computer requires four timing pulses to perform the FETCH operation of an instruction. The instruction set with the associated timing pulses is given in the table below. Design a sequencer which uses a flexible counter to generate the timing sequences required for the automatic fetch and execution of the computer instructions.

Instruction	*Number of timing pulses*
ADD/SUB	3
COMPLEMENT	2
MULTIPLY	15
DIVIDE	15
SHIFT	1

11.12. A very simple model of the control part of a computer is shown in Fig. P11.12. The instruction register is 2 bits, and the address register is 3 bits.

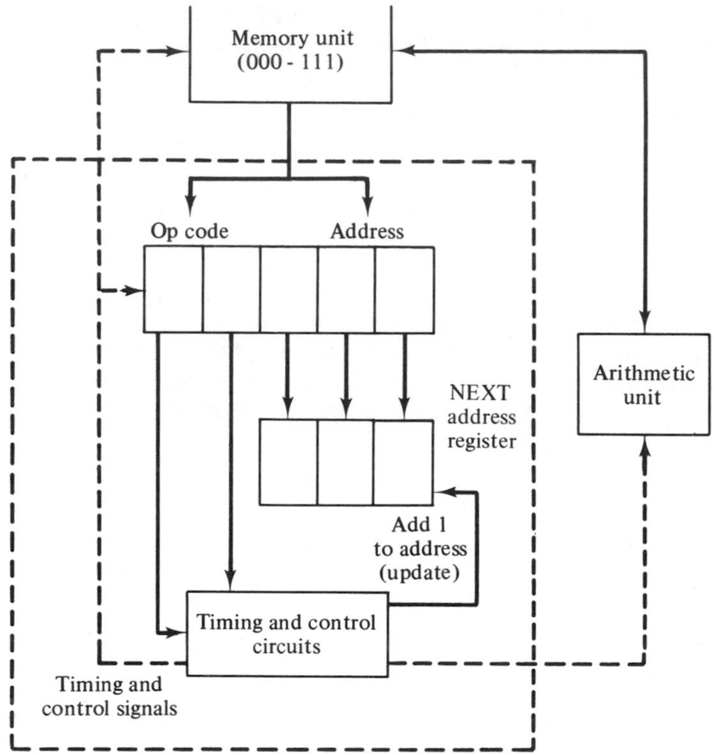

Figure P11.12 Simple control unit.

For 2 bits only four commands are possible. Use the following for this example:

00 = Add

01 = Subtract

10 = Branch if minus

11 = Store in memory

With only 3 bits there are eight different memory cells which can be specified. For this simple system, prepare a detailed description of what occurs during the FETCH and during the EXECUTE part of the cycle for each step of the following program. Only binary numbers are used in this simple setup. Rather than add more bits to allow more operation codes, halt will be indicated by the word HALT in place of any real instruction. The program should be started at location 000, and the accumulator is assumed here to be clear.

Location	Instruction	
	Operation	Address
000	00	101
001	01	110
010	10	111
011	11	110
100	HALT	
101	01	111
110	11	111
111	HALT	

11.13. Repeat Problem 11.12 for the following program. Again assume that the accumulator is clear when the program starts at 000.

Location	Operation	Address
000	00	000
001	00	100
010	11	000
011	HALT	
100	00	001

11.14. Compare horizontal and vertical microprogramming with respect to
(a) Control memory size.
(b) Control memory word length.
(c) Speed of operation.

11.15. Repeat Problem 11.14 for minimally decoded and highly decoded microprogramming.

12

Design of a Small Digital Computer

In this chapter, we shall design a small digital computer using the principles of microprogramming to define the machine. This computer will illustrate only the principles of digital computer design and will not be a complete computer which the student can find in the marketplace. However, it should give the student a good idea of how one goes about the design of a microprogrammed machine.

After surveying the marketplace, we decided to design a fixed word length binary digital computer with a 24-bit word. We came to this conclusion because there are not many 24-bit minicomputers, and, most importantly, a 24-bit computer is easier to design than a 16-bit machine, since with a 16-bit machine the method of addressing becomes more complicated due to the limited size of the direct address field in an instruction. A 24-bit word length computer is also a good compromise from a pedagogical viewpoint. It is larger than a 16-bit minicomputer but not so large as a 32-bit or 36-bit general-purpose large computer. Our machine is therefore entitled to be called a midicomputer.

12.1 DATA, INSTRUCTION FORMAT, AND REGISTERS

12.1.1 Data Formats

The computer will have two data formats. The format used for mathematical calculations will be a 24-bit fractional 2's complement binary number. Figure 12.1 shows this data format.

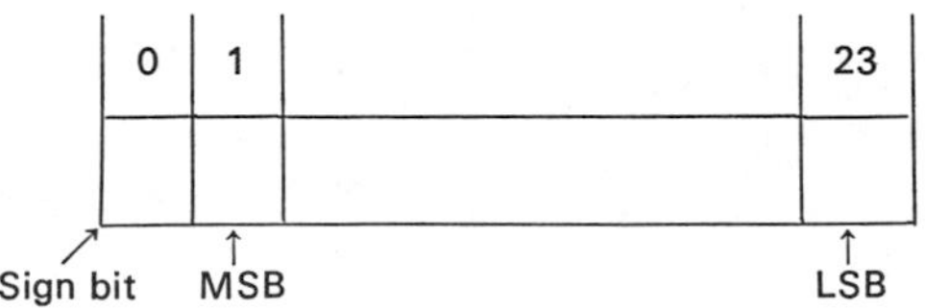

Figure 12.1 Number format.

The data format used for input/output operations will be a 6-bit Hollerith alphanumeric code to represent a character. There will be four characters to a 24-bit word. Figure 12.2 shows the I/O data format.

0 5	6 11	12 17	18 23
Character	Character	Character	Character

Figure 12.2 I/O Data format.

12.1.2 Instruction Formats

The computer has two different instruction formats. The normal format is shown in Fig. 12.3. Bits 0–4 constitute a 5-bit op-code field, bit 5 is an index register addressing flag, bit 6 is a self-relative addressing flag, bit 7 is an indirect address flag field, and bits 8–23 are a 16-bit direct address field. The following abbreviations will be used when referring to the fields of the normal instruction format:

0 4	5	6	7	8 23
Op-code	X	S	I	DA

Figure 12.3 Normal instruction format.

OP = op-code field, bits 0–4
X = index register flag field, bit 5
S = self-relative addressing flag field, bit 6
I = indirect addressing flag field, bit 7
DA = direct address field, bits 8–23

If bits 0–4, i.e., the op-code field of the instruction, contain 0s, op-code = 00000, then the second type of instruction format will be used. This is called the augmented instruction format. The 24-bit instruction word is interpreted into the fields shown in Fig. 12.4. Bits 0–4 contain 0s. Bits 4, 5, and 6 are still the X, S, and I fields but are not used in this type of instruction

0	1	2	3	4	5	6	7	8 11	12 23
0	0	0	0	0	X	S	I	Augmented op-code	N

Figure 12.4 Augmented instruction format.

format. Bits 8–11 are a 4-bit augmented op-code field, and bits 12–23 are a 12-bit field which contains a positive binary number. The abbreviations used to refer to the fields of this instruction format are

OP = op-code field, bits 0–4, all 0s
X = index register field, set to 0
S = self-relative addressing field, set to 0
I = indirect addressing flag field, set to 0
AOP = augmented op-code field, bits 8–11
N = positive binary number, bits 12–23

The normal instruction format is used with those instructions which require a reference to an address of main memory. It permits the use of a single 16-bit index register to obtain an effective address. It also allows for indirect addressing or self-relative addressing. The direct address field, DA, is 16 bits in length, as is the effective address, and therefore the computer has a maximum memory size of 2^{16} (65,536) 24-bit words.

The 5-bit op-code of this instruction permits a maximum of 32 different instructions to use the normal instruction format. However, the op-code 00000 has been used to identify the augmented instruction format, and thus there are only 31 instructions using this normal format.

The augmented instruction format has 4 bits in the augmented op-code field. It can therefore provide 16 additional instructions. The number of instructions for this computer is 31 normal instructions plus 16 augmented instructions for a total of 47 instructions. The augmented instruction format does not reference an address of main memory but is used for instructions whose address is implicit in the op-code, such as left-shift the accumulator N places; or for an instruction with immediate addressing such as load the index register immediate with the number N; or for designating the I/O device and operation to be performed by the I/O device.

12.1.3 Registers and Memory

The computer has a separate instruction unit (I unit) and execution unit (E unit). Figure 12.5 shows a block diagram of the registers in the I unit, while Fig. 12.6 shows a block diagram of the registers in the E unit. The

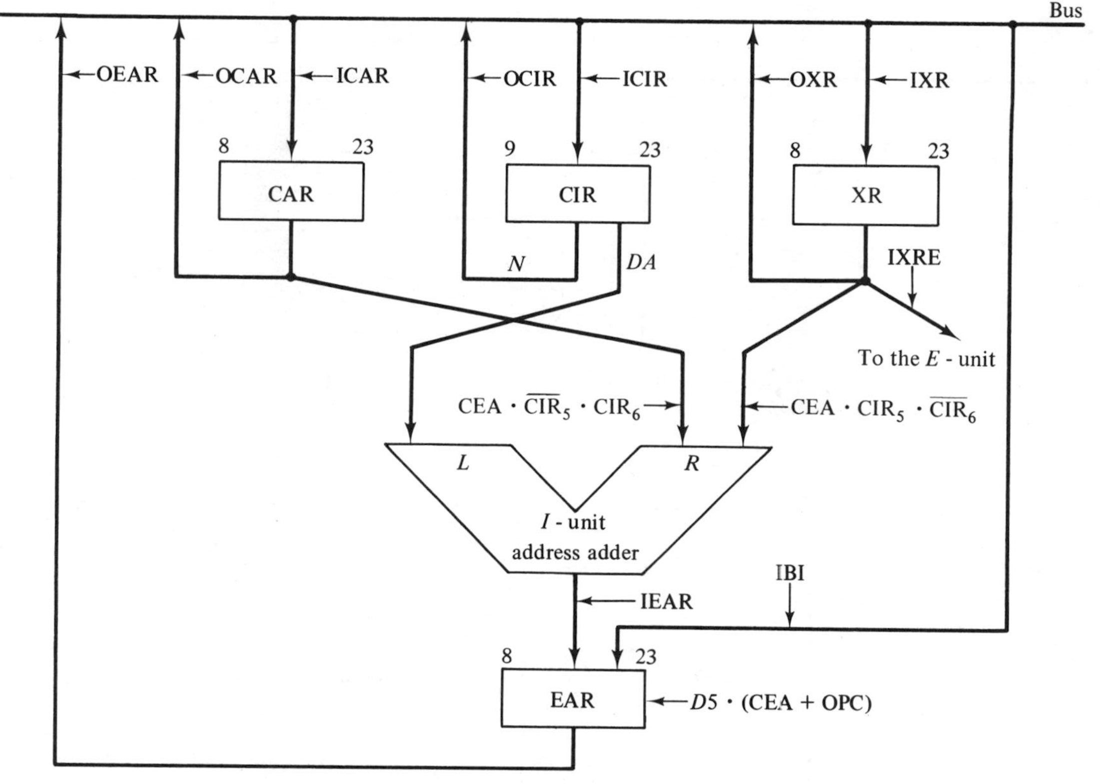

Figure 12.5 The *I*-unit information flow.

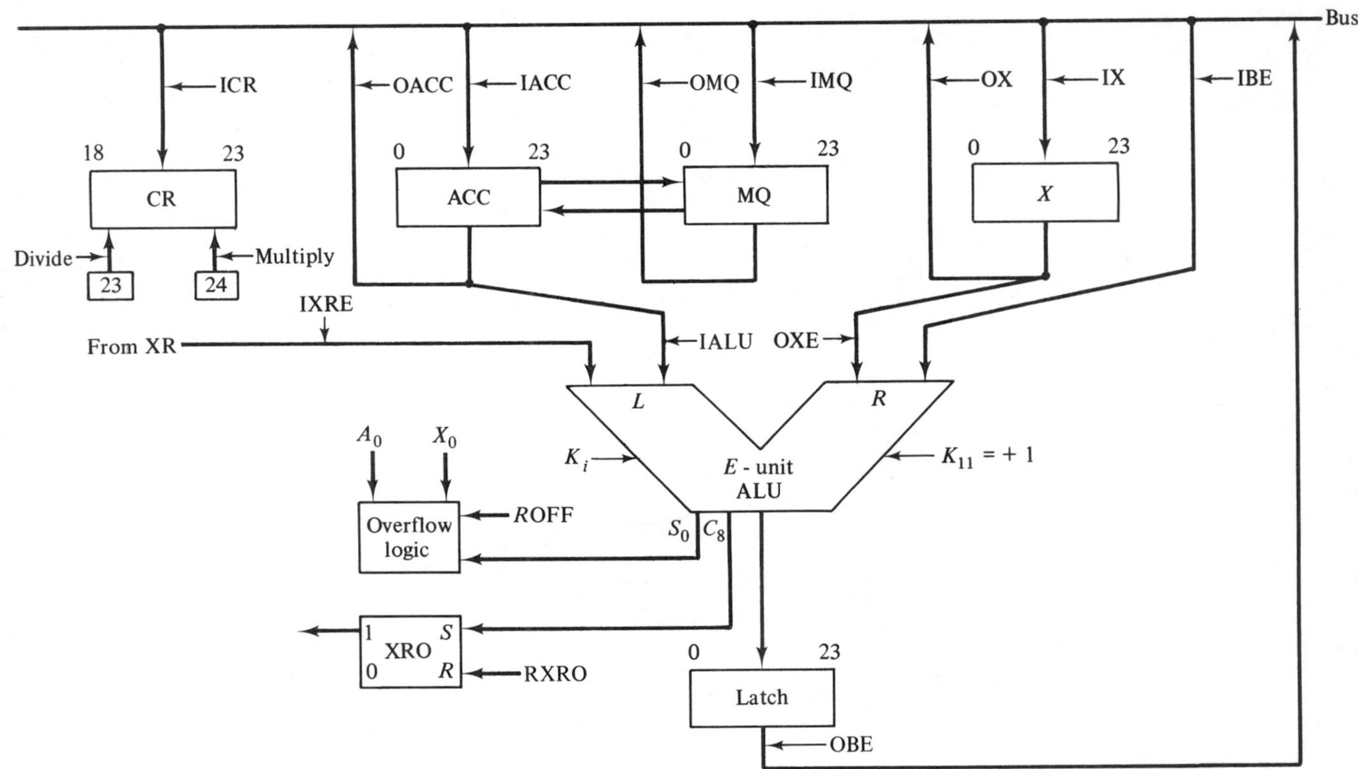

Figure 12.6 *E*-unit information flow.

timing signals which control the various transfers are indicated by arrows and are explained in later sections. The main memory of the computer is composed of 65,536 words of 24 bits each. It is a 3D magnetic core memory with a 1.024-microsecond cycle time. The main memory works asynchronously from the CPU.

12.1.3.1 The I Unit

The instruction unit contains the following registers:

1. *CAR* (current address register), 16 bits long. This register contains the address of the current instruction.
2. *CIR* (current instruction register), 24 bits long. This register holds the current instruction.
3. *XR* (index register), 16 bits long.
4. *EAR* (effective address register), 16 bits long. This register contains the computed effective address of the data.

The *CIR* can be subdivided into several subregisters which are defined as

1. *OP* (*CIR*): The op-code position, bits 0–4.
2. *X* (*CIR*): The indexing flag bit, bit 5.
3. *S* (*CIR*): The self-relative address flag bit, bit 6.
4. *I* (*CIR*): The indirect addressing flag bit, bit 7.
5. *DA* (*CIR*): The direct address field, bits 8–23.

The above subregisters are for the normal instruction format. When the augmented instruction format is being used, the *CIR* can be considered to be subdivided into the following subregisters:

1. *OP* (*CIR*): Op-code portion, bits 0–4.
2. *X* (*CIR*): Indexing flag bit, bit 5.
3. *S* (*CIR*): The self-relative flag bit, bit 6.
4. *I* (*CIR*): The indirect address flag bit, bit 7.
5. *AOP* (*CIR*): The augmented op-code portion, bits 8–11.
6. *N* (*CIR*): A binary number or immediate data, bits 12–23.
7. *SH* (*CIR*): A binary number used for shifting bits 18–23.

To compute the effective address, the contents of two registers must be summed. In this machine a separate 16-bit parallel binary adder is provided in the I unit for this purpose. It is called the address adder.

12.1.3.2 The E Unit

The E unit contains three registers and a 24-bit parallel binary arithemetic and logic unit. The registers are

1. *ACC*: A 24-bit accumulator.
2. *MQ*: A 24-bit multiplier-quotient register.
3. *X*: A 24 bit register used to hold the multiplicand for multiplication and the divisor in division. It is also used to hold an operand once it has been fetched from memory for use with the arithmetic and logic unit (ALU).

The E Unit also contains a special 6-bit counter register, *CR*, which is used to control the number of shifts performed in a shifting instruction or the number of iterations performed in a multiply or divide instruction.

The arithmetic and logic unit (ALU) can perform 23 different operations which depend on the control signals it receives. Besides performing the addition and subtraction of two binary numbers, it can form the 16 different Boolean functions of 2 Boolean variables.

The accumulator overflow flip-flop (*OFF*) is a single-bit flip-flop which is set at the output of the ALU if an overflow condition is detected during an arithmetic operation involving the accumulator.

The index register overflow flip-flop (*XRO*) is a single-bit register which is set at the output of the ALU if a carry is produced from bit location 8 during an instruction whose execution involves the index register.

12.1.3.3 The Memory Unit

The main memory is a random access magnetic core memory with a cycle time of 1024 nanoseconds and an access time of 384 nanoseconds. It contains $2^{16} = 65{,}536$, 24-bit words of memory. The memory has two main registers connected to the bus. One is the *MAR* (memory address register), which contains the effective address where information is being transferred. The other is the *MIR* (memory information register), which holds the information during the write into a word location or receives the information from a readout of a location.

The memory unit is designed to operate in an asynchronous manner from the CPU. That is, once the CPU initiates a transfer, the main memory contains sufficient timing and logic circuitry to control its own sequencing. Thus, the memory operates similarly to an I/O device. A hand-shaking mechanism must be established between the CPU and the main memory subsystem with regard to the reception and sending of addresses and data. The CPU can adapt to the speed of the memory. Since it is desirable to operate the computer at the fastest possible speed, the memory should be designed to have a fast cycle time so as not to slow down the entire computer system. Therefore, it is desirable to have the memory synchronized to the CPU to achieve a maximum memory transfer rate.

Fig. 12.7 shows a block diagram of the informatian flow in memory. The timing sequence for the operation of the memory is as follows:

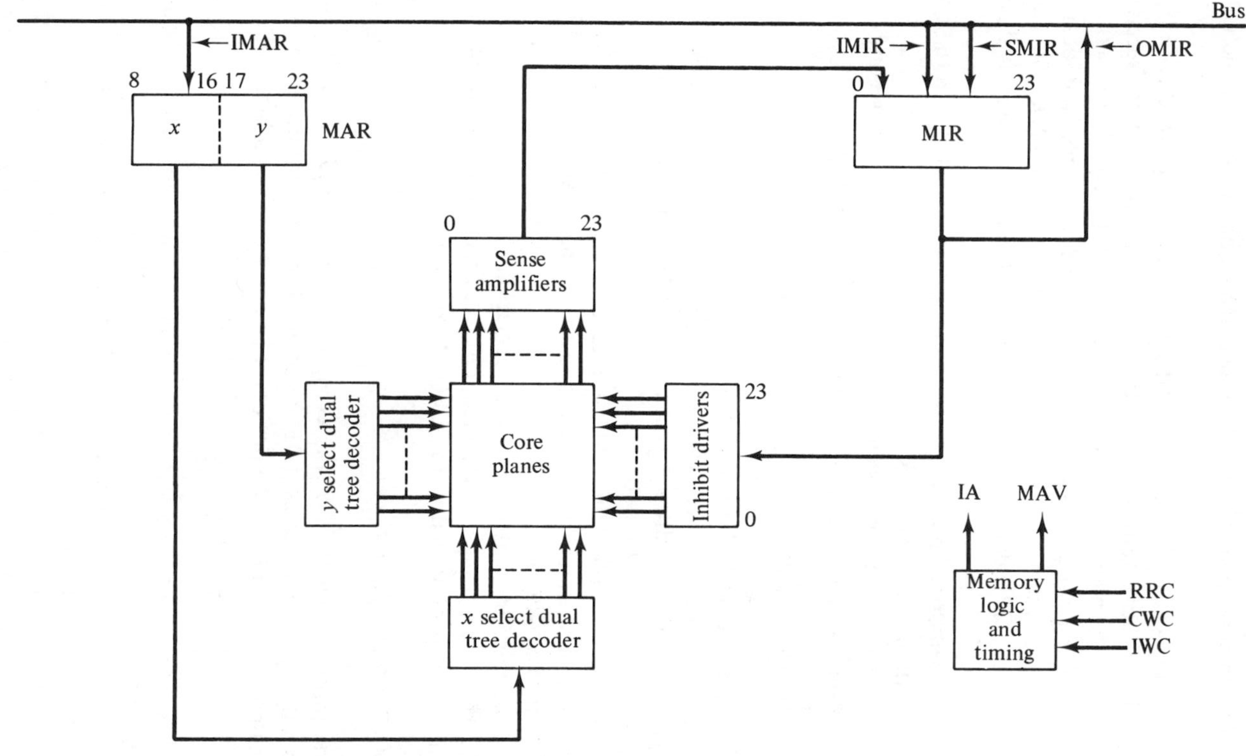

Figure 12.7 Memory unit information flow.

Read-Restore Cycle (RRC)

1. Memory available: A signal is sent from the memory to the CPU, $MAV = 1$.

2. Commence *read* cycle: The RRC signal is sent from the CPU to memory. The effective address is sent to the memory MAR. After receiving the read signal, the memory signals the CPU that it is busy, $MAV = 0$.

3. The memory decodes the address in MAR and cycles the location. After a delay of 384 nanoseconds, the data will arrive in the MIR. The memory will signal the CPU that data are available in MIR (information available, IA).

4. The memory commences the restore part of the cycle, and the CPU transfers the data out of the MIR.

5. When the cycle has been completed, i.e., 1024 nanoseconds, the memory-available signal is presented to the CPU, $MAV = 1$.

Clear-Write Cycle (CWC)

1. Same as item 1 above.

2. The commence *write* cycle signal, CWC, is sent from the CPU to memory. The effective address is sent to the MAR. The memory signals the CPU that it is busy, $MAV = 0$.

3. Memory decodes the address and cycles the location. After 384 nanoseconds, the location is cleared, and the memory signals the CPU to send data into MIR (IA).

4. The memory writes the new information into the location.

5. After the writing is completed, i.e., 1024 nanoseconds, the memory-available signal is presented to the CPU, $MAV = 1$.

12.1.3.4 The Input/Output

The input/output part of the computer is composed of five registers used in a direct memory access (DMA) channel to move information to and from main memory and the I/O devices. Figure 12.8 shows a block diagram of the I/O channel data flow. The registers are

1. $IOIR$: A 24-bit register which buffers data to and from I/O units and main memory.

2. $IOCR$: I/O Control register used to control the I/O channel. This register is 24 bits in length.

3. $IOAR$: A 16-bit register which holds the effective address of the location in main memory from which information is being transferred.

4. $IOWC$: A 16-bit register which initially contains the number of words of information to be transferred between memory and the I/O device.

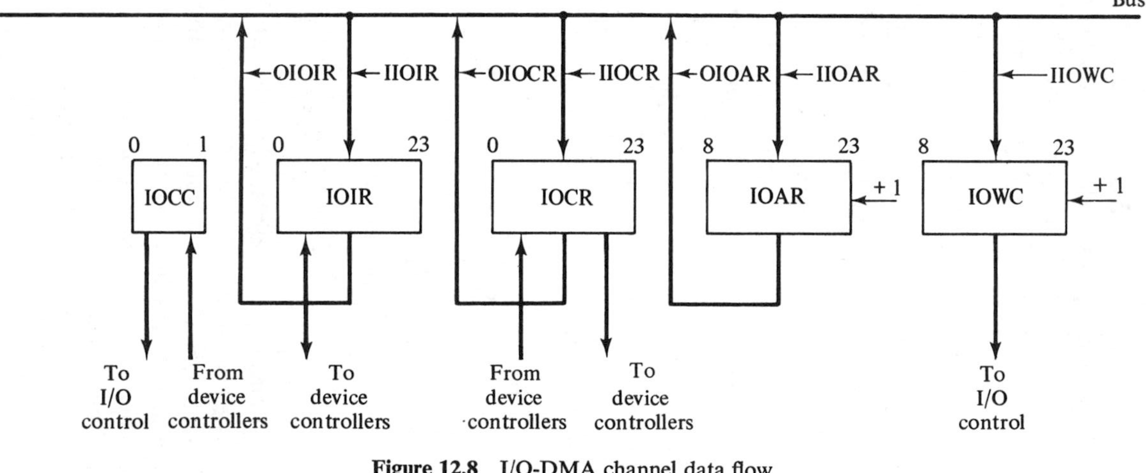

Figure 12.8 I/O-DMA channel data flow.

5. *IOCC*: A 2-bit register which counts the number of characters which are being transferred between the device and the *IOIR*.

Information is moved between the I/O information register (*IOIR*) and main memory as a 24-bit word in parallel. Each word is composed of four 6-bit Hollerith coded characters. Data move between the *IOIR* and I/O devices one 6-bit character at a time. The assembling of a word is controlled by the *IOCC* (I/O character counter).

The start I/O instruction (SIO) moves the effective address of the first word of information to be placed in main memory into the I/O address register (*IOAR*). The load I/O instruction (LIO) loads the number of words of information to be transferred to and from main memory into the I/O word counter (*IOWC*).

The connect device instruction (CND) moves the address of the I/O device into the I/O control register (*IOCR*). It also moves the device command code into the *IOCR*. The command code is the operation which the device is to perform.

Bits 0–11 of the *IOCR* receive the status of the device from the device control unit. The load status instruction (LDS) transfers this status code from the *IOCR* into the accumulator.

The *IOCR* is divided into several subregisters:

1. Status: Bits 0–11.
2. Command code: Bits 12–15.
3. I/O Class: Bits 16–19.
4. Device number: Bits 20–23.

12.1.4 Information Flow

The register diagrams of the I unit and E unit show one-line inputs to or outputs from a register. Since we shall design a parallel machine, there are in fact 16 or 24 lines used to transfer the bits of information into or from the registers. However, only one line is shown in the diagrams.

The output bus lines are constructed from the ORing of all the outputs of the corresponding bits of each of the registers. To permit only the outputs of one register at a time on the bus an AND gate or its logical equivalent is inserted between the output bus OR gate and a particular register. A control line is the other input to the AND gate. Figure 12.9 shows the arrangement for the output bus.

The input to the flip-flops of the registers are also ORed together. The signal is passed to an AND gate where a control signal is applied and usually a timing strobe signal derived from the clock and timing subfunctions. Figure

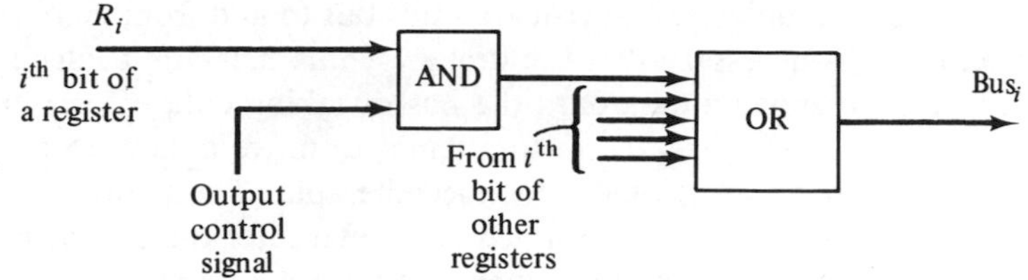

Figure 12.9 Output control of registers.

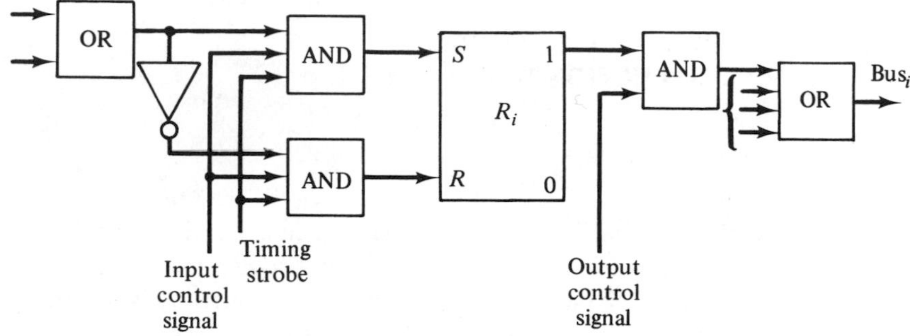

Figure 12.10 Register control and information flow.

12.10 shows the input to the ith bit of a register as well as the output from the register to the output bus.

There are times when the register must be reset to the all-0 state. This is usually accomplished by inserting an OR gate in front of the reset input of the flip-flop in Fig. 12.10. Figure 12.11 shows this configuration.

The register diagram of the machine shown in the previous figures reveals that this computer has a single-bus structure. All information moving between any of its registers and/or memory moves across this single bus. All

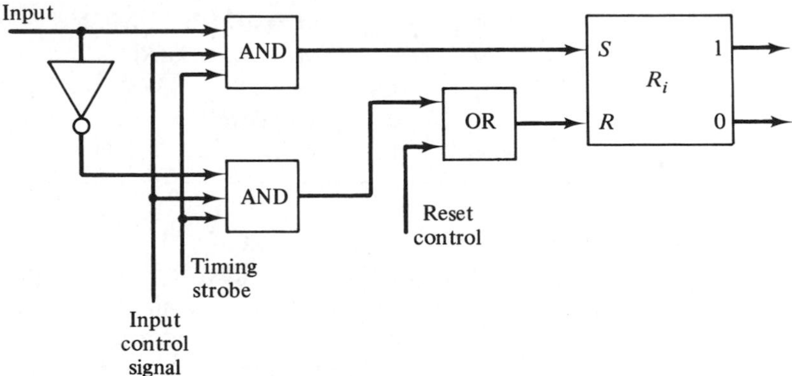

Figure 12.11 Register control with reset.

the I/O data moving to devices move across this bus to and from main memory. Therefore, it is necessary that the design of this machine be such that only one register is moving data onto the bus or taking data off the bus at any given time. If two registers were to attempt to move data onto the bus simultaneously, the results would be undecipherable. The control signals shown in Fig. 12.5–12.8 must be so timed as to guarantee the uniqueness of the data path. The control signals are shown as arrows pointing to the transfer paths. The name of the control signal is also indicated.

12.2 METHODS OF ADDRESSING AND INSTRUCTION SET

12.2.1 Methods of Addressing

Bits 5–7 of the normal instruction format are flags used to specify the type of addressing which will be used with the particular instruction. A 0 in a bit field means that the particular type of addressing method is not being used, while a 1 means that the option is used. Table 12.1 shows how the effective address (*EA*) is obtained and what combinations of addressing methods are permitted.

Table 12.1 *Definition of Addressing Methods*

Name	*X*, *bit 5*	*S*, *bit 6*	*I*, *bit 7*	*Effective address* (*EA*)
Direct address	0	0	0	$EA = DA$
Indirect address	0	0	1	$EA = [DA]$, bits 8–23
Self-relative	0	1	0	$EA = [CAR] + DA$
	0	1	1	$EA = [[CAR] + DA]$, bits 8–23
Indexing	1	0	0	$EA = DA + [XR]$
	1	0	1	$EA = [DA + [XR]]$, bits 8–23
	1	1	0	Not allowed
	1	1	1	Not allowed

In the two cases where there are a combination of addressing methods, the indirect address is obtained after the effective address has been computed.

The effective address is computed in the I unit using an address adder; see Fig. 12.5. The effective address is placed into the effective address register, (*EAR*). If direct addressing is used, the direct address (*DA*) is gated to one input of the adder and 0s are gated to the other input. The output placed into the *EAR* will be equal to the *DA*.

The I unit address adder performs addition on two 16-bit numbers without any indication of an overflow. Thus, to subtract 1 from the direct address, it is necessary to add $2^{16} - 1$. The following example shows this.

Example 12.1

Assume that the *CAR* contains the octal number 000132, i.e., decimal 90. The instruction which is fetched from this location specifies self-relative addressing by having bit 6 equal to 1. The intent of the instruction is to branch back four instructions in the program. Thus, we wish to branch to address $(000126)_8$. This is accomplished by placing the number −4 in 2's complement in the *DA* field of the instruction. The instruction looks like

0 — 4	5	6	7	8 — 23
Op-code	0	1	0	$(177774)_8$

The address adder will perform the following operation, shown in octal notation:

$$\begin{aligned} [CAR] &= 000132 \\ \underline{DA} &= \underline{177774} \\ [CAR] + DA &= 200126 \end{aligned}$$

The contents of the 16 bits in the *DA* field can be represented by six octal characters with the most significant character being either 0 or 1. Therefore, the largest octal location which can be addressed is 177777. The leading 2 in the above will overflow the adder and be discarded. Therefore, the contents of the *EAR* will be

$$[EAR] = 200126 \text{ mod } 2^{16} = 000126$$

12.2.2 The Instruction Set

Table 12.2 is a listing of the instructions which utilize the normal instruction format. Included in the table is the binary op-code, its decimal equivalent, the instruction mnemonic used by the assembler, the name of the instruction, and a short definition of the instruction.

Table 12.3 is a listing of the instructions which utilize the augmented instruction format. The same information is also presented for these instructions.

12.2.3 Instruction Definitions and Use

In this subsection we shall define the instructions of the machine which were listed in Tables 12.2 and 12.3.

Table 12.2 *Instructions Using the Normal Format*

Decimal	*Op-code*	*Mnemonic*	*Name*	*Definition*
0	00000	—		Indicates augmented instruction format
1	00001	ADD	Addition	$[ACC] + [EA]\dagger \longrightarrow ACC$
2	00010	SUB	Subtraction	$[ACC] - [EA] \longrightarrow ACC$
3	00011	MUL	Multiplication	$[ACC] \times [EA] \longrightarrow ACC//MQ$
4	00100	DIV	Division	$[ACC] \div [EA] \longrightarrow MQ$; remainder in ACC
5	00101	STQ	Store MQ	$[MQ] \longrightarrow EA$
6	00110	STA	Store ACC	$[ACC] \longrightarrow EA$
7	00111	STX	Store index	$[XR] \longrightarrow EA_{8-23}$
8	01000	LDA	Load ACC	$[EA] \longrightarrow ACC$
9	01001	LDX	Load index	$[EA]_{8-23} \longrightarrow XR$
10	01010	STC	Store CAR	$[CAR] \longrightarrow EA_{8-23}$
11	01011	TXA	Transfer XR to ACC	$[XR] \longrightarrow ACC_{8-23}; 0 \longrightarrow ACC_{0-7}$
12	01100	TMQ	Transfer MQ to ACC	$[MQ] \longrightarrow ACC$
13	01101	ADX	Add to index	$[XR] + [EA]_{8-23} \longrightarrow XR$
14	01110	SIO	Start I/O	$EA \longrightarrow IOAR$

15	01111	LIO	Load I/O	$EA \longrightarrow IOWC$
16	10000	UNB	Unconditional branch	$[EAR] \longrightarrow CAR$
17	10001	BAO	Branch *ACC* overflow	If $[ACC] > 1\text{–}2^{-23}$, $[EAR] \longrightarrow CAR$
18	10010	BXP	Branch index positive	If $[XR] > 0$, $[EAR] \longrightarrow CAR$
19	10011	BXZ	Branch index zero	If $[XR] = 0$, $[EAR] \longrightarrow CAR$
20	10100	BXN	Branch index negative	If $[XR] < 0$, $[EAR] \longrightarrow CAR$
21	10101	TLD	Tally down	If $[XR] > 0$, $[XR] - 1 \longrightarrow XR$; $[EAR] \longrightarrow CAR$
22	10110	BAP	Branch *ACC* positive	If $[ACC] > 0$, $[EAR] \longrightarrow CAR$
23	10111	BAZ	Branch *ACC* zero	If $[ACC] = 0$, $[EAR] \longrightarrow CAR$
24	11000	BAN	Branch *ACC* negative	If $[ACC] < 0$, $[EAR] \longrightarrow CAR$
25	11001	LOR	Logical-OR	$[ACC]$ or $[EA] \longrightarrow ACC$
26	11010	LPR	Logical product	$[ACC] \cdot [EA] \longrightarrow ACC$
27	11011	LNG	Logical negation	$[\overline{ACC}] \longrightarrow ACC$
28	11100	EOR	Logical exclusive-OR	$[ACC] \oplus [EA] \longrightarrow ACC$
29	11101	SRJ	Subroutine jump	$[CAR] + 1 \longrightarrow EA_{8-23}$; $[EAR] \longrightarrow CAR$
30	11110	BDN	Branch device not available	If I/O device not available, $[EAR] \longrightarrow CAR$
31	11111	NOP	No operation	Take the next sequential instruction

†*EA* is the effective address in memory.

Table 12.3 Instructions Using the Augmented Format

Decimal	Augmented Op-code	Mnemonic	Name	Definition
0	0000	STP	Stop	Computer stopped $0 \longrightarrow RUN$
1	0001	CMA	2's complement *ACC*	$[\overline{ACC}] + 1 \longrightarrow ACC$
2	0010	ALA	Arithmetic left shift *ACC*	$R(ACC) \longrightarrow L(ACC); 0 \longrightarrow ACC_{23}$; N times
3	0011	ARA	Arithmetic right shift *ACC*	$L(ACC) \longrightarrow R(ACC); [ACC_0] \longrightarrow ACC_1$; N times
4	0100	LRQ	Logical right shift *ACC//MQ*	$L(ACC \| MQ) \longrightarrow R(ACC \| MQ); 0 \longrightarrow ACC_0$; N times
5	0101	LLQ	Logical left shift *ACC//MQ*	$R(ACC \| MQ) \longrightarrow L(ACC \| MQ); 0 \longrightarrow MQ_{23}$; N times
6	0110	LLA	Logical left shift *ACC*	$R(ACC) \longrightarrow L(ACC); 0 \longrightarrow ACC_{23}$; N times
7	0111	LRA	Logical right shift *ACC*	$L(ACC) \longrightarrow R(ACC); 0 \longrightarrow ACC_0$; N times
8	1000	LCA	Logical circular left shift *ACC*	$\rho^{-1}[ACC]$; N times
9	1001	LAI	Load *ACC* immediate	$N \longrightarrow ACC$
10	1010	LXI	Load index immediate	$N \longrightarrow XR$
11	1011	INX	Increment index	$[XR] + N \longrightarrow XR$
12	1100	DEX	Decrement index	$[XR] - N \longrightarrow XR$
13	1101	CND	Connect I/O device	$[CIR]_{12-23} \longrightarrow IOCR_{12-23}$
14	1110	ENI	Enable interrupt	Enable the interrupt system
15	1111	LDS	Load status	$[IOCR]_{0-11} \longrightarrow ACC_{12-23}; 0 \longrightarrow ACC_{0-11}$

12.2.3.1 Arithmetic Operations

Addition—01—ADD Add the contents of the memory address computed as the effective address to the contents of the accumulator. Place the sum in the accumulator register.

The architecture of this computer requires that the operands fetched from memory be placed temporarily in the X register. The ALU receives its information from the X register and the accumulator register. The accumulator is designed using master-slave flip-flops so that it may receive input information while it is sending output information and not create a race condition. The addition instruction will be executed in the following phases:

1. $[EA] \longrightarrow MIR$
2. $[MIR] \longrightarrow X$
3. $[ACC] + [X] \longrightarrow ACC$

Subtraction—02—SUB Subtract the contents of the effective address from the accumulator. Place the difference in the accumulator.

The machine uses 2's complement notation and performs the subtraction by addition of the 1's complement of the operand plus 1 to the contents of the accumulator. The register transfers involved are

1. $[EA] \longrightarrow MIR$
2. $[MIR] \longrightarrow X$
3. $[ACC] + [\bar{X}] + 1 \longrightarrow ACC$

This is accomplished by complementing the output of the X register as it enters the arithmetic and logic unit of the execution unit and forcing the input carry to the LSB to be a 1.

Multiplication—03—MUL Multiply the contents of the accumulator by the contents of the effective address. Place the most significant bits of the product in the accumulator and the least significant bits in the MQ register. The product of two 24-bit numbers can be 48 bits long. The algorithm is shown in Section 8.2.2. Initially, the multiplier is in the accumulator, and the multiplicand is in memory. The register transfer equations are

1. $[EA] \longrightarrow MIR$; $[ACC]_0 \longrightarrow Scand$; $24 \longrightarrow CR$;
$[ACC]_0 \cdot \{[\overline{ACC}] + 1\}$ or $[\overline{ACC}]_0 \cdot \{[ACC]\} \longrightarrow MQ$;
magnitude into MQ
2. $[MIR] \longrightarrow X$; $[X]_0 \oplus Scand \longrightarrow Scand$; $0 \longrightarrow ACC$
3. $[X]_0 \cdot \{[\bar{X}] + 1\}$ or $[\bar{X}]_0 \cdot [X] \longrightarrow X$; magnitude into X
4. $[ACC] + [X] \longrightarrow ACC$ if $[MQ]_{23} = 1$; add
5. $L(ACC//MQ) \longrightarrow R(ACC//MQ)$; $0 \longrightarrow ACC_0$; shift

6. $[CR] - 1 \longrightarrow CR$; branch to step 4 if $[CR] \neq 0$
7. $[Scand]\cdot\{[\overline{ACC}] + 1\}$ or $[\overline{Scand}]\cdot[ACC] \longrightarrow ACC$; complement answer
8. $[Scand]\cdot\{[\overline{MQ}] + 1\}$ or $[\overline{Scand}]\cdot[MQ] \longrightarrow MQ$; complement answer

The multiplicand is fetched from memory into the *MIR*. The sign of the multiplier is stored in a 1-bit register, *Scand*. The number of iterations is placed in the counter register (*CR*). This register will count down the number of iterations needed for the add and shift algorithm. While the multiplicand is being fetched, the multiplier is transferred through the ALU to obtain its magnitude and placed into the *MQ* register. The multiplicand is transferred to the *X* register. Its sign is exclusive-ORed with the sign of the multiplier and placed back into *Scand*. The contents of the accumulator are set to 0. The contents of the *X* register (multiplier) are set to its magnitude. If the LSB of *MQ* is a 1, the contents of the *X* register are added to the accumulator. Logical right shift the accumulator and *MQ* register one place, placing ACC_{23} into MQ_0. Decrement the counter register by 1. If it is not 0, branch back to step 4, the add step. After completing 24 add and shift cycles, the magnitude of the product is in the accumulator and *MQ* register. If the answer should be negative, it is necessary to 2's complement both the accumulator and the *MQ* register. The time of the multiplication instruction will be variable depending on the sign of the operands as well as the number of 0s in the multiplier.

Division—04—DIV Divide the contents of the accumulator by the contents of the effective address. Place the quotient in the *MQ* register and the remainder in the accumulator. The quotient is 24 bits long. The algorithm used is the nonrestoring division as shown in Section 8.2.3. The dividend is initially in the accumulator, and the divisor is in memory. The register transfer equations are

1. $[EA] \longrightarrow MIR$; $[ACC]_0 \longrightarrow Send$; $23 \longrightarrow CR$; $[ACC]_0\cdot\{[\overline{ACC}] + 1\}$ or $[\overline{ACC}]_0\cdot[ACC] \longrightarrow ACC$; magnitude in ACC
2. $[MIR] \longrightarrow X$; $[X]_0 \oplus [Send] \longrightarrow Send$; $0 \longrightarrow MQ$;
3. $[X]_0\cdot\{[\bar{X}] + 1\}$ or $[\bar{X}]_0\cdot[X] \longrightarrow X$; magnitude of X
4. $[ACC] - [X] \longrightarrow ACC$; test overflow condition
5. if $[ACC]_0 = 0$, $1 \longrightarrow OFF$; set overflow and stop; if $[ACC]_0 = 1$, $[ACC] + [X] \longrightarrow ACC$; restore if no overflow
6. $R[ACC//MQ] \longrightarrow L[ACC//MQ]$; shift
7. $[ACC] - [X] \longrightarrow ACC$; subtract
8. $R[ACC//MQ] \longrightarrow L[ACC//MQ]$; $[\overline{ACC}]_0 \longrightarrow MQ_{23}$; shift; $[CR] - 1 \longrightarrow CR$; decrement counter

9. $[MQ]_{23} \cdot \{[ACC] - [X]\}$ or $\overline{[MQ]}_{23} \cdot \{[ACC] + [X]\} \longrightarrow ACC$
10. If $[CR] \neq 0$, branch to step 8
11. $[Send] \cdot \{\overline{[MQ]} + 1\}$ or $\overline{[Send]} \cdot [MQ] \longrightarrow MQ$

The divisor is fetched from the memory into the *MIR*. The sign of the dividend is recorded in the *Send* flip-flop and 23 is placed into the counter register (*CR*). The dividend is passed through the ALU and placed into magnitude form and returned to the accumulator. In step 2, the divisor is sent to the *X* register, and its sign is exclusive ORed with the *Send* flip-flop and saved in *Send*. The *MQ* register is cleared. In step 3, the divisor is placed into magnitude form. In step 4, the divisor is subtracted from the accumulator, and the results are stored in the accumulator. This is a test for the divide overflow. If the results are positive, i.e., the dividend is larger than the divisor, the overflow flip-flop is sct and the division instruction will be suppressed. The next sequential instruction will be executed. If the accumulator is negative, the algorithm continues by restoring the accumulator back to the magnitude of the dividend. In step 6, the combination of the accumulator and *MQ* register is shifted left 1 bit. In step 7, the divisor is subtracted from the dividend. In step 8, the combined *ACC* and *MQ* registers are shifted 1 bit left, and the sign bit A_0 is negated and sent to bit 23 of the *MQ* register. This will become the sign bit of the quotient. In step 9, if the bit in MQ_{23} is a 1, the divisor is subtracted from the accumulator, while if it is a 0, the divisor is added. The counter register is decremented by 1. In step 10, the *CR* is tested; if it is not a 0, the algorithm branches back to step 8; if it is a 0, the next sequential step is performed. After 23 iterations of steps 8, 9, and 10, the counter will be 0 and step 11 will be performed. Step 11 places the accumulator into the correct representation based on the sign bit computation retained in the *Send* flip-flop.

Add to Index—13—ADX Add the contents of the index register to bits 8–23 of the contents of the effective address. Place the sum in the index register. The register transfers involved are

1. $[EA] \longrightarrow MIR$
2. $[MIR]_{8\text{-}23} \longrightarrow X_{8\text{-}23}$; $0 \longrightarrow X_{0\text{-}7}$
3. $[XR] + [X]_{8\text{-}23} \longrightarrow XR$; if $C_8 = 1$, $1 \longrightarrow XRO$

Since this instruction operates on the index register, indexing is not permitted in computing the effective address. If a carry is produced out of bit position 8, then the *XRO* flip-flop will be set to 1 to indicate *XR* overflow.

2's Complement the Accumulator—0—01—CMA This is an instruction in the augmented instruction format and hence has no effective address. It is performed by the ALU using the following register transfer:

1. $[\overline{ACC}] + 1 \longrightarrow ACC$

Increment Index Register—0—11—INX This is an arithmetic operation using the augmented instruction format, and therefore no reference is made to an effective address. The contents of the index register (XR) is one operand, and the contents of the current instruction register bits 12–23 are the second operand. This is an immediate address instruction format. The register transfer equations are

1. $[N(CIR)] \longrightarrow X$
2. $[XR] + [X] \longrightarrow XR$; if $C_8 = 1$, $1 \longrightarrow XRO$

The instruction is performed in two parts. First, the immediate operand, N, is moved into the X register. Then the contents of the index register and X register are added together in the ALU, and the sum is stored in the index register. If a carry is produced from bit position 8, the XRO flip-flop will be set to indicate XR overflow.

Decrement Index Register—0—12—DEX This is also an immediate address instruction which uses the augmented instruction format. The contents of the index register are one operand, and bits 12–23 of the CIR, $N(CIR)$, contain the second operand. The register transfer equations are

1. $[N(CIR)] \longrightarrow X$
2. $[XR] + [\bar{X}] + 1 \longrightarrow XR$; if $C_8 = 1$, $1 \longrightarrow XRO$

The subtraction is performed by adding the 2's complement of the immediate operand, which has been transferred to the X register, to the contents of the index register. The operation uses the ALU, and the difference is stored in the index register. If a carry is produced from bit 8 of the ALU, the XRO flip-flop is set.

12.2.3.2 Transfer Operations

Store MQ—05—STQ Store the contents of the MQ register into the effective address of main memory. The register transfer equations are

1. $[MQ] \longrightarrow MIR$
2. $[MIR] \longrightarrow EA$

Store ACC—06—STA Store the contents of the accumulator into the effective address of main memory.

1. $[ACC] \longrightarrow MIR$
2. $[MIR] \longrightarrow EA$

Store XR—07—STX Store the contents of the index register into the effective address of main memory.

1. $[XR] \longrightarrow MIR_{8\text{-}23}$
2. $[MIR] \longrightarrow EA$

The index register is 16 bits in length, and when stored into main memory only bits 8–23 of the memory location are changed. Bits 0–7 remain as they were before the store.

Store CAR—10—STC Store the contents of the *CAR* into the effective address of main memory.

1. $[CAR] \longrightarrow MIR_{8\text{-}23}$
2. $[MIR] \longrightarrow EA$

The *CAR* is 16 bits in length and is stored into bits 8–23 of the *EA*. Bits 0–7 of the *EA* are not affected.

Load ACC—08—LDA Load the contents of the effective address of main memory into the accumulator register.

1. $[EA] \longrightarrow MIR$
2. $[MIR] \longrightarrow X$
3. $[X] \longrightarrow ACC$

Load XR—09—LDX Load the contents of the effective address (bits 8–23) into the index register.

1. $[EA] \longrightarrow MIR$
2. $[MIR] \longrightarrow X$
3. $[X]_{8-23} \longrightarrow XR$

Transfer XR into ACC—11—TXA Transfer the contents of the index register into the accumulator bits 8–23. Bits 0–7 of the accumulator are set to 0. The transfer is performed directly.

1. $[XR] \longrightarrow ACC_{8\text{-}23},\ 0 \longrightarrow ACC_{0\text{-}7}$

This instruction has no effective address and is an exception to the normal instruction format.

Transfer MQ into ACC—12—TMQ Transfer the contents of the MQ register into the accumulator.

1. $[MQ] \longrightarrow ACC$

This instruction is also an exception in the normal instruction format since it has no effective address.

Load Accumulator Immediate—0—09—LAI This is an augmented instruction format with an immediate address operand. The operand is a maximum 12-bit positive binary number contained in bits 12–23 of the CIR and designated as N. It is loaded into the accumulator. The transfer equation is

1. $[CIR]_{12\text{-}23} \longrightarrow ACC_{12\text{-}23};\ 0 \longrightarrow ACC_{0\text{-}11}$

Load Index Register Immediate—0—10—LXI This is similar to the load accumulator immediate instruction and the transfer equation is

1. $[CIR]_{12\text{-}23} \longrightarrow XR_{12\text{-}23};\ 0 \longrightarrow XR_{8\text{-}11}$

12.2.3.3 Branch Instructions

Unconditioned Branch—16—UNB The unconditioned branch places the computed effective address which is held in the effective address register (EAR) into the current address register (CAR).

1. $[EAR] \longrightarrow CAR$

Branch on Accumulator Overflow—17—BAO The accumulator overflow flip-flop is tested to determine if it has been set to 1. If it has, the contents of the EAR are placed into the CAR and an instruction fetch cycle is started. If it has not, the next sequential instruction is fetched.

1. If $[OFF] = 1$, $[EAR] \longrightarrow CAR$; if $[OFF] = 0$, $[CAR] + 1 \longrightarrow CAR$

Branch on Index Register Positive—18—BXP The contents of the index register are tested to determine if they are positive. If they are, $[EAR] \longrightarrow CAR$; if they are not positive, the next sequential instruction is fetched.

1. If $[XR] > 0$, $[EAR] \longrightarrow CAR$; if $[XR] \leq 0$, $[CAR] + 1 \longrightarrow CAR$

Branch on Index Register Zero—19—BXZ The contents of the index register are tested to determine if they are 0. If so, $[EAR] \longrightarrow CAR$, and if they are not, the next sequential instruction is fetched.

1. If $[XR] = 0$, $[EAR] \longrightarrow CAR$; if $[XR] \neq 0$, $[CAR] + 1 \longrightarrow CAR$

Branch on Index Register Negative—20—BXN The contents of the index register cannot become negative because the index register is 16 bits in length and represents possible addresses in the main memory. However, its contents can be changed by such instructions as tally and decrement index register. Therefore, a special 1-bit flip-flop register called *XRO* is set whenever an instruction which increments or decrements the contents of the index causes an overflow in the register, that is, if the contents go through zero and become negative, or go beyond 2^{-7} and overflow positive. The number contained in the index register under either of these two conditions is still a valid address. The *XRO* flip-flop is set to 1 if a carry or borrow is produced out of bit location 8 of the ALU during an add to index instruction, ADX; increment index, INX; or a decrement index, DEX. The BXN instruction therefore performs the function of testing the *XRO* to determine its state. If it is a 1, $[EAR] \rightarrow CAR$, and the instruction fetch cycle is started. If the *XRO* is in the 0 state, $[CAR] + 1 \rightarrow CAR$, and the next sequential instruction is fetched.

1. If $[XRO] = 1$, $[EAR] \longrightarrow CAR$; if $[XRO] = 0$, $[CAR] + 1 \longrightarrow CAR$

Tally Down the Index Register—21—TLD The tally down the index register has several parts to it. First, the contents of the index register are tested to determine if they are zero. If they are 0, the next sequential instruction is fetched. If they are not zero, the contents are decremented by one and a branch is performed to fetch the next instruction from the *EA*. The sequence is

1. If $[XR] = 0$, $[CAR] + 1 \longrightarrow CAR$;
2. If $[XR] > 0$, $[XR] - 1 \longrightarrow XR$, $[EAR] \longrightarrow CAR$

Branch Accumulator Positive—22—BAP The sign bit of the accumulator is tested to determine if it is a 0 and if the contents of the accumulator are not 0. If these two requirements are met, the next instruction is fetched from the contents of the effective address. If they are not met, the next sequential instruction is fetched. The transfer equations are

1. If $[ACC]_0 = 0$ and $[ACC]_{1\text{-}23} \neq 0$, $[EAR] \longrightarrow CAR$;
2. If $[ACC]_0 = 1$ or $[ACC]_{1\text{-}23} = 0$, $[CAR] + 1 \longrightarrow CAR$

Branch on Accumulator Zero—23—BAZ The contents of the accumulator are tested to determine if they are 0. If they are, $[EAR] \rightarrow CAR$; if they are not, $[CAR] + 1 \rightarrow CAR$.

1. If $[ACC] = 0$, $[EAR] \longrightarrow CAR$; if $[ACC] \neq 0$, $[CAR] + 1 \longrightarrow CAR$

Branch on Accumulator Negative—24—BAN The sign bit of the accumulator is tested to determine if it is a 1. If it is, $[EAR] \rightarrow CAR$; if it is not, $[CAR] + 1 \rightarrow CAR$.

1. If $[ACC]_0 = 1$, $[EAR] \longrightarrow CAR$; if $[ACC]_0 \neq 1$, $[CAR] + 1 \longrightarrow CAR$

Subroutine Jump—29—SRJ This is a special type of unconditional branch instruction which leaves a trail of where the branch was executed. It is used to enter a subroutine and leave a path to return to the calling program. It performs as follows:

1. The contents of the *CAR* plus 1 are stored at the effective address.
2. The effective address is the address of the next instruction which is to be executed. The transfer equations for this instruction are

1. $[CAR] + 1 \longrightarrow MIR_{8\text{-}23}$
2. $[MIR]_{8\text{-}23} \longrightarrow EA_{8\text{-}23}$ (store the return address)
3. $[EAR] \longrightarrow CAR$

The contents of the effective address, which is the first instruction of the subroutine, must contain a no-operation instruction since that is where the return address, $[CAR] + 1$, is stored.

No-Operation—31—NOP This is another unusual instruction which is not really a branch instruction but is placed here because of its use with the subroutine jump instruction. The no-operation instruction is just that. No operation is performed, and the next sequential instruction is fetched to be executed. Decoding of an NOP instruction results in no transfer of data.

Stop—0—00—STP This instruction is not a branch instruction but is placed here since it does not seem to fit anywhere. The stop instruction places the computer at idle, and no instructions are fetched to be executed. Execution of a stop instruction prevents the incrementing of the *CAR*, and therefore the STP instruction is being executed continuously. To start the computer, the start button on the console must be depressed. This will increment the *CAR*, and the next sequential instruction will be fetched for execution.

12.2.3.4 Logical Operations

Logical-OR—25—LOR Perform the logical-OR operation on a bit-by-bit basis on the contents of the accumulator and the contents fetched from the effective address. The operation is performed in the ALU; the operand is temporarily stored in the *X* register; place the result in the accumulator.

This instruction has the following transfer equations:

1. $[EA] \longrightarrow MIR$
2. $[MIR] \longrightarrow X$
3. $[ACC] + [X] \longrightarrow ACC$

The + sign in step 3 above is the logical-OR operation.

Logical-AND—26—LPR Perform the logical-AND operation on a bit-by-bit basis between the contents of the accumulator and the contents of the effective address. Place the results in the accumulator. The instruction has the following transfer equations:

1. $[EA] \longrightarrow MIR$
2. $[MIR] \longrightarrow X$
3. $[ACC] \cdot [X] \longrightarrow ACC$

Logical Negation—27—LNG Perform the logical negation on a bit-by-bit basis on the contents of the accumulator and place the results into the accumulator. This is the same as forming the 1's complement of the contents of the accumulator.

1. $[\overline{ACC}] \longrightarrow ACC$

This instruction is an exception to the normal instruction format as it has no effective address.

Logical Exclusive-OR—28—EOR Perform the exclusive-OR operation on a bit-by-bit basis between the contents of the accumulator and the contents of the effective address. Place the results in the accumulator. The transfer equations are

1. $[EA] \longrightarrow MIR$
2. $[MIR] \longrightarrow X$
3. $[ACC] \oplus [X] \longrightarrow ACC$

12.2.3.5 Shift Instructions

The shift instructions all use the augmented instruction format. The binary number contained in bits 18–23 of the instruction determines the number of bits to be shifted for the particular instruction. In a single register shift, the maximum number of shifts needed to change the contents of the register is 23. If the number specified is larger than 23, the shifting will continue until the specified number of shifts has been performed. In a double register shift, the maximum number of needed shifts is 47. The counter register is therefore only 6 bits in length.

The shift instructions are of two types: an arithmetic shift or a logical shift. The logical shift involves the entire 24-bit word. The arithmetic shift does not permit the shifting of bits into the sign bit of the register. Likewise, in an arithmetic right shift, the sign bit is preserved and a copy is shifted into the bit 1 position. This preserves the sign of the number in the register.

Arithmetic Left Shift the Accumulator—0—02—ALA The contents of the accumulator are shifted left by the number of bit places specified by the binary number contained in bits 18–23 of the instruction. The sign bit is not shifted. Zeros are shifted into bit 23 of the accumulator. The accumulator is shown in the accompanying figure, and the transfer equations are

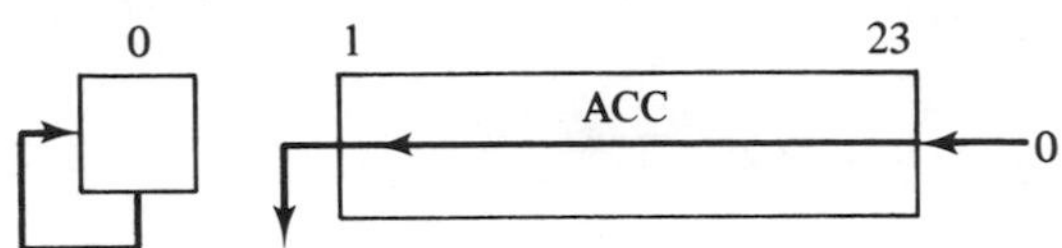

1. $[SH(CIR)] \longrightarrow CR$
2. If $[CR] = 0$, commence the next instruction fetch cycle
3. $[ACC]_{2\text{-}23} \longrightarrow ACC_{1\text{-}22}$; $0 \longrightarrow ACC_{23}$; $[A]_0 \longrightarrow A_0$
4. $[CR] - 1 \longrightarrow CR$; branch to step 2

The contents of the *CIR* bits 18–23 are placed into the counter register (*CR*). If the count is 0, the instruction is completed. If the count is not 0, bits 2–23 of the accumulator are shifted one place left into bits 1–22 of the accumulator. A 0 is shifted into bit 23. The counter register is decremented by 1, and a branch is made to test if the *CR* is 0.

Arithmetic Right Shift the Accumulator—0—03—ARA The contents of the accumulator are shifted right by the number of bit places specified by the binary number contained in bits 18–23 of the instruction. The sign bit is not shifted, but a copy of the sign bit is shifted into bit 1 of the accumulator. The accompanying figure shows the accumulator for this instruction. The transfer equations are

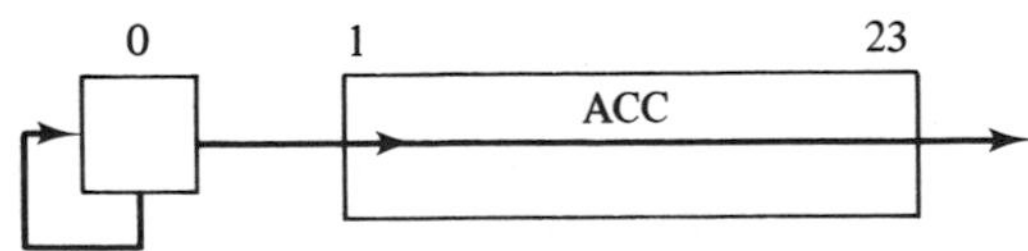

1. $[SH(CIR)] \longrightarrow CR$
2. If $[CR] = 0$, commence the next instruction fetch cycle
3. $[ACC]_{1\text{-}22} \longrightarrow ACC_{2\text{-}23}$; $[A]_0 \longrightarrow ACC_1$; $[A]_0 \longrightarrow A_0$
4. $[CR] - 1 \longrightarrow CR$; branch to step 2.

Logical Right Shift the Accumulator and MQ Register—0—04—LRQ The contents of the accumulator and the MQ register are shifted right by the number of bit places specified by the binary number contained in bits 18–23 of the instruction. Bit 23 of the accumulator is shifted into bit 0 of the MQ register. The sign bit of the accumulator is copied into bit 1 of the accumulator. Bit 23 of the MQ register is shifted out. The shifting is shown by the accompanying figure, and the transfer equations are

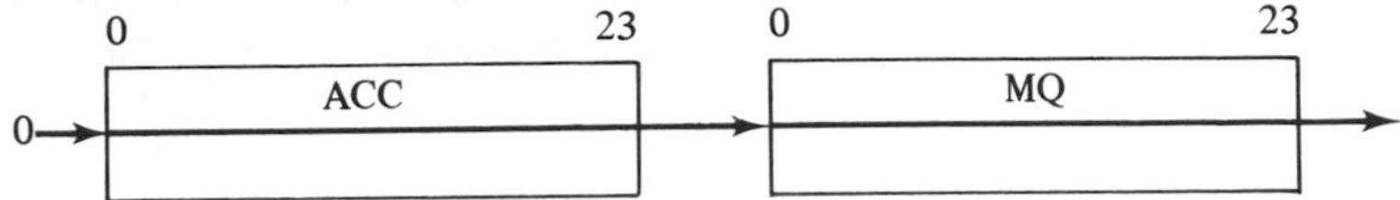

1. $[SH(CIR)] \longrightarrow CR$
2. If $[CR] = 0$, commence the next instruction fetch cycle
3. $[ACC]_{0\text{-}22} \longrightarrow ACC_{1\text{-}23}$; $[ACC]_{23} \longrightarrow MQ_0$;
$[MQ]_{0\text{-}22} \longrightarrow MQ_{1\text{-}23}$; $0 \longrightarrow ACC_0$
4. $[CR] - 1 \longrightarrow CR$; branch to step 2

Logical Left Shift the Accumulator and MQ Register—0—05—LLQ The contents of the accumulator and the MQ register are shifted left by the number of bit places specified by the binary number contained in bits 18–23 of the instruction. Bit 0 of the MQ register is shifted into bit 23 of the accumulator. Zeros are shifted into bit 23 of the MQ register. The shifting is shown by the accompanying figure. The transfer equations are

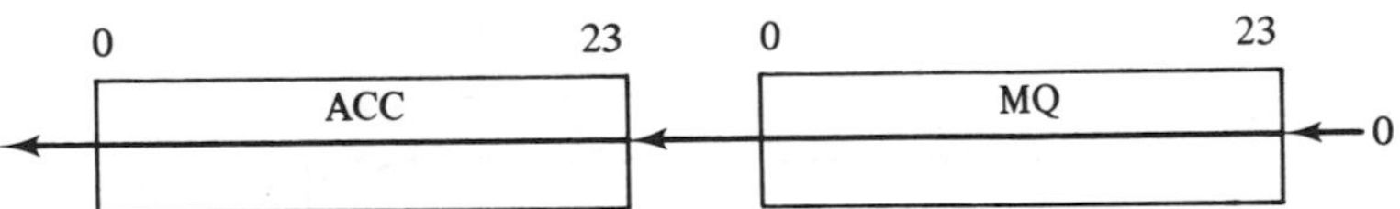

1. $[SH(CIR)] \longrightarrow CR$
2. If $[CR] = 0$, commence the next instruction fetch cycle
3. $[MQ]_{1\text{-}23} \longrightarrow MQ_{0\text{-}22}$; $[ACC]_{1\text{-}23} \longrightarrow ACC_{0\text{-}22}$;
$[MQ]_0 \longrightarrow ACC_{23}$; $0 \longrightarrow MQ_{23}$
4. $[CR] - 1 \longrightarrow CR$; branch to step 2

Logical Left Shift the Accumulator—0—06—LLA The contents of the accumulator are shifted left by the number of bit places specified by the binary number contained in bits 18–23 of the instruction. Zeros are shifted into bit 23 of the accumulator. The accompanying figure shows the shifting. The transfer equations are

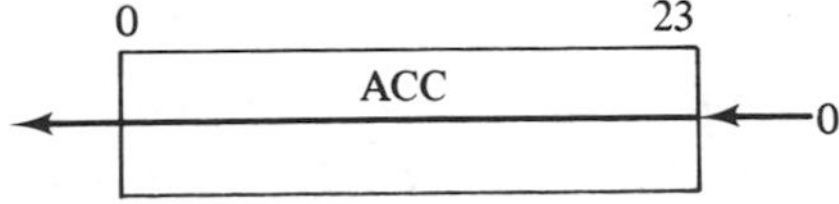

1. $[SH(CIR)] \longrightarrow CR$
2. If $[CR] = 0$, commence the next instruction fetch cycle
3. $R(ACC) \longrightarrow L(ACC); 0 \longrightarrow ACC_{23}$
4. $[CR] - 1 \longrightarrow CR$; branch to step 2

Logical Right Shift the Accumulator—0—07—LRA The contents of the accumulator are shifted right by the number of bit places specified by the binary number contained in bits 18–23 of the instruction. Zeros are shifted into the sign bit, bit 0. The accompanying figure shows the shifting. The register transfer equations are

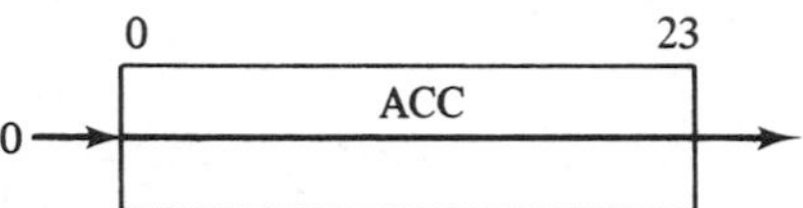

1. $[SH(CIR)] \longrightarrow CR$
2. If $[CR] = 0$, commence the next instruction fetch cycle
3. $L(ACC) \longrightarrow R(ACC); 0 \longrightarrow ACC_0$
4. $[CR] - 1 \longrightarrow CR$; branch to step 2

Logical Left Circular Shift the Accumulator—0—08—LCA The contents of the accumulator are circularly shifted left by the number of bit places specified by the binary number contained in bits 18–23 of the instruction. The accompanying figure shows the shifting. The register transfer equations are

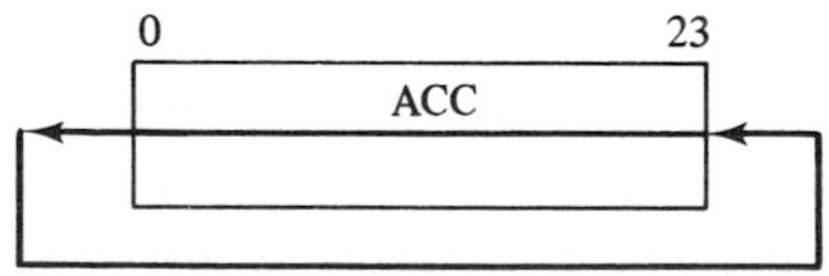

1. $[SH(CIR)] \longrightarrow CR$
2. If $[CR] = 0$, commence the next instruction fetch cycle
3. $\rho^{-1}(ACC) \longrightarrow ACC$
4. $[CR] - 1 \longrightarrow CR$; branch to step 2

12.2.3.6 Input/Output Operations

The computer has a single DMA channel to which all its I/O devices can be connected. Although this is not a large amount of I/O bandwidth, it is sufficient to demonstrate the principles of I/O programming and hardware. There are six instructions utilized in the I/O process.

Connect Device—0—13—CND This instruction uses the augmented instruction format and defines three pieces of information needed to perform an I/O operation. Bits 12–15 define the command code (CC) which the I/O device specified in bits 16–23 is to perform. The device is specified by two fields. Bits 16–19 specify the class of the device, i.e., typewriter, line printer, magnetic tape, etc. Bits 20–23 specify which of the 16 devices of the class is being addressed.

Figure 12.12 shows the format for this instruction.

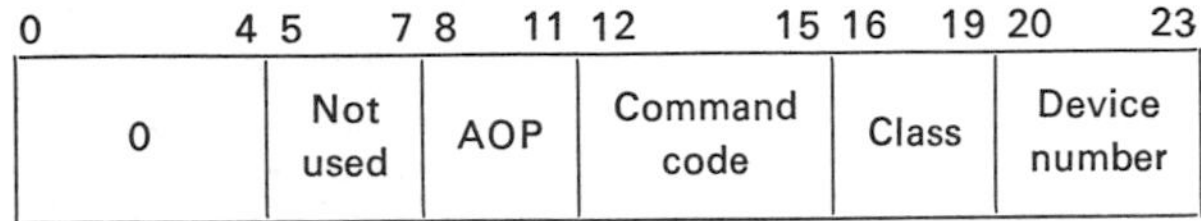

Figure 12.12 CND Instruction format.

The command codes specified in bits 12–15 are interpreted differently for each different class of device. Table 12.4 lists the class codes and command codes for the classes of devices.

The register transfer equation for the instruction is

1. $[CIR]_{12\text{-}23} \longrightarrow IOCR_{12\text{-}23}$

Load Status—0—15—LDS This instruction in the augmented instruction format loads the status of a device or the channel into the accumulator. The status is contained in bits 0–11 of the *IOCR*. The transfer equation for this instruction is

1. $[IOCR]_{0\text{-}11} \longrightarrow ACC_{12\text{-}23};\ 0 \longrightarrow ACC_{0\text{-}11}$

The status is placed in the low-order 12 bits of the accumulator. Each device class has a unique meaning for the 12 possible status codes. Table 12.5 shows the status codes for the device classes.

Branch on Device Not Available—30—BDN This is an instruction in the normal format. It is a special branch instruction which is used to rapidly test the status of the channel and I/O device which has been connected by a CND instruction. If the channel or device is not available—i.e., some other status bits are 1 and the available bit, bit 11, of the *IOCR* is a 0—then a branch will be taken to the effective address to obtain the next instruction. Otherwise, the next sequential instruction is executed since the device is available. The transfer equations are

1. If $[IOCR]_{11} = 0$, $[EAR] \longrightarrow CAR$; if $[IOCR]_{11} = 1$, $[CAR] + 1 \longrightarrow CAR$

Table 12.4 *I/O Class Codes and Command Codes*

Class code in decimal	*Device type*	*Command code in decimal*	*Definition of command*
0	Channel	1	Clear channel
		2	Test channel
		3	Request channel status
1	Typewriter	1	Typewriter output
		2	Typewriter input
		3	Request typewriter status
2	Paper-tape reader	1	Input
		2	Request status
3	Tape punch	1	Output
		2	Request status
4	Card reader	1	Clear
		2	Read one card
		3	Free run
		4	Request status
5	Card punch	1	Clear
		2	Punch one card
		3	Free run
		4	Request status
6	Magnetic tape	1	Write forward
		2	Read forward
		3	Read backward
		4	Write end-of-file mark
		5	Backspace one record
		6	Backspace to end-of-file mark
		7	Search forward to end-of-file mark
		8	Rewind to load point
		9	Rewind and unload
		10	Move forward to next record
		11	Set low density
		12	Set high density
		13	Set even parity
		14	Set odd parity
		15	Request status
7	Magnetic disk	1	Read
		2	Write
		3	Write and verify
		4	Seek to cylinder
		5	Search on key
		6	Request status

***Table 12.5** I/O Status Codes*

Device class	*Status code bit location*	*Definition of status code*
Channel	11	Channel available
	10	Channel busy
	9	Channel malfunctioning
	8	Interrupt pending
Typewriter	11	Channel and device available
	10	Device busy
	9	Power off
	8	Device disconnected
Paper-tape reader	11	Channel and device available
	10	Device busy
	9	Parity error detected
	8	Power off
	7	Device disconnected
Paper-tape punch	11	Channel and device available
	10	Device busy
	9	Power off
	8	Device disconnected
	7	Paper empty
Card reader	11	Channel and device available
	10	Device busy
	9	Power off
	8	Device disconnected
	7	Card hopper empty
	6	Stacker full
	5	Feed failure
	4	Read error
Card punch	11	Channel and device available
	10	Device busy
	9	Power off
	8	Device disconnected
	7	Card hopper empty
	6	Stacker full
	5	Feed failure
	4	Punch stuck
	3	Read comparison error
Magnetic tape	11	Channel and device available
	10	Device busy
	9	Power off
	8	Device disconnected
	7	Tape not loaded
	6	Parity error on read

Table 12.5 *(Cont.)*

Device class	*Status code bit location*	*Definition of status code*
Magnetic tape	5	End-of-file mark read
	4	End of tape sensed
	3	Load point sensed
	2	Parity error on read after write
Magnetic disk	11	Channel and device available
	10	Device busy
	9	Power off
	8	Device disconnected
	7	Disk pack not loaded
	6	Parity error on read
	5	Error on read after write

Enable Interrupt—0—14—ENI This is an instruction in the augmented format. When the DMA channel presents an interrupt to the CPU, the interrupt system is disabled to prevent any additional interrupts from occurring before the return path and register saves can be performed. After the subroutine has performed these "housekeeping" chores, the interrupt system can be enabled. There are no registers or register transfers involved in this instruction.

Start Input/Output—14—SIO This instruction uses the normal format. It starts information moving across the I/O interface. The effective address is placed into the *IOAR* and is the address from which the first word is transferred to or from the *IOIR*. After the data have been transferred the contents of the *IOAR* are incremented by 1. The transfer equations for this instruction are

1. $[EAR] \longrightarrow IOAR$
2. I/O operations commence over the channel

Load Input/Output Word Count (IOWC)—15—LIO This instruction is also in the normal format. It is used to load the *IOWC* (I/O word counter). The number of words of information to be transferred over the channel is specified in the *DA* portion of the instruction. Bits 5, 6, and 7 are not used and must be 0. The transfer equation for the instruction is

1. $[EAR] \longrightarrow IOWC$

12.2.4 Programming Examples

In this section we shall present some example programs for this machine. These examples are simple and short and used only to present a particular concept. As such, they are not necessarily complete programs.

The assembler for the machine uses the coding format shown in Fig. 12.13. The symbolic names are limited to a maximum of six letters and must begin with a letter. All numbers are shown in decimal and are translated by the assembler. If a number is to be shown in octal, it is preceded by the letter O. Address modification is shown with the letters X, S, and I in the modification space.

Symbolic label	*Op-code mnemonic*	*Address modification*	*Symbolic address or data*	*Comments*

Figure 12.13 Assembler coding form.

12.2.4.1 Looping Using Tally Down

This program finds the sum of 100 numbers. The first number is located at a symbolic address called LIST. The sum is placed in the location ANSWER. The program tests for an overflow of the accumulator. If an overflow is detected, a branch is made to a program located at ERROR. That program is not shown. The assembly language listing and the equivalent binary listing are shown in Figs. 12.14 and 12.15, respectively.

The assembler assembles this program into machine language. Let us assume that the program is placed at location $(012341)_8$ in main memory.

Symbolic label	Op-code mnemonic	Address modification	Symbolic address	Comment
.	.		.	.
.	.		.	.
.	.		.	.
BEGIN	LAI		0	Clear accumulator
	LXI		99	99 ⟶ index register
	ADD	X	LIST	[LIST + [*XR*]] + [*ACC*] ⟶ *ACC*
	BAO		ERROR	If overflow, branch to ERROR
	TLD	S	−2	Tally down; if not 0, go back 2 locations
	STA		ANSWER	Put sum at ANSWER
	.		.	
	.		.	
	.		.	

Figure 12.14 Assembly language program using tally.

$$\text{BEGIN} = (012341)_8$$
$$\text{LIST} = (001443)_8$$
$$\text{ERROR} = (014727)_8$$
$$\text{ANSWER} = (013314)_8$$
$$(99)_{10} = (143)_8$$

Address in octal	24-Bit instructions							
.								
.								
.								
012341	00000	000	1001		000	000	000	000
012342	00000	000	1010		000	001	100	011
012343	00001	100	0	000	001	100	100	011
012344	10001	000	0	001	100	111	010	111
012345	10101	010	1	111	111	111	111	110
012346	00110	000	0	001	011	011	001	100
.			.					
.			.					
.			.					

Figure 12.15 Machine language program.

The symbolic variables are assigned absolute addresses. Figure 12.15 shows the symbol table and the machine language program.

12.2.4.2 Looping Without Tally

This section of a program shows an alternative method for adding the list of 100 numbers which was performed in the previous subsection. See Fig. 12.16.

Symbolic label	*Op-code mnemonic*	*Address modification*	*Symbolic address*	*Comment*
BEGIN	LAI		0	Clear the accumulator
	LDX	0	NUMBER	Load index register with 100
	DEX		1	Decrement index register
	ADD	X	LIST	[*ACC*] + [LIST + [*XR*]] ⟶ *ACC*
	BAO		ERROR	Branch if accumulator overflows
	BXP	S	−3	Branch back 3 if [*XR*] > 0
	STA		ANSWER	Store sum

[NUMBER] = $(100)_{10} = (143)_8$

Figure 12.16 Assembly language program II.

12.2.4.3 Subroutine Entry and Return

This programming example shows how the entry to a subroutine and the return from the subroutine is made in this computer. There are two programs. Program A is the calling program, while Program B is the called subroutine. See Fig. 12.17.

The main program (calling program) calls for a subroutine located at location FIX with the subroutine jump instruction (SRJ) located at JUMP.

Symbolic label	*Op-code mnemonic*	*Adress modification*	*Symbolic address*	*Comment*
.	.			
.	.			
.	.			
JUMP	SRJ	0	FIX	Subroutine jump to routine starting at address FIX
.	.			
.	.			
.	.			

Program A, calling program

Symbolic label	*Op-code mnemonic*	*Address modification*	*Symbolic address*	*Comment*
FIX	NOP		0	
.	.		.	
.	.		.	
.	.		.	
END	UNB	I	FIX	

Program B, called Subroutine

Figure 12.17 Subroutine entry and return

The computer branches to location FIX and places the address (JUMP + 1) into the address portion of the instruction at FIX. The op-code at that location must be an NOP. After storing the return address (JUMP + 1) at FIX, the computer executes the instruction at FIX. Since this is a no-operation instruction, nothing is performed. The body of the subroutine follows the address FIX. When the subroutine wishes to return to the calling program, it performs an unconditional branch with indirect addressing to location FIX. This will cause the direct address to be the contents of the address portion of FIX, which is the proper return address (JUMP + 1).

12.2.4.4 Normal I/O

This programming example shows how a typical I/O request using the computer's DMA channel is programmed. The request is to read 12 words of data from a paper-tape reader. Twelve words of data are 48 characters. The paper-tape reader is class 2, and the selected unit is number 1. The data are to be stored in memory starting at location 3015. See Fig. 12.18.

Symbolic label	*Op-code mnemonic*	*Address modification*	*Symbolic address*	*Comment*
BEGIN	CND	0	3, 0, 0	Request channel status
	BDN	0	FIXO1	Branch to FIXO1 if channel not available
	CND	0	1, 2, 1	Connect paper tape unit 1 to channel
	BDN	0	FIXO2	Branch to FIXO2 if device and channel not available
	LIO	0	12	12 ⟶ *IOWC*
	SIO	0	3015	3015 ⟶ *IOAR*
				Start I/O operation on DMA
				} Remainder of the program

Figure 12.18 I/O Programming.

12.3 THE ARITHMETIC AND LOGIC UNIT (ALU)

This section shows the logical design of an arithmetic and logic unit (ALU) which is well suited to microprogramming control. The design will permit the ALU to generate all 16 Boolean functions of two variables. In addition, the ALU will perform the arithmetic operation of SUM and DIFFERENCE of two binary numbers as well as the increment, the decrement, and the 2's complement of the accumulator.

The inputs to the ALU are shown in Fig. 12.6. One side of the ALU receives inputs from either the accumulator (*ACC*) or the index register (*XR*). The other side has the possible inputs of the *X* register or the main bus. Each of the inputs are 24 bits except for the index register, which supplies inputs only to positions 8–23. Figure 12.19 shows a block diagram of the registers and the control signals used to gate data to the ALU. The output of the ALU is used to gate the output latches. The diagram shows only one line, but it represents the 24 lines. The logic diagrams that are shown are only for a single bit of the ALU. However, it is general, and by cascading this bit for 24 bit locations, the entire ALU can be constructed.

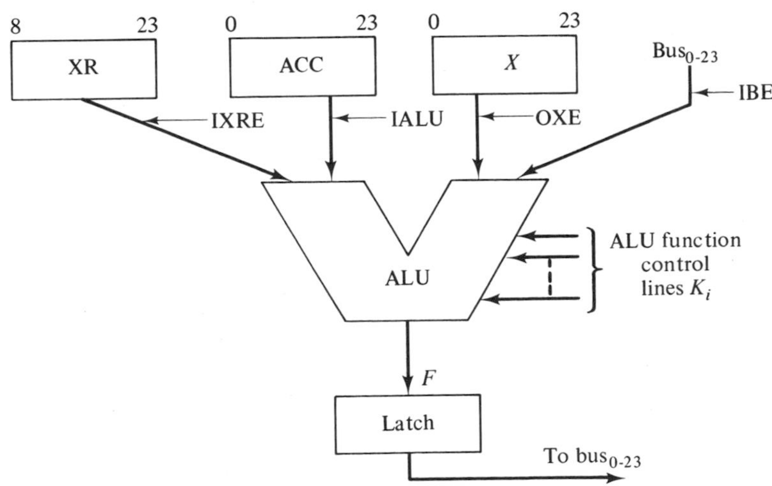

Figure 12.19 Block diagram of ALU and inputs.

12.3.1 Logic Design of the ALU

Before we begin the design, let us consider all the output functions (F) we wish to produce. Table 12.6 shows the functions and their names.

To design the ALU, a type of logic must be chosen. The authors have chosen to implement in terms of NAND gates. Let us now investigate some of the simple functions which can be generated using NAND gates and control input signals. The control inputs are designated as K_i.

Case 1: Consider the simple two-input NAND gate with inputs labeled *1* and *2* shown in Fig. 12.20. Input *1* can be one of four possible literals (A_i, $\bar{A}_i$, 0, 1); input *2* can also be one of four possible inputs (B_i, $\bar{B}_i$, 0, 1). The control signal K_1 can only

Table 12.6 *ALU Functions*

Function	*Name*
$f_0 = 0$	Zero
$f_1 = A_iB_i$	And
$f_2 = \bar{A}_iB_i$	And not A
$f_3 = B_i$	Identity
$f_4 = A_i\bar{B}_i$	And not B
$f_5 = A_i$	Identity
$f_6 = A_i\bar{B}_i + \bar{A}_iB_i$	Exclusive-OR
$f_7 = A_i + B_i$	Or
$f_8 = \bar{A}_i \cdot \bar{B}_i$	Nor
$f_9 = \bar{A}_i\bar{B}_i + A_iB_i$	Coincidence
$f_{10} = \bar{A}_i$	Not
$f_{11} = \bar{A}_i + B_i$	Or not A
$f_{12} = \bar{B}_i$	Not
$f_{13} = A_i + \bar{B}_i$	Or not B
$f_{14} = \bar{A}_i + \bar{B}_i$	Nand
$f_{15} = 1$	Unity
$A_i \oplus B_i \oplus C_{i-1}$	Sum or difference $= S_i$
$(A_i \oplus B_i)C_{i-1} + A_iB_i$	Carry out $= C_i$
$\bar{A}$ plus 1	2's Complement of A
A plus 1	Increment A
A minus 1	Decrement A
B plus 1	Increment B
B minus 1	Decrement B

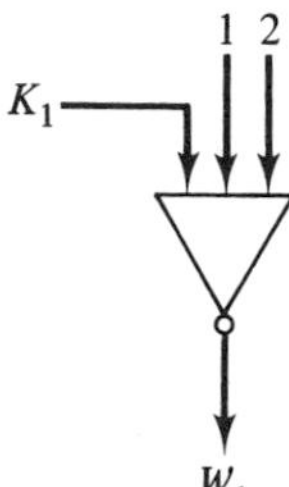

Figure 12.20 Case 1 gate.

be a 0 or a 1. This circuit can produce the logic functions shown in Table 12.7, where d is a don't care condition and can be any of the possible inputs. This circuit cannot produce an exclusive-OR output. If two-level NAND logic is utilized, then an equivalent AND-OR logic will be obtained. This should be able to produce the needed exclusive-OR function and many other functions with the proper control signals.

Table 12.7 *Logic Functions Produced by Case 1*

K_1	*1*	*2*	W_1
0	d	d	$1 = f_{15}$
1	0	d	$1 = f_{15}$
1	d	0	$1 = f_{15}$
1	1	1	$\underline{0 = f_0}$
1	$\bar{A}_i$	$\bar{B}_i$	$\overline{\bar{A}_i\bar{B}_i} = A_i + B_i = f_7$
1	$\bar{A}_i$	B_i	$\overline{\bar{A}_iB_i} = A_i + \bar{B}_i = f_{13}$
1	A_i	$\bar{B}_i$	$\overline{A_i\bar{B}_i} = \bar{A}_i + B_i = f_{11}$
1	A_i	B_i	$\overline{A_iB_i} = \bar{A}_i + \bar{B}_i = f_{14}$

Case 2: For Case 2, consider the circuit shown in Fig. 12.21. Once again, inputs *1* and *2* can present the same four variables as in Case 1, and each of the control signals K_1 and K_2 can present 0 or 1. This circuit can produce the Boolean functions shown in Table 12.8. An inspection of Table 12.8 shows that the circuit cannot produce f_7, f_{11}, f_{13}, and f_{14}.

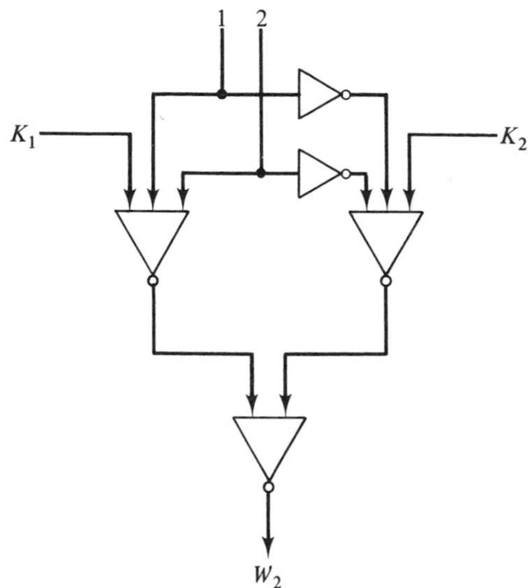

Figure 12.21 Case 2 gates.

Case 3: The circuit of Case 2 required three NAND gates. A logic which uses two NAND gates with the output of each gate strapped together produces the logic family NAND-AND. This logic function is also called AND-OR-INVERT or AND-NOR logic. All three of these functions are logically equivalent. This results in a savings of a gate and also its delay. The circuit shown in Fig. 12.22 produces NAND-AND and can produce the logic functions shown in Table 12.9. An analysis

Table 12.8 *Logic Functions Produced by Case 2*

K_1	K_2	*1*	2	W_2
0	0	d	d	$0 = f_0$
0	1	0	0	$1 = f_{15}$
0	1	$\bar{A}_i$	$\bar{B}_i$	$A \cdot B = f_1$
0	1	$\bar{A}_i$	B_i	$A \cdot \bar{B} = f_4$
0	1	A_i	$\bar{B}_i$	$\bar{A} \cdot B = f_2$
0	1	A_i	B_i	$\bar{A} \cdot \bar{B} = f_8$
1	0	$\bar{A}_i$	$\bar{B}_i$	$\bar{A} \cdot \bar{B} = f_8$
1	0	$\bar{A}_i$	B_i	$\bar{A} \cdot B = f_2$
1	0	A_i	$\bar{B}_i$	$A \cdot \bar{B} = f_4$
1	0	A_i	B_i	$A \cdot B = f_1$
1	1	$\bar{A}_i$	$\bar{B}_i$	$\bar{A}\bar{B} + AB = f_9$
1	1	$\bar{A}_i$	B_i	$\bar{A}B + A\bar{B} = f_6$
1	1	A_i	$\bar{B}_i$	$A\bar{B} + \bar{A}B = f_6$
1	1	A_i	B_i	$AB + \bar{A}\bar{B} = f_9$
1	0	A_i	1	$A = f_5$
1	0	1	B_i	$B = f_3$
0	1	A_i	0	$\bar{A} = f_{10}$
0	1	0	B_i	$\bar{B} = f_{12}$

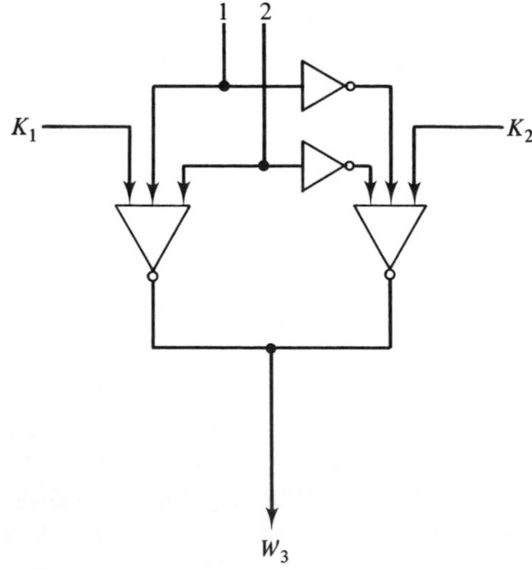

Figure 12.22 NAND-AND gates for Case 3.

Table 12.9 *Logic Functions Produced by Case 3*

K_1	K_2	*1*	*2*	W_3
0	0	d	d	$1 = f_{15}$
0	1	0	0	$0 = f_0$
0	1	$\bar{A}$	$\bar{B}$	$\overline{A \cdot B} = \bar{A} + \bar{B} = f_{14}$
0	1	$\bar{A}$	B	$\overline{A \cdot \bar{B}} = \bar{A} + B = f_{11}$
0	1	A	$\bar{B}$	$\overline{\bar{A} \cdot B} = A + \bar{B} = f_{13}$
0	1	A	B	$\overline{\bar{A} \cdot \bar{B}} = A + B = f_7$
1	0	$\bar{A}$	$\bar{B}$	$\overline{\bar{A} \cdot \bar{B}} = A + B = f_7$
1	0	$\bar{A}$	B	$\overline{\bar{A} \cdot B} = A + \bar{B} = f_{13}$
1	0	A	$\bar{B}$	$\overline{A \cdot \bar{B}} = \bar{A} + B = f_{11}$
1	0	A	B	$\overline{A \cdot B} = \bar{A} + \bar{B} = f_{14}$
1	1	$\bar{A}$	$\bar{B}$	$\overline{\bar{A} \cdot \bar{B}} \cdot \overline{A \cdot B} = \overline{\bar{A} \cdot \bar{B} + A \cdot B} = \overline{A \odot B} = A \oplus B = f_6$
1	1	$\bar{A}$	B	$\overline{\bar{A} \cdot B} \cdot \overline{A \cdot \bar{B}} = \overline{\bar{A} \cdot B + A \cdot \bar{B}} = \overline{A \oplus B} = A \odot B = f_9$
1	1	A	$\bar{B}$	$= \overline{A \oplus B} = f_9$
1	1	A	B	$= A \oplus B = f_6$
1	d	A	1	$\bar{A} = f_{10}$
1	1	A	0	$A = f_5$
1	d	1	B	$\bar{B} = f_{12}$
1	1	0	B	$B = f_3$
1	d	1	1	$0 = f_0$
1	0	0	0	$1 = f_{15}$

of Table 12.9 shows that all the functions can be produced with $K_1 = 1$, and therefore they are independent of K_1. It can be eliminated from the circuit. This circuit cannot produce f_1, f_2, f_4, and f_8. However, the circuit can produce the functions involving the OR operation, exclusive-OR, coincidence, and most important it can behave as a selectable inverter of A or B.

Case 4: Since we wish to produce the function SUM $= A_i \oplus B_i \oplus C_{i-1}$, let us consider another case in which two Case 3 NAND-AND circuits are cascaded as shown in Fig. 12.23. A second-level input *3* has been added. It can be any of four values C_{i-1}, $\overline{C_{i-1}}$, 1, or 0; also, K_4, a control signal, can be a 0 or a 1. Notice that the control signal K_1 has been eliminated. The truth table for this circuit is shown in Table 12.10. An analysis of the truth table shows that all the functions which were needed can be produced with the circuit. The functions are independent of K_4, which can be eliminated. The only control signal is K_2. Input *1* must be able to produce a logical signal which represents A_i, $\bar{A}_i$, or a 1. Input *2* must be able to produce B_i, $\bar{B}_i$, or a 1, and input *3* must present C_{i-1}, 0, or 1.

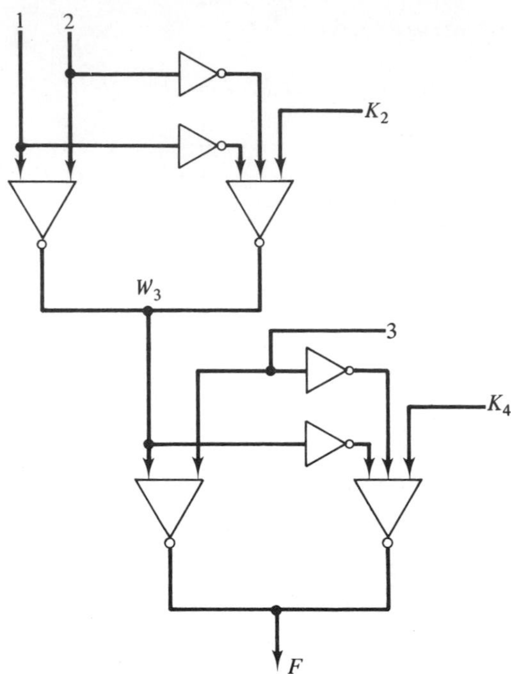

Figure 12.23 NAND-AND gates for Case 4.

Table 12.10 *Truth Table for Case 4*

K_2	*1*	*2*	W_3	*3*	K_4	F
d	1	1	$0 = f_0$	0	0	$0 = f_0$
0	A	B	$\overline{A \cdot B}$	1	1	$A \cdot B = f_1$
0	$\bar{A}$	B	$\overline{\bar{A} \cdot B}$	1	1	$\bar{A} \cdot B = f_2$
0	1	B	$\bar{B}$	1	1	$B = f_3$
0	A	$\bar{B}$	$\overline{A \cdot \bar{B}}$	1	1	$A \cdot \bar{B} = f_4$
d	A	1	$\bar{A}$	1	1	$A = f_5$
1	A	B	$\overline{(A \cdot B)} \cdot \overline{(\bar{A} \cdot \bar{B})}$	0	1	$\overline{(A \cdot B) + (\bar{A} \cdot \bar{B})} = A \oplus B = f_6$
0	$\bar{A}$	$\bar{B}$	$\overline{\bar{A} \cdot \bar{B}}$	0	1	$\overline{\bar{A} \cdot \bar{B}} = A + B = f_7$
0	$\bar{A}$	$\bar{B}$	$\overline{\bar{A} \cdot \bar{B}}$	1	1	$\bar{A} \cdot \bar{B} = f_8$
1	A	B	$A \oplus B$	1	1	$\overline{A \oplus B} = f_9$
d	A	1	$\bar{A}$	0	1	$\bar{A} = f_{10}$
0	A	$\bar{B}$	$\overline{A\bar{B}}$	0	1	$\bar{A} + B = f_{11}$
d	1	B	$\bar{B}$	0	1	$\bar{B} = f_{12}$
0	$\bar{A}$	B	$\overline{\bar{A}B}$	0	1	$A + \bar{B} = f_{13}$
0	A	B	$\overline{AB}$	0	1	$\bar{A} + \bar{B} = f_{14}$
d	1	1	0	1	1	$1 = f_{15}$
1	A	B	$A \oplus B$	C_{i-1}	1	$A \oplus B \oplus C_{i-1} = \text{sum}$
d	$\bar{A}$	1	$A \oplus 0$	C_{i-1}	1	$A \oplus C_{i-1} = A + 1;$ $C_n = 1$

If input *1* is connected through logic to the *i*th bit of a flip-flop accumulator which has both A_i and $\bar{A}_i$ outputs, then the circuit shown in Fig. 12.24 will produce the desired inputs. The signals K_5 and K_6 control the signal presented to input *1*. The truth table of Table 12.11 shows the signal presented to input *1*. If the input is

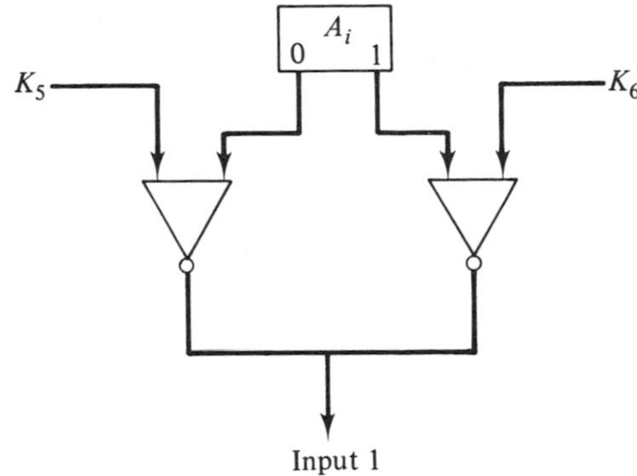

Figure 12.24 Steering circuit for input 1.

Table 12.11 Truth Table for Fig. 12.24

K_5	K_6	*Input 1*
0	0	1
0	1	$\bar{A}_i$
1	0	A_i
1	1	0

connected through logic to the input bus, which only has single rail logic, it will be necessary to produce the inverse of the signal on the *i*th bit line B_i. The circuit in Fig. 12.25 can perform this function with control lines K_7 and K_8.

The circuitry used to produce the carry out from the *i*th bit still needs to be

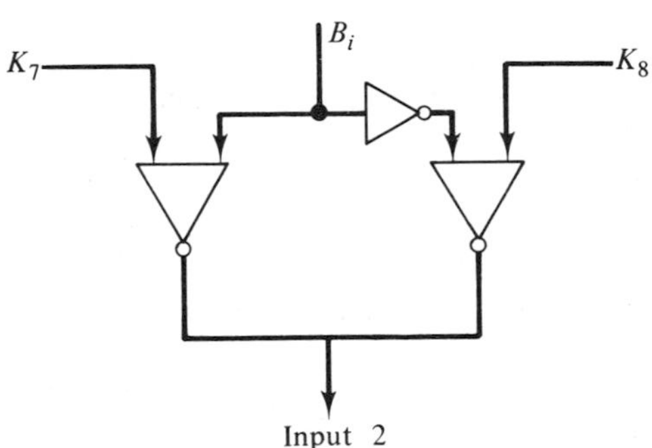

Figure 12.25 Steering circuit from the bus to input 2.

developed. The carry out, C_i, can be written as

$$C_i = A_iB_i + (A_i \oplus B_i)C_{i-1}$$

The signal at input *1* is A_i; the signal at input *2* is B_i; K_2 is a 1; W_3 is $A_i \oplus B_i$; and the signal on input 3 is C_{i-1}. The carry out, C_i, can be produced with the circuit shown in Fig. 12.26.

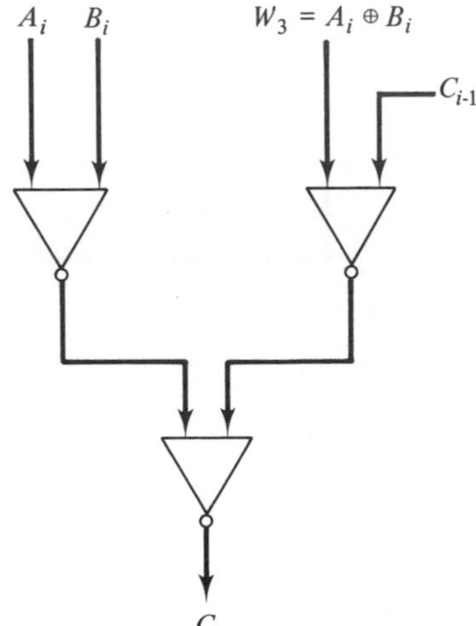

Figure 12.26 NAND-AND logic producing the carry-out.

The logic circuit for the *i*th bit of the ALU with its associated control signals is shown in Fig. 12.27. Input *1* comes from either the accumulator or the index register. Both of these are flip-flop registers so that double rail logic is available. Input *2* also has two inputs. One is from the flip-flop *X* register, while the other is from the main bus, which has only single rail logic.

There are two sets of control signals shown in the diagram. One set is used to gate the required signals into terminals *1* and *2*. These signals are also shown in Fig. 12.6. They are IXRE, which gates the index register to the ALU, and IALU, which gates the accumulator to terminal *1*. The index register, *XR*, is connected only to bits 8–23 of the ALU since it is a 16-bit register. The signals IBE and OXE gate the bus and the *X* register, respectively, to terminal *2*.

The other set of control signals are the K_i used to produce the desired arithmetic and logical functions. The complete truth table showing the states of the control signals needed to produce all the functions is shown in Table 12.12. The control signal K_{11} is used to inject a 1 or a 0 into the carry in of the LSB of the ALU. This signal is required because we are designing a 2's complement parallel adder.

12.3.2 Signal Propagation Through the ALU

How long does it take to add two binary numbers together using the ALU shown in Fig. 12.27? Assume that one number is contained in the

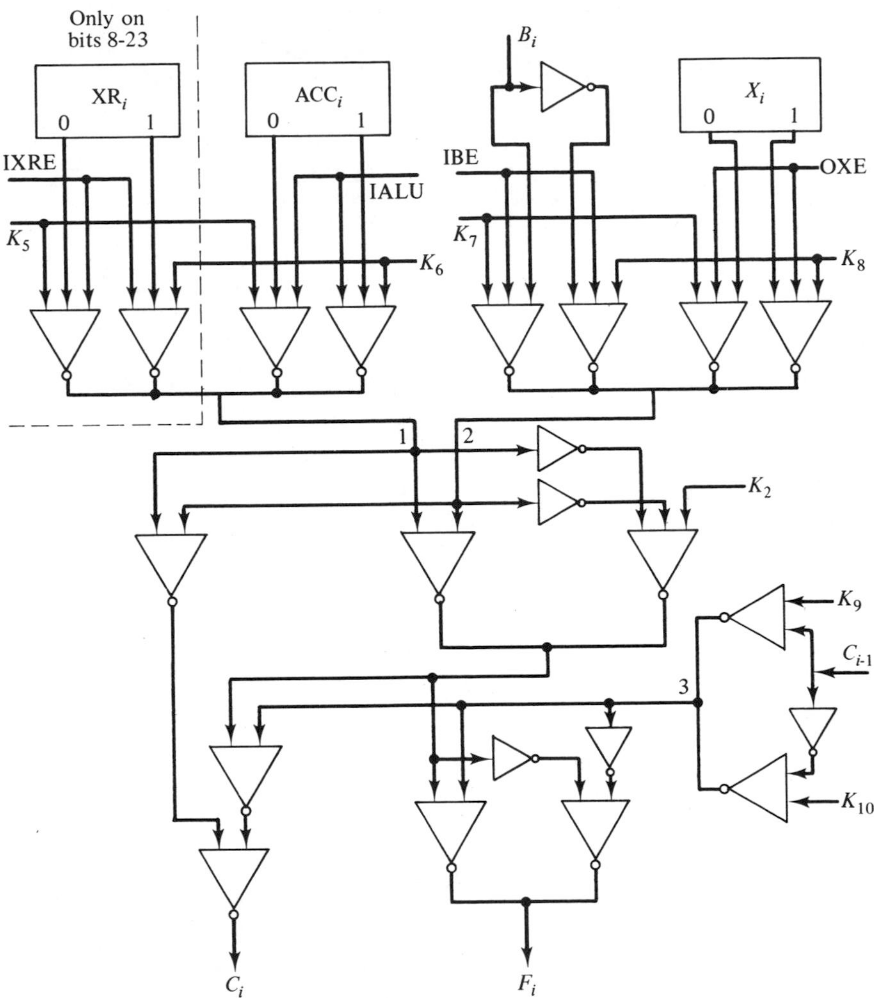

Figure 12.27 Logic diagram of the ith bit of the ALU.

accumulator A and the other is available on the bus B. The signals to inputs *1* and *2* will be available after 2 gate delay times on all 24 bits of the ALU. The C_i signals must propagate down the parallel adder. Since this a *ripple adder*, each C_i will require 4 gate delay times before it can appear. Thus, the carry out of bit 23 will appear after 6 gate delay times, the carry from bit 22 after 10 delay times, and each successive bit will add 4 gate delay times. Thus, the sum will appear at F after $(24 \times 4) + 2$ gate delay times or 98 gate delay times. The carry out of bit 0 will also appear after 98 gate delay

Table 12.12 Control Signals for the ALU

Function f	Control signals							
	K_2	K_5	K_6	K_7	K_8	K_9	K_{10}	K_{11}
$f_0 = 0$	d	0	0	0	0	1	1	d
$f_1 = A_iB_i$	0	1	0	1	0	0	0	d
$f_2 = \bar{A}_iB_i$	0	0	1	1	0	0	0	d
$f_3 = B_i$	0	0	0	1	0	0	0	d
$f_4 = A_i\overline{B_i}$	0	1	0	0	1	0	0	d
$f_5 = A_i$	d	1	0	0	0	0	0	d
$f_6 = A_i \oplus B_i$	1	1	0	1	0	1	1	d
$f_7 = A_i + B_i$	0	0	1	0	1	1	1	d
$f_8 = \bar{A}_i \cdot \overline{B_i}$	0	0	1	0	1	0	0	d
$f_9 = A_i \odot B_i$	1	1	0	1	0	0	0	d
$f_{10} = \overline{A_i}$	d	1	0	0	0	1	1	d
$f_{11} = \bar{A}_i + B_i$	0	1	0	0	1	1	1	d
$f_{12} = \bar{B}_i$	d	0	0	1	0	1	1	d
$f_{13} = A_i + \bar{B}_i$	0	0	1	1	0	1	1	d
$f_{14} = \bar{A}_i + \overline{B_i}$	0	1	0	1	0	1	1	d
$f_{15} = 1$	d	0	0	0	0	0	0	d
$S_i = A_i \oplus B_i \oplus C_{i-1}$	1	1	0	1	0	1	0	0
$A - B$	1	1	0	0	1	1	0	1
A plus 1	1	1	0	1	1	1	0	1
A minus 1	1	1	0	0	0	1	0	0
B plus 1	1	1	1	1	0	1	0	1
B minus 1	1	0	0	1	0	1	0	0
2's Complement of A	1	0	1	1	1	1	0	1
2's Complement of B	1	1	1	0	1	1	0	1

times. If each NAND gate has a maximum delay time of 8 nanoseconds, the output will appear in a maximum of 784 nanoseconds.

12.3.3 Overflow Detection in the ALU

The computer utilizes 2's complement notation to represent the binary numbers. The overflow condition for addition and subtraction is

$$(\overline{ACC_0 \oplus X_0}) \cdot Sum_0$$

This can be implemented by the logic circuit of Fig. 12.28.

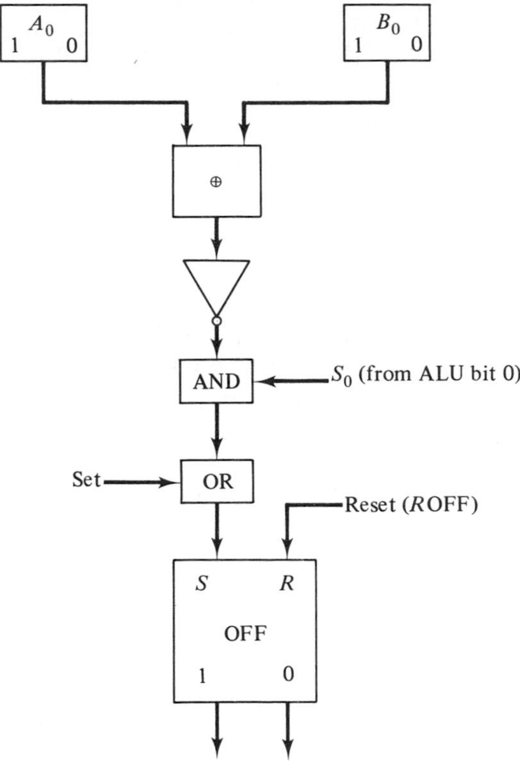

Figure 12.28 Overflow detection circuit.

12.4 COMPUTER PHASES, TIMING, AND INSTRUCTION UNIT

12.4.1 Phases and Clock Time

The stored program, single-address, parallel arithmetic unit, 2's complement arithmetic, general-purpose, microprogrammed digital computer which is designed herein operates using two basic phases:

1. The instruction phase.
2. The execution phase.

Each of these two phases is further divided into one or more CPU cycles. Each CPU cycle is 512 nanoseconds long, which is one-half of the main memory cycle time. Since the access time of main memory is 384 nanoseconds, a word can be fetched from main memory during a CPU cycle. However, during the next CPU cycle, the main memory must restore the address, in a read-restore memory cycle, and therefore another word may not be fetched. Through a hand-shaking convention, the main memory is synchronized with the CPU major clock cycles.

12.4.1.1 Major CPU Clock Time

The CPU major clock period is 512 nanoseconds. Two of these cycles constitute the instruction phase or fetch phase, while one or more of these cycles constitutes the execution phase. The length of the execution phase is a function of the particular operation to be performed.

The computer is microprogrammed using an integrated circuit bipolar transistor control memory. The control memory word length is 48 bits, and there are 256 words in this memory. The access time to fetch a microinstruction from this memory and place it into the *CMIR* is 64 nanoseconds. The control memory is synchronized to the CPU clock cycle. At the beginning of each 512-nanosecond CPU cycle, a microinstruction is fetched. During an instruction phase, two microinstructions are fetched and executed, one for each CPU cycle which constitutes this phase of operation. During an execution phase, one or more microinstructions would be fetched, depending on the operation to be performed in the execution unit.

12.4.1.2 Minor Clock Cycle

Each major CPU cycle is subdivided into eight minor cycles of 64 nanoseconds. These minor time periods are obtained from a 3-bit binary counter and decoder. The periods are named D0, D1, D2, . . . , D6, D7. This is a modulo-8 counter and recycles back to D0 from D7. During each D0 period, the next microinstruction is fetched from the control memory (*CM*) and placed into the control memory instruction register (*CMIR*). During each period D7, the address of the next microinstruction is determined and placed into the control memory address register (*CMAR*). The major CPU cycle is obtained from the 3-bit *D* clock when it recycles back to a 0 count.

12.4.2 Timing Sequences in the Instruction Unit

During the execution of each microinstruction, there may occur several different gating operations which move information into and out of several registers. Because this machine has only a single bus, only one register may be placing information onto the bus during any given 64-nanosecond minor clock period. Therefore, the microinstruction is divided into its eight 64-nanosecond subintervals. Transfer equations are written for any minor cycle interval for which a transfer is to occur. In this subsection we shall develop the timing sequences of the microcode for the instruction unit.

12.4.2.1 Instruction Phase (Fetch Cycle)

The instruction phase (fetch cycle) is composed of two microinstructions. The first of these microinstructions is given the symbolic control memory address of FETCH∅. The second is called FETCH1. Figure 12.29 shows the register transfer needed during these two cycles.

Major cycle	Minor cycle	Instruction unit	Main memory unit	
FETCH∅	D0	Fetch microinstruction	$MAV = 1$	
	D1	$[CAR] \longrightarrow MAR$; $1 \longrightarrow RRC$	$0 \longrightarrow MAV$	READ (D1–D6)
	D2			
	D3			
	D4			
	D5			
	D6		$1 \longrightarrow IA$	
	D7	$[MIR] \longrightarrow CIR$		
FETCH1	D0	Fetch microinstruction		RESTORE (D0–D6)
	D1			
	D2	Obtain the effective address (D2–D6)		
	D3			
	D4			
	D5			
	D6	$EA \longrightarrow EAR$; $[CAR] + 1 \longrightarrow CAR$	$0 \longrightarrow IA$	
	D7	If indirect, $IND1 \longrightarrow CMAR$; if operand, $OP\ CODE \longrightarrow CMAR$	$1 \longrightarrow MAV$	

Figure 12.29 Fetch phase register transfers.

The FETCH∅ cycle will be fetched from the control memory only if several conditions are met. These conditions will be discussed in detail in a later section, but some are presented here to show some conditional requirements needed to enter the instruction phase. Three such conditions are

1. Interrupt not active = $\overline{INT}$
2. Memory available = MAV
3. Automatic mode or single instruction mode = $AUTO + SIN$

Let us assume that the conditions are met. During the minor cycle D0, the microinstruction FETCH∅ is fetched from the control memory and placed into the *CMIR*. It is decoded, and the transfers are commenced at the minor cycle times shown in Fig. 12.29. During D1, the contents of the *CAR* are gated into the *MAR*; a 1 is placed on the read-restore cycle line (*RRC*) to the memory, and the memory places a 0 on the memory available line, *MAV* = 0. The memory completes its read part of the read-restore cycle at the end of D6, and the instruction is in the *MIR*. The memory signals the CPU that data are available by placing a 1 on the *IA* line. During cycle D7, the control unit gates the contents of the *MIR* onto the bus and then into the *CIR*. The contents of the *CMAR* are incremented by 1 during D7, and during the subsequent D0 the next sequential microinstruction, FETCH1, is placed in the *CMIR*. The memory commences its restore cycle. During periods D1–D6, the effective address is computed in the I unit using the address adder. At the end of D6, the effective address (*EA*) is gated into the effective address register (*EAR*). Also, the contents of the *CAR* are incremented by 1. The information-available flip-flop is reset in the memory. During D7, the memory sets its memory-available flip-flop to a 1. The control unit tests to see if indirect addressing is specified with this instruction. If it is, the address of the indirect addressing microprogram is placed into the *CMAR*. If it is not, the address of the microprogram for the particular operation specified by the op-code is placed into the *CMAR*, and the machine enters the execution phase.

12.4.2.2 Indirect Addressing Phase

During D7 of the FETCH1 cycle of the microprogram for the instruction phase a test is made to determine the status of the indirect bit flag of the instruction. If this bit, CIR_7, is a 1, indirect addressing is required. The microprogram branches to a special two-cycle microprogram which performs the transfers for the indirect addressing phase.

During D0 of IND1, the microprogram step is fetched from the control memory and placed into the *CMIR*. During D1 the contents of the *EAR* are transferred into the *MAR* and a read-restore cycle is initiated in the main memory. At D6, the memory signals that the direct address has been placed into the *MIR*. During D7 the contents of the *MIR* are transferred into the *X* register and the *CMAR* is incremented by 1.

During D0 of IND2, the microprogram step is fetched from the control memory. During D1 the contents of the *X* register are moved into the *EAR*. During D7 of IND2, a special command, OPC, moves the contents of the op-code field of the *CIR* into the *CMAR* in order to begin the execution phase during the subsequent cycle. Figure 12.30 shows the timing and register transfers for the indirect addressing phase.

12.4.2.3 Interrupt Handling Phase

The computer has an interrupt interface with its I/O devices. This interface is handled through the DMA channel. The CPU control unit contains a flip-flop called *INT*. When an I/O device wishes to interrupt the CPU it sets

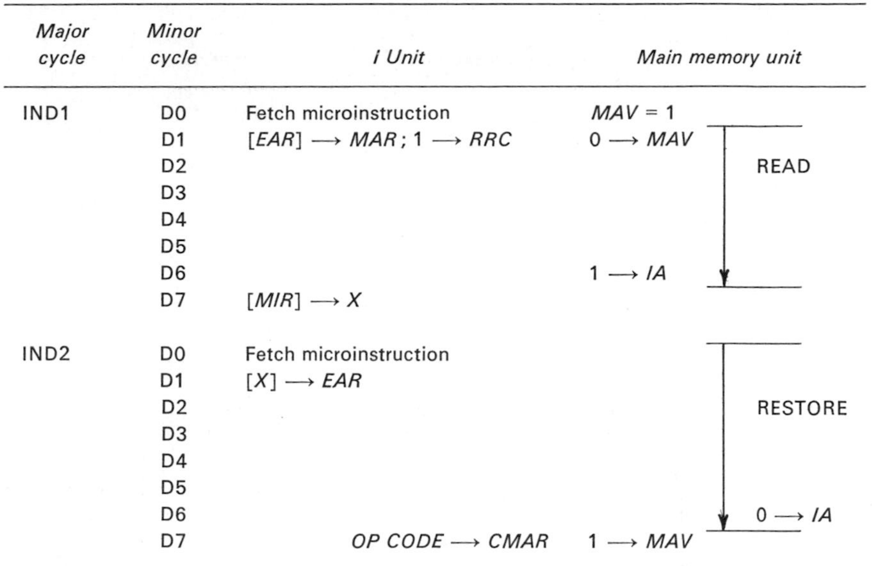

Major cycle	Minor cycle	*I* Unit	Main memory unit	
IND1	D0	Fetch microinstruction	$MAV = 1$	
	D1	$[EAR] \longrightarrow MAR$; $1 \longrightarrow RRC$	$0 \longrightarrow MAV$	
	D2			READ
	D3			
	D4			
	D5			
	D6		$1 \longrightarrow IA$	
	D7	$[MIR] \longrightarrow X$		
IND2	D0	Fetch microinstruction		
	D1	$[X] \longrightarrow EAR$		
	D2			RESTORE
	D3			
	D4			
	D5			
	D6			$0 \longrightarrow IA$
	D7	OP CODE $\longrightarrow$ CMAR	$1 \longrightarrow MAV$	

Figure 12.30 Indirect addressing phase register transfers.

the interrupt flip-flop in its I/O device control unit, which, in turn, sets the *INT* flip-flop to a 1. At the completion of the execution phase of an instruction, the microinstruction performs a test to determine the state of the *INT* flip-flop. If it is set to a 1, a branch in the microcode is made to the microprogam which handles the interrupt. If the flip-flop is in the 0 state, a branch is made to FETCH∅ in the microcode to start the next instruction phase.

Some operations cannot test for an interrupt at the completion of their execution phase because another test is performed at that time. These instructions branch to a special microinstruction upon their completion. This microinstruction is labeled INT∅. It tests for an interrupt. If the interrupt is set, it permits the next sequential microinstruction to be fetched, which is INT1, which handles the interrupt. If the interrupt is 0, INT∅ branches to FETCH∅ to start the next instruction phase.

At the completion of performing the register transfers needed to force a jump to a programmed subroutine to handle the interrupt, the microinstruction INT2 resets the flip-flop INT and fetches the next sequential microinstruction, which is FETCH∅.

Figure 12.31 shows the register transfers needed to force a jump to a location to handle the interrupt. The interrupt procedure for the computer is the following:

1. Each class of I/O device has a unique class code; when the device sets its interrupt flip-flop and therefore the INT flip-flop, it also places its class code into a register called *CLASS* in the DMA channel.

Major cycle	Minor cycle	Instruction unit	Main memory unit	
INTØ	D0	Fetch microinstruction	*MAV* = 1	
	D1			
	D2			
	D3			
	D4			
	D5			
	D6			
	D7	Test *INT* flip-flop: if 1, INT1 ⟶ *CMAR*; if 0, FETCHØ ⟶ *CMAR*		
INT1	D0	Fetch microinstruction	*MAV* = 1	
	D1	0 ⟶ *MAR*; 1 ⟶ *IWC*	0 ⟶ *MAV*	CLEAR
	D2			
	D3			
	D4			
	D5			
	D6			
	D7	[*CAR*] ⟶ $MIR_{8\text{-}23}$	1 ⟶ *IA*	
INT2	D0	Fetch microinstruction		WRITE
	D1			
	D2			
	D3			
	D4			
	D5			
	D6	[*CLASS*] ⟶ *CAR*	0 ⟶ *IA*	
	D7	*RINT*	1 ⟶ *MAV*	

Figure 12.31 Interrupt handling phase register transfers.

2. Upon testing the interrupt flip-flop, the microcode branches to microinstruction INT1.

3. At D1 of INT1, the contents of the *MAR* are cleared to 0 and a special clear-write cycle (IWC) is initiated. Since the contents of the *MAR* are 0, this will clear location 0 in main memory.

4. At D7 of INT1, the contents of the *CAR* are transferred into bits 8–23 of the *MIR*.

5. During INT2, the contents of the *MIR* are written into location 0. This saves the return location in address 0 of main memory.

6. At D6 of INT2, the contents of the register *CLASS* are transferred into the *CAR*. Thus, the next instruction will be fetched from the address in main memory corresponding to the class of the I/O device causing the interrupt.

7. At D7 of INT2, the *INT* flip-flop is reset to 0. This does not enable the interrupt system, which was disabled when the device caused the interrupt.

At the location corresponding to the class of the device causing the interrupt, the programmer places an unconditional branch instruction to a location of a subprogram which serves that class of I/O device. This type of interrupt structure and the software to support it were demonstrated in Section 10.3.4.

12.4.3 The I Unit Address Adder

The I unit of this machine has its own adder for use in computing the effective address. The input to the left side of the adder as shown in Fig. 12.5 is the direct address (*DA*) which comes from the *CIR*. The input to the right side of the adder is either the *CAR*, when self-relative addressing is used, or the index register (*XR*), when indexing is requested. The effective address must be computed in five minor clock periods or 384 nanoseconds, i.e., D(1)–D(6). The I unit adder adds positive binary numbers of 16-bit length. Because of the need for speed, it will be designed using two-level NAND gates with a maximum delay of 8 nanoseconds per gate. Thus, the carry will require 16 nanoseconds $\times$ 16 bits or 256 nanoseconds to propagate. This will yield an effective address in the *EAR* well within the 384-nanosecond limitation.

Figure 12.32 shows the logic circuits required for a single bit of the I unit address adder. Included in the diagram are all the control signals required for the transfer of data into and out of the registers and the address adder. The *CAR* and *XR* are 16 bits in length. The *CIR* is 24 bits in length, but only bits 8–23 are involved in determining the effective address.

The effective address (*EA*) is computed during the second cycle of the instruction phase, FETCH1. The computation is performed during periods D1–D6. The sum is latched into the effective address register (*EAR*) during the minor cycle D6. During periods D1–D6, the control signal, compute effective address, (*CEA*) is present during FETCH1. If bit 5 of the *CIR* is a 1, indicating that indexing is used, the path from the index register to the address adder is open. If bit 6 of the *CIR* is a 1, indicating self-relative addressing, the path from the *CAR* to the address adder is open. The two conditions are mutually exclusive.

At D5 of either FETCH1 or IND2, the *EAR* is reset to 0. At D6 of FETCH1, the effective address is clocked into the *EAR*. At D6 of IND2, if indirect addressing is operative, the direct address fetched from memory is clocked into the *EAR*.

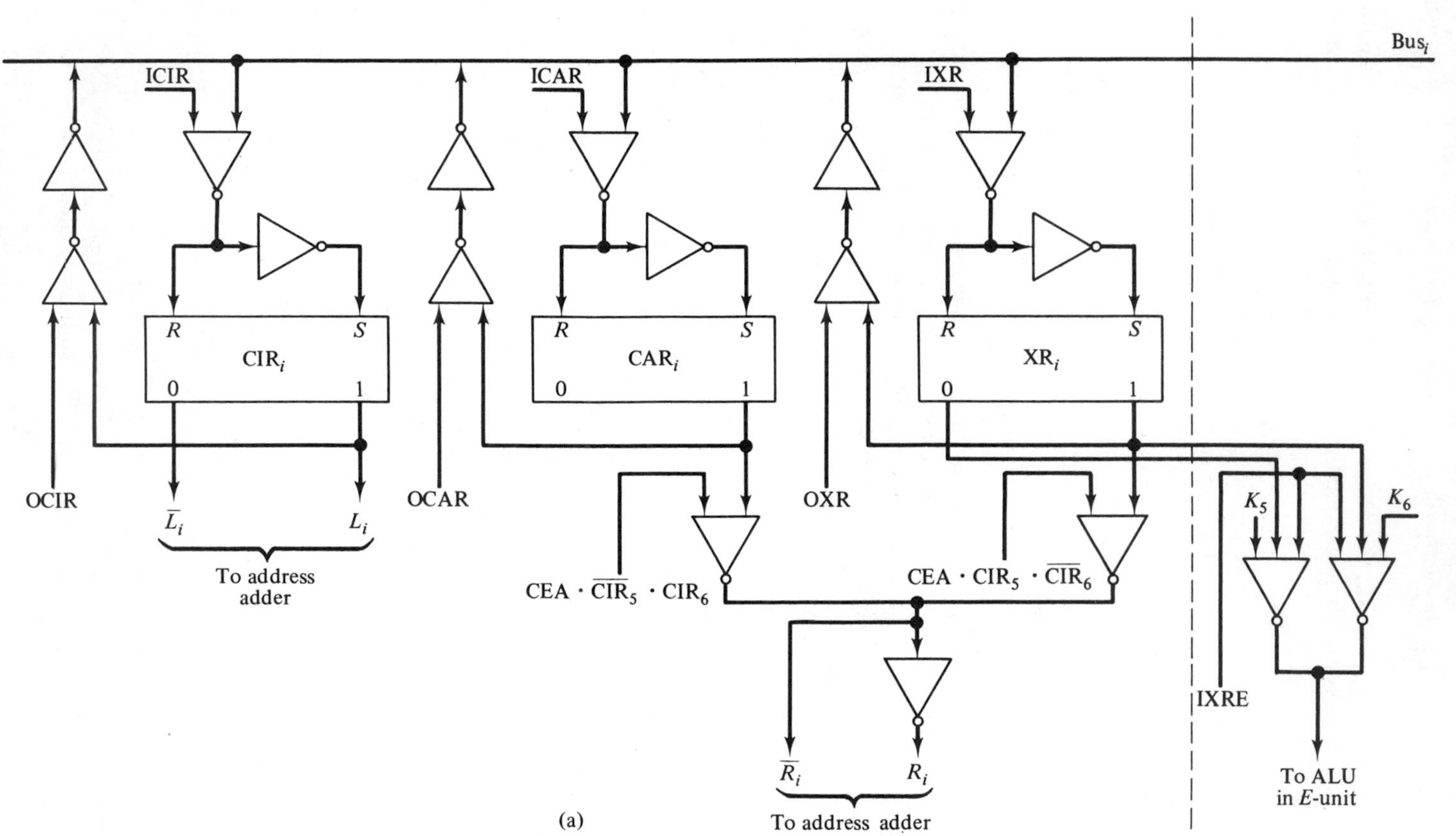

Bus_i
ICIR
ICAR
IXR
R
S
CIR_i
CAR_i
XR_i
0
1
OCIR
OCAR
OXR
$\overline{L_i}$
L_i
To address
adder
$\text{CEA} \cdot \overline{\text{CIR}_5} \cdot \text{CIR}_6$
$\text{CEA} \cdot \text{CIR}_5 \cdot \overline{\text{CIR}_6}$
K_5
K_6
IXRE
$\overline{R_i}$
R_i
To address adder
To ALU
in E-unit

(a)

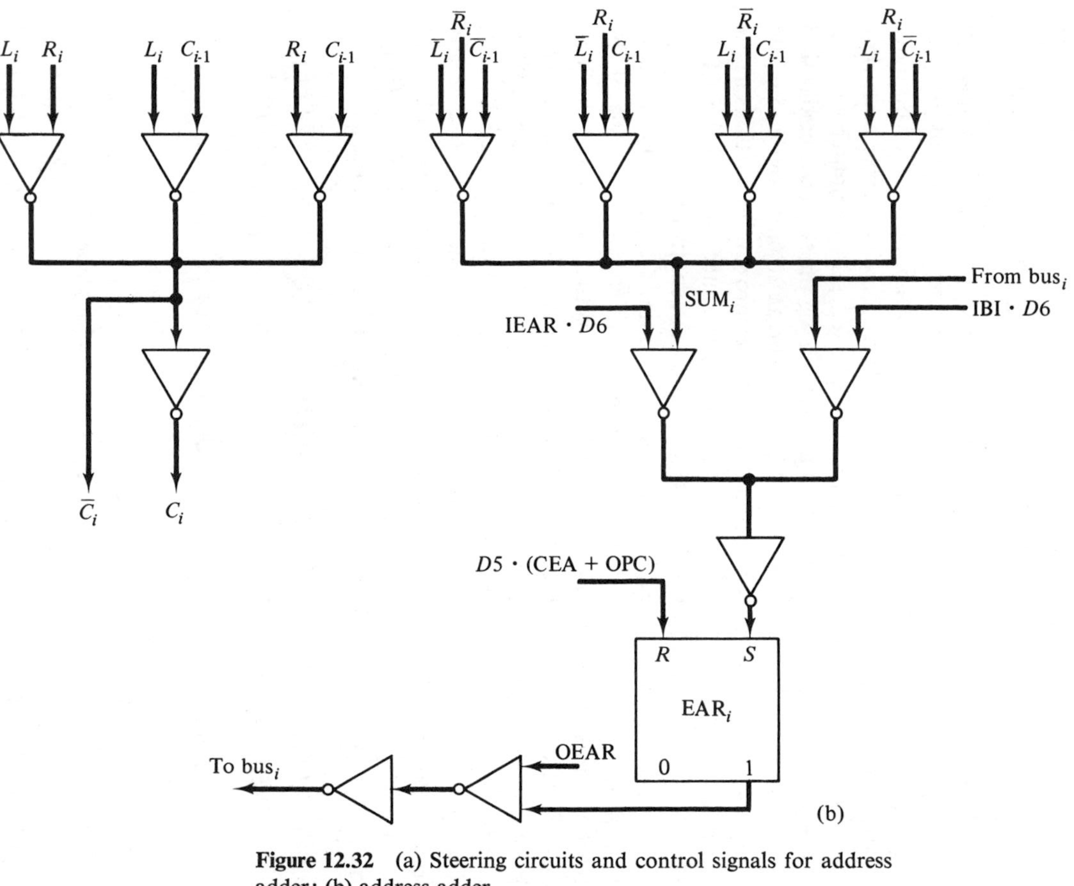

Figure 12.32 (a) Steering circuits and control signals for address adder; (b) address adder.

12.4.4 Operating Modes

The computer has three modes of operation under which it can operate:

1. Single-instruction operation.
2. Single-cycle operation.
3. Automatic continuous run.

The single-instruction operation allows the computer to fetch and execute only one instruction each time a *STEP* button on the control panel of the computer is depressed. The single-cycle operation allows the computer to perform the necessary transfers of a single major clock cycle each time the *STEP* button on the control panel is depressed. The computer will execute one instruction after another, but each major phase of an instruction requires the depression of the *STEP* button. The automatic continuous run mode of operation permits the computer to cycle through instructions at the normal clock rate until a STOP instruction is executed or until the *RUN-HALT* switch is placed into the *HALT* state on the control panel.

The control panel contains switches to determine the mode of operation. One switch is a three-position *MODE* selector which can select one of the three possible modes. A second switch is a two-position switch with *RUN* or *HALT*. A third switch is called a *STEP* button and is used for stepping instructions or cycles depending on the setting of the *MODE* selector. A fourth switch is a system reset and initial program load switch called *RESET*. The fifth switch is a *START* button. The mode of the computer, at any instant, is determined by the state of the control flip-flops in the CPU.

A *RUN* flip-flop determines if the computer is in the *RUN* or *HALT* mode. It is an *R-S* flip-flop with the following excitation equations:

$$RUNFF_S = \mathrm{D6} \cdot RUN \cdot [STARTB \cdot AUTO + (SCY + SIN) \cdot STEPFF]$$
$$RUNFF_R = \mathrm{D6} \cdot [RRUN + HALT \cdot END + (SCY + SIN) \cdot \overline{STEPFF}]$$

The *RUN* flip-flop will be placed into the *RUN* state if

1. The minor clock is in time D6.
2. The *RUN-HALT* switch on the control panel is in *RUN*.
3. The *START* button on the control panel is depressed.
4. The *MODE* switch on the control panel is in either the automatic mode *AUTO* or single-cycle or single-instruction mode and the *STEP* button has been depressed.

The *RUN* flip-flop will be reset if

1. The clock is in time D6.

2. A reset the *RUN* flip-flop (*RRUN*) command is given by a STOP instruction.

3. The *RUN-HALT* switch is placed into *HALT* and the machine has completed execution of the instruction.

4. The machine is in a single-cycle or single-instruction mode but the *STEP* button has not been depressed.

The *STEP* flip-flop (*STEPFF*) is also an *R-S* flip-flop with the following excitation equations:

$$STEPFF_S = RUN \cdot \mathrm{D7} \cdot STEPB \cdot (SCY + SIN)$$
$$STEPFF_R = HALT \cdot \mathrm{D7} + RUN \cdot \mathrm{D7} \cdot (SCY + SIN \cdot END) \cdot \overline{STEPB}$$

The *STEP* flip-flop will be set only at time D7 in the *RUN* mode and a single-cycle or single-instruction mode and when the *STEP* button has been pressed. It will be reset if the *RUN-HALT* switch is placed into *HALT* at time D7 or if while running at time D7 the *STEP* button has not been depressed and the computer is in the single-cycle mode or in the single-instruction mode and the *END* of an excitation cycle has been detected.

The *RUN* flip-flop and *STEP* flip-flop are used together with the *MODE* switch to control the change in the control memory address register (*CMAR*). The *CMAR* will be permitted to change during clock time D7 in order to fetch the next microinstruction from the control memory if the following conditions are met:

$$CHANGE = \mathrm{D7} \cdot RUNFF \cdot [AUTO + STEPFF \cdot (SCY + SIN)]$$

If the conditions are not met, the contents of the *CMAR* will not change, the contents of the *CMIR* will remain the same, but no register transfers will be permitted or arithmetic operations performed.

12.5 THE CONTROL MEMORY

The control unit of this machine uses microprogramming to define the register transfers and arithmetic operations to be performed for each instruction. The machine must therefore contain a control memory which stores the microprogram. This machine will use a read-only semiconductor memory with an access time of 64 nanoseconds. Each word is 48 bits, and there are 256 words of control memory. This requires an 8-bit register for the *CMAR* and a 48-bit register for the *CMIR*. Each microinstruction for this machine determines all transfers and operations for a 512-nanosecond major CPU cycle. Thus, one microinstruction will be fetched from the control memory every 512 nanoseconds with an access time of 64 nanoseconds. The microinstruction is fetched during the minor cycle period D0.

12.5.1 Control Word Format

The 48-bit microinstruction is composed of 11 fields. Figure 12.33 shows the microinstruction format and the fields. The fields are

S_1, 3 bits, the control for the source register of a transfer during time period D1.

D_1, 2 bits, the control for the destination register of a transfer occurring during D1.

S_2, 4 bits, the control for the source register of a transfer occurring during D6.

D_2, 4 bits, the control for the destination register of a transfer occurring during D6.

3	2	4	4	3	4	3	4	5	8	8
S_1	D_1	S_2	D_2	S_3	D_3	C_1	C_2	TEST	ALU	NA

List of Control Signal Mnemonics

S_1	D_1	S_2	D_2	S_3	D_3	C_1	C_2	*TEST*
ZERO	ZERO	ZERO	ZERO	ZERO	ZERO	ZERO	ZERO	ZERO
OCAR	IMAR	OBE	IACC	OMIR	ICIR	RRC	DCR	UNB
OEAR	OXE	OMQ	IXR	OEAR	ICAR	CWC	SOFF	TINT
IALU	ICR	OCIR	IMQ	OIOIR	IACC	MUL	SXRO	TAS
IXRE		OX	IEAR	OACC	IMIR	DIV	ROFF	TXS
OIOAR		OXR	IAS	OMQ	SMIR	SHT	RXRO	TQ23
OCIR		OACC	IX	OXR	IX	IWC	DXR	TCR
		NSI	IIOCR	OCAR	IIOIR	END	RRUN	TSD
		OIOCR	IIOAR		ICR		ENI	TAO
		OIOIR	IIOWC				RACC	TXP
		OEAR	ICAR				RMQ	TXZ
		OCL	IBI				AQ23	TXRO
			SGN				RINT	TAP
							OPC	TAZ
							CEA	TRUN
							RCS	TDA
								TIND

Figure 12.33 Microinstruction format and control signal mnemonics.

S_3, 3 bits, the control for the source register of a transfer occurring during D7.

D_3, 4 bits, the control for the designation register of a transfer occurring during D7.

C_1, 3 bits, a control signal which causes a specific command to occur, usually during D1.

C_1, 4 bits, another control signal which causes a specific command to occur, usually at D6 or D7.

TEST, 5 bits, specifies the code of a particular test which will be made during D7.

ALU, 8 bits, specifies the particular operations which the ALU will perform during the major cycle. It is the K_i signals to the ALU.

NA, 8 bits, is the next address field of the microinstruction. If the test of the TEST field is positive, this field specifies the address in the control memory from which the next microinstruction will be fetched.

Each of the S or D fields contains a binary code which specifies a particular control signal which opens the AND gates between a given register and the single main bus. The mnemonics used to define these control signals in the design are explained in the next subsection.

12.5.2 Register Transfer Control Signals

All the transfer paths within the computer have control signals which open the AND gates and permit the data to flow on the appropriate data lines. These control signals are shown for the I unit in Fig. 12.5 and for the E unit in Fig. 12.6. In addition, the control signals K_i have been used to produce the desired functions of the ALU.

The control signals for the arithmetic unit were developed for a single bit. However, except for the carry into the least significant bit (LSB), all the control lines for each of the 24 bits of the ALU will be connected in parallel. That is, all the bit positions of the ALU perform the same operation at any given time. The carry out from each bit is applied to the next higher-order bit as its carry in when addition (or subtraction) is being performed. The carry out of the most significant bit (MSB) of the adder will be used to determine overflow conditions and certain other test conditions.

The control signals are applied to AND gates, or their logical equivalents, to permit the flow of the logic signal onto or out of the bus. The mnemonics used to designate these signals will adhere to the following convention. Control signals which gate signals out of a register and onto the main bus will be designated by an O followed by the register's abbreviation, the O standing for output. Thus, the signal used to gate the output of the accumulator onto the bus is called OACC. Control signals which gate signals from the main bus into a register are designated with an I followed by the register's abbreviation. Thus, IACC is a control signal used to gate data from the bus into the accumulator. The control signals which are exceptions to the above rule are

1. IALU: Gates the output of the accumulator into the ALU during D1–D7.
2. IBE: Gates the bus directly into the ALU during D1–D7.
3. IXRE: Gates the output of the index register directly to the ALU (bits 8–23) during D1–D7.
4. OXE: Gates the output of the X register directly into the ALU during D1–D7.
5. OBE: Gates the output of the ALU latches onto the bus.
6. IBI: Gates the bus directly into the *EAR*.
7. NSI: Is really a command which causes the *CAR* to increment by 1.
8. IEAR: Latches the *EAR* register with the effective address.
9. IAS: Places ACC_0 into the *Scand* flip-flop.
10. SMIR: Gates signals from bits 8–23 of the bus into bits 8–23 of the *MIR* without affecting bits 0–7 of the *MIR*.
11. SGN: Computes $[X]_0 \oplus [Scand]$ and places the results back into *Scand.*

If a specific bit of a register is to be controlled for a transfer without affecting the rest of the register, then the number of that position is appended to the nomenclature. Thus, IMQ23 is a control signal used to gate bit 23 of the bus into bit 23 of the *MQ* register. Table 12.13 shows the mnemonics of the control signals, the source and destination of the data, and the bit positions of the registers which are being opened by the control signal.

12.5.3 Command Signals

In addition to control signals used to gate data signals to and from the registers, the microinstruction has two fields designated as C_1 and C_2 which are command signals. The fields provide signals used to perform specific functions. Some of these signals are timed with a specific output from the minor clock and are effective only during that period. Others are applied throughout the major cycle of the microinstruction. Table 12.14 lists the command mnemonics, their definitions, and the time during which they are applied to the registers.

The command signals are present in two fields, and therefore two commands may be given during a major cycle. However, the commands of each field are mutually exclusive and are encoded in the command fields.

Table 12.13 Control Signal Mnemonics

Mnemonic	*Source*	*Destination*	*Bits*
ICAR	BUS	*CAR*	8–23
ICIR	BUS	*CIR*	0–23
IXR	BUS	*XR*	8–23
IEAR	AA	*EAR*	8–23
IBI	BUS	*EAR*	8–23
IXRE	*XR*	ALU	8–23
OCAR	*CAR*	BUS	8–23
OCIR	*CIR*	BUS	12–23
OXR	*XR*	BUS	8–23
OEAR	*EAR*	BUS	8–23
ICR	BUS	*CR*	18–23
IACC	BUS	*ACC*	0–23
IMQ	BUS	*MQ*	0–23
IX	BUS	*X*	0–23
IALU	*ACC*	ALU	0–23
IBE	BUS	ALU	0–23
OACC	*ACC*	BUS	0–23
OMQ	*MQ*	BUS	0–23
OX	*X*	BUS	0–23
OBE	ALU	BUS	0–23
IMIR	BUS	*MIR*	0–23
IMAR	BUS	*MAR*	8–23
OMIR	*MIR*	BUS	0–23
IIOIR	BUS	*IOIR*	0–23
IIOAR	BUS	*IOAR*	8–23
IIOCR	BUS	*IOCR*	12–23
IIOWC	BUS	*IOWC*	8–23
OIOIR	*IOIR*	BUS	0–23
OIOAR	*IOAR*	BUS	8–23
OIOCR	*IOCR*(0–11)	BUS(12–23)	12–23
SMIR	BUS	*MIR*(8–23)	8–23
OXE	*X*	ALU	0–23
NSI	*CAR*	*CAR*	8–23
OCL	*CLASS*	BUS	8–23
IAS	ACC_0	*Scand*	0
SGN	X_0	*Scand*	—

Table 12.14 List of Command Signals

Mnemonic	*Definition*	*Time*
	A. Command Signals Specified in the C_1 Field	
Zero	No signal	—
CWC	Clear-write cycle CPU to memory	D1
RRC	Read-restore cycle CPU to memory	D1
MUL	Multiply; 24 $\longrightarrow$ *CR* (counter register)	D1
DIV	Division; 23 $\longrightarrow$ *CR*	D1
SHT	Indicates a shift operation	D1–D7
IWC	Interrupt clear-write cycle	D1
END	Last microinstruction in execution phase	D6–D7
	B. Command Signals Specified in the C_2 Field	
Zero	No signal	—
DCR	Decrement counter register by 1	D6
SOFF	Set overflow flip-flop	D6
ROFF	Reset overflow flip-flop	D6
SXRO	Set *XR* overflow flip-flop	D6
RXRO	Reset *XR* overflow flip-flop	D6
DXR	Decrement index register by 1	D6
RRUN	Reset *RUN* flip-flop (STOP)	D6
ENI	Enable the interrupt	D6
RACC	Reset the accumulator to 0	D7
RMQ	Reset the *MQ* register to 0	D7
AQ23	$\overline{ACC_0} \longrightarrow MQ23$	D6
RINT	Reset the interrupt flip-flop	D7
OPC	Move the op-code into the *CMAR*	D1–D7
CEA	Compute the effective address	D1–D7
RCS	Reset the cycle steal flip-flop	D7

12.5.4 Control Memory Unit

The control memory unit is shown in the block diagram of Fig. 12.34. The memory is a semiconductor, linear select memory with 256 words of 48 bits each. The control memory address register (*CMAR*) is 8 bits long, and the control memory instruction register (*CMIR*) is 48 bits long. Each of the source and destination fields of the *CMIR* is connected to a decoder which produces a unique line output for the coded control signals. This is also true for the two command signal fields C_1 and C_2. The test field also

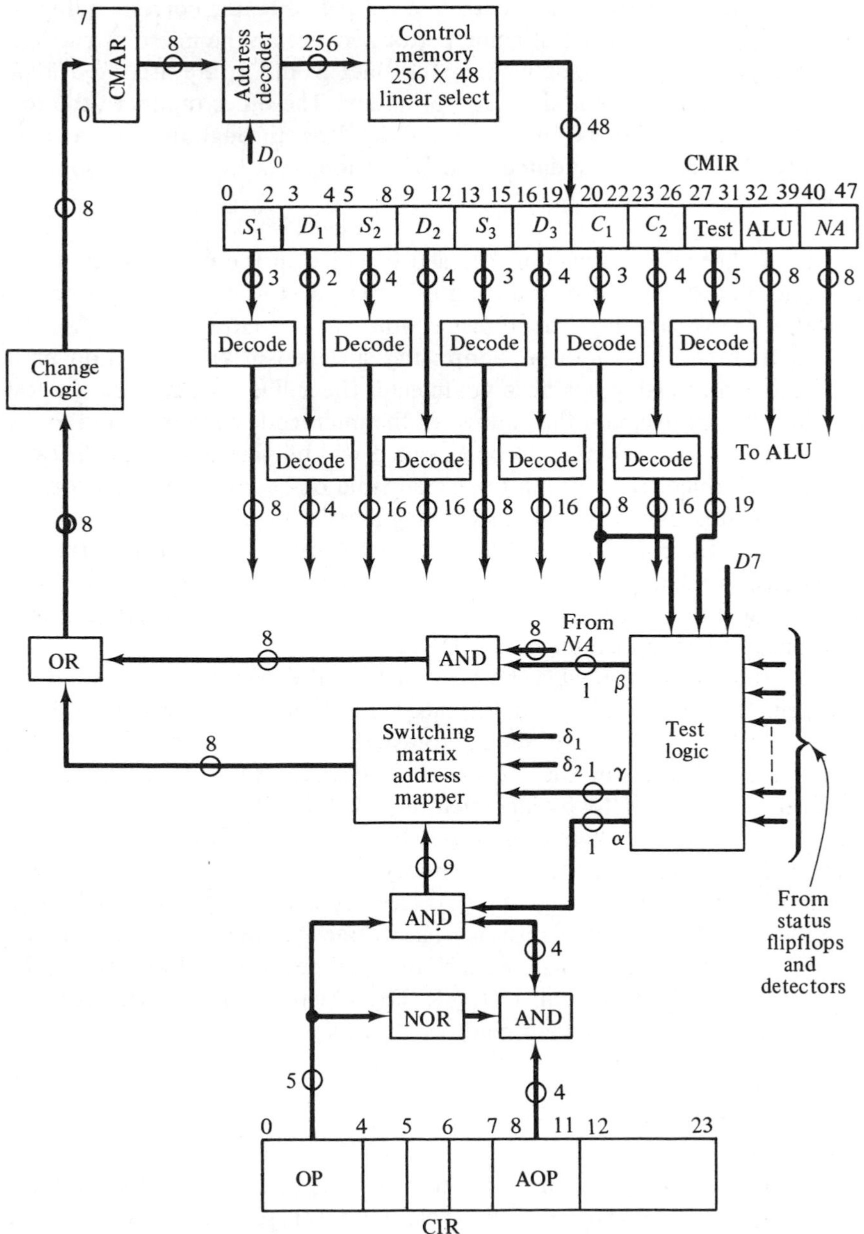

Figure 12.34 Control memory unit.

has a decoder which produces a signal on the unique line corresponding to the code of the particular test being performed during the microinstruction. The test lines, as well as some command lines from C_1, are used in a logic section to produce output signals γ, β, and α. The other inputs to the test logic block are the status of flip-flops and detectors throughout the machine.

The *CMAR* can be updated each D7 period from five different sources, which are

1. If the microinstruction has a 0 specified in the next address (*NA*) field and no test is performed during D7, then the *CMAR* is updated to the address of next sequential microinstruction by $[CMAR] + 1 \longrightarrow CMAR$.

2. If there is no indirect addressing specified in an instruction or if the indirect addressing has been completed, the α line output from the test logic will be positive, and the address of the microcode corresponding to the instruction op-code which is to be executed will be placed into the *CMAR*. This is performed by mapping the 47 possible op-codes onto the 8 control memory address lines through a switching matrix map.

3. If at the completion of the execution cycle of an instruction the test logic detects that an interrupt has occurred, the γ line from the output of the test logic will be positive, and the address of the interrupt handling microprogram will be mapped into the *CMAR*.

4. If a test is specified in the TEST field of the microinstruction and the test is made by the test logic and is successful, then the address specified in the next address (*NA*) field of the microinstruction is placed into the *CMAR*. If the test is unsuccessful, the address of the next sequential microinstruction is placed into the *CMAR* by incrementing the *CMAR*.

5. If the DMA channel indicates that it is ready for a *cycle steal* to store or fetch data from the main memory by setting the *CSFF* (cycle steal flip-flop) and the interrupt signal is 0, indicating that no interrupt is pending, then the δ_1 or δ_2 line will be positive. The δ_1 line indicates a READ from the I/O device, which means that the memory will perform a clear-write cycle, while the δ_2 line indicates an I/O WRITE, which means that the memory must fetch data and perform a read-restore cycle.

12.5.5 CMAR Change Logic

There are conditions for which the contents of the *CMAR* should not change during D7 of a major CPU cycle. Under these conditions, when the next microinstruction is fetched from the control memory during D0 of the next major CPU cycle, it will be the same microinstruction as is presently in the *CMIR*. Thus, the *CMAR* will be stuck at the same addıess, and the control unit will not progress through the microprogram but will remain at one location. The logic which prevents the *CMAR* from changing is labeled CHANGE in Fig. 12.34.

The condition for which the *CMAR* should not change during D7 of a particular microinstruction is dependent on the settings of the operating mode switches on the control panel. In addition it is dependent on the availability of memory during a microinstruction which specifies an RRC, CWC, or IWC command signal during the D1 minor cycle. If the memory is not available, i.e., MAV = 0, then the microinstruction will not be permitted to execute and the *CMAR* will not be permitted to change at D7 time. Thus, the machine will "hang up" until the memory indicates it is available, i.e., MAV = 1. The logic equation is

$$\text{CHANGE} = \text{D7} \cdot MSFF$$

where *MSFF* is a memory synchronization flip-flop. The excitation equations for the MSFF are

$$MSFF_S = \text{MAV} \cdot RUNFF \cdot \text{D1} \cdot (\text{RRC} + \text{CWC} + \text{IWC})$$

$$MSFF_R = \overline{\text{MAV}} \cdot RUNFF \cdot \text{D1} \cdot (\text{RRC} + \text{CWC} + \text{IWC})$$

If the *MSFF* = 0 at D1, then no register transfers are permitted, and the *CMAR* will not be changed at D7. Thus, the microinstruction will not change or be executed until the memory signals it is available to execute a read-restore or clear-write cycle.

Combining this CHANGE equation with the CHANGE equation from the operating mode switches yields

$$\text{CHANGE} = \text{D7} \cdot [RUNFF \cdot [AUTO + STEPFF(SCY + SIN)]] \cdot MSFF$$

12.5.6 Test Condition Logic

The execution of the microprogram is similar to the execution of the main memory program. The control memory unit fetches microinstructions into the *CMIR* from sequential control memory addresses unless the order is changed by a test condition which is successful. The particular test which is to be performed during the D7 minor cycle of the particular microinstruction is specified by a 5-bit code in the TEST field of the microinstruction. Only one test may be specified for each microinstruction. The test mnemonics and their definitions are listed in Table 12.15.

The test code is decoded and fed into a section of logic called the *test logic*, as shown in Fig. 12.34. The output from this logic is three lines labeled α, β, and γ. These lines control the sequencing of the microcode which will be executed.

Table 12.15 *Definition of Test Mnemonics*

Mnemonic	*Definition*
ZERO	No test condition
UNB	Unconditional branch (test always successful)
TINT	Test for interrupt condition
TIND	Test for indirect addressing
TAS	Test accumulator sign
TXS	Test *X* register sign
TQ23	Test bit 23 of the *MQ* register
TCR	Test the zero detector on the counter register
TSD	Test the *Scand* flip-flop
TAO	Test the accumulator overflow flip-flop
TXP	Test the index register positive
TXZ	Test the index register zero detector
TXRO	Test the index register overflow
TAP	Test the accumulator positive
TAZ	Test the accumulator zero detector
TRUN	Test the *RUN* flip-flop
TDA	Test the device-not-available bit

The α line is used to place the address of the start of the microcode for a particular operation into the *CMAR* to begin the execution phase. It can be positive only at D7 of the microinstruction FETCH1 or at D7 of IND2. The logic for this test is

$$\alpha = \mathrm{D7}\cdot(\mathrm{TIND}\cdot\overline{CIR_7} + \mathrm{OPC})$$

where TIND is the test for the indirect addressing mnemonics. This signal appears in the microinstruction FETCH1. CIR_7 is the indirect addressing flag in the *CIR*, and OPC is a command signal to move the microcode address corresponding to the op-code in the *CIR* into the *CMAR* via the map switch matrix. This command signal appears in the microinstruction IND2.

The γ line is used to cause the control memory to jump to the microprogram which handles the interrupt, i.e., INT1. It can be positive at the end of each execution phase, where a test is made to see if an interrupt has occurred during the execution of an instruction. The logic for this test is

$$\gamma = \mathrm{D7}\cdot\mathrm{TINT}\cdot\mathrm{INT}\cdot\mathrm{END}$$

where TINT is the test for interrupt mnemonic, INT is the set output of the interrupt flip-flop, and END is a command signal which signifies the last minor cycle of an execution phase. If the test is successful, the address of the

Table 12.16 *Test Condition Logic for Line* β

Test condition	*Location of occurrence*	*NA*
UNB	DIV17; DIV22	DIV20; DIV 24
$\mathrm{TIND} \cdot CIR_7$	FETCH1	IND1
$\mathrm{TINT} \cdot \overline{\mathrm{INT}} \cdot \mathrm{END} \cdot \overline{CSFF}$	Last microcode of execution phase	FETCH∅
$\mathrm{TAS} \cdot \overline{ACC_0}$	MUL2; DIV1; DIV8; BAN1	MUL5; DIV4; DIV23; INT∅
$\mathrm{TXS} \cdot \bar{X}_0$	MUL5; DIV4	MUL8; DIV7
$\mathrm{TQ23} \cdot \overline{MQ23}$	DIV14; MUL8	DIV18; MUL11
$\mathrm{TCR} \cdot \overline{CR \text{ zero detection}}$	MUL11; DIV16; DIV19	MUL8; DIV14; DIV14
$\mathrm{TSD} \cdot \overline{Scand}$	MUL12; DIV20	MUL17; DIV24
$\mathrm{TAO} \cdot \overline{OFF}$	BAO1	INT∅
$\mathrm{TXP} \cdot (XR \text{ zero detect} \dotplus XRO)$	BXP1	INT∅
$\mathrm{TXZ} \cdot (\overline{XR \text{ zero detect}}) \cdot \mathrm{BXZ}$	BXZ1	INT∅
$\mathrm{TXZ} \cdot (XR \text{ zero detect}) \cdot \mathrm{TLD}$	TLD1	INT∅
$\mathrm{TXRO} \cdot \overline{XRO}$	BXN1	INT∅
$\mathrm{TAP} \cdot (ACC \text{ zero detect} + ACC_0)$	BAP1	INT∅
$\mathrm{TAZ} \cdot (\overline{ACC \text{ zero detect}})$	BAZ1	INT∅
$\mathrm{TRUN} \cdot RUN$	STP2	INT∅
$\mathrm{TCR} \cdot (CR \text{ zero detect}) \cdot \mathrm{SHT}$	Step 1 of all shift instructions	INT∅
$\mathrm{TDA} \cdot IOCR_{11}$	BDN1	INT∅

start of the microinstruction which handles the interrupt, INT1, is placed into the *CMAR*. If the test is unsuccessful, i.e., no interrupt has occurred, then this is a condition under the β line tests which will be successful, and the address of the start of the fetch cycle FETCH∅ will be placed into the *CMAR*.

The β line is used for 18 different tests, as specified by the contents of the TEST field in the microcode. All the tests are made during minor cycle time D7. Table 12.16 shows the test conditions, the mnemonics of the microinstruction where the tests occur, and the next address of the microcode, which is placed into the *CMAR* if the test is successful. If the test is unsuccessful, i.e., the condition has not occurred, the *CMAR* is incremented by 1, except for the case of testing for indirect addressing and interrupt. These two cases are handled by lines α and γ, as explained above. The request for a DMA channel cycle steal is made via the two lines δ_1 and δ_2. These two requests are not made as tests but are a result of conditions which exist in the machine at a specific time.

12.6 THE EXECUTION UNIT AND INSTRUCTION TIMING

In this section, the register transfer equations for each of the 47 operations which the computer can execute will be developed in detail. Since the structure of the control memory has been developed in the previous section, the basic timing features inherent in its design will be utilized in specifying the register transfer equations. The microprogrammed control permits register transfers to occur only at minor cycle times D1, D6, and D7. It also requires that all tests be made at time D7 and that only one test can be requested during each microinstruction, i.e., major CPU cycle. If the test is affirmative, a branch may be made in the microprogram; i.e., the microinstruction specified by the address in the NA field is executed. If the test is negative, the next sequential microinstruction is executed. If no test is specified, the next sequential microinstruction is executed. If the control field C_1 contains the command END, this signifies the last microinstruction for the execution of the operation, and the next microinstruction executed is FETCH∅, the beginning of the fetch cycle, unless an interrupt has occurred and then the interrupt phase is entered.

Since register transfers occur only during D1, D6, and D7, only these times will be shown in the succeeding descriptions of the operations.

12.6.1 General Timing Considerations

Before each operation is examined in detail, let us develop some general timing concepts for the execution phase.

Case 1: Operand Required: This case assumes a normal format instruction which requires either the accessing of an operand or the storing of an operand. At the end of the fetch cycle, FETCH1, or the end of the indirect addressing cycle, IND2, the effective address of the operand has been stored in the *EAR*. At D7 of either of these two microinstructions the α test line will be positive and cause the op-code to be mapped into the address of the first microinstruction for that operation which is placed into the *CMAR*. At the subsequent D0 time, the microinstruction will be accessed from the CM and thus the execution phase starts.

Since an operand is required, the first microinstruction must request a memory cycle. At time D1, the contents of the *EAR* are sent to the *MAR*. The CPU signals the main memory that it requires either a read-restore cycle (RRC), i.e., an operand is required, or a clear-write cycle (CWC), i.e., the contents of a register are to be stored. The memory responds by resetting the memory-available flip-flop. At the end of D6, the operand will be available in the *MIR*, or the memory location will be cleared. The memory signals the CPU by setting the information-available flip-flop. During minor cycle time D7, the contents of the *MIR* are transferred to the *X* register, or the contents of the register to be stored is placed into the *MIR*. During the next D0, the next sequential microinstruction is fetched from the CM. During this microinstruction the memory must be restoring the location or writing new information into a cleared location. This will not be completed until D6. At that time, the memory will reset the *IA* flip-flop. At minor cycle time D7 it will set the memory-available flip-flop.

During this microinstruction, while the memory is in the restore phase, it is possible to specify register transfers since the operand is in the *X* register. Thus, the execution of the instruction can be overlapped with the memory restore. Figure 12.35 shows the timing for the fetching of an operand.

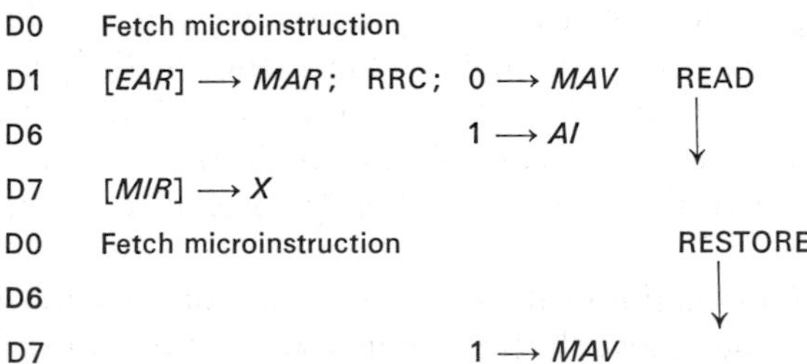

D0 Fetch microinstruction
D1 [EAR] ⟶ MAR; RRC; 0 ⟶ MAV
READ
D6 1 ⟶ AI
D7 [MIR] ⟶ X
D0 Fetch microinstruction
RESTORE
D6
D7 1 ⟶ MAV

Figure 12.35 Register transfers for an operand fetch.

Case 2: No Operand Required: This case covers the normal format instructions which do not involve the fetch or store of an operand. These are the branch instructions and the transfer instructions.

Case 2a: Branch Instructions: The microprogram for the conditional branch instructions requires two microinstructions. No register transfers are performed during the first microinstruction, but the test as specified by the TEST field is performed during D7. The test mnemonics and the test conditions are listed in Tables 12.15 and 12.16. If the test is successful, the microprogram branches to INT∅ to test for the occurrence of an interrupt before entering the fetch cycle. If the test is unsuccessful, the next sequential microinstruction is fetched at D0. Again, no transfers are made until D7, when the contents of the *EAR* are placed into the *CAR* and the software branch will be made.

Case 2b: Transfer Instructions: The instructions TXA and TMQ, which are exceptions in the normal format since they require no operands or address computations, are executed in one major execution cycle. At time D6, the contents of the registers are transferred.

Case 3: Immediate Instructions: The instructions which are in the augmented format never require the use of memory for operands or the need to compute an effective address. Some of these instructions require only one major CPU cycle to execute:

1. LAI: Load accumulator immediate
2. LXI: Load index register immediate
3. STP: Stop
4. ENI: Enable interrupt
5. LDS: Load status
6. CND: Connect I/O device

The register transfers involved in these instructions are generally performed during minor cycle time D6. Two of these executions, STP and ENI, require no transfers but only command signals which reset the *RUN* flip-flop and enable the interrupt flip-flop, respectively. The remainder of the instructions in this format will be discussed separately.

12.6.2 Register Transfers and Microprogramming

In this subsection we shall specify all the register transfers which must occur for each instruction as well as the timing of their occurrence. The commands and tests needed for each instruction will be shown. From these detailed specifications, the microcode for the instruction can be stated. We shall deal only with the mnemonics already developed in earlier sections of this chapter.

The transfer equations will be shown for minor cycle times D1, D6, and D7. The mnemonics above the transfer arrows are the control signals needed to open the gates to permit the transfer to occur. These, in turn, become the signals which are specified in the microprogram. Each microinstruction will be assigned a symbolic address in the control memory. This symbolic address is specified and used to signify microaddresses for the branch under test conditions.

12.6.2.1 Instruction Phase Revisited

To keep this section complete and comprehensive, we shall restate the detailed register transfers for the fetch phase, interrupt phase, and indirect phase. This time we shall also present the microcode which these equations represent.

Interrupt, fetch, and indirect timing

INTØ	D0			
	D1			
	D6			
	D7		END; TINT;	FETCHØ
INT1	D0			
	D1	$0 \xrightarrow{\text{IWC}} MAR$	IWC	
	D6			
	D7	$[CAR] \xrightarrow{\text{OCAR}} \text{bus} \xrightarrow{\text{SMIR}} MIR$		
INT2	D0			
	D1			
	D6	$[CLASS] \xrightarrow{\text{OCL}} \text{bus} \xrightarrow{\text{ICAR}} CAR$		
	D7		END; RINT	
FETCHØ	D0			
	D1	$[CAR] \xrightarrow{\text{OCAR}} \text{bus} \xrightarrow{\text{IMAR}} MAR;$	RRC	
	D6			
	D7	$[MIR] \xrightarrow{\text{OMIR}} \text{bus} \xrightarrow{\text{ICIR}} CIR$		
FETCH1	D0			
	D1	$[SH(CIR)] \xrightarrow{\text{OCIR}} \text{bus} \xrightarrow{\text{ICR}} CR;$	CEA	
	D6	$EA \xrightarrow{\text{IEAR}} EAR;\ [CAR] + 1 \xrightarrow{\text{NSI}} CAR$		
	D7		TIND	
IND1	D0			
	D1	$[EAR] \xrightarrow{\text{OEAR}} \text{bus} \xrightarrow{\text{IMAR}} MAR;$	RRC	
	D6			
	D7	$[MIR] \xrightarrow{\text{OMIR}} \text{bus} \xrightarrow{\text{IX}} X$		
IND2	D0			
	D1			
	D6	$[X] \xrightarrow{\text{OX}} \text{bus} \xrightarrow{\text{IBI}} EAR$		
	D7		OPC	

The microprogram for these operations performed during the instruction phase is

Symbolic address	S_1	D_1	S_2	D_2	S_3	D_3	C_1	C_2	*TEST*	*ALU*	*NA*
INTØ	0	0	0	0	0	0	END	0	TINT	0	FETCHØ
INT1	0	0	0	0	OCAR	SMIR	IWC	0	0	0	0
INT2	0	0	OCL	ICAR	0	0	END	RINT	0	0	0
FETCHØ	OCAR	IMAR	0	0	OMIR	ICIR	RRC	0	0	0	0
FETCH1	OCIR	ICR	IEAR	NSI	0	0	0	CEA	TIND	0	0
IND1	OEAR	IMAR	0	0	OMIR	IX	RRC	0	0	0	0
IND2	0	0	OX	IBI	0	0	0	OPC	0	0	0

Some comments about the above microcode are necessary.

INT∅ : Some execution phases cannot test for an interrupt at the end of the phase. They perform a branch to INT∅ where the test is performed.

INT1 and INT2: These handle the register transfers needed if an interrupt is being processed.

FETCH1: At S_2 and D_2 are special control signals. IEAR strobes the output from the address adder into the EAR. NSI increments the *CAR* by 1.

12.6.2.2 Arithmetic Operations

Addition Operation

ADD1	D0		
	D1	$[EAR] \xrightarrow{\text{OEAR}} \text{bus} \xrightarrow{\text{IMAR}} MAR;$	RRC
	D6		
	D7	$[MIR] \xrightarrow{\text{OMIR}} \text{bus} \xrightarrow{\text{IX}} X$	
ADD2	D0		
	D1	$[ACC] \xrightarrow{\text{IALU}} \text{ALU}; [X] \xrightarrow{\text{OXE}} \text{ALU};$	ADD
	D6		ROFF
	D7		
ADD3	D0		
	D1	$[ACC] \xrightarrow{\text{IALU}} \text{ALU}; [X] \xrightarrow{\text{OXE}} \text{ALU};$	ADD
	D6	$[ALU] \xrightarrow{\text{OBE}} \text{bus} \xrightarrow{\text{IACC}} ACC$	
	D7		END; TINT; FETCH∅

The addition operation starts by fetching the operand from main memory and placing it into the X register at time D7. While the memory is performing the restore, the addition is started at D1 of ADD2. The addition continues during ADD3, and the sum is transferred into the accumulator at D6 of ADD3. At ADD3·D7 a test is made to see if an interrupt has occurred. If it has, a branch will be made to INT1 because the γ line will be positive (see Section 12.5.4). If no interrupt is pending, the next microinstruction to be executed is located at FETCH∅ (see Section 12.5.6). The microcode for the above transfers is

Symbolic address	S_1	D_1	S_2	D_2	S_3	D_3	C_1	C_2	*TEST*	*ALU*	*NA*
ADD1	OEAR	IMAR	0	0	OMIR	IX	RRC	0	0	0	0
ADD2	IALU	OXE	0	0	0	0	0	ROFF	0	ADD	0
ADD3	IALU	OXE	OBE	IACC	0	0	END	0	TINT	ADD	FETCH∅

Subtraction Operation The subtraction operation is just like the addition operation, with the exception that the K_i control signals to the ALU will indicate subtraction instead of addition. The timing and register transfers are identical and will not be repeated. The microcode for subtraction is

Symbolic address	S_1	D_1	S_2	D_2	S_3	D_3	C_1	C_2	*TEST*	*ALU*	*NA*
SUB1	OEAR	IMAR	0	0	OMIR	IX	RRC	0	0	0	0
SUB2	IALU	OXE	0	0	0	0	0	ROFF	0	SUB	0
SUB3	IALU	OXE	OBE	IACC	0	0	END	0	TINT	SUB	FETCHØ

Add to Index—ADX The add to index operation is also similar to the addition operation except that the index register is used in place of the accumulator. The transfer equations are

ADX1	D0			
	D1	$[EAR] \xrightarrow{\text{OEAR}} \text{bus} \xrightarrow{\text{IMAR}} MAR;$	RRC	
	D6			
	D7	$[MIR] \xrightarrow{\text{OMIR}} \text{bus} \xrightarrow{\text{IX}} X$		
ADX2	D0			
	D1	$[XR] \xrightarrow{\text{IXRE}} \text{ALU}; [X] \xrightarrow{\text{OXE}} \text{ALU};$	ADD	
	D6		RXRO	
	D7			
ADX3	D0			
	D1	$[XR] \xrightarrow{\text{IXRE}} \text{ALU}; [X] \xrightarrow{\text{OXE}} \text{ALU};$	ADD	
	D6	$[\text{ALU}] \xrightarrow{\text{OBE}} \text{bus} \xrightarrow{\text{IXR}} XR$		
	D7		END; TINT;	FETCHØ

The microcode derived from the above transfers is

Symbolic address	S_1	D_1	S_2	D_2	S_3	D_3	C_1	C_2	*TEST*	*ALU*	*NA*
ADX1	OEAR	IMAR	0	0	OMIR	IX	RRC	0	0	0	0
ADX2	IXRE	OXE	0	0	0	0	0	RXRO	0	ADD	0
ADX3	IXRE	OXE	OBE	IXR	0	0	END	0	TINT	ADD	FETCHØ

2's Complement the Accumulator This operation does not require the fetch of an operand. If can therefore be completed in only two major CPU cycles. The register transfers are

CMA1	D0		
	D1	$[ACC] \xrightarrow{\text{IALU}} \text{ALU}; 0 \longrightarrow \text{ALU};$	CMA
	D6		
	D7		
CMA2	D0		
	D1	$[ACC] \xrightarrow{\text{IALU}} \text{ALU}; 0 \longrightarrow \text{ALU};$	CMA
	D6	$[\text{ALU}] \xrightarrow{\text{OBE}} \text{bus} \xrightarrow{\text{IACC}} ACC$	
	D7		END; TINT; FETCH∅

This operation requires only one implicitly defined operand. Therefore, one of the operands into the ALU is 0. The control signal to the ALU specifies a CMA, 2's complement the accumulator operation. The microcode for the above is

Symbolic address	S_1	D_1	S_2	D_2	S_3	D_3	C_1	C_2	*TEST*	*ALU*	*NA*
CMA1	IALU	0	0	0	0	0	0	0	0	CMA	0
CMA2	IALU	0	OBE	IACC	0	0	END	0	TINT	CMA	FETCH∅

Increment Index Register This operation is very similar to the add to index operation. The difference is that this operation uses immediate data and does not reference main memory. The transfer equations are

INX1	D0		
	D1		
	D6	$[N(CIR)] \xrightarrow{\text{OCIR}} \text{bus} \xrightarrow{\text{IX}} X$	
	D7		
INX2	D0		
	D1	$[XR] \xrightarrow{\text{IXRE}} \text{ALU}; [X] \xrightarrow{\text{OXE}} \text{ALU};$	ADD
	D6		RXRO
	D7		
INX3	D0		
	D1	$[XR] \xrightarrow{\text{IXRE}} \text{ALU}; [X] \xrightarrow{\text{OXE}} \text{ALU};$	ADD
	D6	$[\text{ALU}] \xrightarrow{\text{OBE}} \text{bus} \xrightarrow{\text{IXR}} XR$	
	D7		END; TINT; FETCH∅

The microcode for the above transfers is

Symbolic address	S_1	D_1	S_2	D_2	S_3	D_3	C_1	C_2	*TEST*	*ALU*	*NA*
INX1	0	0	OCIR	IX	0	0	0	0	0	0	0
INX2	IXRE	OXE	0	0	0	0	0	RXRO	0	ADD	0
INX3	IXRE	OXE	OBE	IXR	0	0	END	0	TINT	ADD	FETCH∅

Decrement Index Register This operation is similar to the increment index register except that a subtraction command is given to the ALU instead of an addition. The register transfers are the same, and the microcode is

Symbolic address	S_1	D_1	S_2	D_2	S_3	D_3	C_1	C_2	*TEST*	*ALU*	*NA*
DEX1	0	0	OCIR	IX	0	0	0	0	0	0	0
DEX2	IXRE	OXE	0	0	0	0	0	RXRO	0	SUB	0
DEX3	IXRE	OXE	OBE	IXR	0	0	END	0	TINT	SUB	FETCHØ

12.6.2.3 Multiply Operation

The multiply operation requires that the multiplier be fetched from memory and placed into magnitude form. The multiplicand must also be placed into magnitude form. Of course the signs must be used to determine the sign of the product. If the product is negative, the results must be complemented. The product is double length in the accumulator and MQ register. The following register transfers describe the process:

MUL1	D0				
	D1	$[EAR] \xrightarrow{\text{OEAR}} \text{bus} \xrightarrow{\text{IMAR}} MAR;$	RRC		
	D6	$[ACC]_0 \xrightarrow{\text{IAS}} Scand$			
	D7	$[MIR] \xrightarrow{\text{OMIR}} \text{bus} \xrightarrow{\text{IX}} X$			
MUL2	D0				
	D1		MUL;	$24 \longrightarrow CR$	
	D6	$[X]_0 \oplus [Scand] \xrightarrow{\text{SGN}} Scand$			
	D7		TAS;		MUL5
MUL3	D0				
	D1	$[ACC] \xrightarrow{\text{IALU}} \text{ALU}; 0 \longrightarrow \text{ALU};$	CMA		
	D6				
	D7				
MUL4	D0				
	D1	$[ACC] \xrightarrow{\text{IALU}} \text{ALU}; 0 \longrightarrow \text{ALU};$	CMA		
	D6	$[\text{ALU}] \xrightarrow{\text{OBE}} \text{bus} \xrightarrow{\text{IACC}} ACC$			
	D7				
MUL5	D0				
	D1				
	D6	$[ACC] \xrightarrow{\text{OACC}} \text{bus} \xrightarrow{\text{IMQ}} MQ$			
	D7		RACC;	TXS;	MUL8
MUL6	D0				
	D1	$[X] \xrightarrow{\text{OXE}} \text{ALU}; 0 \longrightarrow \text{ALU};$	CMX		
	D6				
	D7				

MUL7	D0			
	D1	$[X] \xrightarrow{\text{OXE}} \text{ALU};\ 0 \longrightarrow \text{ALU};$	CMX	
	D6	$[\text{ALU}] \xrightarrow{\text{OBE}} \text{bus} \xrightarrow{\text{IX}} X$		
	D7			
MUL8	D0			
	D1			
	D6	$[CR] - 1 \longrightarrow CR$	DCR	
	D7		TQ23;	MUL11
MUL9	D0			
	D1	$[ACC] \xrightarrow{\text{IALU}} \text{ALU};\ [X] \xrightarrow{\text{OXE}} \text{ALU};$	ADD	
	D6			
	D7			
MUL10	D0			
	D1	$[ACC] \xrightarrow{\text{IALU}} \text{ALU};\ [X] \xrightarrow{\text{OXE}} \text{ALU};$	ADD	
	D6	$[\text{ALU}] \xrightarrow{\text{OBE}} \text{bus} \xrightarrow{\text{IACC}} ACC$		
	D7			
MUL11	D0			
	D1		SHT	
	D6	$[ACC]_{0\text{-}22} \longrightarrow ACC_{1\text{-}23};\ [MQ]_{0\text{-}22} \longrightarrow MQ_{1\text{-}23};$	LRQ	
	D7		TCR;	MUL8
MUL12	D0			
	D1			
	D6			
	D7		TSD;	MUL17
MUL13	D0			
	D1	$[ACC] \xrightarrow{\text{IALU}} \text{ALU};\ 0 \longrightarrow \text{ALU};$	CMA	
	D6			
	D7			
MUL14	D0			
	D1	$[ACC] \xrightarrow{\text{IALU}} \text{ALU};\ 0 \longrightarrow \text{ALU};$	CMA	
	D6	$[\text{ALU}] \xrightarrow{\text{OBE}} \text{bus} \xrightarrow{\text{IACC}} ACC$		
	D7	$[MQ] \xrightarrow{\text{OMQ}} \text{bus} \xrightarrow{\text{IX}} X$		
MUL15	D0			
	D1	$[X] \xrightarrow{\text{OXE}} \text{ALU};\ 0 \longrightarrow \text{ALU};$	CMX	
	D6			
	D7			
MUL16	D0			
	D1	$[X] \xrightarrow{\text{OXE}} \text{ALU};\ 0 \longrightarrow \text{ALU};$	CMX	
	D6	$[\text{ALU}] \xrightarrow{\text{OBE}} \text{bus} \xrightarrow{\text{IMQ}} MQ$		
	D7			

MUL17 D0
D1
D6
D7 END, TINT; FETCHØ

The microcode for the above transfers is

Symbolic address	S_1	D_1	S_2	D_2	S_3	D_3	C_1	C_2	*TEST*	*ALU*	*NA*
MUL1	OEAR	IMAR	0	IAS	OMIR	IX	RRC	0	0	0	0
MUL2	0	0	0	SGN	0	0	MUL	0	TAS	0	MUL5
MUL3	IALU	0	0	0	0	0	0	0	0	CMA	0
MUL4	IALU	0	OBE	IACC	0	0	0	0	0	CMA	0
MUL5	0	0	OACC	IMQ	0	0	0	RACC	TXS	0	MUL8
MUL6	OXE	0	0	0	0	0	0	0	0	CMX	0
MUL7	OXE	0	OBE	IX	0	0	0	0	0	CMX	0
MUL8	0	0	0	0	0	0	0	DCR	TQ23	0	MUL11
MUL9	IALU	OXE	0	0	0	0	0	0	0	ADD	0
MUL10	IALU	OXE	OBE	IACC	0	0	0	0	0	ADD	0
MUL11	0	0	0	LRQ	0	0	SHT	0	TCR	0	MUL8
MUL12	0	0	0	0	0	0	0	0	TSD	0	MUL17
MUL13	IALU	0	0	0	0	0	0	0	0	CMA	0
MUL14	IALU	0	OBE	IACC	OMQ	IX	0	0	0	CMA	0
MUL15	0	OXE	0	0	0	0	0	0	0	CMX	0
MUL16	0	OXE	OBE	IMQ	0	0	0	0	0	CMX	0
MUL17	0	0	0	0	0	0	END	0	TINT	0	FETCHØ

12.6.2.4 Divide Operation

Division, like multiplication, requires a long process and microcode. The divisor must be fetched from memory and placed into magntiude form if negative. The dividend must also be placed into magnitude form if negative. The nonrestoring method is utilized with a test made for overflow prior to beginning the shift and either add or subtract iteration. If the results should be negative, the MQ register must be complemented. The register transfer equations which describe the process are

DIV1 D0
D1 $[EAR] \xrightarrow{\text{OEAR}} \text{bus} \xrightarrow{\text{IMAR}} MAR;$ RRC
D6 $[ACC]_0 \xrightarrow{\text{IAS}} Scand$
D7 $[MIR] \xrightarrow{\text{OMIR}} \text{bus} \xrightarrow{\text{IX}} X;$ TAS; DIV4
DIV2 D0
D1 $[ACC] \xrightarrow{\text{IALU}} \text{ALU}; \ 0 \longrightarrow \text{ALU};$ CMA
D6
D7

DIV3	D0			
	D1	$[ACC] \xrightarrow{\text{IALU}} \text{ALU};\ 0 \longrightarrow \text{ALU};$	CMA	
	D6	$[\text{ALU}] \xrightarrow{\text{OBE}} \text{bus} \xrightarrow{\text{IACC}} ACC$		
	D7			
DIV4	D0			
	D1		DIV; $23 \longrightarrow CR$	
	D6	$[X]_0 \oplus [Scand] \xrightarrow{\text{SGN}} Scand$		
	D7		RMQ; TXS;	DIV7
DIV5	D0			
	D1	$[X] \xrightarrow{\text{OXE}} \text{ALU};\ 0 \longrightarrow \text{ALU};$	CMX	
	D6			
	D7			
DIV6	D0			
	D1	$[X] \xrightarrow{\text{OXE}} \text{ALU},\ 0 \longrightarrow \text{ALU};$	CMX	
	D6	$[\text{ALU}] \xrightarrow{\text{OBE}} \text{bus} \xrightarrow{\text{IX}} X$		
	D7			
DIV7	D0			
	D1	$[ACC] \xrightarrow{\text{IALU}} \text{ALU}; [X] \xrightarrow{\text{OXE}} \text{ALU};$	SUB	
	D6		ROFF	
	D7			
DIV8	D0			
	D1	$[ACC] \xrightarrow{\text{IALU}} \text{ALU}; [X] \xrightarrow{\text{OXE}} \text{ALU};$	SUB	
	D6	$[\text{ALU}] \xrightarrow{\text{OBE}} \text{bus} \xrightarrow{\text{IACC}} ACC$		
	D7		TAS;	DIV23
DIV9	D0			
	D1	$[ACC] \xrightarrow{\text{IALU}} \text{ALU}; [X] \xrightarrow{\text{OXE}} \text{ALU};$	ADD	
	D6			
	D7			
DIV10	D0			
	D1	$[ACC] \xrightarrow{\text{IALU}} \text{ALU}; [X] \xrightarrow{\text{OXE}} \text{ALU};$	ADD	
	D6	$[\text{ALU}] \xrightarrow{\text{OBE}} \text{bus} \xrightarrow{\text{IACC}} ACC$		
	D7			
DIV11	D0			
	D1		SHT	
	D6		LLQ; AQ23	
	D7			
DIV12	D0			
	D1	$[ACC] \xrightarrow{\text{IALU}} \text{ALU}; [X] \xrightarrow{\text{OXE}} \text{ALU};$	SUB	
	D6	$[CR] - 1 \longrightarrow CR$	DCR	
	D7			

DIV13	D0			
	D1	$[ACC] \xrightarrow{\text{IALU}} \text{ALU}; [X] \xrightarrow{\text{OXE}} \text{ALU};$	SUB	
	D6	$[\text{ALU}] \xrightarrow{\text{OBE}} \text{bus} \xrightarrow{\text{IACC}} ACC$		
	D7			
DIV14	D0			
	D1		SHT	
	D6		LLQ; AQ23	
	D7		TQ23;	DIV18
DIV15	D0			
	D1	$[ACC] \xrightarrow{\text{IALU}} \text{ALU}; [X] \xrightarrow{\text{OXE}} \text{ALU};$	SUB	
	D6	$[CR] - 1 \longrightarrow CR$	DCR	
	D7			
DIV16	D0			
	D1	$[ACC] \xrightarrow{\text{IALU}} \text{ALU}; [X] \xrightarrow{\text{OXE}} \text{ALU};$	SUB	
	D6	$[\text{ALU}] \xrightarrow{\text{OBE}} \text{bus} \xrightarrow{\text{IACC}} ACC$		
	D7		TCR;	DIV14
DIV17	D0			
	D1			
	D6			
	D7		UNB;	DIV20
DIV18	D0			
	D1	$[ACC] \xrightarrow{\text{IALU}} \text{ALU}; [X] \xrightarrow{\text{OXE}} \text{ALU};$	ADD	
	D6	$[CR] - 1 \longrightarrow CR$	DCR	
	D7			
DIV19	D0			
	D1	$[ACC] \xrightarrow{\text{IALU}} \text{ALU}; [X] \xrightarrow{\text{OXE}} \text{ALU};$	ADD	
	D6	$[\text{ALU}] \xrightarrow{\text{OBE}} \text{bus} \xrightarrow{\text{IACC}} ACC$		
	D7		TCR;	DIV14
DIV20	D0			
	D1			
	D6			
	D7	$[MQ] \xrightarrow{\text{OMQ}} \text{bus} \xrightarrow{\text{IX}} X;$	TSD;	DIV24
DIV21	D0			
	D1	$[X] \xrightarrow{\text{OXE}} \text{ALU}; \; 0 \longrightarrow \text{ALU};$	CMX	
	D6			
	D7			
DIV22	D0			
	D1	$[X] \xrightarrow{\text{OXE}} \text{ALU}; \; 0 \longrightarrow \text{ALU};$	CMX	
	D6	$[\text{ALU}] \xrightarrow{\text{OBE}} \text{bus} \xrightarrow{\text{IMQ}} MQ$		
	D7		UNB;	DIV24

DIV23 D0
D1
D6 $1 \longrightarrow OFF$ SOFF
D7
DIV24 D0
D1
D6
D7 END; TINT; FETCHØ

The microcode which corresponds to these register transfers is

Symbolic address	S_1	D_1	S_2	D_2	S_3	D_3	C_1	C_2	*TEST*	*ALU*	*NA*
DIV1	OEAR	IMAR	0	IAS	OMIR	IX	RRC	0	TAS	0	DIV4
DIV2	IALU	0	0	0	0	0	0	0	0	CMA	0
DIV3	IALU	0	OBE	IACC	0	0	0	0	0	CMA	0
DIV4	0	0	0	SGN	0	0	DIV	RMQ	TXS	0	DIV7
DIV5	0	OXE	0	0	0	0	0	0	0	CMX	0
DIV6	0	OXE	OBE	IX	0	0	0	0	0	CMX	0
DIV7	IALU	OXE	0	0	0	0	0	ROFF	0	SUB	0
DIV8	IALU	OXE	OBE	IACC	0	0	0	0	TAS	SUB	DIV23
DIV9	IALU	OXE	0	0	0	0	0	0	0	ADD	0
DIV10	IALU	OXE	OBE	IACC	0	0	0	0	0	ADD	0
DIV11	0	0	0	LLQ	0	0	SHT	AQ23	0	0	0
DIV12	IALU	OXE	0	0	0	0	0	DCR	0	SUB	0
DIV13	IALU	OXE	OBE	IACC	0	0	0	0	0	SUB	0
DIV14	0	0	0	LLQ	0	0	SHT	AQ23	TQ23	0	DIV18
DIV15	IALU	OXE	0	0	0	0	0	DCR	0	SUB	0
DIV16	IALU	OXE	OBE	IACC	0	0	0	0	TCR	SUB	DIV14
DIV17	0	0	0	0	0	0	0	0	UNB	0	DIV20
DIV18	IALU	OXE	0	0	0	0	0	DCR	0	SUB	0
DIV19	IALU	OXE	OBE	IACC	0	0	0	0	TCR	SUB	DIV14
DIV20	0	0	0	0	OMQ	IX	0	0	TSD	0	DIV24
DIV21	0	OXE	0	0	0	0	0	0	0	CMX	0
DIV22	0	OXE	OBE	IMQ	0	0	0	0	UNB	CMX	DIV24
DIV23	0	0	0	0	0	0	0	SOFF	0	0	0
DIV24	0	0	0	0	0	0	END	0	TINT	0	FETCHØ

12.6.2.5 Load, Transfer, and Store Operations

Store the MQ

STQ1 D0
D1 $[EAR] \xrightarrow{\text{OEAR}} \text{bus} \xrightarrow{\text{IMAR}} MAR$; CWC
D6
D7 $[MQ] \xrightarrow{\text{OMQ}} \text{bus} \xrightarrow{\text{IMIR}} MIR$
STQ2 D0
D1
D6
D7 END; TINT; FETCHØ

The above register transfers show a typical store. At D1, the address and a command for a clear-write cycle are sent to main memory. At D7, the data to be stored are sent to main memory. However, the CPU must wait one main cycle until the memory has completed its write cycle. The microprogram is

Symbolic address	S_1	D_1	S_2	D_2	S_3	D_3	C_1	C_2	*TEST*	*ALU*	*NA*
STQ1	OEAR	IMAR	0	0	OMQ	IMIR	CWC	0	0	0	0
STQ2	0	0	0	0	0	0	END	0	TINT	0	FETCHØ

Store the Accumulator This operation is similar to the above with the accumulator substituted for the *MQ* register. The microprogram is

Symbolic address	S_1	D_1	S_2	D_2	S_3	D_3	C_1	C_2	*TEST*	*ALU*	*NA*
STA1	OEAR	IMAR	0	0	OACC	IMIR	CWC	0	0	0	0
STA2	0	0	0	0	0	0	END	0	TINT	0	FETCHØ

Store the Index Register This operation is also similar to the store the *MQ* with the index register substituted for the *MQ*. The major change is that at D7 a special signal SMIR is used to transfer data into the *MIR*. This control signal changes the data in bit positions 8–23 without affecting bits 0–7. The microcode is

Symbolic address	S_1	D_1	S_2	D_2	S_3	D_3	C_1	C_2	*TEST*	*ALU*	*NA*
STX1	OEAR	IMAR	0	0	OXR	SMIR	CWC	0	0	0	0
STX2	0	0	0	0	0	0	END	0	TINT	0	FETCHØ

Store the CAR This operation is very similar to the store the index register with the *CAR* substituted for the index register. The microcode is

Symbolic address	S_1	D_1	S_2	D_2	S_3	D_3	C_1	C_2	*TEST*	*ALU*	*NA*
STC1	OEAR	IMAR	0	0	OCAR	SMIR	CWC	0	0	0	0
STC2	0	0	0	0	0	0	END	0	TINT	0	FETCHØ

Load the Accumulator The load the accumulator operati n, as all load operations, is performed using the *X* register as an intermedia. . Thus, data from memory are first sent to the X register and then to the destination register. The transfer equations are

LDA1	D0		
	D1	$[EAR] \xrightarrow{\text{OEAR}} \text{bus} \xrightarrow{\text{IMAR}} MAR;$	RRC
	D6		
	D7	$[MIR] \xrightarrow{\text{OMIR}} \text{bus} \xrightarrow{\text{IX}} X$	
LDA2	D0		
	D1		
	D6	$[X] \xrightarrow{\text{OX}} \text{bus} \xrightarrow{\text{IACC}} ACC$	
	D7		END; TINT; FETCH∅

The microcode is

Symbolic address	S_1	D_1	S_2	D_2	S_3	D_3	C_1	C_2	TEST	ALU	NA
LDA1	OEAR	IMAR	0	0	OMIR	IX	RRC	0	0	0	0
LDA2	0	0	OX	IACC	0	0	END	0	TINT	0	FETCH∅

Load the Index Register This operation is similar to the load the accumulator with the index register substituted for the accumulator. The microcode is

Symbolic address	S_1	D_1	S_2	D_2	S_3	D_3	C_1	C_2	TEST	ALU	NA
LDX1	OEAR	IMAR	0	0	OMIR	IX	RRC	0	0	0	0
LDX2	0	0	OX	IXR	0	0	END	0	TINT	0	FETCH∅

Load the Accumulator Immediate Since no memory reference must be made for this immediate operation, only one major CPU cycle is needed for the register transfers. The register transfers are

LAI	D0		
	D1		
	D6	$[N(CIR)] \xrightarrow{\text{OCIR}} \text{bus} \xrightarrow{\text{IACC}} ACC$	
	D7		END; TINT; FETCH∅

The microcode is

Symbolic address	S_1	D_1	S_2	D_2	S_3	D_3	C_1	C_2	TEST	ALU	NA
LAI1	0	0	OCIR	IACC	0	0	END	0	TINT	0	FETCH∅

Load the Index Register Immediate This operation is similar to the above with the index register substituted for the accumulator. The microcode is

Symbolic address	S_1	D_1	S_2	D_2	S_3	D_3	C_1	C_2	*TEST*	*ALU*	*NA*
LXI1	0	0	OCIR	IXR	0	0	END	0	TINT	0	FETCHØ

Transfer XR into ACC The transfer equations also do not reference memory, and hence the transfers can be completed within one major CPU cycle. The register transfers are

TXA1 D0
D1
D6 $[XR] \xrightarrow{\text{OXR}} \text{bus} \xrightarrow{\text{IACC}} ACC$
D7 END; TINT; FETCHØ

The microcode is

Symbolic address	S_1	D_1	S_2	D_2	S_3	D_3	C_1	C_2	*TEST*	*ALU*	*NA*
TXA1	0	0	OXR	IACC	0	0	END	0	TINT	0	FETCHØ

Transfer MQ into ACC This instruction is very similar to the above with the MQ substituted for the XR. The microcode is

Symbolic address	S_1	D_1	S_2	D_2	S_3	D_3	C_1	C_2	*TEST*	*ALU*	*NA*
TMQ1	0	0	OMQ	IACC	0	0	END	0	TINT	0	FETCHØ

12.6.2.6 Branch Operations

Unconditional Branch This operation requires no test in order to perform the instruction. Therefore, it can be accomplished in one major CPU cycle. The register transfer equations are

UNB1 D0
D1
D6
D7 $[EAR] \xrightarrow{\text{OEAR}} \text{bus} \xrightarrow{\text{ICAR}} CAR$; END; TINT; FETCHØ

The microcode is

Symbolic address	S_1	D_1	S_2	D_2	S_3	D_3	C_1	C_2	TEST	ALU	NA
UNB1	0	0	0	0	OEAR	ICAR	END	0	TINT	0	FETCHØ

Branch on Accumulator Overflow This operation, as well as all other conditional branch operations, requires two major CPU cycles: one to test for the condition, and the other to transfer the registers if a branch is to be performed. The register transfer equations are

BAO1 D0
D1
D6
D7 TAO; INTØ
BAO2 D0
D1
D6
D7 $[EAR] \xrightarrow{\text{OEAR}} \text{bus} \xrightarrow{\text{ICAR}} CAR$; END; TINT; FETCHØ

The microcode is

Symbolic address	S_1	D_1	S_2	D_2	S_3	D_3	C_1	C_2	TEST	ALU	NA
BAO1	0	0	0	0	0	0	0	0	TAO	0	INTØ
BAO2	0	0	0	0	OEAR	ICAR	END	0	TINT	0	FETCHØ

Branch on Index Register Positive This operation is similar to the above but with a different test condition. The microcode for the operation is

Symbolic address	S_1	D_1	S_2	D_2	S_3	D_3	C_1	C_2	TEST	ALU	NA
BXP1	0	0	0	0	0	0	0	0	TXP	0	INTØ
BXP2	0	0	0	0	OEAR	ICAR	END	0	TINT	0	FETCHØ

Branch on Index Register Zero This operation is also similar to those above. The microcode is

Symbolic address	S_1	D_1	S_2	D_2	S_3	D_3	C_1	C_2	*TEST*	*ALU*	*NA*
BXZ1	0	0	0	0	0	0	0	0	TXZ	0	INTØ
BXZ2	0	0	0	0	OEAR	ICAR	END	0	TINT	0	FETCHØ

Branch on Index Register Negative This instruction tests the state of the *XRO* flip-flop and performs a branch if it has been set. The *XRO* flip-flop can be set from two sources. One source is due to a carry from bit position 8 of the ALU. This will be set by executing an add to index, increment index, or decrement index instruction. The other source is due to a carry out of bit position 8 of the index register itself whenever it is decremented during a tally instruction. This condition will arise if the index register has overflowed positive or has gone through 0 and become negative. The microcode for this instruction is similar to all other conditional branch instructions:

Symbolic address	S_1	D_1	S_2	D_2	S_3	D_3	C_1	C_2	*TEST*	*ALU*	*NA*
BXN1	0	0	0	0	0	0	0	0	TXRO	0	INTØ
BXN2	0	0	0	0	OEAR	ICAR	END	0	TINT	0	FETCHØ

Tally Down Index Register This operation requires that the index register be tested to determine if its contents are zero. If they are, a branch is made to INTØ. If the contents have not reached 0, the contents are decremented by 1 and a branch is made to the effective address. The decrementing of the index register is made by the command signal DXR. The register transfers are

TLD1 D0
D1
D6
D7 TXZ; INTØ
TLD2 D0
D1
D6 $[XR] - 1 \longrightarrow XR$; DXR
D7 $[EAR] \xrightarrow{\text{OEAR}} \text{bus} \xrightarrow{\text{ICAR}} CAR$; END; TINT; FETCHØ

The microcode is

Symbolic address	S_1	D_1	S_2	D_2	S_3	D_3	C_1	C_2	*TEST*	*ALU*	*NA*
TLD1	0	0	0	0	0	0	0	0	TXZ	0	INTØ
TLD2	0	0	0	0	OEAR	ICAR	END	DXR	TINT	0	FETCHØ

Branch on Accumulator Positive This operation is similar to any branch on condition. In this case a test is made on the accumulator. The register transfer equations are

BAP1	D0	
	D1	
	D6	
	D7	TAP; INT∅
BAP2	D0	
	D1	
	D6	
	D7	$[EAR] \xrightarrow{\text{OEAR}} \text{bus} \xrightarrow{\text{ICAR}} CAR$; END; TINT; FETCH∅

The microcode is

Symbolic address	S_1	D_1	S_2	D_2	S_3	D_3	C_1	C_2	*TEST*	*ALU*	*NA*
BAP1	0	0	0	0	0	0	0	0	TAP	0	INT∅
BAP2	0	0	0	0	OEAR	ICAR	END	0	TINT	0	FETCH∅

Branch on Accumulator Zero This operation is similar to branch on accumulator positive except that the test is made for a 0 contents in the accumulator. This is done by connecting each bit of the accumulator to an AND gate. If all the bits are 0, the output of the AND gate is a 1, indicating a 0 content. The microcode is

Symbolic address	S_1	D_1	S_2	D_2	S_3	D_3	C_1	C_2	*TEST*	*ALU*	*NA*
BAZ1	0	0	0	0	0	0	0	0	TAZ	0	INT∅
BAZ2	0	0	0	0	OEAR	ICAR	END	0	TINT	0	FETCH∅

Branch on Accumulator Negative This operation is also similar to the one above except that the test is made on the sign bit of the accumulator. The microcode is

Symbolic address	S_1	D_1	S_2	D_2	S_3	D_3	C_1	C_2	*TEST*	*ALU*	*NA*
BAN1	0	0	0	0	0	0	0	0	TAS	0	INT∅
BAN2	0	0	0	0	OEAR	ICAR	END	0	TINT	0	FETCH∅

Subroutine Jump This operation can be performed in two minor cycles. First, the contents of the *CAR* are stored at the effective address. While the memory is writing the information at that location, the effective address is placed into the *CAR*. The register transfer equations are

$$\begin{array}{lll} \text{SRJ1} & \text{D0} & \\ & \text{D1} & [EAR] \xrightarrow{\text{OEAR}} \text{bus} \xrightarrow{\text{IMAR}} MAR;\quad \text{CWC} \\ & \text{D6} & \\ & \text{D7} & [CAR] \xrightarrow{\text{OCAR}} \text{bus} \xrightarrow{\text{SMIR}} MIR \\ \text{SRJ2} & \text{D0} & \\ & \text{D1} & \\ & \text{D6} & \\ & \text{D7} & [EAR] \xrightarrow{\text{OEAR}} \text{bus} \xrightarrow{\text{ICAR}} CAR;\quad \text{END};\quad \text{TINT};\quad \text{FETCH}\varnothing \end{array}$$

The microcode is

Symbolic address	S_1	D_1	S_2	D_2	S_3	D_3	C_1	C_2	*TEST*	*ALU*	*NA*
SRJ1	OEAR	IMAR	0	0	OCAR	SMIR	CWC	0	0	0	0
SRJ2	0	0	0	0	OEAR	ICAR	END	0	TINT	0	FETCH∅

No Operation This instruction does nothing except test for an interrupt. It performs no register transfers. The microcode is

Symbolic address	S_1	D_1	S_2	D_2	S_3	D_3	C_1	C_2	*TEST*	*ALU*	*NA*
NOP1	0	0	0	0	0	0	END	0	TINT	0	FETCH∅

STOP The stop operation resets the *RUN* flip-flop during D6. This will prevent the *CMAR* from changing at time D7 and prevent the execution of microinstructions and thereby the program; see Section 12.4.4. The microcode is

Symbolic address	S_1	D_1	S_2	D_2	S_3	D_3	C_1	C_2	*TEST*	*ALU*	*NA*
STP1	0	0	0	0	0	0	0	RRUN	0	0	INT∅

12.6.2.7 Logic Operations

The logic operations of AND, OR, and exclusive-OR require an operand to be fetched and the use of the ALU to perform the logic on a bit-by-bit basis. Since no carry propagation is performed, the logic operations of the ALU require only one major CPU cycle to complete.

Logical-OR The register transfer equations are straightforward with the ALU programmed to perform the OR operation f_7 (see Tables 12.6 and 12.12). The transfer equations are

LOR1	D0		
	D1	$[EAR] \xrightarrow{\text{OEAR}} \text{bus} \xrightarrow{\text{IMAR}} MAR;$	RRC
	D6		
	D7	$[MIR] \xrightarrow{\text{OMIR}} \text{bus} \xrightarrow{\text{IX}} X$	
LOR2	D0		
	D1	$[ACC] \xrightarrow{\text{IALU}} \text{ALU}; [X] \xrightarrow{\text{OXE}} ALU;$	OR
	D6	$[ALU] \xrightarrow{\text{OBE}} \text{bus} \xrightarrow{\text{IACC}} ACC$	
	D7		END; TINT; FETCH∅

The microcode corresponding to these transfers is

Symbolic address	S_1	D_1	S_2	D_2	S_3	D_3	C_1	C_2	*TEST*	*ALU*	*NA*
LOR1	OEAR	IMAR	0	0	OMIR	IX	RRC	0	0	0	0
LOR2	IALU	OXE	OBE	IACC	0	0	END	0	TINT	OR	FETCH∅

Logical-AND This is similar to the logical-OR except for the control signals supplied to the ALU which command it to perform the logical-AND. The microcode is

Symbolic address	S_1	D_1	S_2	D_2	S_3	D_3	C_1	C_2	*TEST*	*ALU*	*NA*
LPR1	OEAR	IMAR	0	0	OMIR	IX	RRC	0	0	0	0
LPR2	IALU	OXE	OBE	IACC	0	0	END	0	TINT	AND	FETCH∅

Logical Exclusive-OR The microcode is

Symbolic address	S_1	D_1	S_2	D_2	S_3	D_3	C_1	C_2	*TEST*	*ALU*	*NA*
EOR1	OEAR	IMAR	0	0	OMIR	IX	RRC	0	0	0	0
EOR2	IALU	OXE	OBE	IACC	0	0	END	0	TINT	EOR	FETCH∅

Logical Negation This operation only uses the contents of the accumulator and does not require an access to main memory. It can be accomplished in one CPU major cycle. The microcode is

Symbolic address	S_1	D_1	S_2	D_2	S_3	D_3	C_1	C_2	*TEST*	*ALU*	*NA*
LNG1	IALU	0	OBE	IACC	0	0	END	0	TINT	NOTA	FETCH∅

12.6.2.8 Shift Operations

The shifting operations use the augmented instruction format to specify a shift of a particular register. At time D1 of FETCH1 of each instruction phase, bits 18–23 of the *CIR* are transferred into the counter register (*CR*). Although this is done for every instruction, only the shift instructions utilize the counter register with the number of bits to be shifted now in the counter register.

The shift operations all perform the following steps. First, a test is made to determine if the contents of the counter register have reached 0. If they have, the microprogram branches into INT∅ to test for any pending interrupt. If they have not reached 0, the next microinstruction is fetched. Next, the specified shift of 1 bit is performed, and then the counter register is decremented by 1. An unconditional branch is then made back to the first microinstruction to test the contents of the counter register.

The format of the microinstruction is interpreted in a different manner for shift operations. First, the C_1 field of each microinstruction contains the SHT command, indicating a shift operation. Second, the particular shift which is to be performed is specified by decoding the contents of the D_2 field. Thus, for a shift operation the D_2 field does not specify a destination register control signal but specifies a type of shift. Figure 12.36 shows how this field plays a double role in the microprogram control.

Each shift operation requires two microinstructions. These microinstructions are repeated for each bit position shifted. The total number of times the microinstructions are executed is equal to the binary number specified by bits 18–23 of the shift instruction, which were transferred into the counter register at the time the instruction phase was executed.

Arithmetic Left Shift the Accumulator The transfer equations for this operation are

ALA1	D0				
	D1		SHT		
	D6				
	D7			TCR;	INT∅
ALA2	D0				
	D1		SHT		
	D6	$[ACC]_{2\text{-}23} \longrightarrow ACC_{1\text{-}23}$; $0 \longrightarrow ACC_{23}$;	DCR		
	D7		UNB;		ALA1

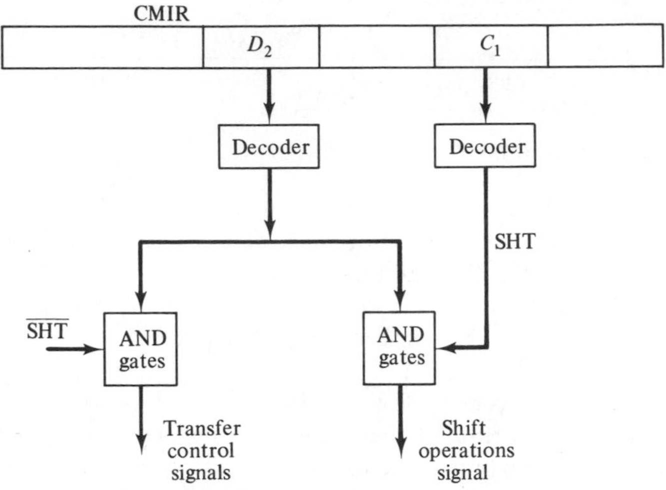

Figure 12.36 Shift control signal logic.

The microcode for these two major cycles is

Symbolic address	S_1	D_1	S_2	D_2	S_3	D_3	C_1	C_2	*TEST*	*ALU*	*NA*
ALA1	0	0	0	0	0	0	SHT	0	TCR	0	INTØ
ALA2	0	0	0	ALA	0	0	SHT	DCR	UNB	0	ALA1

Arithmetic Right Shift the Accumulator The register transfers for this operation are

ARA1	D0			
	D1		SHT	
	D6			
	D7		TCR;	INTØ
ARA2	D0			
	D1		SHT	
	D6	$[ACC]_{1\text{-}22} \longrightarrow ACC_{2\text{-}23}$; $[ACC]_0 \longrightarrow ACC_1$;	DCR	
	D7		UNB;	ARA1

The microcode is

Symbolic address	S_1	D_1	S_2	D_2	S_3	D_3	C_1	C_2	*TEST*	*ALU*	*NA*
ARA1	0	0	0	0	0	0	SHT	0	TCR	0	INTØ
ARA2	0	0	0	ARA	0	0	SHT	DCR	UNB	0	ARA1

Logical Right Shift the Accumulator and MQ Register This operation is performed on two registers concatenated so as to appear as one double-length register. The transfer equations are

LRQ1	D0				
	D1		SHT		
	D6				
	D7			TCR;	INT∅
LRQ2	D0				
	D1		SHT		
	D6	$[ACC]_{0\text{-}22} \longrightarrow ACC_{1\text{-}23}$; $[ACC]_{23} \longrightarrow MQ_0$; $[MQ]_{0\text{-}22} \longrightarrow MQ_{1\text{-}23}$; $0 \longrightarrow ACC_0$;	DCR		
	D7		UNB;		LRQ1

The microcode is

Symbolic address	S_1	D_1	S_2	D_2	S_3	D_3	C_1	C_2	*TEST*	*ALU*	*NA*
LRQ1	0	0	0	0	0	0	SHT	0	TCR	0	INT∅
LRQ2	0	0	0	LRQ	0	0	SHT	DCR	UNB	0	LRQ1

Logical Left Shift the Accumulator and MQ Register This operation is similar to the one above except that it performs a logical left shift on the double-length register. The microcode is

Symbolic address	S_1	D_1	S_2	D_2	S_3	D_3	C_1	C_2	*TEST*	*ALU*	*NA*
LLQ1	0	0	0	0	0	0	SHT	0	TCR	0	INT∅
LLQ2	0	0	0	LLQ	0	0	SHT	DCR	UNB	0	LLQ1

Logical Left Shift the Accumulator This operation is similar to the arithmetic right shift except that the sign bit will be shifted. The transfer equations are

LLA1	D0				
	D1		SHT		
	D6				
	D7			TCR;	INT∅
LLA2	D0				
	D1		SHT		
	D6	$R(ACC) \longrightarrow L(ACC)$; $0 \longrightarrow ACC_{23}$;	DCR		
	D7		UNB;		LLA1

The microcode is

Symbolic address	S_1	D_1	S_2	D_2	S_3	D_3	C_1	C_2	*TEST*	*ALU*	*NA*
LLA1	0	0	0	0	0	0	SHT	0	TCR	0	INTØ
LLA2	0	0	0	LLA	0	0	SHT	DCR	UNB	0	LLA1

Logical Right Shift the Accumulator This operation is similar to the one above except for the direction of the shift. The microcode is

Symbolic address	S_1	D_1	S_2	D_2	S_3	D_3	C_1	C_2	*TEST*	*ALU*	*NA*
LRA1	0	0	0	0	0	0	SHT	0	TCR	0	INTØ
LRA2	0	0	0	LRA	0	0	SHT	DCR	UNB	0	LRA1

Logical Left Circular Shift the Accumulator This is also similar to the logical left shift execept that $ACC_0 \longrightarrow ACC_{23}$ to make a circular shift. The microcode is

Symbolic address	S_1	D_1	S_2	D_2	S_3	D_3	C_1	C_2	*TEST*	*ALU*	*NA*
LCA1	0	0	0	0	0	0	SHT	0	TCR	0	INTØ
LCA2	0	0	0	LCA	0	0	SHT	DCR	UNB	0	LCA1

12.6.2.9 I/O Operations

In this section the operations used to perform I/O in association with the DMA channel will be developed. In addition the transfers associated with the DMA channel when it is working asynchronously from the CPU will be described.

Connect Device This instruction is really a register transfer operation and can be performed in a single major CPU cycle. The register transfer equations are

CND1 D0

D1

D6 $[CIR]_{12\text{-}23} \xrightarrow{\text{OCIR}} \text{bus} \xrightarrow{\text{IIOCR}} IOCR_{12\text{-}23}$

D7 END; TINT; FETCHØ

The microcode is

Symbolic address	S_1	D_1	S_2	D_2	S_3	D_3	C_1	C_2	*TEST*	*ALU*	*NA*
CND1	0	0	OCIR	IIOCR	0	0	END	0	TINT	0	FETCH∅

Load Status This operation is also a register transfer which involves one of the registers in the DMA channel. The transfer equations are

LDS1 D0
D1
D6 $[IOCR]_{0-11} \xrightarrow{\text{OIOCR}} \text{bus}_{12-23} \xrightarrow{\text{IACC}} ACC_{12-23}$
D7 END; TINT; FETCH∅

The microcode is

Symbolic address	S_1	D_1	S_2	D_2	S_3	D_3	C_1	C_2	*TEST*	*ALU*	*NA*
LDS1	0	0	OIOCR	IACC	0	0	END	0	TINT	0	FETCH∅

Branch on Device Not Available This is a special branch operation which tests the status of bit 11 in the *IOCR* to determine the availability of the channel and the addressed device. It is similar to the other branch operations. The register transfer equations are

BDN1 D0
D1
D6
D7 TDA; INT∅
BDN2 D0
D1
D6
D7 $[EAR] \xrightarrow{\text{OEAR}} \text{bus} \xrightarrow{\text{ICAR}} \text{CAR}$; END; TINT; FETCH∅

The microcode for this test and transfer is

Symbolic address	S_1	D_1	S_2	D_2	S_3	D_3	C_1	C_2	*TEST*	*ALU*	*NA*
BDN1	0	0	0	0	0	0	0	0	TDA	0	INT∅
BDN2	0	0	0	0	OEAR	ICAR	END	0	TINT	0	FETCH∅

Enable the Interrupt This operation involves no register transfers but is a command to the interrupt system. The microcode for this operation is

Symbolic address	S_1	D_1	S_2	D_2	S_3	D_3	C_1	C_2	*TEST*	*ALU*	*NA*
ENI1	0	0	0	0	0	0	END	ENI	TINT	0	FETCH∅

Start I/O This operation has two parts. First it transfers data (an address) into the *IOAR*. Then it signals the DMA channel that it can commence the I/O operations. It does this with the command signal RCS at time D7. The register transfer equations are

SIO1 D0
D1
D6 $[EAR] \xrightarrow{\text{OEAR}} \text{bus} \xrightarrow{\text{IIOAR}} IOAR$
D7 END; RCS; TINT; FETCH∅

The microcode is

Symbolic address	S_1	D_1	S_2	D_2	S_3	D_3	C_1	C_2	*TEST*	*ALU*	*NA*
SIO1	0	0	OEAR	IIOAR	0	0	END	RCS	TINT	0	FETCH∅

Load I/O This register transfer loads the *IOWC* with the word count. It is a single transfer operation whose microcode is

Symbolic address	S_1	D_1	S_2	D_2	S_3	D_3	C_1	C_2	*TEST*	*ALU*	*NA*
LIO1	0	0	OEAR	IIOWC	0	0	END	0	TINT	0	FETCH∅

12.6.3 Timing of the Execution Phase

The amount of time in the execution phase is dependent on the operation to be performed. The amount of time is summarized in Table 12.17. The time unit is the major CPU cycle or 512 nanoseconds.

The load and store instructions, op-codes 5–10, require two major CPU cycles in their execution phase. This corresponds to the 1024 nanoseconds needed to cycle the main memory for these two types of operations.

Table 12.17 Execution Phase Cycles

Decimal op-code	*Mnemonic*	*Name*	*Number of cycles*
0	—	Augmented op-code	—
1	ADD	Addition	3
2	SUB	Subtraction	3
3	MUL	Multiplication	Var.
4	DIV	Division	Var.
5	STQ	Store *MQ* register	2
6	STA	Store *ACC* register	2
7	STX	Store index register	2
8	LDA	Load *ACC* register	2
9	LDX	Load index register	2
10	STC	Store *CAR*	2
11	TXA	Transfer *XR* to *ACC*	1
12	TMQ	Transfer *MQ* to *ACC*	1
13	ADX	Add to index register	3
14	SIO	Start I/O	1
15	LIO	Load I/O	1
16	UNB	Unconditional branch	1
17	BAO	Branch on *ACC* overflow	2
18	BXP	Branch on index positive	2
19	BXZ	Branch on index zero	2
20	BXN	Branch on index negative	2
21	TLD	Tally down	2
22	BAP	Branch on *ACC* positive	2
23	BAZ	Branch on *ACC* zero	2
24	BAN	Branch on *ACC* negative	2
25	LOR	Logical-OR	2
26	LPR	Logical-AND	2
27	LNG	Logical negation	1
28	EOR	Logical exclusive-OR	2
29	SRJ	Subroutine jump	2
30	BDN	Branch on device not available	2
31	NOP	No operation	1
0–0	STP	Stop	1
0–1	CMA	2's complement *ACC*	2
0–2	ALA	Arithmetic left shift *ACC*	2/bit
0–3	ARA	Arithmetic right shift *ACC*	2/bit
0–4	LRQ	Logical right shift $ACC \| MQ$	2/bit
0–5	LLQ	Logical left shift $ACC \| MQ$	2/bit

Table 12.17 Cont.

Decimal op-code	*Mnemonic*	*Name*	*Number of cycles*
0–6	LRA	Logical right shift *ACC*	2/bit
0–7	LLA	Logical left shift *ACC*	2/bit
0–8	LCA	Logical circular left shift *ACC*	2/bit
0–9	LAI	Load *ACC* immediate	1
0–10	LXI	Load index register immediate	1
0–11	INX	Increment index	3
0–12	DEX	Decrement index	3
0–13	CND	Connect I/O device	1
0–14	ENI	Enable interrupt	1
0–15	LDS	Load status	1

The transfer instructions, op-codes 11 and 12, and the transfer instructions associated with I/O operations, op-codes 14, 15, 0–13, and 0–15, require only one major CPU cycle since they do not require a memory cycle.

The short arithmetic operations, op-codes 1, 2, 13, 0–11, and 0–12, require three major CPU cycles. The first is the read cycle of the memory to fetch an operand. While the restore cycle in memory is occurring during the second cycle, the arithmetic operation is started. Since the adder requires two cycles to ripple its carry, a third CPU cycle is required to complete the operation.

The logic operations, op-codes 25, 26, and 28, require only two major CPU cycles. The first is the read cycle of the memory to obtain the operand. The second overlaps the memory restore with the logic computation in the ALU.

The branch operations, op-codes 17–24 and 30, require one or two major CPU cycles. The first major CPU cycle tests the particular condition. If it is not met, then a return is made to the instruction phase, and the operation completes in one cycle. If it is met, the second cycle completes the branch by changing the contents of the *CAR*. The unconditional branch, op-code 16, always requires only one major cycle since no test needs to be made. Likewise, the special branch instruction subroutine jump, op-code 29, always requires two major cycles to complete its register transfers and cycle the main memory to store the return address.

The immediate instructions, op-codes 0–9 and 0–10, require only one major CPU cycle to complete the required register transfers.

The shift instructions, op-codes 0–2 through 0–8, require two major CPU cycles per shift. The first tests to see if the shifting is completed; the second performs the shift and branches back to the first.

The logical negation, op-code 27, requires only one cycle since it requires no memory cycle and can complete the logic operation in one cycle time. The commands stop and enable the interrupt also require only one cycle to complete, op-codes 31 and 0–0. However, the instruction 2's complement the accumulator, op-code 0–1, requires two major CPU cycles to complete in order to permit the carries to ripple through the ALU circuit.

The number of cycles required to complete the long instructions multiplication and division, op-codes 3 and 4, is a function of the bits in the operands as well as the signs of the operands. The minimum time for a multiplication will occur if the multiplicand is positive and the multiplier is a 0. For this condition the operation will require 53 major CPU cycles. The maximum time occurs when the multiplicand is negative and the multiplier is positive and all 1s. For this condition the operations will require 107 cycles. The minimum time for a division is 58 major CPU cycles, while the maximum time is 63 cycles.

12.7 I/O AND DMA CHANNEL

A block diagram of the DMA channel is shown in Fig. 12.8. The single channel of this computer is connected directly to the single main bus. Because of this architecture, only one transfer can be made over the bus in any minor cycle. To guarantee that the DMA channel has unique possession of the bus during a transfer to or from main memory while it is operating in the cycle steal mode, the CPU cannot be performing any transfers. Thus, the DMA channel will not only steal cycles from the memory but will steal cycles from the CPU. It will perform this by waiting until the end of the execution phase of each instruction. At that time, the CPU makes a decision with the following priority of entry:

1. Interrupt handling.
2. DMA channel cycle steal.
3. Enter instruction phase.

It will enter a new instruction phase if the condition.

$$\mathrm{D7 \cdot END \cdot TINT} \cdot \overline{\mathrm{INT}} \cdot \overline{CSFF}$$

is true. This will cause the β line to be positive, and a branch to FETCH∅ in the microprogram will occur. It will enter the DMA cycle steal microprogram if the following condition is true:

$$\mathrm{D7 \cdot END} \cdot CSFF \cdot \overline{\mathrm{INT}}$$

This will cause one of the δ lines to be positive and a branch to be made to the appropriate microprogram to handle an I/O read or write. If the condition

$$\text{D7}\cdot\text{END}\cdot\text{TINT}\cdot\text{INT}$$

is true, the γ line will be positive, and the microprogram will branch to INT1 to handle the interrupt. The conditions are mutually exclusive so that only one can be true at any one time.

To handle a DMA cycle steal a second flip-flop, in addition to the *CSFF* (cycle steal flip-flop), is needed. This is the *RWFF* (read-write flip-flop). If it is a 1, it indicates that an I/O write is to be performed. An I/O write requires a memory read-restore cycle. If the *RWFF* is in the 0 state, then an I/O read is to be performed, which means that a memory clear-write cycle is required. The two control signals δ_1 and δ_2 are defined as

$$\delta_1 = \text{D7}\cdot\text{END}\cdot CSFF\cdot\overline{\text{INT}}\cdot\text{READ}$$
$$\delta_2 = \text{D7}\cdot\text{END}\cdot CSFF\cdot\overline{\text{INT}}\cdot\overline{\text{READ}}$$

where $\overline{\text{READ}}$ is equal to an I/O write. Thus, δ_1 indicates an I/O READ, while δ_2 indicates an I/O WRITE.

The register transfers which occur for an I/O read are

IOR1	D0	
	D1	$[IOAR] \xrightarrow{\text{OIOAR}} \text{bus} \xrightarrow{\text{IMAR}} MAR$; CWC
	D6	$1 \longrightarrow IA$
	D7	$[IOIR] \xrightarrow{\text{OIOIR}} \text{bus} \xrightarrow{\text{IMIR}} MIR$
IOR2	D0	
	D1	
	D6	
	D7	$0 \xrightarrow{\text{RCS}} CSFF$; END; TINT; FETCH∅

The register transfers which must be performed for an I/O write which occurs when the δ_2 line is positive are

IOW1	D0	
	D1	$[IOAR] \xrightarrow{\text{OIOAR}} \text{bus} \xrightarrow{\text{IMAR}} MAR$; RRC
	D6	
	D7	$[MIR] \xrightarrow{\text{OMIR}} \text{bus} \xrightarrow{\text{IIOIR}} IOIR$
IOW2	D0	
	D1	
	D6	
	D7	$0 \xrightarrow{\text{RCS}} \text{CSFF}$; END; TINT; FETCH∅

The microcode for these two types of DMA channel operation is

Symbolic address	S_1	D_1	S_2	D_2	S_3	D_3	C_1	C_2	*TEST*	*ALU*	*NA*
IOR1	OIOAR	IMIR	0	0	OIOIR	IMIR	CWC	0	0	0	0
IOR2	0	0	0	0	0	0	END	RCS	TINT	0	FETCHØ
IOW1	OIOAR	IMIR	0	0	OMIR	IIOIR	RRC	0	0	0	0
IOW2	0	0	0	0	0	0	END	RCS	TINT	0	FETCHØ

The automatic branch in the microprogram is made to IOR1 if condition δ_1 is true and to IOW1 if condition δ_2 is true.

A more detailed block diagram of a DMA channel is shown in Fig. 10.13, and the details for a simple channel interrupt structure are shown in Fig. 10.11. The DMA channel logic of our design requests a cycle steal for data transfer by use of the δ_1 and δ_2 lines. It requests an interrupt by the use of the γ line. The DMA channel uses the command signal RCS (reset the cycle steal flip-flop) for two purposes. One is to reset this flip-flop after a memory transfer has been completed. The other purpose is to signal the DMA channel to commence asynchronous operation with the start I/O (SIO) instruction.

Figure 12.37 shows the interrupt interface for this machine. The two command signals RINT and ENI are shown applied to the proper flip-flops. The control signal OCL places the address of the service routine from the *CLASS* register into the *CAR*.

An alternative architecture using a separate memory bus to which both the CPU and DMA channels have access could have been designed for this computer. With such an architecture it would be possible for the DMA channel to cycle steal the memory while the CPU did overlapping computation. To provide a memory with two access ports requires that a priority of access be established. To illustrate such an architecture, we shall assume that the DMA channel has a higher priority access to memory than the CPU.

When the DMA channel wishes to request a memory access, it presents a signal to the main memory called the direct memory access request (DMAR). When the CPU wishes to request a main memory cycle it presents the signal CPUR (CPU request) to main memory, where

$$\mathrm{CPUR} = \mathrm{D1} \cdot (\mathrm{RRC} + \mathrm{CWC} + \mathrm{IWC})$$

The memory will accept a request only if it is available to start a cycle, i.e., MAV = 1. Figure 12.38 shows the priority request line. When the logic accepts a request from the DMA channel, it sets a flip-flop called the direct memory access priority flip-flop (*DMAPFF*). If the logic accepts the request from the CPU, it sets a flip-flop called CPU priority flip-flop (*CPUPFF*).

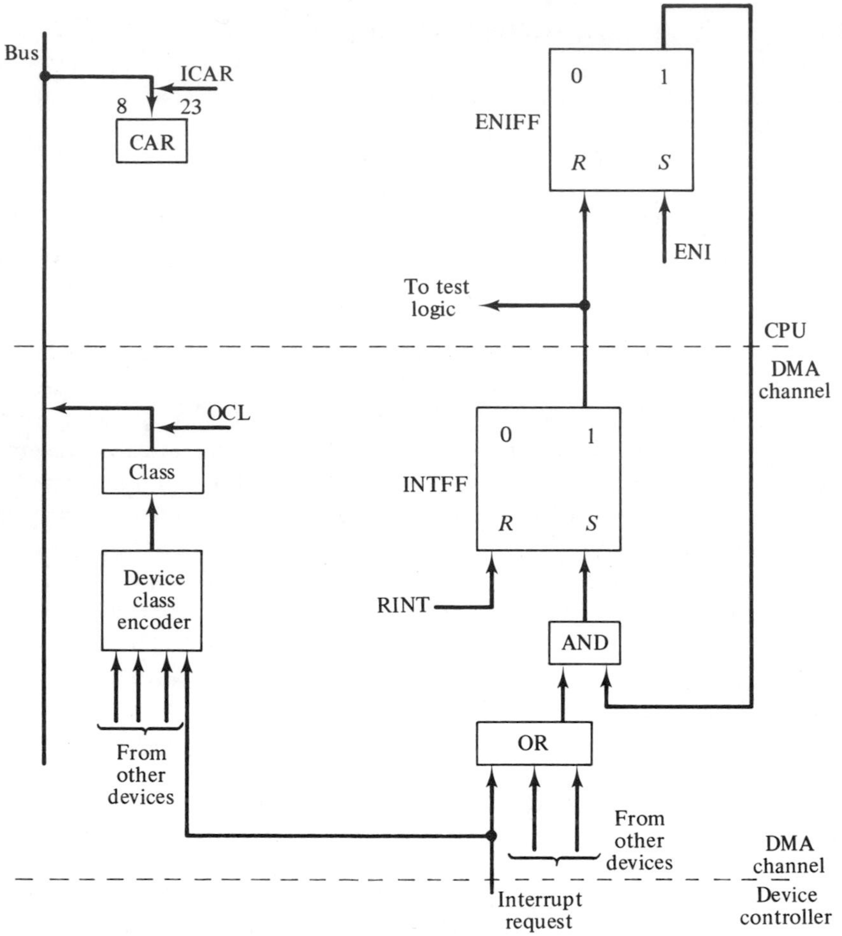

Bus
ICAR
8
23
CAR
ENIFF
0
1
R
S
ENI
To test logic
CPU
DMA channel
OCL
Class
INTFF
0
1
R
S
Device class encoder
RINT
AND
OR
From other devices
From other devices
Interrupt request
DMA channel
Device controller

Figure 12.37 Interrupt interface.

DMAR
MAV
AND
DMAPFF$_S$
OR
To memory logic
CPUR
MAV
AND
CPUPFF$_S$

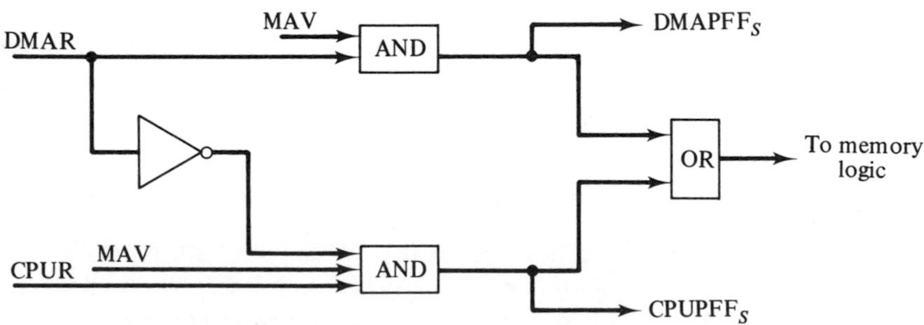

Figure 12.38 Memory request priority logic.

These two actions are mutually exclusive because of the priority design. Upon the acceptance of a memory access request, the memory logic presents a memory-unavailable signal, i.e., MAV = 0.

At the time that either the CPU or the channel requests a memory access, it also sends the address to the *MAR*. The priority logic which gives priority to the DMA channel is shown in Fig. 12.39.

The data lines to and from the *MIR* are shown in Fig. 12.40(a) and (b). The signals are gated by the outputs of the priority flip-flops, which are mutually exclusive.

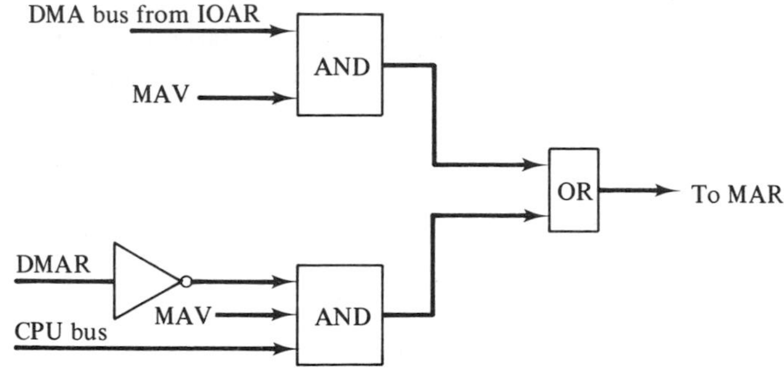

Figure 12.39 Address line priority.

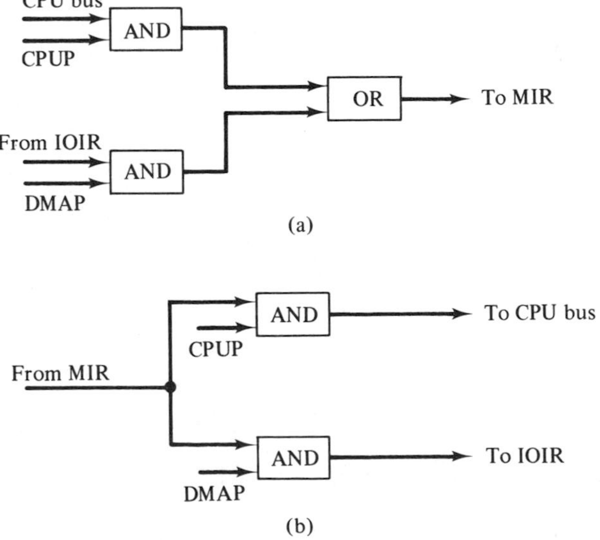

Figure 12.40 Data paths with priority.

At the completion of a memory cycle, the memory logic must indicate that the memory is not busy, i.e., available MAV = 1, and it must reset both of the priority flip-flops to zero, i.e., $DMAP = CPUP = 0$.

12.8 THE MEMORY UNIT

The block diagram of the memory unit is shown in Figure 12.7. The details of core memory design have been developed in Chapter 9 and will not be repeated here. The memory logic and timing will be shown in this section.

The CPU requests a main memory cycle by presenting either an RRC, CWC, or IWC signal to the memory. If the memory is available, the cycle will begin, and the memory will change its status from available to busy. The circuit shown in Fig. 12.41 will perform the function. It is a flip-flop which will be reset to $\overline{MAV}$ when the CPU signals a request for a main memory cycle and will be set to MAV by MEMFIN when the memory completes its cycle. The Δ symbol is a one-minor-cycle delay, i.e., 64 nanoseconds.

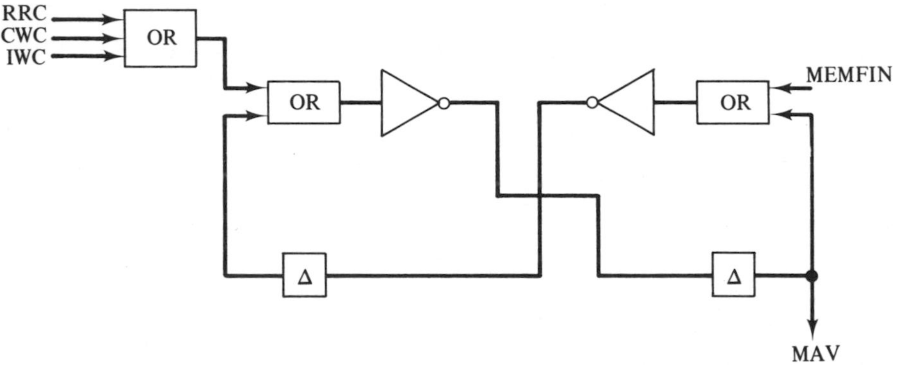

Figure 12.41 Memory available circuit.

The synchronization of the main memory with the CPU is accomplished with the use of the MAV signal and the memory synchronization flip-flop ($MSFF$) in the CPU. The excitation conditions for this flip-flop are

$$\text{MSFF}_S = \text{MAV}\cdot\text{RUNFF}\cdot\text{D1}\cdot(\text{RRC} + \text{CWC} + \text{IWC})$$
$$\text{MSFF}_R = \overline{MAV}\cdot\text{RUNFF}\cdot\text{D1}\cdot(\text{RRC} + \text{CWC} + \text{IWC})$$

Thus, at time D1 if a memory cycle is required but the memory is not available, the MSFF will be reset to 0. This will cause the execute flip-flop (EXECFF) to reset to 0.

$$\text{EXECFF}_R = \text{D1}\cdot(\overline{\text{MSFF}} + \overline{\text{RUNFF}})$$
$$\text{EXECFF}_S = \text{D1}\cdot\text{MSFF}\cdot\text{RUNFF}$$

If the EXECFF is a 0, no register transfers or commands of the microprogram will execute. Likewise, if the memory synchronization flip-flop is a 0 at time D7, the contents of the CMAR will not change.

$$\text{CHANGE} = \text{D7}\cdot\text{RUNFF}\cdot\text{MSFF}\cdot[\text{AUTO} + \text{STEPFF}(\text{SCY} + \text{SIN})]$$

This will cause the same microinstruction to be fetched during the subsequent D0 cycle. If at cycle D1 the memory has signaled that it is available and thus MSFF is a 1, the execute flip-flop will also set and the microprogram will now be permitted to execute. This logic keeps the memory and CPU in synchronization.

Once the memory and CPU are synchronized, the microprogrammed control unit can execute and provide control signals to the memory. The memory is connected to the 24 lines of the main bus. In addition control signals are set on four lines for IMIR, OMIR, SMIR, and IMAR. There are three lines which indicate the type of memory cycle required, i.e., RRC, CWC, and IWC. The memory also supplies two signals to the CPU, i.e., IA and MAV. Thus, there are a total of 33 lines from the memory to the CPU.

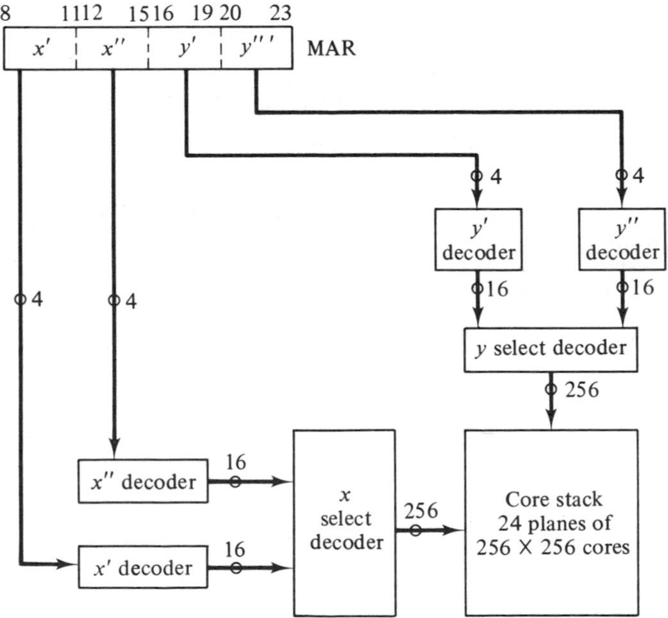

Figure 12.42 Dual tree decoders for memory.

The memory contains 2^{16} words each of 24 bits. The core plane is $2^8 \times 2^8$ cores (256×256). There are 24 planes plus one plane for an odd parity bit check. The address decoders are organized into dual-tree decoders. Figure 12.42 shows the organization.

We shall not dwell on the detailed design of a 3D core memory, since this was shown in Chapter 9. We have shown the details of the design of a midicomputer with emphasis on the microprogrammed control unit, the logic design of the arithmetic and logic unit, the interrupt structure and interfacing of an I/O channel, and the timing and synchronization of all the units which constitute the computer. The design starts with a programmer's description of the machine and then shows the register transfer equations which must be performed. This extends to the microprogramming of the control unit and the required control of the gates on the bus. We did not justify every decision in the design parameters for the machine. To do so would have required another book instead of just a lengthy chapter. It is hoped that this chapter, which is a miniature reference manual for the computer, will give greater insight into the timing and control needed in the design of a digital computer.

REFERENCES

Chu, Y., *Computer Organization and Microprogramming*. Englewood Cliffs, N.J.: Prentice-Hall, 1972.

Davidow, W. H., "General-Purpose Microcontrollers, Part I: Economic Considerations," *Computer Design*, 11, No. 7 (July 1972), pp. 75–79.

Dollhoff, T. L., "Microprogrammed Control for Small Computers," *Computer Design*, 12, No. 5 (May 1973), pp. 91–97.

Hill, F. J., and G. F. Peterson, *Digital Systems: Hardware Organization and Design*. New York: Wiley, 1973.

Husson, S. S., *Microprogramming Principles and Practice*. Englewood Cliffs, N.J.: Prentice-Hall, 1970.

Sobel, H. S., *Introduction to Digital Computer Design*. Reading, Mass.: Addison-Wesley, 1970.

PROBLEMS

12.1. How could the concept of base addressing be included in the design of the computer of this chapter? How many base registers would be needed? Would self-relative and indirect addressing be needed if base addressing is included? Redesign the normal instruction format to include base addressing.

12.2. Is the separate address adder really needed in this machine, or could the ALU be used to compute the effective address? Develop the timing, transfer paths, and control signals needed if it can be implemented.

12.3. Assume that the computer did not provide for a hardware multiply or divide instruction. Write a program which would perform these two operations using the remaining machine instructions. Compare the time needed for the execution of the software multiply with the hardware multiply.

12.4. Redesign the ALU unit so that it performs only 2's complement addition, negation, OR, AND, and exclusive-OR. Compare your gate count with the design in the text.

12.5. Show the design of the computer with four DMA channels instead of only one such channel. Redefine the I/O instructions and show the selection logic as well as the new data paths to the memory for cycle steal.

12.6. What must be added to the I unit to permit both indexing and self-relative addressing to be used in the same instruction? How will this affect the instruction phase timing?

12.7. What are the advantages of providing the capability of indirect addressing before indexing instead of after indexing?

12.8. Discuss the advantages and disadvantages of the wraparound memory method of addressing which the computer provides.

12.9. Is it really necessary to provide the three instructions BAP, BAZ, and BAN? Could the computer be designed with only two branch on accumulator condition instructions?

12.10. If the computer were to have seven general-purpose registers, develop the instruction format needed for register-to-register instructions.

12.11. Develop an algorithm which will permit the direct multiplication of two 2's complement numbers without their conversion to signed magnitude notation. Show the register transfer equations for your algorithm.

12.12. Define a set of instructions for a 24-bit computer which provides seven general-purpose registers that can be used as index registers, base registers, or accumulators.

12.13. Implement the register transfer equations for a rounded multiply instruction where the contents of the accumulator are rounded up if the most significant bit of the MQ register, which holds the least significant half of the product, is a 1.

12.14. Design the logic gating to provide all the shifting instructions of the computer if the accumulator is constructed from S-R-T flip-flops.

12.15. What is the need for a disable the interrupt instruction? Why wasn't one provided in the computer designed in this chapter?

12.16. Redesign the ALU using NOR logic, and compute the time to add two 24-bit numbers.

12.17. Redesign the control memory unit so that a microinstruction is fetched each minor cycle (64 nanoseconds) but designates only one source register and one destination register. Make the new microinstruction 32 bits in length.

12.18. Design the logic needed for the test logic block of the control memory unit.

12.19. Place all the microinstructions in sequential order and assign a binary address to each of the symbolic addresses. Then design the address matrix map.

12.20. Assign binary numbers to the symbolic names used in the microinstructions and show a map of the control memory.

12.21. How could the microprogram be changed in the control memory if it were constructed as a writable control store? Design the required registers and data paths to provide this capability.

Index